Advances in Industrial Control

Other titles published in this series:

Digital Controller Implementation and Fragility
Robert S.H. Istepanian and James F. Whidborne (Eds.)

Optimisation of Industrial Processes at Supervisory Level
Doris Sáez, Aldo Cipriano and Andrzej W. Ordys

Robust Control of Diesel Ship Propulsion
Nikolaos Xiros

Hydraulic Servo-systems
Mohieddine Jelali and Andreas Kroll

Model-based Fault Diagnosis in Dynamic Systems Using Identification Techniques
Silvio Simani, Cesare Fantuzzi and Ron J. Patton

Strategies for Feedback Linearisation
Freddy Garces, Victor M. Becerra, Chandrasekhar Kambhampati and Kevin Warwick

Robust Autonomous Guidance
Alberto Isidori, Lorenzo Marconi and Andrea Serrani

Dynamic Modelling of Gas Turbines
Gennady G. Kulikov and Haydn A. Thompson (Eds.)

Control of Fuel Cell Power Systems
Jay T. Pukrushpan, Anna G. Stefanopoulou and Huei Peng

Fuzzy Logic, Identification and Predictive Control
Jairo Espinosa, Joos Vandewalle and Vincent Wertz

Optimal Real-time Control of Sewer Networks
Magdalene Marinaki and Markos Papageorgiou

Process Modelling for Control
Benoît Codrons

Computational Intelligence in Time Series Forecasting
Ajoy K. Palit and Dobrivoje Popovic

Modelling and Control of Mini-Flying Machines
Pedro Castillo, Rogelio Lozano and Alejandro Dzul

Ship Motion Control
Tristan Perez

Hard Disk Drive Servo Systems (2nd Ed.)
Ben M. Chen, Tong H. Lee, Kemao Peng and Venkatakrishnan Venkataramanan

Measurement, Control, and Communication Using IEEE 1588
John C. Eidson

Piezoelectric Transducers for Vibration Control and Damping
S.O. Reza Moheimani and Andrew J. Fleming

Manufacturing Systems Control Design
Stjepan Bogdan, Frank L. Lewis, Zdenko Kovačić and José Mireles Jr.

Windup in Control
Peter Hippe

Nonlinear H_2/H_∞ Constrained Feedback Control
Murad Abu-Khalaf, Jie Huang and Frank L. Lewis

Practical Grey-box Process Identification
Torsten Bohlin

Control of Traffic Systems in Buildings
Sandor Markon, Hajime Kita, Hiroshi Kise and Thomas Bartz-Beielstein

Wind Turbine Control Systems
Fernando D. Bianchi, Hernán De Battista and Ricardo J. Mantz

Advanced Fuzzy Logic Technologies in Industrial Applications
Ying Bai, Hanqi Zhuang and Dali Wang (Eds.)

Practical PID Control
Antonio Visioli

(continued after Index)

Hugues Garnier • Liuping Wang
Editors

Identification of Continuous-time Models from Sampled Data

Springer

Hugues Garnier, PhD
Centre de Recherche en Automatique
  de Nancy (CRAN)
Nancy-Université
CNRS
Faculté des Sciences et Techniques
54506 Vandoeuvre-les-Nancy
France

Liuping Wang, PhD
RMIT University
School of Electrical and Computing
  Engineering
Swanston Street
Melbourne 3000
Victoria
Australia

ISBN 978-1-84800-160-2   e-ISBN 978-1-84800-161-9

DOI 10.1007/978-1-84800-161-9

Advances in Industrial Control series ISSN 1430-9491

British Library Cataloguing in Publication Data
Identification of continuous-time models from sampled data.
  - (Advances in industrial control)
  1. Linear time invariant systems - Mathematical models -
  Congresses 2. Automatic control - Congresses
  I. Garnier, Hugues II. Wang, Liuping
  629.8'32
ISBN-13: 9781848001602

Library of Congress Control Number: 2007942577

© 2008 Springer-Verlag London Limited

MATLAB® and Simulink® are registered trademarks of The MathWorks, Inc., 3 Apple Hill Drive, Natick, MA 01760-2098, USA. http://www.mathworks.com

Apart from any fair dealing for the purposes of research or private study, or criticism or review, as permitted under the Copyright, Designs and Patents Act 1988, this publication may only be reproduced, stored or transmitted, in any form or by any means, with the prior permission in writing of the publishers, or in the case of reprographic reproduction in accordance with the terms of licences issued by the Copyright Licensing Agency. Enquiries concerning reproduction outside those terms should be sent to the publishers.

The use of registered names, trademarks, etc. in this publication does not imply, even in the absence of a specific statement, that such names are exempt from the relevant laws and regulations and therefore free for general use.

The publisher makes no representation, express or implied, with regard to the accuracy of the information contained in this book and cannot accept any legal responsibility or liability for any errors or omissions that may be made.

Cover design: eStudio Calamar S.L., Girona, Spain

Printed on acid-free paper

9 8 7 6 5 4 3 2 1

springer.com

Advances in Industrial Control

Series Editors

Professor Michael J. Grimble, Professor of Industrial Systems and Director
Professor Michael A. Johnson, Professor (Emeritus) of Control Systems and Deputy Director

Industrial Control Centre
Department of Electronic and Electrical Engineering
University of Strathclyde
Graham Hills Building
50 George Street
Glasgow G1 1QE
United Kingdom

Series Advisory Board

Professor E.F. Camacho
Escuela Superior de Ingenieros
Universidad de Sevilla
Camino de los Descubrimientos s/n
41092 Sevilla
Spain

Professor S. Engell
Lehrstuhl für Anlagensteuerungstechnik
Fachbereich Chemietechnik
Universität Dortmund
44221 Dortmund
Germany

Professor G. Goodwin
Department of Electrical and Computer Engineering
The University of Newcastle
Callaghan
NSW 2308
Australia

Professor T.J. Harris
Department of Chemical Engineering
Queen's University
Kingston, Ontario
K7L 3N6
Canada

Professor T.H. Lee
Department of Electrical Engineering
National University of Singapore
4 Engineering Drive 3
Singapore 117576

Professor Emeritus O.P. Malik
Department of Electrical and Computer Engineering
University of Calgary
2500, University Drive, NW
Calgary
Alberta
T2N 1N4
Canada

Professor K.-F. Man
Electronic Engineering Department
City University of Hong Kong
Tat Chee Avenue
Kowloon
Hong Kong

Professor G. Olsson
Department of Industrial Electrical Engineering and Automation
Lund Institute of Technology
Box 118
S-221 00 Lund
Sweden

Professor A. Ray
Pennsylvania State University
Department of Mechanical Engineering
0329 Reber Building
University Park
PA 16802
USA

Professor D.E. Seborg
Chemical Engineering
3335 Engineering II
University of California Santa Barbara
Santa Barbara
CA 93106
USA

Doctor K.K. Tan
Department of Electrical Engineering
National University of Singapore
4 Engineering Drive 3
Singapore 117576

Professor Ikuo Yamamoto
The University of Kitakyushu
Department of Mechanical Systems and Environmental Engineering
Faculty of Environmental Engineering
1-1, Hibikino,Wakamatsu-ku, Kitakyushu, Fukuoka, 808-0135
Japan

Series Editors' Foreword

The series *Advances in Industrial Control* aims to report and encourage technology transfer in control engineering. The rapid development of control technology has an impact on all areas of the control discipline. New theory, new controllers, actuators, sensors, new industrial processes, computer methods, new applications, new philosophies..., new challenges. Much of this development work resides in industrial reports, feasibility study papers and the reports of advanced collaborative projects. The series offers an opportunity for researchers to present an extended exposition of such new work in all aspects of industrial control for wider and rapid dissemination.

The importance of system models in the current paradigm of advanced control design cannot be overestimated. Three recent volumes in the *Advances in Industrial Control* series: *Model-based Process Supervision* by Arun Samantaray and Belkacem Ould Bouamama, *Wind Turbine Control Systems* by Fernando Bianchi and colleagues and *Soft Sensors for Monitoring and Control of Industrial Processes* by Luigi Fortuna and colleagues testify to the pervasive use of system models in different aspects of control engineering and in different application areas.

This growth in the use of models to accomplish different objectives in the design of industrial control systems has been accompanied by a similar growth in the science of system identification. Today, there is a thriving research community pursuing new developments in system identification that support the use of system models in control design, and for process comprehension. The IFAC Symposium on System Identification (SYSID) is a popular forum for the work of this research community.

System identification is often classed as a white-box problem or a black-box problem, but when the designer is allowed to introduce *a priori* system knowledge into the process then more pragmatic grey-box methods emerge. For the interested reader, the *Advances in Industrial Control* monograph *Practical Grey-box Process Identification* by Torsten Bohlin describes the fundamentals of, some new software for and some applications of the grey-box identification approach.

A mainstay of the control system modelling paradigm are continuous-time models because they arise naturally when describing the physical phenomena of systems and processes. These models of physical systems usually involve differential equations that stem from the application of physical and chemical laws. However, the widespread use of digital computing technology and the concomitant sampled data led to an emphasis on the use of discrete system models, discrete control designs and sampled-data-based system identification algorithms from the 1980s onward. In an attempt to address this imbalance in technical methods, Hugues Garnier and Liuping Wang initiated international conference sessions and research activities to re-establish interest in the techniques for the identification of continuous-time models. One outcome of their endeavours is this entry in the *Advances in Industrial Control* series. Hugues Garnier and Liuping Wang are not only the Editors of this fourteen-contributed-chapter book but have also, along with many other leading international researchers in the system identification field, contributed to several chapters as authors.

Identification of Continuous-time Models from Sampled Data covers a wealth of material from this field. Usefully, the opening survey chapter defines the basic identification problem, reviews the issues arising from the continuous-time aspect of the problem and then provides a road map to the very substantial set of contributed chapters that follows. The range of topics covered includes: experimental design, model structure determination, closed-loop identification, and software aspects along with a generous number of practical examples. This list is by no means exhaustive of the breadth of the contents and subject matter of the book.

The encyclopedic and highly focussed nature of the book means that it is likely to become a standard reference text for this particular aspect of the system identification subject. It is, after all, the first major book contribution to this field for nearly fifteen years and as such is a very welcome addition to the *Advances in Industrial Control* series.

Industrial Control Centre *M.J. Grimble*
Glasgow *M.A. Johnson*
Scotland, UK
2007

Preface

It is often true that a book is developed through a long journey that consists of many tiny steps and interactions with many people. This book started in July 2004 when we, Hugues Garnier and Liuping Wang, met for the first time during the fifth Asian Control Conference in Melbourne. We decided to organise an invited session on continuous-time system identification for the 16th World IFAC Congress in Prague (2005). The invitation was first presented to Professor Graham Goodwin, and was accepted without any hesitation. Our invited session in Prague was successful, with support and contributions from Professors Lennart Ljung, Torsten Söderström, Graham Goodwin, Peter Young, Peter Gawthrop, Tomas McKelvey, Johan Schoukens and Rik Pintelon. The special session was well received. From the World Congress, we then decided to organise another three invited sessions for the IFAC Symposium on System Identification (SYSID'2006) in Newcastle, with the same authors from the World IFAC Congress, in which one was devoted to theoretical study and algorithmic development and one was devoted to application of continuous-time system identification. The majority of the authors in this monograph were the contributors to the invited sessions.

Although a broad overview of the different techniques available for direct continuous-time model identification has been given in the books by Unbehauen and Rao (1987) [1] and Sinha and Rao (1991) [2], more than fifteen years has passed since the publication of the last book on continuous-time system identification. Interest in continuous-time approaches to system identification has indeed been growing in the last ten years. Apart from the joint activities in organising the more recent invited sessions, the first editor (Hugues Garnier) has organised invited sessions for the 5th European Control Conference (ECC'1999) in Karlsruhe, for the 15th World IFAC Congress in Barcelona (2002) and for the SYSID'2003 Symposium in Rotterdam. The large number of publications in recent years reflects the intensive effort devoted to the development of theory, software, and applications of these techniques. We felt that it was time to reflect on the recent development of this area. Thereby arose our intention of editing this book with the contributors who have been working with us for the past many years.

It has been a privilege for us to have the opportunity to work with them. Our thanks go to all the contributors of this book who have supported us over the years. Indeed, both of us enjoyed the time we spent interacting with the contributors and editing this book. Our special gratitude goes to our friends Professor Graham Goodwin and Professor Peter Young who have guided us in various aspects of our careers.

Finally, but not least, we give deepest gratitude to our families: Liuping's husband (Jianshe) and son (Robin); Hugues's wife (Nathalie) and children (Elliot, Victoria and Marie-Sarah) for their endless understanding, caring, patience and support.

It is to them all that we dedicate this book.

Nancy, France and Melbourne, Australia *Hugues Garnier*
March 2007 *Liuping Wang*

References

1. H. Unbehauen and G.P. Rao. *Identification of Continuous Systems*. North-Holland, Systems and Control Series, Amsterdam, 1987.
2. N.K. Sinha and G.P. Rao (eds). *Identification of Continuous-time Systems. Methodology and Computer Implementation*. Kluwer Academic Publishers, Dordrecht, 1991.

Contents

List of Abbreviations and Symbols xviii

List of Contributors xxiii

1 Direct Identification of Continuous-time Models from Sampled Data: Issues, Basic Solutions and Relevance
Hugues Garnier, Liuping Wang and Peter C. Young 1
1.1 Introduction 1
1.2 System Identification Problem and Procedure 2
1.3 Basic Discrete-time Model Identification 5
    1.3.1 Difference Equation Models 5
    1.3.2 The Traditional Least Squares Method 5
    1.3.3 Example: First-order Difference Equation 6
    1.3.4 Models for the Measurement Noise 7
1.4 Issues in Direct Continuous-time Model Identification 7
    1.4.1 Differential Equation Models 7
    1.4.2 Input–Output Time Derivatives 8
    1.4.3 Models for the Measurement Noise 8
1.5 Basic Direct Continuous-time Model Identification 9
    1.5.1 The Traditional State-variable Filter Method 9
    1.5.2 Example: First-order Differential Equation 11
1.6 Motivations for Identifying Continuous-time Models Directly from Sampled Data 11
    1.6.1 Physical Insight into the System Properties 12
    1.6.2 Preservation of *a priori* Knowledge 12
    1.6.3 Inherent Data Filtering 13
    1.6.4 Non-uniformly Sampled Data 13
    1.6.5 Transformation between CT and DT Models 13
    1.6.6 Sensitivity Problems of DT Models at High Sampling Rates 14
    1.6.7 Stiff Systems 14

1.7	Specialised Topics in System Identification	15
	1.7.1 Identification of the Model Structure	15
	1.7.2 Identification of Pure Time (Transportation) Delay	15
	1.7.3 Identification of Continuous-time Noise Models	15
	1.7.4 Identification of Multi-variable Systems	16
	1.7.5 Identification in Closed Loop	16
	1.7.6 Identification in the Frequency Domain	16
	1.7.7 Software for Continuous-time Model Identification	16
1.8	Historical Review	16
1.9	Outline of the Book	20
1.10	Main References	25
References		26

2 Estimation of Continuous-time Stochastic System Parameters

Erik K. Larsson, Magnus Mossberg, Torsten Söderström 31

2.1	Background and Motivation	31
2.2	Modelling of Continuous-time Stochastic Systems	33
2.3	Sampling of Continuous-time Stochastic Models	34
	2.3.1 Sampling of CARMA Systems	35
	2.3.2 Sampling of Systems with Inputs	37
2.4	A General Approach to Estimation of Continuous-time Stochastic Models	38
	2.4.1 Direct and Indirect Methods	40
2.5	Introductory Examples	42
2.6	Derivative Approximations for Direct Methods	46
	2.6.1 Non-uniformly Sampled Data	52
2.7	The Cramér–Rao Bound	54
	2.7.1 The Cramér–Rao Bound for Irregularly Sampled CARMA Models	55
2.8	Numerical Studies of Direct Methods	58
2.9	Conclusions	62
References		63

3 Robust Identification of Continuous-time Systems from Sampled Data

Juan I. Yuz, Graham C. Goodwin 67

3.1	Overview	68
3.2	Limited-bandwidth Estimation	69
	3.2.1 Frequency-domain Maximum Likelihood	72
3.3	Robust Continuous-time Model Identification	75
	3.3.1 Effect of Sampling Zeros in Deterministic Systems	75
	3.3.2 Effect of Sampling Zeros in Stochastic Systems	79
	3.3.3 Continuous-time Undermodelling	82
	3.3.4 Restricted-bandwidth FDML Estimation	84

3.4	Conclusions		86
References			87

4 Refined Instrumental Variable Identification of Continuous-time Hybrid Box–Jenkins Models
Peter C. Young, Hugues Garnier, Marion Gilson 91

4.1	Introduction		91
4.2	Problem Formulation		93
4.3	Optimal RIVC Estimation: Theoretical Motivation		96
	4.3.1	The Hybrid Box–Jenkins Estimation Model	96
	4.3.2	RIVC Estimation	97
4.4	The RIVC and SRIVC Algorithms		100
	4.4.1	The RIVC Algorithm	100
	4.4.2	The SRIVC Algorithm	101
	4.4.3	Multiple-input Systems	103
	4.4.4	Non-uniformly Sampled Data	103
4.5	Theoretical Background and Statistical Properties of the RIVC Estimates		104
	4.5.1	Optimality of RIVC Estimation	104
	4.5.2	The Asymptotic Independence of the System and Noise Model Parameter Estimates	105
4.6	Model Order Identification		108
4.7	Simulation Examples		109
	4.7.1	The Rao–Garnier Test System	109
	4.7.2	Noise-free Case	110
	4.7.3	Noisy-output Case	112
4.8	Practical Examples		119
	4.8.1	Hadley Centre Global Circulation Model (GCM) Data	120
	4.8.2	A Multiple-input Winding Process	122
4.9	Conclusions		127
References			129

5 Instrumental Variable Methods for Closed-loop Continuous-time Model Identification
Marion Gilson, Hugues Garnier, Peter C. Young, Paul Van den Hof ... 133

5.1	Introduction		133
5.2	Problem Formulation		135
5.3	Basic Instrumental Variable Estimators		138
	5.3.1	Consistency Properties	138
	5.3.2	Accuracy Analysis	139
5.4	Extended Instrumental Variable Estimators		139
	5.4.1	Consistency Properties	140
	5.4.2	Accuracy Analysis	140
5.5	Optimal Instrumental Variable Estimators		140
	5.5.1	Main Results	140

	5.5.2	Implementation Issues 141
	5.5.3	Multi-step Approximate Implementations of the Optimal IV Estimate 144
	5.5.4	Iterative Implementations of the Optimal IV Estimate ... 147
5.6	Summary 152	
5.7	Numerical Examples 153	
	5.7.1	Example 1: White Noise 154
	5.7.2	Example 2: Coloured Noise 155
5.8	Conclusions 159	
References 159		

6 Model Order Identification for Continuous-time Models
Liuping Wang, Peter C. Young 161

6.1	Introduction 161	
6.2	Instrumental Variable Identification 162	
6.3	Instrumental Variable Estimation using a Multiple-model Structure . 166	
	6.3.1	Augmented Data Regressor 166
	6.3.2	Instrumental Variable Solution Using UDV Factorisation . 168
	6.3.3	Computational Procedure 172
6.4	Model Structure Selection Using PRESS 174	
6.5	Simulation Studies 179	
6.6	Conclusions 185	
References 186		

7 Estimation of the Parameters of Continuous-time Systems Using Data Compression
Liuping Wang, Peter J. Gawthrop 189

7.1	Introduction 189	
7.2	Data Compression Using Frequency-sampling Filters 189	
	7.2.1	FSF Model 190
	7.2.2	FSF Model in Data Compression 192
	7.2.3	Estimation Using FSF Structure 195
7.3	Data Compression with Constraints 197	
	7.3.1	Formulation of the Constraints 197
	7.3.2	Solution of the Estimation Problem with Constraints 198
	7.3.3	Monte Carlo Simulation Study 199
7.4	Physical-model-based Estimation 201	
7.5	Example: Inverted Pendulum 203	
	7.5.1	FSF Estimation 205
	7.5.2	PMB Estimation 207
7.6	Conclusions 210	
References 212		

8 Frequency-domain Approach to Continuous-time System Identification: Some Practical Aspects
Rik Pintelon, Johan Schoukens, Yves Rolain 215
8.1 Introduction 215
8.2 The Inter-sample Behaviour and the Measurement Setup 216
    8.2.1 Plant Modelling 216
    8.2.2 Noise Modelling 220
    8.2.3 Summary 222
8.3 Parametric Models 223
    8.3.1 Plant Models 223
    8.3.2 Noise Models 225
    8.3.3 Summary 226
8.4 The Stochastic Framework 227
    8.4.1 Periodic Excitations 227
    8.4.2 Arbitrary Excitations 228
8.5 Identification Methods 229
    8.5.1 Asymptotic Properties of the Frequency-domain Gaussian Maximum Likelihood Estimators 231
    8.5.2 Periodic Excitations 231
    8.5.3 Arbitrary Excitations: Generalised Output Error 234
    8.5.4 Arbitrary Excitations: Errors-in-variables 236
8.6 Real Measurement Examples 237
    8.6.1 Operational Amplifier 237
    8.6.2 Flight-flutter Analysis 240
8.7 Guidelines for Continuous-time Modelling 241
    8.7.1 Prime Choice: Uniform Sampling, Band-limited Measurement Setup, Periodic Excitation 241
    8.7.2 Second Choice: Uniform Sampling, Band-limited Measurement Setup, Arbitrary Excitation 242
    8.7.3 Third Choice: Uniform Sampling, Zero-order-hold Measurement Setup 242
    8.7.4 Last Resort: Non-uniform Sampling 243
    8.7.5 To be Avoided 243
8.8 Conclusions 243
References 243

9 The CONTSID Toolbox: A Software Support for Data-based Continuous-time Modelling
Hugues Garnier, Marion Gilson, Thierry Bastogne, Michel Mensler 249
9.1 Introduction 249
9.2 General Procedure for Continuous-time Model Identification 250
9.3 Overview of the CONTSID Toolbox 250
    9.3.1 Parametric Model Estimation 250
    9.3.2 Model Order Selection and Validation 256
9.4 Software Description 260

		9.4.1	Introductory Example to the Command Mode 261

9.4.1 Introductory Example to the Command Mode 261
9.4.2 The Graphical User Interface 267
9.5 Advantages and Relevance of the CONTSID Toolbox Methods ... 271
9.6 Successful Application Examples 275
    9.6.1 Complex Flexible Robot Arm 275
    9.6.2 Uptake Kinetics of a Photosensitising Agent into Cancer Cells.. 278
    9.6.3 Multi-variable Winding Process 283
9.7 Conclusions.. 285
References 287

10 Subspace-based Continuous-time Identification
Rolf Johansson 291
10.1 Introduction 291
10.2 Problem Formulation 292
    10.2.1 Discrete-time Measurements 292
    10.2.2 Continuous-time State-space Linear System............. 293
10.3 System Identification Algorithms 296
    10.3.1 Theoretical Remarks on the Algorithms 299
    10.3.2 Numerical Example................................... 301
10.4 Statistical Model Validation 302
10.5 Discussion 306
10.6 Conclusions... 308
References 309

11 Process Parameter and Delay Estimation from Non-uniformly Sampled Data
Salim Ahmed, Biao Huang, Sirish L. Shah 313
11.1 Introduction 313
11.2 Estimation of Parameters and Delay 315
    11.2.1 Second-order Modelling 315
    11.2.2 Higher-order Modelling 318
    11.2.3 Treatment of Initial Conditions 320
    11.2.4 Parameter Estimation 321
    11.2.5 Non-minimum Phase Processes......................... 323
    11.2.6 Choice of $\hat{A}_0(s)$ and $\hat{\tau}_0$ 324
11.3 Identification from Non-uniformly Sampled Data 324
    11.3.1 The Iterative Prediction Algorithm 324
    11.3.2 Input-only Modelling Using Basis-function Model 325
    11.3.3 Choice of Basis-function Parameters 327
    11.3.4 Criterion of Convergence 328
11.4 Simulation Results... 328
    11.4.1 Estimation from Uniformly Sampled Data 329
    11.4.2 Estimation from Non-uniformly Sampled Data 330
11.5 Experimental Evaluation 331

	11.5.1	Identification of a Dryer 331
	11.5.2	Identification of a Mixing Process 332
11.6	Conclusions 333
References 335

12 Iterative Methods for Identification of Multiple-input Continuous-time Systems with Unknown Time Delays
Zi-Jiang Yang 339
12.1 Introduction 339
12.2 Statement of the Problem 341
12.3 Approximate Discrete-time Model Estimation 342
12.4 SEPNLS Method 343
12.5 GSEPNLS Method 347
12.6 GSEPNIV Method 351
12.7 Numerical Results 355
    12.7.1 GSEPNLS Method in the Case of Low Measurement Noise . 357
    12.7.2 GSEPNIV Method 358
12.8 Conclusions 360
References 361

13 Closed-loop Parametric Identification for Continuous-time Linear Systems via New Algebraic Techniques
Michel Fliess, Hebertt Sira-Ramírez 363
13.1 Introduction 363
13.2 A Module-theoretic Approach to Linear Systems: a Short Summary . 364
    13.2.1 Some Basic Facts about Modules over Principal Ideal Rings 364
    13.2.2 Formal Laplace Transform 365
    13.2.3 Basic System-theoretic Definitions 366
    13.2.4 Transfer Matrices 367
13.3 Identifiability 368
    13.3.1 Uncertain Parameters 368
    13.3.2 The Algebraic Derivative and a New Module Structure ... 368
    13.3.3 Linear Identifiability 368
    13.3.4 An Elementary Example 369
13.4 Perturbations 370
    13.4.1 Structured Perturbations 370
    13.4.2 Unstructured Perturbations 370
    13.4.3 Linear Identifier 371
    13.4.4 Robustness 371
13.5 First Example: Dragging an Unknown Mass in Open Loop 371
    13.5.1 Description and First Results 371
    13.5.2 Denoising 374
    13.5.3 A Comparison with an Adaptive-observer Approach 376
13.6 Second Example: A Perturbed First-order System 377

	13.6.1	Presentation 377
	13.6.2	A Certainty Equivalence Controller 378
	13.6.3	Parameter Identification 378
	13.6.4	Noise-free Simulation Results 380
	13.6.5	Noisy Measurements and Plant Perturbations 381
	13.6.6	Simulation Results with Noises 381
	13.6.7	Comparison with Adaptive Control 381
	13.6.8	Simulations for the Adaptive Scheme 383
13.7	Third Example: A Double-bridge Buck Converter 383
	13.7.1	An Input–Output Model 384
	13.7.2	Problem Formulation 385
	13.7.3	A Certainty Equivalence Controller 385
	13.7.4	Closed-loop Behaviour 385
	13.7.5	Algebraic Determination of the Unknown Parameters 386
	13.7.6	Simulation Results 387
13.8	Conclusion 388
References 389

14 Continuous-time Model Identification Using Spectrum Analysis with Passivity-preserving Model Reduction
Rolf Johansson 393
14.1 Introduction 393
14.2 Preliminaries 394
 14.2.1 Continuous-time Model Identification 394
 14.2.2 Spectrum Analysis and Positivity 396
 14.2.3 Spectral Factorisation and Positivity 399
 14.2.4 Balanced Model Reduction 399
14.3 Problem Formulation 400
14.4 Main Results 401
14.5 Discussion 405
14.6 Conclusions 406
References 406

Index 409

List of Abbreviations and Symbols

arg min	Minimising argument
A	State matrix of state-space model
B	Input-to-state matrix of state-space model
C	State-to-output matrix of state-space model
D	Direct feedthrough matrix of state space model
(A, B, C, D)	State-space realisation
ARMA	Autoregressive moving average
	ARMA model structure $D(\cdot)v(t_k) = C(\cdot)e(t_k)$
ARMAX	Autoregressive moving average with external input
	ARMAX model structure $A(\cdot)y(t_k) = B(\cdot)u(t_k) + C(\cdot)e(t_k)$
ARX	Autoregressive with external input
	ARX model structure $A(\cdot)y(t_k) = B(\cdot)u(t_k) + e(t_k)$
BCLS	Bias-compensated least squares
BJ	Box–Jenkins
	BJ model structure $y(t_k) = \frac{B(\cdot)}{A(\cdot)}u(t_k) + \frac{C(\cdot)}{D(\cdot)}e(t_k)$
Cov{}	Covariance
CRB	Cramér–Rao Bound
$\mathbb{C}$	Set of complex numbers
δ	delta operator $\delta = \frac{q-1}{T_s}$
$\delta(t_k)$	Discrete-time pulse function
δ_{ij}	Kronecker delta function
$e(t_k)$	White-noise stochastic process
$e_o(t_k)$	'True' driving disturbance acting on a given system $\mathcal{S}$
ε	One-step-ahead prediction error
E{}	Expectation (expected value of a random variable)
$\bar{\text{E}}${}	Generalised expectation of a quasi-stationary process
FIR	Finite impulse response
$\boldsymbol{\varphi}(t_k)$	Regression vector at time instant t_k
$\boldsymbol{\Phi}_N$	Matrix of regression vectors $[\boldsymbol{\varphi}(t_1) \ldots \boldsymbol{\varphi}(t_N)]^T$

List of Abbreviations and Symbols

$\Phi_v(\omega)$	Spectrum of v, Fourier transform of $R_v(\tau)$
$\Phi_{sw}(\omega)$	Cross spectrum between s and w, Fourier transform of $R_{sw}(\tau)$
$\hat{\Phi}_u^N(\omega)$	Estimate of the spectrum of u based on N finite samples of $u(t_k)$
$G(s)$	Transfer function from u to y
$G(s, \boldsymbol{\theta})$	Transfer function in a model structure, corresponding to the parameter value $\boldsymbol{\theta}$
$G_o(s)$	'True' data generating input–output system
$G^*(s)$	Limiting estimate of $G(s)$ as $N \to \infty$
$\hat{G}_N(s)$	Estimate of $G(s)$ based on N samples of the input and output signals
$\mathcal{G}$	Set of input–output models (deterministic part only in data-generating system)
$\eta(t_k)$	Filtered measurement noise at time instant t_k
$H(\cdot)$	Transfer function from e to y
$H(\cdot, \boldsymbol{\theta})$	Transfer function in a model structure, corresponding to the parameter value $\boldsymbol{\theta}$
$H_o(\cdot)$	'True' noise shaping filter in data-generating system
$\hat{H}_N(\cdot)$	Estimate of $H(\cdot)$ based on N samples of the input and output signals
i	Complex number, $\mathrm{i} = \sqrt{-1}$
I_n	$n \times n$ identity matrix
Im[ ]	Imaginary part
IPM	Instrumental product matrix
IV	Instrumental variable
$L(\cdot)$	Transfer function of a filter
LS	Least squares
LTI	Linear time-invariant
m	Transfer function numerator order
$\mathcal{M}$	Model set (including deterministic and stochastic part)
$\mathcal{M}(\boldsymbol{\theta})$	Particular model corresponding to the parameter value $\boldsymbol{\theta}$
MIMO	Multiple input, multiple output
MISO	Multiple input, single output
ML	Maximum likelihood
MSE	Mean square error
n	State-space dimension; model order
N	Number of data samples
NSR	Noise-to-signal ratio
$\mathbb{N}$	Set of natural numbers
$\mathcal{N}$	Gaussian or normal distribution
ω	Radial frequency
ω_s	Sampling (radial) frequency

OE	Output error
	OE model structure $y(t_k) = \frac{B(\cdot)}{A(\cdot)} u(t_k) + e(t_k)$
$\mathcal{O}(x)$	Big ordo of x, $\mathcal{O}(x)/|x|$ bounded as $x \to 0$
p	Order of the AR part in an ARMA model; order of Laguerre and Kautz models; time-domain differential operator
$\mathbf{P}_\theta$	Asymptotic covariance matrix of $\boldsymbol{\theta}$
PDF	Probability density function
PEM	Prediction error method
q	Forward shift operator; order of the MA part in an ARMA model
q^{-1}	Backward shift or delay operator
Q^{-1}	Inverse of the matrix Q
Q^T	Transpose of the matrix Q
Q^{-T}	Transpose of the inverse of the matrix Q
$\mathbb{R}$	Set of real numbers
Re[ ]	Real part
$R_u(\tau)$	Autocovariance function of u, $\bar{\mathsf{E}}\{u(t)u^T(t-\tau)\}$
$R_{yu}(\tau)$	Cross-covariance function of y and u, $\bar{\mathsf{E}}\{y(t)u^T(t-\tau)\}$
s	Laplace variable
$\mathcal{S}$	Data-generating system (including both deterministic and stochastic parts)
SDE	Stochastic differential equation
SISO	Single input single output
σ	Singular value
σ_e^2	Variance of stochastic process e
SLS	Shifted least squares
SNR	Signal-to-noise ratio
t_k	Time instant $t_k = kT_s$
τ	Fractional time delay
τ_d	Time delay as an integral multiple of the sampling period $\tau_\mathrm{d} = dT_\mathrm{s}$
T_s	Sampling interval
$\boldsymbol{\theta}$	Vector used to parameterise models
$\hat{\boldsymbol{\theta}}_N$	Estimate of the parameter vector based on Z^N
$\boldsymbol{\theta}_o$	'True' parameter vector
$\hat{\boldsymbol{\theta}}^j$	Estimate of the parameter vector at the jth iteration
$\boldsymbol{\theta}^*$	Limiting estimate of $\boldsymbol{\theta}$ as $N \to \infty$
Θ	Set of parameter vectors
$u(t_k)$	Input signal at time instant t_k
$\mathring{u}(t_k)$	Noise-free (deterministic) input signal at time instant t_k
$U_N(\omega)$	Fourier transform of u from N finite samples
Var{}	Variance
$V_N(\boldsymbol{\theta}, Z^N)$	Sum of squares loss function

List of Abbreviations and Symbols

V_N	Vector of noise samples $[v(t_1)\ldots v(t_N)]^T$
$v(t_k)$	Output noise at time instant t_k
$w(t_k)$	Disturbance signal at time instant t_k
w.p.	with probability
w.p. 1	with probability one
w.r.t.	with respect to
WLS	Weighted least squares
$x(t_k)$	State vector; noise-free (deterministic) output signal at time instant t_k
$\xi(t_k)$	Coloured noise variable at time instant t_k
$y(t_k)$	Measured output signal at time instant t_k
$y^{(i)}(t)$	ith time derivative of a continuous-time signal $y(t)$
$y_\mathrm{f}(t_k)$	Filtered signal $y_\mathrm{f}(t_k) = L(\cdot)y(t_k)$
Y_N	Vector of output samples $[y(t_1)\ldots y(t_N)]^T$
$\boldsymbol{\zeta}(t_k)$	Instrumental variable vector, the 'correlation vector'
$\boldsymbol{\zeta}(t_k, \boldsymbol{\theta})$	Instrumental variable vector with respect to $\boldsymbol{\theta}$
$\boldsymbol{\Psi}_N$	Matrix of instrumental variable vectors $[\boldsymbol{\zeta}(t_1)\ldots\boldsymbol{\zeta}(t_N)]^T$
$\mathcal{Z}$	Set of integer numbers
Z^N	Set of input and output data samples $\{u(t_1), y(t_1)\ldots u(t_N), y(t_N)\}$
$(\cdot)^T$	Transpose of a matrix
$(\bar{\cdot})$	Complex conjugate for scalars; Element-wise complex conjugate for matrices
$\|\cdot\|$	(Euclidean) norm of a vector
$\|\cdot\|$	(Frobenius) norm of a matrix
$Vec(\cdot)$	Vector-operation on a matrix, stacking its columns on top of each other
$\star$	convolution
$\otimes$	Kronecker matrix product
$\odot$	Element-wise multiplication

List of Contributors

Salim Ahmed
Department of Chemical and Materials Engineering
University of Alberta
Edmonton, AB, Canada T6G 2G6
salim.ahmed@ualberta.ca

Thierry Bastogne
Centre de Recherche en Automatique de Nancy
Nancy-Université, CNRS
BP 239, F-54506 Vandoeuvre-lès-Nancy Cedex, France
thierry.bastogne@cran.uhp-nancy.fr

Michel Fliess
Projet ALIEN, INRIA Futurs & Équipe MAX
LIX, CNRS UMR 7161
École Polytechnique, F-91128 Palaiseau, France
michel.fliess@polytechnique.fr

Hugues Garnier
Centre de Recherche en Automatique de Nancy
Nancy-Université, CNRS
BP 239, F-54506 Vandoeuvre-lès-Nancy Cedex, France
hugues.garnier@cran.uhp-nancy.fr

Peter J. Gawthrop
Center for Systems and Control and Department of Mechanical Engineering
University of Glasgow
Glasgow G12 8QQ, UK
p.gawthrop@eng.gla.ac.uk

Marion Gilson
Centre de Recherche en Automatique de Nancy
Nancy-Université, CNRS
BP 239, F-54506 Vandoeuvre-lès-Nancy Cedex, France
marion.gilson@cran.uhp-nancy.fr

Graham C. Goodwin
School of Electrical Engineering and Computer Science
The University of Newcastle
NSW 2308, Australia
graham.goodwin@newcastle.edu.au

Biao Huang
Department of Chemical and Materials Engineering
University of Alberta
Edmonton, AB, Canada T6G 2G6
biao.huang@ualberta.ca

Rolf Johansson
Department of Automatic Control
Lund University
PO Box 118, SE221 00 Lund, Sweden
rolf.johansson@control.lth.se

Erik K. Larsson
Ericsson AB, Ericsson Research
SE-164 80 Stockholm, Sweden
erik.larsson@ericsson.com

Michel Mensler
Direction de la Recherche, Etudes Avancées, Matériaux
Renault Technocentre
1 avenue du Golf, F-78288 Guyancourt Cedex, France
michel.mensler@renault.com

Magnus Mossberg
Department of Electrical Engineering
Karlstad University
SE-651 88 Karlstad, Sweden
magnus.mossberg@kau.se

Rik Pintelon
Dept. ELEC
Vrije Universiteit Brussel
Pleinlaan 2, 1050 Brussels, Belgium
rik.pintelon@vub.ac.be

Yves Rolain
Dept. ELEC
Vrije Universiteit Brussel
Pleinlaan 2, 1050 Brussels, Belgium
yves.rolain@vub.ac.be

Johan Schoukens
Dept. ELEC
Vrije Universiteit Brussel
Pleinlaan 2, 1050 Brussels, Belgium
johan.schoukens@vub.ac.be

Sirish L. Shah
Department of Chemical and Materials Engineering
University of Alberta
Edmonton, AB, Canada T6G 2G6
sirish.shah@ualberta.ca

Hebertt Sira-Ramírez
Cinvestav-IPN, Depto. de Ing. Eléctrica
Av. IPN, No. 2508, Col. San Pedro Zacatenco AP 14740
07300 México D.F., México
hsira@cinvestav.mx

Torsten Söderström
Department of Information Technology
Uppsala University
POBox 337, SE-751 05 Uppsala, Sweden
torsten.soderstrom@it.uu.se

Paul Van den Hof
Delft Center for Systems and Control
Delft University of Technology
Mekelweg 2, 2628 CD Delft, The Netherlands
p.m.j.vandenhof@tudelft.nl

Liuping Wang
School of Electrical and Computer Engineering
RMIT University
Melbourne, Victoria 3000, Australia
liuping.wang@rmit.edu.au

Zi-Jiang Yang
Department of Electrical and Electronic Systems Engineering
Kyushu University
744 Motooka, Nishi-ku, Fukuoka 819-0395, Japan
yoh@ees.kyushu-u.ac.jp

Peter C. Young
Centre for Research on Environmental Systems and Statistics
Lancaster University
Lancaster LA1 4YQ, UK
p.young@lancaster.ac.uk

Juan I. Yuz E.
Electronics Department
Universidad Técnica Federico Santa María
Casilla 110-V, Valparaíso, Chile
juan.yuz@elo.utfsm.cl

1

Direct Identification of Continuous-time Models from Sampled Data: Issues, Basic Solutions and Relevance

Hugues Garnier[1], Liuping Wang[2] and Peter C. Young[3]

[1] Centre de Recherche en Automatique de Nancy, Nancy-Université, CNRS, France
[2] RMIT University, Melbourne, Australia
[3] Lancaster University, UK & Australian National University

1.1 Introduction

Mathematical models of dynamic systems are required in most areas of scientific enquiry and take various forms, such as differential equations, difference equations, state-space equations and transfer functions. The most widely used approach to mathematical modelling involves the construction of mathematical equations based on physical laws that are known to govern the behaviour of the system. Amongst the drawbacks to this approach are that the resulting models are often complex and not easily estimated directly from the available data because of identifiability problems caused by over-parameterisation. This complexity also makes them difficult to use in applications such as control system design.

If sufficient experimental or operational data are available, an alternative to physically-based mathematical modelling is data-based 'system identification', which can be applied to virtually any system and typically yields relatively simple models that can well describe the system's behaviour within a defined operational regime. Such models can be either in a 'black-box' form, which describes only the input–output behaviour, or in some other, internally descriptive form, such as state-space equations, that can be interpreted in physically meaningful terms. This book presents some recent developments in system identification applied to the modelling of continuous-time systems.

Dynamic systems in the physical world are naturally described in continuous-time (CT), differential equation terms because the physical laws, such as conservation equations, have been evolved mainly in this form. Paradoxically, however, the best known system identification schemes have been based on discrete-time (DT) models (sometimes referred to as sampled-data models), without much concern for the merits of natural continuous-time model descriptions and their associated identification methods. In fact, the development of CT system identification techniques occurred in the the last century, before the development of the DT techniques, but was overshadowed by the more extensive DT developments. This was mainly due to the 'go completely discrete-time' trend that was spurred by parallel developments in digital computers.

Much less attention has been devoted to CT modelling from DT data and many practitioners appear unaware that such alternative methods not only exist but may be better suited to their modelling problems. The identification of continuous-time models is indeed a problem of considerable importance that has applications in virtually all disciplines of science and engineering. This book presents an up-to-date view of this active area of research and describes methods and software tools recently developed in this field.

This chapter is organised as follows. In the first section, the general procedure for system identification is reviewed. Thereafter, the basic features for fitting DT and CT models to sampled data are presented with the objective of highlighting issues in CT model identification. Basic solutions to the main issues are then presented. The main motivations for identifying CT models directly from sampled data are then discussed, before we present some specialised topics in system identification that deserve special attention. At the same time, this introductory chapter aims at tying together the different contributions of the book. In this regard, the outline of the book is presented in the last section.

1.2 System Identification Problem and Procedure

A linear time-invariant continuous-time system with input u and output y can always be described by

$$y(t) = G(p)u(t) + \xi(t) \tag{1.1}$$

where G is the transfer function, p the time-domain differential operator and the additive term $\xi(t)$ represents errors and disturbances of all natures. The source of $\xi(t)$ could be measurement errors, unmeasured process disturbances, model inadequacy, or combinations of these. It is assumed that the input signal $\{u(t), t_1 < t < t_N\}$ is applied to the system, with $u(t)$ and the output $y(t)$ both sampled at discrete times $t_1, \cdots, t_N$. The sampled signals are denoted by $\{u(t_k); y(t_k)\}$.

The identification problem can be stated as follows: determine a continuous-time model for the original CT system from N sampled measurements of the input and output $Z^N = \{u(t_k); y(t_k)\}_{k=1}^N$.

There are three different kinds of parameterised models:

- **grey-box models**, where the model is constructed in continuous-time from basic physical principles and the parameters represent unknown values of the system coefficients that, at least in principle, have a direct physical interpretation. Such models are also known as physically parameterised or tailor-made models;
- **black-box models**, which are families of flexible models of general applicability. The parameters in such models, which can be continuous time or discrete time, have no direct physical interpretation (even though the CT version is closer to the physically parameterised model than the DT version), but are used as vehicles to describe the properties of the input–output relationships of the system. Such models are also known as ready-made models;
- **data-based mechanistic (DBM) models**, which are effectively models identified initially in a black-box, generic model form but only considered credible if they can be interpreted in physically meaningful terms.

In this book, we restrict our attention to black-box model identification. The reader is referred, for instance, to [4] and the references therein, for grey-box model identification; and [53] and the references therein, for DBM model identification.

The basic ingredients for the system identification problem are as follows

- the data set;
- a model description class (the model structure);
- a criterion of fit between data and models;
- a way to evaluate the resulting models.

System identification deals with the problem of determining mathematical models of dynamical, continuous-time systems using measured input–output data. Basically this means that a set of candidate models is chosen and then a criterion of fit between model and data is developed. Finally, the model that best describes the data according to the criterion, within the model set, is computed using some suitable algorithm.

There are two fundamentally different time-domain approaches to the problem of obtaining a black-box CT model of a natural CT system from its sampled input–output data:

- the *indirect approach*, which involves two steps. First, a DT model for the original CT system is obtained by applying DT model estimation methods to the available sampled data; and then the DT model is transformed into the required CT form. This indirect approach has the advantage that

it uses well-established DT model identification methods [23, 39, 52]. Examples of such methods, which are known to give consistent and statistically efficient estimates under very general conditions, are prediction error methods optimal instrumental variable techniques;
- the **direct approach**, where a CT model is obtained immediately using CT model identification methods, such as those discussed in this book. Without relying any longer on analogue computers, the present techniques exploit the power of the digital tools. In this direct approach, the model remains in its original CT form.

Independent of how the identification problem is approached, a model parametrisation will lead to the definition of a predictor

$$\hat{y}(t_k, \boldsymbol{\theta}) = g(\boldsymbol{\theta}, Z^{k-1}) \tag{1.2}$$

that depends on the unknown parameter vector $\boldsymbol{\theta}$, and past data Z^{k-1}. The general procedure for estimating a parameterised model from sampled data, regardless of whether it is a CT or DT model, is as follows:

1. from observed data and the predictor $\hat{y}(t_k, \boldsymbol{\theta})$, form the sequence of prediction errors

$$\varepsilon(t_k, \boldsymbol{\theta}) = y(t_k) - \hat{y}(t_k, \boldsymbol{\theta}) \quad k = 1, \ldots, N \tag{1.3}$$

2. filter the prediction errors through a linear filter $F(\bullet)$ to enhance or attenuate interesting or unimportant frequency bands in the signals

$$\varepsilon_{\mathrm{f}}(t_k, \boldsymbol{\theta}) = F(\bullet)\varepsilon(t_k, \boldsymbol{\theta}) \tag{1.4}$$

where $\bullet$ can be the shift operator if the filter is in discrete time or the differential operator when the filter is in continuous time;

3. choose a scalar-valued, positive function $l(\cdot)$ to measure the size or norm of the filtered prediction error

$$l(\varepsilon_{\mathrm{f}}(t_k, \boldsymbol{\theta})) \tag{1.5}$$

4. minimise the sum of these norms

$$\hat{\boldsymbol{\theta}} = \arg\min_{\boldsymbol{\theta}} V_N(\boldsymbol{\theta}) \tag{1.6}$$

where

$$V_N(\boldsymbol{\theta}) = \frac{1}{N} \sum_{k=1}^{N} l(\varepsilon_{\mathrm{f}}(t_k, \boldsymbol{\theta})) \tag{1.7}$$

This procedure is general and pragmatic, in the sense that it is independent of the particular CT or DT model parametrisation used, although this parametrisation will affect the minimisation procedure. Indeed, as we will see, some peculiarities occur in CT model identification that do not occur in DT model identification. We deal with these aspects of the estimation problem in the following three sections. For simplicity of presentation, the formulation and basic solution of both CT and DT model identification problems will be restricted to the case of a linear, single-input, single-output system.

1.3 Basic Discrete-time Model Identification

1.3.1 Difference Equation Models

Perhaps the simplest model of a linear, discrete-time system is the linear difference equation

$$y(t_k) + a_1 y(t_{k-1}) + \ldots + a_{n_a} y(t_{k-n_a}) = b_1 u(t_{k-1}) + \ldots + b_{n_b} u(t_{k-n_b}) + v(t_k) \quad (1.8)$$

where the relationship between the input and output is expressed in terms of the sampled sequences $u(t_k)$ and $y(t_k)$ for $k = 1, 2, \ldots, N$.
Equation (1.8) can also be written as

$$A(q^{-1})y(t_k) = B(q^{-1})u(t_k) + v(t_k) \quad (1.9)$$

or

$$y(t_k) = \frac{B(q^{-1})}{A(q^{-1})}u(t_k) + \xi(t_k); \quad \xi(t_k) = \frac{1}{A(q^{-1})}v(t_k) \quad (1.10)$$

with

$$B(q^{-1}) = b_1 q^{-1} + \cdots + b_{n_b} q^{-n_b},$$
$$A(q^{-1}) = 1 + a_1 q^{-1} + \cdots + a_{n_a} q^{-n_a}$$

where q^{-1} is the backward shift operator, *i.e.*, $q^{-1}x(t_k) = x(t_{k-1})$. Equation (1.8) can be expressed in a vector form that is linear in the model parameters

$$y(t_k) = \boldsymbol{\varphi}^T(t_k)\boldsymbol{\theta} + v(t_k) \quad (1.11)$$

with

$$\boldsymbol{\varphi}^T(t_k) = \begin{bmatrix} -y(t_{k-1}) \cdots -y(t_{k-n_a}) \; u(t_{k-1}) \cdots u(t_{k-n_b}) \end{bmatrix} \quad (1.12)$$
$$\boldsymbol{\theta} = \begin{bmatrix} a_1 \; \ldots \; a_{n_a} \; b_1 \; \ldots \; b_{n_b} \end{bmatrix}^T \quad (1.13)$$

In this case, the predictor defined in (1.2) takes the form

$$\hat{y}(t_k, \boldsymbol{\theta}) = \boldsymbol{\varphi}^T(t_k)\boldsymbol{\theta} \quad (1.14)$$

1.3.2 The Traditional Least Squares Method

A traditional way of determining $\boldsymbol{\theta}$ is to minimise the sum of the squares of the prediction error $\varepsilon(t_k, \boldsymbol{\theta})$ by defining the criterion function

$$V_N(\boldsymbol{\theta}) = \frac{1}{N}\sum_{k=1}^{N}(y(t_k) - \hat{y}(t_k, \boldsymbol{\theta}))^2 \quad (1.15)$$

then minimising with respect to $\boldsymbol{\theta}$. In the present case, $\hat{y}(t_k, \boldsymbol{\theta})$ is linear in $\boldsymbol{\theta}$ and the criterion V_N is quadratic, so that $V_N(\boldsymbol{\theta})$ can be minimised analytically to give the least squares (LS) estimate

$$\hat{\boldsymbol{\theta}}_{\text{LS}} = \left[\frac{1}{N}\sum_{k=1}^{N}\boldsymbol{\varphi}(t_k)\boldsymbol{\varphi}^T(t_k)\right]^{-1} \frac{1}{N}\sum_{k=1}^{N}\boldsymbol{\varphi}(t_k)y(t_k) \qquad (1.16)$$

Once the regression vector $\boldsymbol{\varphi}(t_k)$ is constructed (from the measured sampled input–output data), the solution can be computed easily. In the statistical literature, this approach is usually referred to as 'linear regression analysis' and the linear, simultaneous equations that yield the solution (1.16) are termed the 'normal equations'. It is important to realise, however, that this is not a classical regression problem because the elements of the regression vector $\boldsymbol{\varphi}(t_k)$ are not exactly known, as required in regression analysis, but are measured variables that can be contaminated by noise. This has deleterious effects on the parameter estimates that are considered later in the book. It should also be noted that this basic LS method is a special case of the more general prediction error method discussed in Section 1.2, where the analytical solution does not exist and recourse has to be made to other optimisation approaches, such as gradient optimisation or iterative 'relaxation' estimation.

1.3.3 Example: First-order Difference Equation

The traditional LS method is outlined below for the case of a simple first-order DT model

$$y(t_k) + a_1 y(t_{k-1}) = b_1 u(t_{k-1}) + v(t_k) \qquad (1.17)$$

which can be written in regression form as

$$y(t_k) = [-y(t_{k-1})\; u(t_{k-1})] \begin{bmatrix} a_1 \\ b_1 \end{bmatrix} + v(t_k) \qquad (1.18)$$

Now, according to (1.16), from N available samples of the input and output signals observed at discrete times $t_1, \ldots, t_N$, uniformly spaced, the linear LS parameter estimates are given by

$$\begin{bmatrix} \hat{a}_1 \\ \hat{b}_1 \end{bmatrix} = \begin{bmatrix} \frac{1}{N}\sum_{k=1}^{N} y^2(t_{k-1}) & -\frac{1}{N}\sum_{k=1}^{N} y(t_{k-1})u(t_{k-1}) \\ -\frac{1}{N}\sum_{k=1}^{N} y(t_{k-1})u(t_{k-1}) & \frac{1}{N}\sum_{k=1}^{N} u^2(t_{k-1}) \end{bmatrix}^{-1} \begin{bmatrix} -\frac{1}{N}\sum_{k=1}^{N} y(t_k)y(t_{k-1}) \\ \frac{1}{N}\sum_{k=1}^{N} y(t_k)u(t_{k-1}) \end{bmatrix}$$

It is well known that, except in the special case when $v(t_k)$ in (1.8) is a white noise, simple LS estimation is unsatisfactory. Solutions to this problem led to the development of various approaches, as documented in many books (see e.g., [23, 39, 52]).

The simple difference equation model (1.8) and the well-known LS estimator (1.16) represent the simplest archetype of DT model identification.

1.3.4 Models for the Measurement Noise

In the previous section, we parameterised the description of dynamical systems in a particular form. There are many other possibilities that depend on the method used to model the measurement noise. A common approach used in DT model identification is to assume that the additive disturbance $\xi(t_k)$, contaminating the output measurement has a rational spectral density and can be represented as a DT white noise source $e(t_k)$ passed through a linear filter $H(q^{-1})$, i.e.,

$$\xi(t_k) = H(q^{-1})e(t_k) \tag{1.19}$$

When the system and noise models are combined, the standard discrete-time model of a linear dynamic system then takes the form

$$y(t_k) = G(q^{-1})u(t_k) + H(q^{-1})e(t_k) \tag{1.20}$$

In general, the estimation of the parameters in this model is a non-linear statistical estimation problem that can be solved in several ways: *e.g.*, gradient optimisation procedures, such as the maximum likelihood and prediction error methods; and iterative procedures, such as optimal instrumental variables.

1.4 Issues in Direct Continuous-time Model Identification

1.4.1 Differential Equation Models

A continuous-time model of the system takes the form of a constant coefficient differential equation

$$\frac{d^n y(t)}{dt^n} + a_1 \frac{d^{n-1} y(t)}{dt^{n-1}} + \cdots + a_n y(t) = b_0 \frac{d^m u(t)}{dt^m} + \cdots + b_m u(t) + v(t) \tag{1.21}$$

where $\frac{d^i x(t)}{dt^i}$ denotes the ith time derivative of the continuous-time signal $x(t)$. Equation (1.21) can be written alternatively as

$$y^{(n)}(t) + a_1 y^{(n-1)}(t) + \cdots + a_n y(t) = b_0 u^{(m)}(t) + \cdots + b_m u(t) + v(t) \tag{1.22}$$

where $x^{(i)}(t)$ denotes the ith time derivative of the continuous-time signal $x(t)$. Equation (1.21) or (1.22) can be written in the alternative time-domain differential operator form

$$A(p)y(t) = B(p)u(t) + v(t) \tag{1.23}$$

or

$$y(t) = \frac{B(p)}{A(p)} u(t) + \xi(t); \quad \xi(t) = \frac{1}{A(p)} v(t) \tag{1.24}$$

with

$$B(s) = b_0 p^m + b_1 p^{m-1} + \cdots + b_m$$
$$A(s) = p^n + a_1 p^{n-1} + \cdots + a_n, \quad n \geq m$$

and p denoting the differential operator, *i.e.*, $px(t) = \frac{dx(t)}{dt}$.
At any time instant $t = t_k$, (1.22) can be rewritten in regression-like form as

$$y^{(n)}(t_k) = \boldsymbol{\varphi}^T(t_k)\boldsymbol{\theta} + v(t_k) \qquad (1.25)$$

where the regressor and parameter vectors are now defined by

$$\boldsymbol{\varphi}^T(t_k) = \begin{bmatrix} -y^{(n-1)}(t_k) \cdots - y(t_k) \, u^{(m)}(t_k) \cdots u(t_k) \end{bmatrix} \qquad (1.26)$$
$$\boldsymbol{\theta}^T = [a_1 \; \ldots \; a_n \; b_0 \; \ldots \; b_m] \qquad (1.27)$$

However, unlike the difference equation model, where only sampled input and output data appear, the differential equation model (1.25) contains input and output time derivatives that are not available as measurement data in most practical cases. When compared with DT model identification, direct CT model identification raises several technical issues that are discussed in the following sections.

1.4.2 Input–Output Time Derivatives

The first difficulty in handling CT models is due to the need for the (normally unmeasured) time derivatives of the input–output signals. Various methods have been devised to deal with the reconstruction of the time derivatives [8, 37, 40–42, 45, 56]. Each method is characterised by specific advantages, such as mathematical convenience, simplicity in numerical implementation and computation, handling of initial conditions, physical insight, accuracy and others.

One traditional approach that dates from the days of analogue computers [47] is known as the state-variable filter (SVF) method. This method will be reviewed in Section 1.5.1, with the objective to highlight the differences from DT model identification.

1.4.3 Models for the Measurement Noise

Another difficulty with CT model identification is due to continuous-time stochastic processes. Although the noise model can be given in a CT form, difficulties arise in the estimation because of the theoretical and practical problems associated with the use of CT white noise and its derivatives. A noise model in an equivalent discrete-time form is much more flexible and easier to implement in the estimation problem. Thus, a hybrid model parametrisation method has evolved that involves the identification of a CT model for the process and a DT model for the noise [16, 29, 58]. The continuous-time hybrid model of a linear dynamic system then takes the following form,

$$x(t) = G(p)u(t) \tag{1.28a}$$

$$\xi(t_k) = H(q^{-1})e(t_k) \tag{1.28b}$$

$$y(t_k) = x(t_k) + \xi(t_k) \tag{1.28c}$$

or,

$$y(t_k) = G(p)u(t_k) + H(q^{-1})e(t_k) \tag{1.29}$$

where the operators have been mixed informally here in order to illustrate the nature of the estimation model. This approach alleviates the practical difficulties that may be encountered in the parameter estimation of the fully stochastic CT model.

1.5 Basic Direct Continuous-time Model Identification

1.5.1 The Traditional State-variable Filter Method

Let us first consider the transfer function (TF) model (1.23) in the simple noise-free case (the noise-free output is denoted as $x(t)$), *i.e.*,

$$A(p)x(t) = B(p)u(t) \tag{1.30}$$

Assume now that a SVF filter with operator model $F(p)$ is applied to both sides of (1.30). Then, ignoring transient initial conditions

$$A(p)F(p)x(t) = B(p)F(p)u(t) \tag{1.31}$$

The minimum-order SVF filter is typically chosen to have the following operator model form[4]

$$F(p) = \frac{1}{(p+\lambda)^n} \tag{1.32}$$

where λ is the parameter that can be used to define the bandwidth of the filter.

Equation (1.31) can then be rewritten, in expanded form, as

$$\left(\frac{p^n}{(p+\lambda)^n} + a_1 \frac{p^{n-1}}{(p+\lambda)^n} + \ldots + a_n \frac{1}{(p+\lambda)^n} \right) x(t)$$
$$= \left(b_0 \frac{p^m}{(p+\lambda)^n} + \ldots + b_m \frac{1}{(p+\lambda)^n} \right) u(t) \tag{1.33}$$

Let $F_i(p)$ for $i = 0, 1, \ldots, n$ be a set of filters defined as

---
[4] The filter dc gain can be made unity if this is thought desirable.

$$F_i(p) = \frac{p^i}{(p+\lambda)^n} \tag{1.34}$$

By using the filters defined in (1.34), Equation (1.33) can be rewritten, as

$$(F_n(p) + a_1 F_{n-1}(p) + \ldots + a_n F_0(p))\, x(t) = (b_0 F_m(p) + \ldots + b_m F_0(p))\, u(t) \tag{1.35}$$

Equation (1.35) can also be written as

$$x_f^{(n)}(t) + a_1 x_f^{(n-1)}(t) + \ldots + a_n x_f^{(0)}(t) = b_0 u_f^{(m)}(t) + \ldots + b_m u_f^{(0)}(t) \tag{1.36}$$

with

$$x_f^{(i)}(t) = f_i(t) * x(t)$$
$$u_f^{(i)}(t) = f_i(t) * u(t)$$

where $f_i(t)$, for $i = 0,\ldots,n$ represent the impulse responses of the filters defined in (1.34) and $*$ denotes the convolution operator. The filter outputs $x_f^{(i)}(t)$ and $u_f^{(i)}(t)$ provide *prefiltered* time derivatives of the inputs and outputs in the bandwidth of interest, which may then be exploited for model parameter estimation.

Consider now the situation where there is an additive noise on the output measurement. Then, at time instant $t = t_k$, substituting $x_f(t)$ for $y_f(t)$, (1.36) can be rewritten in standard linear regression-like form as

$$y_f^{(n)}(t_k) = \varphi_f^T(t_k)\boldsymbol{\theta} + \eta(t_k) \tag{1.37}$$

where $\eta(t_k)$ is a filtered noise term arising from the output measurement noise $\xi(t_k)$ and the filtering operations, while

$$\varphi_f^T(t_k) = \left[-y_f^{(n-1)}(t_k) \cdots -y_f^{(0)}(t_k)\, u_f^{(m)}(t_k) \cdots u_f^{(0)}(t_k)\right] \tag{1.38}$$

$$\boldsymbol{\theta} = [a_1 \,\ldots\, a_n \; b_0 \,\ldots\, b_m]^T \tag{1.39}$$

Now, from N available samples of the input and output signals observed at discrete times $t_1,\ldots,t_N$, not necessarily uniformly spaced, the linear least-squares (LS)-based SVF parameter estimates are given by

$$\hat{\boldsymbol{\theta}}_{\text{LSSVF}} = \left[\frac{1}{N}\sum_{k=1}^{N}\varphi_f(t_k)\varphi_f^T(t_k)\right]^{-1} \frac{1}{N}\sum_{k=1}^{N}\varphi_f(t_k)y_f^{(n)}(t_k) \tag{1.40}$$

It is well known that, except in the special case where $\eta(t_k)$ in (1.37) is zero mean and uncorrelated (white noise), LS-based SVF estimation although simple, is unsatisfactory. For example, even if the noise term $\xi(t_k)$ in (1.24) is white, the resultant parameter estimates are asymptotically biased and inconsistent. Solutions to this problem are the subject of various chapters in this book.

1.5.2 Example: First-order Differential Equation

The LS-based SVF method is outlined below for the case of a simple first-order differential model given by

$$y^{(1)}(t) + a_1 y(t) = b_0 u(t) + v(t) \qquad (1.41)$$

Applying a first-order SVF filter to both sides yields

$$\left(\frac{p}{p+\lambda} + a_1 \frac{1}{p+\lambda}\right) y(t) = b_0 \frac{1}{p+\lambda} u(t) + \frac{1}{p+\lambda} v(t) \qquad (1.42)$$

which can be rewritten as

$$(F_1(p) + a_1 F_0(p)) y(t) = b_0 F_0(p) u(t) + F_0(p) v(t) \qquad (1.43)$$

Equation (1.43) can be expressed for $t = t_k$ as

$$y_\mathrm{f}^{(1)}(t_k) + a_1 y_\mathrm{f}(t_k) = b_0 u_\mathrm{f}(t_k) + \eta(t_k) \qquad (1.44)$$

and written in regression-like form as

$$y_\mathrm{f}^{(1)}(t_k) = [-y_\mathrm{f}(t_k) \; u_\mathrm{f}(t_k)] \begin{bmatrix} a_1 \\ b_0 \end{bmatrix} + \eta(t_k) \qquad (1.45)$$

Now, according to (1.40), from N available samples of the input and output signals observed at discrete times $t_1, \ldots, t_N$, not necessarily uniformly spaced, the linear LS-based SVF parameter estimates are given by

$$\begin{bmatrix} \hat{a}_1 \\ \hat{b}_0 \end{bmatrix} = \begin{bmatrix} \frac{1}{N} \sum_{k=1}^{N} y_\mathrm{f}^2(t_k) & -\frac{1}{N} \sum_{k=1}^{N} y_\mathrm{f}(t_k) u_\mathrm{f}(t_k) \\ -\frac{1}{N} \sum_{k=1}^{N} y_\mathrm{f}(t_k) u_\mathrm{f}(t_k) & \frac{1}{N} \sum_{k=1}^{N} u_\mathrm{f}^2(t_k) \end{bmatrix}^{-1} \begin{bmatrix} -\frac{1}{N} \sum_{k=1}^{N} y_\mathrm{f}^{(1)}(t_k) y_\mathrm{f}(t_k) \\ \frac{1}{N} \sum_{k=1}^{N} y_\mathrm{f}^{(1)}(t_k) u_\mathrm{f}(t_k) \end{bmatrix}$$

The differential equation model (1.21) and the traditional LS-based SVF estimator (1.40) represent the simplest archetype of direct CT model identification.

1.6 Motivations for Identifying Continuous-time Models Directly from Sampled Data

There are many advantages in describing a physical system using CT models and also in identifying the CT models directly from sampled data. Here, we implicitly assume that the sampling rate is sufficiently fast to permit the identification of a continuous-time model from sampled data. It is true that DT models may be better suited to the design and simulation of control systems in a digital environment. However, because a DT model is estimated from sampled data with a fixed sampling rate, it is only really valid for this chosen sampling rate in its later applications. On the other hand, if a CT model is obtained using data collected at a fast sampling rate, this CT model may be discretised into a DT model with any sampling rate (not necessarily related to the original sampling rate). This is particularly advantageous in the situation where the issue is one of choosing the appropriate sampling rate for discrete-time system modelling and control system design.

The following subsections provide a discussion of the various issues of continuous-time versus discrete-time modelling.

1.6.1 Physical Insight into the System Properties

Most physical phenomena are more transparent in a CT setting, as the models of a physical system obtained from the application of physical laws are naturally in a CT form, such as differential equations. A continuous-time model is preferred to its discrete-time counterpart in the situation where one seeks a model that represents an underlying CT physical system, and wishes to estimate parameter values that have a physical meaning, such as time constants, natural frequencies, reaction times, elasticities, mass values, *etc*. While these parameters are directly linked to the CT model, the parameters of DT models are a function of the sampling interval and do not normally have any direct physical interpretation. For example, consider a mechanical system represented by the following second order CT transfer function

$$\frac{1}{ms^2 + bs + k} \quad (1.46)$$

where the parameters represent the mass, elasticity and friction that have a direct physical meaning. Now, a DT model of the same process will take the following form

$$\frac{b_0 z + b_1}{a_0 z^2 + a_1 z + a_2} \quad (1.47)$$

where z denotes the Z-transform variable. The parameters of the corresponding DT model do not have a direct physical meaning.

In many areas such as, for example, astrophysics, economics, mechanics, environmental science and biophysics, one is interested in the analysis of the physical system [3, 18, 27, 56]. In these areas, the direct identification of CT models has definite advantages.

1.6.2 Preservation of *a priori* Knowledge

The *a priori* knowledge of relative degree (the difference between the orders of the denominator and numerator) is easy to accommodate in CT models and, therefore, allows for the identification of more parsimonious models than in discrete time. This is obvious in the example of the second-order mechanical system, where additional parameters are introduced in the numerator of the DT transfer function by the sampling process.

1.6.3 Inherent Data Filtering

Explicit prefiltering strategies are recommended in the general approach to system identification [23, 52], where it is shown that these strategies improve the statistical efficiency of the parameter estimates. However, the prefiltering strategy is not inherent in DT model identification and the user is, therefore, confronted with a choice of whether to add prefiltering. This scenario is contrasted with the situation in CT identification, where the prefiltering is inherent and has two roles: in addition to its original use for reconstructing the filtered time derivatives within the bandwidth of the system to be identified, it became clear [58] that it can perform the same, statistically meaningful prefiltering role as in DT identification.

1.6.4 Non-uniformly Sampled Data

In some situations, it is difficult to obtain equidistant sampled data. This problem arises in medicine, environmental science, transport and traffic systems, astrophysics and other areas, where measurement is not under the control of the experimenter or where uniform sampling is practically impossible. For these non-uniformly sampled data systems, the standard DT linear, time-invariant models will not be applicable because the assumption of a uniformly sampled environment, as required for the existence of such discrete-time models, is violated. On the other hand, the coefficients of CT models are assumed to be independent of the sampling period and so they have a built-in capability to cope with the non-uniformly sampled data situation. With a small modification of the data handling procedure, the measurements are considered as points on a continuous line, which do not need to be equidistantly spaced.

1.6.5 Transformation between CT and DT Models

The parameter transformation between DT and CT representations is well studied [32]. The poles of a DT model are mapped according to the poles in the continuous-time model via the relation: $p_\mathrm{d} = \mathrm{e}^{p_\mathrm{c} T_\mathrm{s}}$, where p_d is the discrete-time pole, p_c is the continuous-time pole and T_s is the sampling interval. However, the zeros of the DT model are not as easily mapped as the poles. Even if the continuous-time system is minimum phase (*i.e.*, all zeros in the left half-plane), the corresponding discrete-time model can be non-minimum phase (*i.e.*, possesses zeros outside of the unit circle). In addition, due to the discrete nature of the measurements, the discrete-time models do not capture all of the information about the CT signals.

Moreover, in order to describe the signals *between* the sampling instants, some additional assumptions have to be made: for example, assuming that the excitation signal is constant within the sampling intervals (the zero-order hold assumption). However, violation of these assumptions may lead to estimation errors [35].

1.6.6 Sensitivity Problems of DT Models at High Sampling Rates

It is well known that discrete-time models encounter difficulties when the sampling frequency is too high in relation to the dominant frequencies of the system under study [1]. In this situation, the DT poles lie too close to the unit circle in the complex domain and the parameter estimates can become statistically ill-defined.

1.6.7 Stiff Systems

Stiff systems are systems with eigenvalues that are of a different order of magnitude, *i.e.*, the system contains both slow and fast dynamics. Since a DT model is related to a single sampling rate, it is often difficult in such situations to select a sampling rate that captures the complete dynamics of the system without any compromise. In order to illustrate this scenario, suppose that there are two time constants in a stiff system, the fast time constant is 1 (s) and the slow time constant is 100 (s). Typically, the sampling interval T_s is selected approximately in the range of 0.1 to 0.25 of the time constant in order to capture the dynamics associated with this time constant. Assume that $T_s = 0.25$ of the fast time constant, the poles in the discrete-time model are then $e^{-0.01T_s} = 0.9975$ and $e^{-T_s} = 0.7788$; and the slow pole is now very close to the unit circle in the complex plane (see previous subsection). As a result, a small estimation error could cause the estimated model to become unstable. However, if we now reduce the sampling rate, in order to avoid this difficulty, the dynamics associated with the fast time constant can become poorly identified. For example, suppose that $T_s = 10$ (s), then poles in the discrete model are $e^{-0.1} = 0.9048$ and $e^{-10} = 4.54 \times 10^{-5}$; so that, although the slow pole moves away from the unit circle, the fast pole virtually disappears from the model structure. Thus, we see that DT models find it difficult, at a specified sampling interval, to deal with both the quick and slow dynamics. In contrast to this, a stiff system can be better captured by a continuous-time model estimated from rapidly sampled data and the coefficients of this model are independent of the sampling rate.

1.7 Specialised Topics in System Identification

The general framework of parameter estimation for linear, time-invariant CT models has to include the consideration of additional factors, such as the identification of the model structure (the order of the transfer function polynomials and the size of any pure time delay), the possible non-integral nature of any pure time delay, identification from data collected during closed-loop experiments, *etc*. The following subsections briefly introduce these other factors and how they are discussed in the present book.

1.7.1 Identification of the Model Structure

Data-based modelling of a continuous-time model consists of model structure identification and the estimation of the parameters that characterise this structure. A continuous-time model structure is prescribed by its model order: *i.e.*, the order of its denominator polynomial and a relative degree. Due to the relative degree, there are many candidate model structures for a given model order. The objective of model structure identification is to select the 'best' model structure among all candidates, based on performance indices, which are often the sum of squares of prediction errors, the statistical properties of the errors and numerous statistical identification criteria. Model structure identification will be discussed in Chapter 6.

1.7.2 Identification of Pure Time (Transportation) Delay

An important additional part of the model structure is the existence of a pure time delay parameter. Unlike the situation in DT identification, where the time delay is assumed to be an integral number of sampling intervals and is often absorbed into the definition of the numerator polynomial (as leading zero-valued parameters), the time-delay parameter for CT system models is normally associated directly with the input signal and can have a non-integral value. As a result, the estimation of the time-delay parameter in CT identification deserves special attention. The interesting issues, in this regard, include simultaneously identifying the continuous-time model parameters and time-delays. Identification of systems with unknown time-delay is discussed in Chapters 11 and 12.

1.7.3 Identification of Continuous-time Noise Models

Identification of the system characteristics from output observations only is referred to as time-series analysis in econometrics, blind identification in signal processing, noise modelling in system identification, and operational modal analysis in mechanical engineering. A fundamental problem is how to model a continuous-time stochastic process based on sampled measurements. Several solutions are possible. One of the key issues is how to sample a continuous-time stochastic system. These issues are discussed in Chapter 2.

1.7.4 Identification of Multi-variable Systems

Systems with many input signals and/or many output signals are called multi-variable. Such systems are often more challenging to model. In particular, systems with several outputs can be difficult. A basic reason for the difficulties is that the couplings between several inputs and outputs lead to more complex models. The structures involved are richer and more parameters are required to obtain a good fit. A class of multi-variable system identification schemes, based on the subspace estimation and state-space realisations have emerged since the late 1980s. The use of these subspace methods to identify CT state-space models is discussed in Chapter 10.

1.7.5 Identification in Closed Loop

Many systems have feedback that cannot be interrupted for an identification experiment, as for example when an existing controller cannot safely be disconnected from an industrial process. In this situation, special procedures are necessary to avoid identifiability problems that can be induced by the feedback connection. Closed-loop identification schemes are described in Chapters 5 and 13.

1.7.6 Identification in the Frequency Domain

Linear dynamic systems have equivalent and complementary descriptions: in the time domain and in the frequency domain. Although the two descriptions are basically equivalent to each other, the formulation of the identification problem leads to different methods in the two domains. In many practical situations, parameter estimation in the frequency domain is of considerable interest [30]. Practical aspects of frequency-domain parametric identification methods are discussed in Chapter 8.

1.7.7 Software for Continuous-time Model Identification

System identification is typically an iterative procedure, where the insights and judgements of the user are mingled with formal considerations, extensive data handling and complex algorithms. To make the application of the identification procedure successful, it is almost always necessary to have some user-friendly software tools to facilitate the user's modelling. These software aspects are discussed in Chapters 8 and 9.

1.8 Historical Review

In contrast to the present day, the control world of the 1950s and 1960s was dominated by CT models as most control system design was concerned with

CT systems and most control system implementations employed analogue techniques. Moreover, almost all CT identification methods were largely deterministic, in the sense that they did not explicitly model the additive noise process nor attempt to quantify the statistical properties of the parameter estimates. Nevertheless, it is fascinating to see that some of these early papers introduced interesting concepts that foreshadowed later, important developments of a very similar nature. For instance, Valstar [43] and Young [47, 48] suggested the use of prefilters to solve the derivative measurement problem and this same 'state-variable filter' (SVF) approach[5] was re-discovered, some 20 years afterwards [34], under the title of 'Poisson-moment functionals' (PMF). Most early research also used completely analogue implementation, with both the prefilters and the estimation algorithm implemented in an analogue manner (*e.g.*, [14, 22, 48]); while some, adumbrating developments to come, utilised hybrid implementations where analogue prefiltering was combined with a digital identification algorithm [47, 49]. Indeed, two of these references [14, 22] also consider non-linear system identification, using a purely deterministic 'state-dependent parameter' approach that would emerge, many years later, in a stochastic, purely digital form (*e.g.*, [59]).

Also in the 1960s, it was realised that measurement noise could cause asymptotic bias on the parameter estimates when linear least squares (regression) methods were used to estimate the parameters in dynamic systems. Within a largely deterministic setting, papers appeared (*e.g.*, [20, 46, 49, 50]) that graphically demonstrated the value of the instrumental variable (IV) modification to both the recursive and *en-bloc* least squares algorithms that had been used for CT identification prior to this. Here, the instrumental variables were generated as the output of a recursively updated 'auxiliary model' of the system that, together with the prefilters, was implemented directly in continuous time.

Perhaps because of the dominant interest in DT identification and estimation since 1970, a stochastic formulation of the CT estimation problem did not appear until 1980. Then, Young and Jakeman [58], following the optimal prefiltering and recursive-iterative estimation procedures for DT systems (first presented in [51]), suggested an optimal 'hybrid' refined instrumental variable solution to the CT identification problem (RIVC). This involves a CT model of the system and a discrete-time ARMA model for the noise. However, at that time, it was only implemented in a simplified form (SRIVC) that yields consistent and statistically efficient parameter estimates when the additive noise $\xi(t)$ in (1.24) is white.

Responding to the research on RIVC estimation, Huang *et al.* [15] implemented an alternative hybrid solution that allowed for coloured noise and

---

[5] Also called the 'method of multiple filters' [49].

utilised a gradient optimisation algorithm, rather than the iterative solution used in the SRIVC algorithm and proposed in the RIVC algorithm. However, they chose to convert the problem into an entirely DT form and so did not implement the prefilters and auxiliary model explicitly in continuous time. Also, they did not present any stochastic simulation results and, as such, it is not possible to reach any clear conclusions about the statistically efficiency of the algorithm.

Despite these excursions into stochastic systems and full statistical estimation, most publications on CT identification during the 1970s and 1980s were deterministic in concept and suggested various methods of implementing prefilters (see [8] for a recent overview for example). Most of the deterministic approaches are available in the continuous-time system identification (CONTSID) toolbox[6] for MATLAB® (see Chapter 9 in this book). Since the deterministic methods have been documented so fully, it will suffice here merely to outline the main features of each approach.

Linear Filter Methods

These methods originated from the third author's early research in this area [47, 48, 50] where the method was referred to as the *'method of multiple filters'* (MMF). It involves passing the input and output signals through a chain of (usually identical) first-order prefilters with user-specified bandwidth, normally selected so that it spans the anticipated bandwidth of the system being identified (see Section 1.5.2 for the simplest example of this method). More recently this MMF approach has been re-named the *generalised Poisson moment functionals* (GPMF) approach [34, 40]. Recent MMF/GPMF developments have been proposed by the first author and his co-workers [2, 7, 9–11].

Integration-based Methods

The main idea of these methods is to avoid the differentiation of the data by performing an order n integration. These integral methods can be roughly divided into two groups. The first group, using numerical integration and orthogonal function methods, performs a basic integration of the data and special attention has to be paid to the initial condition issue. The second group includes the *linear integral filter* (LIF: [33]) and the *re-initialised partial moments* (RPM: [6]) approaches. Here, advanced integration methods are used to avoid the initial condition problem either by exploiting a moving integration window (LIF) or a time-shifting window (RPM).

---

[6] http://www.cran.uhp-nancy.fr/contsid: the CONTSID toolbox also contains the SRIVC and RIVC algorithms.

Modulating Function Methods

This approach was first suggested almost half a century ago by Shinbrot in order to estimate the parameters of linear and non-linear systems [36]. Further developments have been based on different modulating functions. These include the Fourier-based functions [26], in either trigonometric or complex exponential form; spline-type functions; Hermite functions and, more recently, Hartley-based functions [42]. A very important advantage of using Fourier- and Hartley-based modulating functions is that the model estimation can be formulated entirely in the frequency domain, making it possible to use efficient DFT/FFT techniques.

Other Methods

Several other approaches have been suggested that cannot be classified directly into any of the categories discussed in the previous subsections. An interesting approach is reported in [16] where the idea is to replace the differentiation as represented by the Laplace operator s by the operator w. These operators are related via the bilinear relationship $w = \frac{s-a}{s+a}$. The new w-domain model can be estimated directly from sampled data, using filtered signals. Afterwards, the parameters of this model are translated back to the parameters of the ordinary continuous-time model, using simple algebraic relations. The w operator can be an all-pass filter. In this case, the filter does not alter the frequency content of the signals and only affects the phase. This setup is closely related to the SVF method (see also [5] for a related scheme where the filters take the form of CT Laguerre functions). Two other approaches that have attracted a lot of attention in the identification community in the 1990s are subspace-based methods (see [2, 13, 17, 21, 24, 25] but also Chapter 10 in this book) and finite difference methods [19, 28, 38]. This latter approach, which is based on replacing the differentiation operator with finite differences, will be considered in some depth in this book (see Chapters 2 and 3).

Most recently, attention has re-focused on stochastic model identification and statistically optimal CT estimation procedures. First, in discussing a paper on optimal CT identification by Wang and Gawthrop [44], Young [54] drew attention to the virtues of the existing SRIVC estimation algorithm and demonstrated its superiority in a simulation example. This encouraged, at last, the implementation of the full hybrid RIVC algorithm [57] that is presented and evaluated in Chapter 4 of this book, as well as the development of an associated closed-loop version of the algorithm [12], which is described in Chapter 5. A useful by-product of this renewed interest in these optimal algorithms is that optimal RIV algorithms are now available for Box–Jenkins-type stochastic transfer function models of CT and DT systems, providing a unified approach to the identification and estimation of transfer function models [55].

Table 1.1. Organisation of the book

Topics	User's choice
Experiment design	Chapter 8
Model structure determination	Chapter 6
Model parameter estimation	Chapters 2, 3, 4, 7, 14
Model validation	Chapter 6, 10
Closed-loop identification	Chapters 5, 13
Subspace identification	Chapter 10
Frequency-domain identification	Chapter 8
Identification of noise models	Chapter 2
Time-delay identification	Chapters 11, 12
Identification from non-uniformly sampled data	Chapters 2, 9, 11
Software aspects	Chapters 8, 9
Practical examples	Chapters 4, 7, 8, 9, 11, 13

1.9 Outline of the Book

The aim of this book is to bring together contributions from well-known experts in the field of continuous-time system identification from sampled data. The book is written as a research monograph with a survey focus. It is meant to be interesting for a broad audience, including researchers and graduate students in systems and control, as well as in signal processing. It also comprehensively covers material suitable for specialised graduate courses in these areas.

Table 1.1 illustrates the book's structure in relation to the system identification procedure and specialised topics of direct CT model identification from sampled data.

The book begins with the work by Erik Larsson, Magnus Mossberg, and Torsten Söderström. In Chapter 2, the authors describe identification of continuous-time systems using discrete-time data via the approximation of the derivatives in the continuous-time description. They focus on continuous-time stochastic systems and study the effect of random noise in the data on the estimated parameters. The chapter begins with the concepts of modelling and sampling of continuous-time stochastic systems, followed by several simple examples that illustrate how and why first-order approximations of the derivatives may result in bias errors of the estimated continuous-time system parameters. The authors proceed to generalise the approximations of the derivatives by imposing conditions on the coefficients of the approximation equations, which create a bandpass effect for the differential operators, instead of a high-pass one. The end result of this modification is a class of more robust estimation algorithms, particularly in the presence of unmodelled wide-band noise. Interestingly, the authors extend the results obtained for the case of uniformly sampled data systems to non-uniformly sampled data systems by modifying the coefficients of the approximation equations to include the variation of sampling interval. The authors have also shown how to compute the Cramér–Rao bound for both uniformly sampled and non-uniformly sampled data systems.

In Chapter 3, Juan Yuz and Graham Goodwin continue the discussion of identification of continuous-time models using discrete-time data via the approximation of continuous-time derivatives. More precisely, the sampled data models expressed using the δ operator are used to estimate the parameters of the underlying continuous-time system. The authors analyse the potential problems, information loss and the robustness issues associated with this approach to continuous-time system identification using discrete-time data. The authors argue that one always needs to define a bandwidth of validity relative to the factors that include sampling rate, nature of input between samples, nature of sampling process, the system relative degree and high-frequency poles and zeros, ensuring the analysis is restricted to the bandwidth defined. The authors then describe time and frequency-domain methods for ensuring insensitivity to high-frequency folded artifacts in the identification of continuous-time systems from sampled data.

In Chapter 4, Peter Young, Hugues Garnier and Marion Gilson depart from the approaches using approximation of continuous-time derivatives. Instead, they focus on the identification and estimation of continuous-time hybrid Box–Jenkins transfer function models from discrete-time data. In this chapter, the model of the dynamic system is estimated in continuous-time, differential equation form, while the associated additive noise model is estimated as a discrete-time, autoregressive-moving average process. The differential operator in the continuous-time dynamic model is replaced by the filtered differential operator, which effectively overcomes the high-pass nature of the differential operator via the appropriate choice of the filters. Their approach involves concurrent discrete-time noise model estimation and uses this estimated noise model in the iterative-adaptive design of statistically optimal prefilters that effectively attenuate noise outside the passband of the system. As a result, the optimal prefilters prewhiten the noise remaining within the bandwidth. Instrumental variable (IV) methods are used in conjunction with the prefiltering and noise modelling to achieve optimal estimation results. The evaluation of the developed algorithms is based on comprehensive Monte Carlo simulation analysis, as well as on two practical examples selected from environmental and electro-mechanical fields.

In Chapter 5, Marion Gilson, Hugues Garnier, Peter Young and Paul Van den Hof continue the discussion on instrumental variable identification of continuous-time systems. However, they focus their attention on closed-loop systems. The authors argue that closed-loop identification of continuous-time models is still an issue that has not received adequate attention, which needs to be addressed. In a general framework of instrumental variable identification, the authors present several instrumental variable and instrumental variable-related methods. Furthermore, a statistically efficient method of closed-loop IV estimation is proposed together with several design variables that are required for the estimation. In order to achieve minimum variance in the estimation, *a priori* information about the noise model is required. The authors propose several bootstrap methods for extracting this required information approximately from the measurement data. A comparison between the proposed methods is illustrated, with a simulation example showing that the optimal estimator can be accurately approximated by an appropriate choice of the design parameters.

In Chapter 6, Liuping Wang and Peter Young present new results in instrumental variable identification to address the potential singularity issue, the selection of the best model structure and the computation of prediction errors in the spirit of cross-validation. By assuming *a priori* knowledge of the relative degree and a maximum model order for the continuous-time system structure, the authors propose an instrumental variable solution of all candidate models within this frame. The instrumental variable solution is based on UDV matrix factorisation, where the higher-order model parameter estimates will not affect the lower-order parameter estimates, allowing the natural truncation of model orders. Another important contribution of this chapter is the derivation of the mathematically compact formulae for the computation of prediction errors based on the instrumental variable identification algorithm. The prediction errors that are in the spirit of cross-validation, calculated for all candidate models in a systematic manner, provides an effective tool for the selection of a continuous-time model structure.

In Chapter 7, Liuping Wang and Peter Gawthrop propose a two-step approach to continuous-time model identification. In the first step, the authors use the frequency-sampling filters model to compress a given set of experimental data for a dynamic system to a non-parametric model in the form of a discrete-step response/frequency response with a link to their continuous-time counter-part. The authors point out that in the context of continuous-time system identification when the system is operating in a fast sampling environment, the frequency sampling filters model has the advantage over the traditional finite impulse response (FIR) model, and yet maintains the same fundamental features of an FIR model. Because the data compression is model based, discrete-time noise models are naturally embedded to allow optimal estimation. The authors also show that the nature of the frequency-sampling filters model permits the incorporation of *a priori* knowledge such as gain, frequency response, and time delays, into the estimation algorithm through equality or inequality constraints. The solution is in the form of quadratic programming with extension to multiple-input and multiple-output systems. Having obtained the relatively noise-free step-response data, the methods proposed from Chapters 2–5 in this book could be applied to obtain a continuous-time transfer function model as the second step of the integrated approach. However, as an alternative, the authors choose to focus on the identification of a partially known continuous-time system; this is essentially a non-linear optimisation problem. This optimisation is simplified by the fact that the step response is a relatively short and noise-free representation of the original data. As an illustration, a partially known system consisting of an unknown unstable system with a known stabilising controller in the feedback loop is identified using the methods of this chapter.

In Chapter 8, Rik Pintelon, Johan Schoukens and Yves Rolain present some practical aspects of frequency-domain approaches to continuous-time system identification in a tutorial style. The chapter begins with the description of inter-sample behaviour and the measurement setup, particularly in the frequency domain. They specifically assume a band-limited continuous-time white noise as the source of disturbances. Then, the authors move on to discuss identification of both continuous-time and discrete-time parametric models using frequency-response analysis. This is particularly useful in continuous-time modelling as the Laplace variable becomes a simple complex variable $i\omega$ without the complication of using implementation filters or approximation of the derivatives. The leakage errors and residual alias errors are quantified accordingly. The authors point out that with periodic excitation signals, both errors will vanish. In the stochastic framework, the authors extend the study scope to include identification in closed-loop systems. Within this framework, the authors discuss the properties of frequency-domain Gaussian maximum likelihood estimators, and their solutions when using either a periodic excitation signal or an arbitrary excitation signal. To demonstrate the practical aspects of the frequency-domain approaches, the authors present two real-world applications: one is the modelling analysis for an operational amplifier, while the other is a flight-flutter analysis. The chapter is concluded with guidelines for continuous-time modelling, based on the authors' experience in the past two decades.

In Chapter 9, as a followup to Rik Pintelon, Johan Schoukens and Yves Rolain's practical aspects of the frequency-domain approach to continuous-time system identification, Hugues Garnier, Marion Gilson, Thierry Bastogne and Michel Mensler present software-development aspects of time-domain continuous-time system identification. The authors have developed a continuous-time system identification (CONTSID) toolbox that was intended to fill the gap between theoretical and algorithmic development and to provide handy tools that engineers can use in a day-to-day operation. The toolbox is a successful implementation of the time-domain methods developed over the last three decades for estimating continuous-time linear transfer function or state-space models directly from uniformly and non-uniformly sampled input–output data. This chapter gives an overview of the toolbox, discusses the advantages of direct continuous-time model identification methods from sampled data and illustrates them on three practical examples selected from robotic, biological and electro-mechanical fields.

In Chapter 10, Rolf Johansson presents the theory, algorithms and validation results for system identification of continuous-time state-space models from finite input–output sequences. The algorithms developed are methods of subspace model identification and stochastic realisation adapted to the continuous-time context. The resulting model can be decomposed into an input–output model and a stochastic innovation model. Using the Riccati equation, the author has designed a procedure to provide a reduced-order stochastic model that is minimal with respect to system order as well as the number of stochastic inputs thereby avoiding several problems appearing in standard application of stochastic realisation to the model validation problem.

Identification of continuous-time systems with time delay is an important issue. The book devotes the next two chapters to present new approaches in this area. In Chapter 11, Salim Ahmed, Biao Huang and Sirish Shah discuss how to estimate a continuous-time time-delay system from measurements with an extension to non-uniformly sampled data. Unlike the majority of the more traditional approaches in which time delay is estimated as part of a non-linear optimisation scheme, their approach embeds this parameter into the linear-in-the-parameter regression. Hence, it can be estimated directly as part of the parameters in the continuous-time transfer function model. With this modification, the existing continuous-time estimation techniques in the framework of linear regression, such as the results presented in Chapters 3 to 8, can be naturally applied to estimate a continuous-time transfer function model along with its time delay. The authors extend their estimation results to include the case of non-uniformly sampled data, where continuous-time basis-function models are used as part of the iterative scheme for predicting and interpreting the missing data during the sampling period. The authors also present two applications: one is the continuous-time model identification of a dryer while the other is a mixing process.

In Chapter 12, Zi-Jiang Yang takes a different approach to identification of continuous-time systems with multiple delays. The author formulates the objective of estimating both the parameters of the transfer function model along with time-delay parameters as a non-linear optimisation problem. Along the lines of gradient methods, the author presents an unseparable non-linear least squares (UNSEPNLS) method and a separable non-linear least squares (SEPNLS) method. These approaches are modified to include instrumental variables in the optimisation procedure to eliminate bias errors due to measurement noise. Simulation studies for SISO systems are provided. It is shown that if the initial parameters are suitably chosen, the UNSEPNLS and SEPNLS algorithms work quite well in the case of low measurement noise, and their modified versions with instrumental variables yield consistent estimates in the presence of high measurement noise. Strategies are also incorporated to avoid the local convergence of the iterative algorithms.

The last two chapters of this book are devoted to two special approaches in continuous-time system identification. In Chapter 13, Michel Fliess and Hebertt Sira-Ramírez use algebraic tools stemming from module theory, operational calculus and differential algebra to tackle the problems of parametric identification of linear continuous-time systems. Their solutions are robust in the presence of noise and can be implemented in real time. One of the applicable areas is in identification of closed-loop systems. The authors carefully explain the mathematical machinery used in the development, and provide several case studies to illustrate the applicability of their work.

The book concludes with Chapter 14 where Rolf Johansson uses systems-analysis tools in continuous-time system identification. There are three stages involved in the estimation of a continuous-time model. The first stage provides discrete-time spectral estimation with an unbiased estimate of the input–output transfer function in the case of uncorrelated noise and control input. The second stage provides an unbiased, overparameterised continuous-time linear model. Finally, the third stage of the algorithm provides a passivity preserving model order reduction resulting in a reduced-order continuous-time state-space model maintaining the same spectral properties and interpretations.

1.10 Main References

Early research on system identification focused on identification of CT models from CT data (*e.g.*, [47]). Subsequently, however as previously said, rapid developments in digital data acquisition and computers have resulted in attention being shifted to the identification of discrete-time models from sampled data, as documented in many books (see, *e.g.*, [23, 39, 52]).
The first significant survey to the subject of identification of CT models written by Young [45] appeared in 1981. Subsequently, Unbehauen and Rao tracked further developments in the field [40, 41]. A book was also devoted to the subject of identification of CT models [37]. Recent surveys can be found in [8, 31, 42, 56].

References

1. K. J. Aström. On the choice of sampling rates in parametric identification of time series. *Information Science*, 1(1):273–278, 1969.
2. T. Bastogne, H. Garnier, and P. Sibille. A PMF-based subspace method for continuous-time model identification. Application to a multivariable winding process. *International Journal of Control*, 74(2):118–132, 2001.
3. A.R. Bergström. *Continuous-time Econometric Modeling*. Oxford University Press, Oxford, UK, 1990.
4. T. Bohlin. *Practical Grey-box Process Identification. Theory and Applications*. Springer, Series: Advanced in Industrial Control, 2006.
5. C.T. Chou, M. Verhaegen, and R. Johansson. Continuous-time identification of SISO systems using Laguerre functions. *IEEE Transactions on Signal Processing*, 47(2):349–362, 1999.
6. M. Djamai, E. Tohme, R. Ouvrard, and S. Bachir. Continuous-time model identification using reinitialised partial moments. Application to power amplifier modeling. *14th IFAC Symposium on System Identification (SYSID'2006)*, Newcastle, Australia, March 2006.
7. H. Garnier, M. Gilson, and W.X. Zheng. A bias-eliminated least-squares method for continuous-time model identification of closed-loop systems. *International Journal of Control*, 73(1):38–48, 2000.
8. H. Garnier, M. Mensler, and A. Richard. Continuous-time model identification from sampled data. Implementation issues and performance evaluation. *International Journal of Control*, 76(13):1337–1357, 2003.
9. H. Garnier, P. Sibille, and T. Bastogne. A bias-free least-squares parameter estimator for continuous-time state space models. *36th IEEE Conference on Decision and Control (CDC'97)*, volume 2, pages 1860–1865, San Diego, California, USA, December 1997.
10. H. Garnier, P. Sibille, and A. Richard. Continuous-time canonical state-space model identification via Poisson moment functionals. *34th IEEE Conference on Decision and Control (CDC'95)*, volume 2, pages 3004–3009, New Orleans, USA, December 1995.
11. H. Garnier, P. Sibille, and T. Spott. Influence of the initial covariance matrix on recursive LS estimation of continuous models via generalised Poisson moment functionals. *10th IFAC Symposium on System Identification (SYSID'94)*, pages 3669–3674, Copenhagen, Denmark, 1994.
12. M. Gilson, H. Garnier, P.C. Young, and P. Van den Hof. A refined IV method for closed-loop system identification. *14th IFAC Symposium on System Identification (SYSID'2006)*, pages 903–908, Newcastle, Australia, March 2006.
13. B. Haverkamp. *State Space Identification. Theory and Practice*. PhD thesis, TU Delft, The Netherlands, 2001.
14. L.L. Hoberock and R.H. Kohr. An experimental determination of differential equations to describe simple nonlinear systems. *Joint Automatic Control Conference*, Washington, Seattle, pages 616–623, 1966.
15. H. Huang, C. Chen, and Y. Chao. Augmented hybrid method for continuous process identification from sampled data with coloured noise. *International Journal of Control*, 46:1373–1390, 1987.
16. R. Johansson. Identification of continuous-time models. *IEEE Transactions on Signal Processing*, 42(4):887–896, 1994.

17. R. Johansson, M. Verhaegen, and C.T. Chou. Stochastic theory of continuous-time state-space identification. *IEEE Transactions on Signal Processing*, 47(1):41–50, 1999.
18. R.E. Kearney and I.W. Hunter. System identification of human joint dynamics. *Critical Reviews in Biomedical Engineering*, 18(1):55–87, 2000.
19. E. Larsson. *Identification of Stochastic Continuous-time Systems. Algorithms, Irregular Sampling and Cramér-Rao Bounds*. PhD thesis, Uppsala University, Sweden, 2004.
20. V.S. Levadi. Parameter estimation of linear systems in the presence of noise. *International Conference on Microwaves, Circuit Theory and Information Theory*, Tokyo, Japan, 1964.
21. W. Li, H. Raghavan, and S. Shah. Subspace identification of continuous time models for process fault detection and isolation. *Journal of Process Control*, 13(5):407–421, 2003.
22. P.M. Lion. Rapid identification of linear and nonlinear systems. *Joint Automatic Control Conference, Washington, Seattle*, pages 605–615, 1966.
23. L. Ljung. *System Identification. Theory for the User*. Prentice Hall, Upper Saddle River, USA, 2nd edition, 1999.
24. G. Mercère, R. Ouvrard, M. Gilson, and H. Garnier. Subspace based methods for continuous-time model identification of MIMO systems from filtered sampled data. *9th European Control Conference (ECC'07)*, Kos, Greece, July 2007.
25. A. Ohsumi, K. Kameyama, and K. Yamaguchi. Subspace identification for continuous-time stochastic systems via distribution-based approach. *Automatica*, 38(1):63–79, 2002.
26. A.E. Pearson, Y. Shen, and V. Klein. Application of Fourier modulating function to parameter estimation of a multivariable linear differential system. *10th IFAC Symposium on System Identification (SYSID'94)*, pages 49–54, Copenhagen, Denmark, 1994.
27. M.S. Phadke and S.M. Wu. Modelling of continuous stochastic processses from discrete observations with application to sunspots data. *Journal American Statistical Association*, 69(346):325–329, 1974.
28. D.T. Pham. Estimation of continuous-time autoregressive model from finely sampled data. *IEEE Transactions on Signal Processing*, 48:2576–2584, 2000.
29. R. Pintelon, J. Schoukens, and Y. Rolain. Box-Jenkins continuous-time modeling. *Automatica*, 36(7):983–991, 2000.
30. R. Pintelon and J. Schoukens. *System Identification: a Frequency Domain Approach*. IEEE Press, Piscataway, USA, 2001.
31. G.P. Rao and H. Unbehauen. Identification of continuous-time systems. *IEE Proceedings - Control Theory and Applications*, 153(2):185–220, 2006.
32. K.J. Åström, P. Hagander, and J. Sternby. Zeros of sampled systems. *Automatica*, 20(1):31–38, 1984.
33. S. Sagara and Z.Y. Zhao. Numerical integration approach to on-line identification of continuous-time systems. *Automatica*, 26(1):63–74, 1990.
34. D.C. Saha and G.P. Rao. *Identification of Continuous Dynamical Systems - The Poisson Moment Functionals (PMF) approach*. Springer-Verlag, Berlin, 1983.
35. J. Schoukens, R. Pintelon, and H. Van Hamme. Identification of linear dynamic systems using piecewise constant excitations: use, misuse and alternatives. *Automatica*, 30(7):1953–1169, 1994.
36. M. Shinbrot. On the analysis of linear and non linear systems. *Transactions of the ASME*, 79:547–552, 1957.

37. N.K. Sinha and G.P. Rao (eds). *Identification of Continuous-time Systems. Methodology and Computer Implementation.* Kluwer Academic Publishers, Dordrecht, 1991.
38. T. Söderström, H. Fan, B. Carlsson, and S. Bigi. Least squares parameter estimation of continuous-time ARX models from discrete-time data. *IEEE Transactions on Automatic Control*, 42(5):659–673, 1997.
39. T. Söderström and P. Stoica. *System Identification.* Series in Systems and Control Engineering. Prentice Hall, Englewood Cliffs, 1989.
40. H. Unbehauen and G.P. Rao. *Identification of Continuous Systems.* Systems and control series. North-Holland, Amsterdam, 1987.
41. H. Unbehauen and G.P. Rao. Continuous-time approaches to system identification - a survey. *Automatica*, 26(1):23–35, 1990.
42. H. Unbehauen and G.P. Rao. A review of identification in continuous-time systems. *Annual Reviews in Control*, 22:145–171, 1998.
43. J.E. Valstar. In flight dynamic checkout. *IEEE Transactions on Aerospace*, AS1(2), 1963.
44. L. Wang and P. Gawthrop. On the estimation of continuous-time transfer functions. *International Journal of Control*, 74(9):889–904, 2001.
45. P. Young. Parameter estimation for continuous-time models - a survey. *Automatica*, 17(1):23–39, 1981.
46. P. C. Young. Regression analysis and process parameter estimation: a cautionary message. *Simulation*, 10:125–128, 1968.
47. P.C. Young. In flight dynamic checkout - a discussion. *IEEE Transactions on Aerospace*, 2:1106–1111, 1964.
48. P.C. Young. The determination of the parameters of a dynamic process. *Radio and Electronic Engineering (Journal of IERE)*, 20:345–361, 1965.
49. P.C. Young. Process parameter estimation and self adaptive control. In P.H. Hammond (ed), *Theory of Self Adaptive Control Systems*, Plenum Press, pages 118–140, 1966.
50. P.C. Young. An instrumental variable method for real-time identification of a noisy process. *Automatica*, 6:271–287, 1970.
51. P.C. Young. Some observations on instrumental variable methods of time-series analysis. *International Journal of Control*, 23:593–612, 1976.
52. P.C. Young. *Recursive Estimation and Time-series Analysis.* Springer-Verlag, Berlin, 1984.
53. P.C. Young. Data-based mechanistic modeling of environmental, ecological, economic and engineering systems. *Journal of Modelling & Software*, 13:105–122, 1998.
54. P.C. Young. Optimal IV identification and estimation of continuous-time TF models. *15th IFAC World Congress on Automatic Control*, Barcelona, Spain, July 2002.
55. P.C. Young. The refined instrumental variable method: unified estimation of discrete and continuous-time transfer function models. *Journal Européen des Systèmes Automatisés*, 2008.
56. P.C. Young and H. Garnier. Identification and estimation of continuous-time data-based mechanistic (DBM) models for environmental systems. *Environmental Modelling and Software*, 21(8):1055–1072, August 2006.
57. P.C. Young, H. Garnier, and M. Gilson. An optimal instrumental variable approach for identifying hybrid continuous-time Box–Jenkins model. *14th IFAC*

Symposium on System Identification (SYSID'2006), pages 225–230, Newcastle, Australia, March 2006.
58. P.C. Young and A.J. Jakeman. Refined instrumental variable methods of time-series analysis: Part III, extensions. *International Journal of Control*, 31:741–764, 1980.
59. P.C. Young, P. McKenna, and J. Bruun. Identification of nonlinear stochastic systems by state dependent parameter estimation. *International Journal of Control*, 74:1837–1857, 2001.

2

Estimation of Continuous-time Stochastic System Parameters

Erik K. Larsson[1], Magnus Mossberg[2], and Torsten Söderström[3]

[1] Ericsson Research, Stockholm, Sweden[†]
[2] Karlstad University, Sweden
[3] Uppsala University, Sweden

2.1 Background and Motivation

Identification of continuous-time (CT) systems is a fundamental problem that has applications in virtually all disciplines of science. Examples of mathematical models of CT phenomena appear in such diverse areas as biology, economics, physics, and signal processing. A small selection of references are cited below. Models in the economics of renewable resources, *e.g.*, in biology, is discussed in [9]. Sunspot data modelling by means of CT ARMA models is carried out in [39]. Aspects of economic growth models is the topic of [59]. Models for stock-price fluctuations are discussed in [48] and stochastic volatility models of the short-term interest rate can be found in [2]. The use of Ito's calculus in modern financial theory with applications in financial decision making is presented in [36]. Continuous-time models for the heat dynamics of a building is described in [35]. Modelling of random fatigue crack growth in materials can be found in [50], and models of human head movements appear in [20]. Identification of ship-steering dynamics by means of linear CT models and the maximum likelihood (ML) method is considered in [6]. Numerous other examples of applications of stochastic differential equations (SDEs) can be found in the literature. See, for example, [25, Chapter 7] where various modelling examples, including population dynamics, investment finance, radio-astronomy, biological waste treatment, *etc.* can be found.

A major reason for using CT models is that most physical systems or phenomena are CT in nature, for example in many control applications. Due to the advent of digital computers, research for control and identification of these CT systems and processes has concentrated on their discretised models with samples from the underlying CT system inputs and outputs. Recently interest in identification of CT systems and processes has arisen (see the above

---

[†] This work was performed during Erik Larsson's PhD studies at Uppsala University.

references). The parameters in CT models are strongly correlated with the physical properties of the systems; something that is very appealing to an engineer. One particularly interesting and practical scenario is identification of CT systems using discrete-time (DT) data.

Most results on identification of CT models are developed in a deterministic perspective, [43]. Sometimes features are included to make the estimates less sensitive to measurement noise in general. The theme of this chapter is rather to describe identification methods that are well suited to cope with random disturbances. Hence, we focus on stochastic systems and study in particular what effects random noise in the data has on the parameter estimates.

Lately, the problem of system identification from irregularly sampling time instants has received some attention. Irregular sampling is common in such diverse applications as time-series analysis [1,44], radar imaging [22], medical imaging [24], and biomedicine [37]. One obvious special case when the measurements are irregularly spaced is when the underlying sampling process is uniform but some samples are occasionally missing, which can happen because of sensor failures, or some other inability to observe the system. However, the sampling process can also be of a more general form, *i.e.*, not necessarily associated with a missing-data scenario. Furthermore, the sampling process can be deterministic or random, and the random times of observations may be dependent on or independent of the actual process.

A main concern when dealing with irregularly observed data is that the (computational) complexity of conventional methods, such as the prediction error method (PEM), increases substantially. Another intriguing question relevant to irregular sampling is how a particular sampling scheme will affect the properties of the corresponding estimation problem. Inevitably, a missing-data scenario will result in loss of information; the question is how this information loss will impact the parameter estimation problem. Although the original data process is ergodic and/or stationary, the subprocess, due to the missing data, may not be ergodic and/or stationary. In general, if the sampling process is nonuniform, a sampled representation of the system may be hard (or impossible) to obtain. In this case, it may be better to work with a CT representation of the dynamic system, and estimate its parameters directly from the (unevenly sampled) measured data, without going via a sampled representation of the system. Even though some of the mathematics associated with CT stochastic dynamic systems is more complicated than the corresponding theory for DT systems, a CT description of a dynamic system is usually easy to obtain since many physical systems are naturally modelled in continuous time. Consequently, CT modelling has been shown to be a way forward in a number of applications.

The chapter is organised as follows. In the next section we describe basic models of CT stochastic systems, where SDEs play an important role. Estimation methods are typically based on DT measurements, so the transformation into DT models is important, and Section 2.3 deals with sampling of stochastic CT models. Section 2.4 describes two basic approaches for identifying CT systems. One is the direct approach, where the CT model parameters are estimated directly. The second is the indirect approach, where first a DT model is fit to data, and then transformed into CT. Some simple examples, showing that there are many traps when using simple estimation techniques, and that the presence of CT noise can produce large errors in the estimates are given in Section 2.5. More systematic approaches for direct methods are given and analysed in Section 2.6. The Cramér–Rao lower bound (CRB) gives a performance limit in many estimation problems. How to compute such bounds in the case of identifying stochastic CT systems is described in Section 2.7. Section 2.8 is devoted to some numerical studies of direct methods, while Section 2.9 contains conclusions of the chapter.

The notations used in the chapter follow basically the general conventions for this book. A few exceptions are that the sampling interval is denoted by h, t_k denotes a general kth time point, particularly for irregularly sampled data, t may denote either continuous or discrete time, and $\triangleq$ denotes a definition.

2.2 Modelling of Continuous-time Stochastic Systems

We consider processes with rational spectra. Note that this condition on the spectrum is natural and non-restrictive for many CT phenomena, see Section 2.1. More exactly, the spectrum is described as

$$\Phi_c(s) = \sigma^2 \frac{D(s)D(-s)}{A(s)A(-s)} \tag{2.1}$$

where

$$A(s) = \prod_{i=1}^{n}(s - \rho_i) = s^n + a_1 s^{n-1} + \cdots + a_n \tag{2.2}$$

$$D(s) = \prod_{i=1}^{m}(s - \eta_i) = s^m + d_1 s^{m-1} + \cdots + d_m \tag{2.3}$$

with $m < n$. By assumption, $A(s)$ and $D(s)$ are coprime, and the zeros of $A(s)$ and $D(s)$, here denoted by ρ_i and η_i, respectively, are in the left half-plane. Note that the roll-off rate of the spectrum, *i.e.*, how fast the spectrum tends to zero for high frequencies, is given by $2(n - m)$.

The process with the spectrum (2.1) is represented in the time domain as

$$A(p)y(t) = D(p)e_c(t) \tag{2.4}$$

where p denotes the differentiation operator and where $e_c(t)$ is CT white noise with intensity σ^2. The process $y(t)$, which is $n-m-1$ times differentiable, is a CT ARMA, or CARMA, process, and the case $D(s) = 1$ is referred to as a CT AR, or CAR, process. Continuous-time white noise, see, e.g., [4], is defined as a stationary Gaussian process with zero mean and constant spectrum. However, such a continuous-time white noise process does not exist in the traditional sense because its variance is infinite. To define CT white noise mathematically we have to rely on generalised functions. In other words, CT white noise is a mathematical trick used when modelling a given spectrum, and not a physically present signal of infinite variance.

Formally, the linear SDE

$$\begin{aligned} d\boldsymbol{x}(t) &= \boldsymbol{A}\boldsymbol{x}(t)\,dt + d\boldsymbol{w}(t) \\ y(t) &= \boldsymbol{C}\boldsymbol{x}(t) \end{aligned} \tag{2.5}$$

is used instead of (2.4). Here, $\boldsymbol{x}(t)$ is an n-dimensional state vector, $\boldsymbol{w}(t)$ a Wiener process with incremental covariance $\boldsymbol{\Sigma}\,dt$, and the initial value $\boldsymbol{x}(t_0)$ is a Gaussian random variable with zero mean and covariance matrix $\boldsymbol{P}_0$, and independent of $\boldsymbol{w}(t)$. By including an input signal $u(t)$ in (2.5), the resulting SDE

$$\begin{aligned} d\boldsymbol{x}(t) &= \boldsymbol{A}\boldsymbol{x}(t)\,dt + \boldsymbol{B}u(t)\,dt + d\boldsymbol{w}(t) \\ y(t) &= \boldsymbol{C}\boldsymbol{x}(t) \end{aligned} \tag{2.6}$$

can describe a CT ARMAX, or CARMAX, process

$$A(p)y(t) = B(p)u(t) + D(p)e_c(t) \tag{2.7}$$

where

$$B(s) = b_1 s^{n-1} + \cdots + b_n \tag{2.8}$$

The special case $D(s) = 1$ gives a CT ARX, or CARX, process. More material on modelling of CT stochastic systems can be found in [4, 53].

2.3 Sampling of Continuous-time Stochastic Models

A main characteristic of CT models is that the signals are functions of a CT variable. In practice, however, it is obviously impossible to observe a process continuously over any given time period, due to, for example, limitations on the precision of the measuring device or due to unavailability of observations at every time point. Consequently, it is very important to know what happens when a CT process is observed at DT instances. The process of converting a CT system into a corresponding DT system is referred to as sampling. One should be aware that there exist different notions of sampling. For instance, in the control literature deterministic (or zero-order hold) sampling is often assumed, which is characterised by the input being kept constant between the sampling instants. In the context of CT stochastic models sampling means that the obtained DT stochastic process is a sequence of stochastic variables with the same second-order statistical properties as the original CT system at the sampling instants. Inevitably, when sampling a system we may lose information about the system since we only observe the system at certain DT points. In particular, a complete probabilistic description of the original system conditioned on the observations can in general not be obtained; the linear Gaussian case being an important exception.

Reconsider the linear SDE (2.6),

$$\begin{aligned} \mathrm{d}\boldsymbol{x}(t) &= \boldsymbol{A}\boldsymbol{x}(t)\,\mathrm{d}t + \boldsymbol{B}u(t)\,\mathrm{d}t + \mathrm{d}\boldsymbol{w}(t) \\ y(t) &= \boldsymbol{C}\boldsymbol{x}(t) \end{aligned} \quad (2.9)$$

Let us assume that $u(t)$ and $y(t)$ are sampled (observed) at the DT instants $t_1, t_2, \ldots, t_N$, and we want equations that relate the values of the state vector $\boldsymbol{x}(t)$ and the measured signals $u(t)$ and $y(t)$ at the sampling instants. It holds that the solution of the SDE (2.9) is given by (see, e.g. [4, 53])

$$\boldsymbol{x}(t) = e^{\boldsymbol{A}(t-t_0)}\boldsymbol{x}(t_0) + \underbrace{\int_{t_0}^{t} e^{\boldsymbol{A}(t-s)}\boldsymbol{B}u(s)\,\mathrm{d}s}_{\boldsymbol{I}_1} + \underbrace{\int_{t_0}^{t} e^{\boldsymbol{A}(t-s)}\,\mathrm{d}\boldsymbol{w}(s)}_{\boldsymbol{I}_2} \quad (2.10)$$

Clearly, the explicit solution of (2.10) depends on the two integrals $\boldsymbol{I}_1$ and $\boldsymbol{I}_2$. In particular, we notice two things:

- the solution of $\boldsymbol{I}_1$ depends on the properties of the input signal $u(t)$. Consequently, we cannot solve (2.10) without knowledge or assumptions about $u(t)$. We will come back to this issue in Section 2.3.2;
- the second integral, $\boldsymbol{I}_2$, is a so-called stochastic integral. This integral cannot be attributed a rigourous meaning by using classical integration theory. To fully understand its characteristics we need tools from the theory of SDEs, see, e.g. [4]. The solution to $\boldsymbol{I}_2$ will be treated in more detail in Section 2.3.1 below.

In order to simplify the presentation we will begin by considering the stochastic part ($\boldsymbol{I}_2$) in more detail. Hence, for now we assume that $u(t) \equiv 0$ so the contribution from ($\boldsymbol{I}_1$) is zero.

2.3.1 Sampling of CARMA Systems

Consider the CARMA process $y(t)$ defined in Section 2.2. Let us represent the process $y(t)$ in state-space form (see (2.5)) as

$$dx(t) = Ax(t)\,dt + dw(t)$$
$$y(t) = Cx(t) \tag{2.11}$$

where $w(t)$ is a Wiener process with incremental covariance matrix $\Sigma\,dt$. Then it can be shown (see, e.g. [4, 53]) that the sampled version of (2.11) satisfies

$$x(t_{k+1}) = F(h_k)x(t_k) + v(t_k)$$
$$y(t_k) = Cx(t_k) \tag{2.12}$$

where $h_k \triangleq t_{k+1} - t_k$,

$$F(h_k) = e^{Ah_k} \tag{2.13}$$

and where $v(t_k)$ is DT white noise with zero mean and covariance matrix

$$R_d(h_k) = \int_0^{h_k} e^{As} \Sigma e^{A^T s}\,ds \tag{2.14}$$

The DT system (2.12) is referred to as the sampled counterpart of the CT system (2.11). This sampling procedure is commonly referred to as exact or instantaneous sampling in the literature. The sampled system (2.12) will have the same second-order statistical properties as the original system (2.11), but, in general, it is time varying and non-stationary.

Let us now turn to the more specific situation with equidistant sampling, i.e., when $h_k = h$ is fixed. In this case (2.12) becomes a time-invariant system

$$x(kh + h) = Fx(kh) + v(kh)$$
$$y(kh) = Cx(kh) \tag{2.15}$$

where $F = e^{Ah}$, and where $v(kh)$ is DT white noise with zero mean and covariance matrix

$$R_d = \int_0^h e^{As} \Sigma e^{A^T s}\,ds \tag{2.16}$$

Also, we would like to point out that in order to generate DT data from a linear CT stochastic system, the correct way is to go via the sampled counterpart. In other words, determine the sampled version of the CT system (as outlined above), and use the sampled system to generate data at the desired sampling instants. Note that there is no approximation of the underlying SDE involved in this approach.

Another common way of representing the system (2.15) is via the DT ARMA model

$$D_d(q)y(kh) = C_d(q)e(kh) \tag{2.17}$$

where $e(kh)$ is DT white noise with zero mean and variance λ^2, and $C_\mathrm{d}(z)$ and $D_\mathrm{d}(z)$ are stable polynomials. The variance λ^2, and the polynomials $D_\mathrm{d}(z)$ and $C_\mathrm{d}(z)$ can be found by means of spectral factorisation

$$\lambda^2 \frac{C_\mathrm{f}(z)C_\mathrm{d}(z^{-1})}{D_\mathrm{d}(z)D_\mathrm{d}(z^{-1})} = \boldsymbol{C}(z\boldsymbol{I}-\boldsymbol{F})^{-1}\boldsymbol{R}_\mathrm{d}(z^{-1}\boldsymbol{I}-\boldsymbol{F}^T)^{-1}\boldsymbol{C}^T \qquad (2.18)$$

It turns out that $C_\mathrm{d}(z)$ and $D_\mathrm{d}(z)$ have the following structure

$$D_\mathrm{d}(z) = \prod_{i=1}^{n}(z-\kappa_i), \qquad C_\mathrm{d}(z) = \prod_{i=1}^{n-1}(z-\zeta_i) \qquad (2.19)$$

where

$$\kappa_i = e^{\rho_i h}, \quad i=1,\ldots,n \qquad (2.20)$$

and ρ_i (see (2.2)) is a CT pole and κ_i (see (2.19)) is the corresponding DT pole. A similar relation between the CT and DT zeros can in general not be described by a closed-form expression. However, for small sampling intervals, some explicit but approximate results exist. It holds that

$$C_\mathrm{d}(z) = C_1(z)C_2(z) \qquad (2.21)$$

where the $n-m-1$ zeros of $C_1(z)$ are usually referred to as sampling zeros, and where the m zeros of $C_2(z)$ are often called intrinsic zeros. Furthermore, it can be shown that as the sampling interval tends to zero, the zeros of $C_1(z)$ converge to the zeros of a constant polynomial that does not depend on the CT parameters, and the zeros of $C_2(z)$ converge to $z=1$, see [5, 61]. In particular, it follows that sampling of a CAR process yields a DT ARMA process with $C_\mathrm{d}(z) = C_1(z)$. This will have a profound impact when estimating CAR models using DT data, see Section 2.5. Further results and extensions can be found in [27, 28].

2.3.2 Sampling of Systems with Inputs

Reconsider the linear SDE (2.6),

$$\begin{aligned} \mathrm{d}\boldsymbol{x}(t) &= \boldsymbol{A}\boldsymbol{x}(t)\,\mathrm{d}t + \boldsymbol{B}u(t)\,\mathrm{d}t + \mathrm{d}\boldsymbol{w}(t) \\ y(t) &= \boldsymbol{C}\boldsymbol{x}(t) \end{aligned} \qquad (2.22)$$

where $\boldsymbol{x}(t)$, $y(t)$, and $u(t)$ are the state vector, and the output and input signals, respectively. As mentioned earlier, the solution of (2.22) depends on the input signal $u(t)$. Up to this point we have focused on the stochastic part of (2.22) and assumed that the input is zero ($u(t) \equiv 0$). We will now show how to deal with the input for two cases of particular interest.

Example 2.1. The first example targets situations when we do not have full knowledge about the input $u(t)$. In such cases it is common to assume that the input is a CT stochastic process with random realisations. This is, for example, achieved by assuming that $u(t)$ is described by a linear SDE

$$\begin{aligned} \mathrm{d}\boldsymbol{x}_2(t) &= \boldsymbol{A}_2\boldsymbol{x}_2(t)\,\mathrm{d}t + \mathrm{d}\boldsymbol{w}_2(t) \\ u(t) &= \boldsymbol{C}_2\boldsymbol{x}_2(t) \end{aligned} \qquad (2.23)$$

where $\boldsymbol{w}_2(t)$ is a Wiener process with incremental covariance matrix $\boldsymbol{\Sigma}_2\,\mathrm{d}t$. In order to sample (2.22) where the input $u(t)$ is described by (2.23), it is convenient to write (2.22) and (2.23) jointly as

$$\begin{aligned} \begin{bmatrix} \mathrm{d}\boldsymbol{x}(t) \\ \mathrm{d}\boldsymbol{x}_2(t) \end{bmatrix} &= \begin{bmatrix} \boldsymbol{A} & \boldsymbol{B}\boldsymbol{C}_2 \\ \boldsymbol{0} & \boldsymbol{A}_2 \end{bmatrix} \begin{bmatrix} \boldsymbol{x}(t) \\ \boldsymbol{x}_2(t) \end{bmatrix} \mathrm{d}t + \begin{bmatrix} \mathrm{d}\boldsymbol{w}(t) \\ \mathrm{d}\boldsymbol{w}_2(t) \end{bmatrix} \\ \begin{bmatrix} y(t) \\ u(t) \end{bmatrix} &= \begin{bmatrix} \boldsymbol{C} & \boldsymbol{0} \\ \boldsymbol{0} & \boldsymbol{C}_2 \end{bmatrix} \begin{bmatrix} \boldsymbol{x}(t) \\ \boldsymbol{x}_2(t) \end{bmatrix} \end{aligned} \qquad (2.24)$$

Then, we can simply apply the techniques discussed in Section 2.3.1 in order to sample (2.24). ∎

Example 2.2. Another common alternative is to assume that the input is kept constant between consecutive samples. This assumption is natural in cases when we have full knowledge of $u(t)$ and its behaviour. For example, in a computer-controlled system, the input to the process is kept constant until a new value is calculated in the control algorithm. The procedure can be described as that of using a sample-and-hold circuit between the control algorithm in the computer and the CT process to be controlled. In this case it can be shown that the sampled system (2.22) satisfies

$$\begin{aligned} \boldsymbol{x}(t_{k+1}) &= \boldsymbol{F}(h_k)\boldsymbol{x}(t_k) + \boldsymbol{\Gamma}(h_k)u(t_k) + \boldsymbol{v}(t_k) \\ y(t_k) &= \boldsymbol{C}\boldsymbol{x}(t_k) \end{aligned} \qquad (2.25)$$

where $h_k = t_{k+1} - t_k$, $\boldsymbol{F}(h_k)$ is given in (2.13), $\boldsymbol{v}(t_k)$ is DT white noise with zero mean and covariance matrix $\boldsymbol{R}_\mathrm{d}(h_k)$ defined in (2.14), and

$$\boldsymbol{\Gamma}(h_k) = \int_0^{h_k} e^{\boldsymbol{A}s}\,\mathrm{d}s\,\boldsymbol{B} \qquad (2.26)$$

∎

2.4 A General Approach to Estimation of Continuous-time Stochastic Models

There exist several different methods for estimating CT stochastic models from DT measurements. We will begin by describing a general framework applicable to a large class of CT models, including the models treated in Section 2.2. This framework is based on a prediction error methodology [3,33, 57] and results presented in Section 2.3.

For simplicity, let us consider the model (2.5),

$$\begin{aligned} d\boldsymbol{x}(t) &= \boldsymbol{A}\boldsymbol{x}(t)\,dt + d\boldsymbol{w}(t) \\ y(t) &= \boldsymbol{C}\boldsymbol{x}(t) \end{aligned} \quad (2.27)$$

where $\boldsymbol{w}(t)$ is a Wiener process with incremental covariance $\boldsymbol{\Sigma}\,dt$. The model (2.27) is parameterised by $\boldsymbol{\theta}$, which may appear in the matrices $\boldsymbol{A}$, $\boldsymbol{C}$, and $\boldsymbol{\Sigma}$. The aim is to estimate $\boldsymbol{\theta}$ from possibly irregular DT measurements $y(t_1), \ldots, y(t_N)$.

From Section 2.3 we know that sampling of the model (2.27) yields the time-varying DT model (2.12). Needless to say, the model matrices $\boldsymbol{F}(h_k)$, $\boldsymbol{C}$, and $\boldsymbol{R}_d(h_k)$ in (2.12) may now depend on $\boldsymbol{\theta}$ and h_k in a rather intricate manner. The basic prediction error estimate of $\boldsymbol{\theta}$ is then obtained as

$$\hat{\boldsymbol{\theta}} = \arg\min_{\boldsymbol{\theta}} V_{\text{PEM}}(\boldsymbol{\theta}), \qquad V_{\text{PEM}}(\boldsymbol{\theta}) = \frac{1}{N}\sum_{k=1}^{N}\varepsilon^2(t_k;\boldsymbol{\theta}) \quad (2.28)$$

where the prediction errors $\varepsilon(t_k;\boldsymbol{\theta})$ are found from the DT model (2.12) by means of the Kalman filter, see, e.g. [3,53]. Some comments are in order

- the minimisation in (2.28) is performed using some standard numerical minimisation scheme. This requires repeated evaluations of the cost function $V_{\text{PEM}}(\boldsymbol{\theta})$, in which the sampling procedure and the Kalman filtering step have to be carried out each time t_k. Consequently, this approach can become computationally very intense;
- by modifying the cost function $V_{\text{PEM}}(\boldsymbol{\theta})$ according to

$$V_{\text{ML}}(\boldsymbol{\theta}) = \frac{1}{N}\sum_{k=1}^{N}\left(\frac{\varepsilon^2(t_k;\boldsymbol{\theta})}{r(t_k;\boldsymbol{\theta})} + \log\{r(t_k;\boldsymbol{\theta})\}\right) \quad (2.29)$$

where $r(t_k;\boldsymbol{\theta})$ is the variance of $\varepsilon(t_k;\boldsymbol{\theta})$, we obtain the ML estimate, which is known to be consistent and statistically efficient under rather weak assumptions (identifiability, *etc.*).
- It is straightforward to include an input component in the model (2.27). However, in this case the sampling procedure will depend on the current assumption regarding the input, cf. Section 2.3.2. Furthermore, DT measurement noise can easily be introduced into this framework.
- For equidistant sampling, the computational load will be substantially reduced since both the sampling process and the Kalman filter algorithm are simplified.

2.4.1 Direct and Indirect Methods

Traditionally, methods for estimating CT models are often classified into two broad categories

- *direct methods*. These methods can work either with approximate or exact DT models. The methods work in two steps. For approximate DT models the first step consists of using DT approximations for signals and operators in the CT model. The result is, in general, an approximate DT model where the original parameterisation is kept. The main role of the first step is to handle the non-measurable input and output time derivatives. In the second step, the CT parameters are estimated from the model obtained in the first step using some identification scheme. The main point with these direct methods is that the desired parameters are estimated directly without any need of introducing additional auxiliary parameters. Traditionally, the DT approximations in the first step are considered to be linear operations, which can be interpreted as input and output signal preprocessing.

  In contrast to the previous point, the first step can also be made exact. For example, the framework described previously in (2.27) and (2.28), can be considered to be a direct approach. In that case, the first step would correspond to the sampling procedure, which yields an exact DT model (parameterised using the CT model parameters). Note, though, that the sampling process is not a linear operation. It is important to realise this distinction between approximate and exact direct methods, since they generally have different characteristics. In-depth studies of some approximate direct methods for estimation of CARX models will be treated in Sections 2.5 and 2.6;

- *indirect methods*. In indirect methods the CT model is transformed into a corresponding DT model that is identified. The desired CT system parameters are then obtained by transferring the DT model back into CT. As an example, let us reconsider the model (2.27) and assume that observations are equidistant. Then, the corresponding sampled version of the CT model is given by a DT ARMA model, see (2.17). The parameters of the DT model (the parameters of $D_d(q)$ and $C_d(q)$) are then estimated using some standard technique and the original parameters are finally obtained by reverse sampling of the estimated DT model. One thing to note is that the DT parameters enter linearly (w.r.t. q) in the model (2.17), which makes it fairly easy to calculate prediction errors, gradients, and Hessians. This is important in order to improve the performance of the involved minimisation algorithms and to decrease the computational load. On the other hand, it is evident that indirect methods are not suited for irregular sampling since the DT model becomes time varying. Finally, we would like to point out that the transformation from CT into DT is generally assumed to be exact for indirect methods. However, one can also envision indirect methods that use approximate transformations from CT into DT. Consequently, just as for direct methods we can distinguish between exact and approximate indirect methods.

Let us conclude this section by listing some advantages/disadvantages with direct and indirect methods (mainly applicable to the CT models discussed in Section 2.2). The discussion targets primarily approximate direct methods and exact indirect methods. For a more comprehensive discussion we refer to [30].

- Estimates from indirect methods are usually consistent, whereas estimates from direct methods are often biased. Needless to say, this statement depends on what estimation scheme is used. However, since the first step in direct methods in general produces an approximate DT model this approach has an inherent bias source. Also, the bias in direct methods is often proportional to the sampling interval.
- The choice of the sampling interval is generally crucial and difficult for indirect methods, whereas direct methods are less sensitive to the choice of the sampling interval. The main concern for indirect methods is the variance, which can increase substantially if too low or too a high sampling interval is chosen. Direct methods usually benefit from fast sampling since the bias is proportional to the sampling interval. On the other hand, the variance usually increases with decreasing values of the sampling interval for direct methods.
- Many estimation schemes rely on some form of numerical minimisation. It has been shown that the initialisation procedure of the optimisation scheme can be a key factor to obtain satisfactory estimation results, see [34, 42]. Indirect methods seem to be more sensitive than direct methods in this respect.
- We have seen that sampling of CT models with an input requires knowledge or assumptions regarding the nature of the input. Violation of these assumptions may lead to severe estimation errors, see [41, 47]. In this respect direct methods may be more robust compared to indirect methods. Nevertheless, this statement is once again coupled with the hypothesis that direct methods often rely on approximations of the involved operators, whereas indirect methods use exact sampling.
- It is well known [49] that indirect methods suffer from numerical problems due to ill-conditioning when the sampling interval becomes small. The main reason is that all DT poles (and some zeros) cluster around the point $z = 1$ when the sampling interval tends to zero. Another way to understand these numerical problems is to realise that the shift operators (q or q^{-1}) have very short memories (one sample), see [62]. A way to partly cure this problem is to restructure the DT representation by using, for instance, the delta operator.

2.5 Introductory Examples

We consider here simple direct identification techniques with constant sampling rate, using CARX models.

For illustration purposes, let us consider a first-order CT system

$$(p + a)y(t) = bu(t) + e_c(t) \tag{2.30}$$

where p is the differentiation operator, $y(t)$ is the output signal, $u(t)$ is a measurable input signal, and $e_c(t)$ a (non-measurable) noise source with intensity σ^2. We notice that the output $y(t)$ can be written as a sum of a deterministic and a stochastic term

$$y(t) = \frac{b}{p+a}u(t) + \frac{1}{p+a}e_c(t) \triangleq y_u(t) + y_e(t) \tag{2.31}$$

see Section 2.2 for further discussions of such models.

We are interested in estimating the parameter vector

$$\boldsymbol{\theta}_o = \begin{bmatrix} a & b \end{bmatrix}^T \tag{2.32}$$

from DT data $\{y(kh), u(kh)\}_{k=1}^N$, where h denotes the sampling interval.

As a first step, we need to decide how to treat the non-measurable derivative of $y(t)$. A natural approach is to approximate the differentiation operator p by a difference operator. Two simple examples are the delta forward operator δ_f and the delta backward operator δ_b, defined as

$$\delta_f \triangleq \frac{q-1}{h}, \qquad \delta_b \triangleq \frac{1-q^{-1}}{h}, \tag{2.33}$$

where q and q^{-1} are the forward and backward shift operators, respectively. The delta operators are known to have an approximation error of order $\mathcal{O}(h)$ (usually referred to as first-order approximations). By substituting the differentiation operator p in (2.30) by the delta forward operator δ_f, we obtain a DT model

$$\begin{aligned} \delta_f y(kh) &= \boldsymbol{\varphi}^T(kh)\boldsymbol{\theta}_f + \varepsilon_1(kh) \\ \boldsymbol{\varphi}^T(kh) &= \begin{bmatrix} -y(kh) & u(kh) \end{bmatrix} \end{aligned} \tag{2.34}$$

where $\varepsilon_1(kh)$ is an equation error. It is of interest to examine whether a least squares approach

$$\hat{\boldsymbol{\theta}}_f(N) = \left(\frac{1}{N}\sum_{k=1}^N \boldsymbol{\varphi}(kh)\boldsymbol{\varphi}^T(kh)\right)^{-1} \left(\frac{1}{N}\sum_{k=1}^N \boldsymbol{\varphi}(kh)\delta_f y(kh)\right) \tag{2.35}$$

for estimating $\boldsymbol{\theta}_f$ gives feasible results. Consider an asymptotic ($N \to \infty$) analysis of the estimates. Due to ergodicity assumptions the asymptotic estimate (2.35) equals

$$\hat{\boldsymbol{\theta}}_f \triangleq \lim_{N \to \infty} \hat{\boldsymbol{\theta}}_f(N) = \left(\mathsf{E}\boldsymbol{\varphi}(kh)\boldsymbol{\varphi}^T(kh)\right)^{-1}\left(\mathsf{E}\boldsymbol{\varphi}(kh)\delta_f y(kh)\right) \qquad (2.36)$$

where E denotes the expectation operator. Similarly, by substituting the differentiation operator in (2.30) by the delta backward operator δ_b, we obtain the DT model

$$\delta_b y(kh) = \boldsymbol{\varphi}^T(kh)\boldsymbol{\theta}_b + \varepsilon_2(kh) \qquad (2.37)$$

where $\varepsilon_2(kh)$ is another equation error. The asymptotic least squares estimate of $\boldsymbol{\theta}_b$ then reads

$$\hat{\boldsymbol{\theta}}_b = \left(\mathsf{E}\boldsymbol{\varphi}(kh)\boldsymbol{\varphi}^T(kh)\right)^{-1}\left(\mathsf{E}\boldsymbol{\varphi}(kh)\delta_b y(kh)\right) \qquad (2.38)$$

At this point, it is important to realise that (2.34) and (2.37) are heuristic models.

Example 2.3. Assume that the input $u(t)$ to (2.30) is a sinusoid of angular frequency $\omega = 0.5$. The asymptotic parameter estimates of a and b were computed for delta forward and delta backward approximations of the differentiation operator (via (2.36) and (2.38)). The true values of the parameters are $a = b = 1$. Two different values of the sampling interval h were considered. The estimates were computed as functions of the signal-to-noise ratio

$$\text{SNR} \triangleq \mathsf{E}y_u^2(t)/\mathsf{E}y_e^2(t) \qquad (2.39)$$

In other words, the intensity σ^2 was chosen so that a certain SNR was obtained. The results are shown in Figure 2.1. It is evident from the figure that the delta forward approximation gives estimates with a small bias. Furthermore, the bias decreases with decreasing h. When a delta backward approximation is used, the bias is still small for high enough SNRs, but becomes very pronounced for moderate and small SNRs.

The results from this example clearly indicate that the estimation problem is considerably more difficult to solve for low SNRs. Although, the delta forward and the delta backward operators both have approximation errors of order $\mathcal{O}(h)$, the least squares estimate is highly dependent on the chosen operator for low SNRs. This is an intriguing observation. For high SNRs the result is intuitively more appealing. Both operators provide estimates with an error of the same magnitude, and the error depends on the sampling interval, *i.e.*, on how accurately the differentiation operator is approximated. The pure stochastic case (SNR $\to 0$) will be treated in the next example. ∎

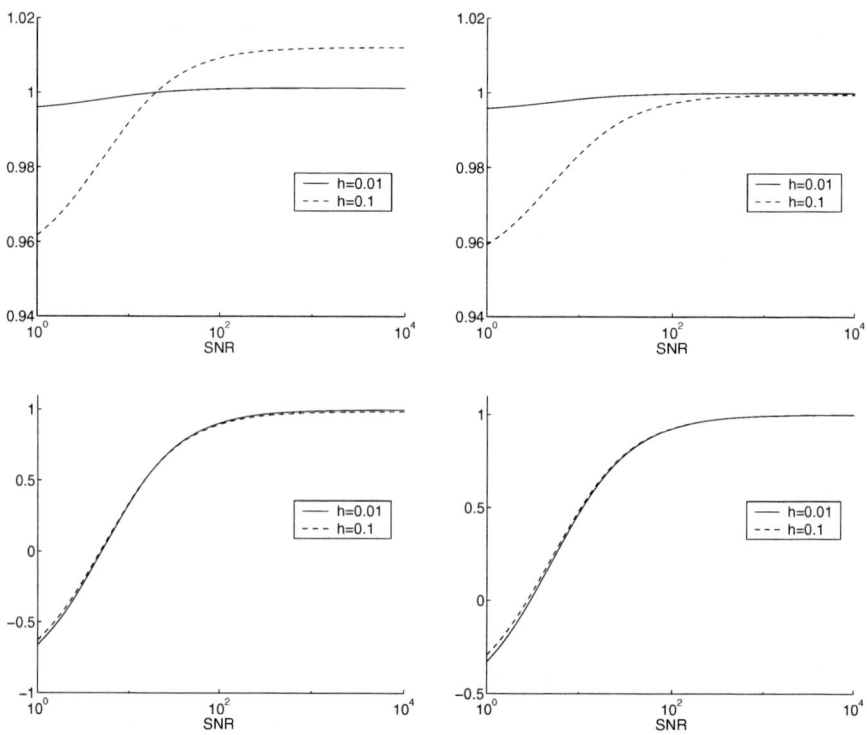

Fig. 2.1. Asymptotic ($N \to \infty$) parameter estimates as functions of the SNR. *Upper figures* – delta forward approximation, *lower figures* – delta backward approximation. *Left figures* – estimate of a, *right figures* – estimate of b. Note that different subfigures have different scales.

Example 2.4. This example treats the stochastic case (SNR $\to 0$). Hence, the system (2.30) reduces to

$$(p+a)y(t) = e_c(t) \tag{2.40}$$

Consider estimating the parameter a by using the methodology outlined above.

As a first approach, we use the delta backward operator $\delta_b y(t)$ as an approximation of the differentiation operator. By substituting $py(t)$ by $\delta_b y(t)$ in (2.40) for $t = kh$, the following DT regression model can be formed

$$\delta_b y(kh) = -y(kh) a_b + \varepsilon_2(kh). \tag{2.41}$$

Asymptotically ($N \to \infty$) the least squares estimate of a_b becomes (cf. (2.38))

$$\hat{a}_b = -\frac{\frac{1}{h}\mathsf{E}\big(y(kh) - y(kh-h)\big)y(kh)}{\mathsf{E}y^2(kh)} = \frac{1}{h}\left(-1 + \frac{r(h)}{r(0)}\right) \tag{2.42}$$

where the covariance function, $r(\tau) = \mathsf{E}y(t)y(t+\tau)$, satisfies

$$r(\tau) = \frac{\sigma^2}{2a}e^{-a|\tau|} \tag{2.43}$$

Hence, we get

$$\hat{a}_b = \frac{1}{h}\left(-1+e^{-ah}\right) = -a + \mathcal{O}(ah) \tag{2.44}$$

which is far from its true value even for the ideal case, N large and h small.

Next, we consider the delta forward operator $\delta_f y(t)$. By substituting (2.33) into (2.40) we obtain the DT model

$$\delta_f y(kh) = -y(kh)a_f + \varepsilon_1(kh) \tag{2.45}$$

The asymptotic least squares estimate of a_f then reads (cf. (2.36))

$$\begin{aligned}\hat{a}_f &= -\frac{\frac{1}{h}\mathsf{E}\bigl(y(kh+h) - y(kh)\bigr)y(kh)}{\mathsf{E}y^2(kh)} = \frac{1}{h}\left(1 - \frac{r(h)}{r(0)}\right) \\ &= \frac{1}{h}\left(1 - e^{-ah}\right) = a + \mathcal{O}(ah)\end{aligned} \tag{2.46}$$

which means that the estimate is accurate for sufficiently large N and sufficiently small h. Actually, it is the product ah that needs to be small. The quantity $1/a$ is the time-constant of the process $y(t)$. Hence, for an accurate estimate the sampling interval h should be small compared to the time constant of the process.

Finally, one could try the central difference operator

$$\delta_c y(kh) = \frac{1}{2h}\bigl(y(kh+h) - y(kh-h)\bigr) \tag{2.47}$$

which has the advantage that the approximation error is of order $\mathcal{O}(h^2)$ (for a smooth enough function). Remember that the delta backward and the delta forward operators have approximation errors of order $\mathcal{O}(h)$. Substituting (2.47) into (2.40), forming a DT regression model similar to (2.41), and estimating a_c using the least squares method gives the asymptotic estimate

$$\hat{a}_c = \frac{r(h) - r(h)}{2hr(0)} = 0 \tag{2.48}$$

which clearly is a completely erroneous estimate. ∎

The above example shows that the estimate of a is crucially dependent on the choice of the differentiation approximation scheme. One might believe that a general recipe is to use the delta forward operator. Unfortunately, this is not the case. It is sufficient to consider a second-order process to see that the delta forward operator approach will give biased estimates.

Example 2.5. For a second-order system using

$$p^2 y(kh) \approx \delta_f^2 y(kh) = \frac{1}{h^2}\big(y(kh+2h) - 2y(kh+h) + y(kh)\big)$$
$$py(kh) \approx \delta_f y(kh) = \frac{1}{h}\big(y(kh+h) - y(kh)\big)$$

one gets, see [54],

$$\hat{a}_1 = \frac{2}{3}a_1 + \mathcal{O}(h)$$
$$\hat{a}_2 = a_2 + \mathcal{O}(h)$$

Hence, good quality of the estimates is not guaranteed by using delta forward approximations. ∎

To understand why the result depends on the choice of the differentiation approximation scheme, it is necessary to analyse this issue in more detail. We will do so in the next section.

2.6 Derivative Approximations for Direct Methods

In this section we generalise the findings from the previous section to more general linear regression models.

To this aim, consider CARX models to be estimated from DT data. Rewrite the output $y(t)$ as a sum of a deterministic and a stochastic term as

$$y(t) = \frac{B(p)}{A(p)}u(t) + \frac{1}{A(p)}e_c(t) \triangleq y_u(t) + y_e(t) \qquad (2.49)$$

We assume the process operates in open loop, so the two terms in (2.49) are independent. We write (2.49) as a linear regression model

$$p^n y(t) = \boldsymbol{\varphi}^T(t)\boldsymbol{\theta} + \varepsilon(t) \qquad (2.50)$$

where

$$\boldsymbol{\varphi}(t) = \begin{bmatrix} -p^{n-1}y(t) & \ldots & -y(t) & p^{n-1}u(t) & \ldots & u(t) \end{bmatrix}^T \qquad (2.51)$$
$$\boldsymbol{\theta} = \begin{bmatrix} a_1 & \ldots & a_n & b_1 & \ldots & b_n \end{bmatrix}^T \qquad (2.52)$$

To discretise the model, we substitute the jth-order differentiation operator p^j by a discrete approximation $\boldsymbol{\mathcal{D}}^j$ as

$$p^j \approx \boldsymbol{\mathcal{D}}^j = \frac{1}{h^j}\sum_{\mu}\beta_{j,\mu}q^\mu \qquad (2.53)$$

For example, when the delta forward operator

$$\mathcal{D} = \delta_f \triangleq \frac{q-1}{h} \qquad (2.54)$$

is used as an approximation of the differentiation operator of order zero, one, and two, the β-weights are those shown in Table 2.1.

Table 2.1. The β-weights when the delta forward operator δ_f is used as an approximation for the differentiation operator of order zero, one, and two

μ	0	1	2
$\beta_{0,\mu}$	1		
$\beta_{1,\mu}$	-1	1	
$\beta_{2,\mu}$	1	-2	1

In general, we have to impose some conditions on the β-weights in order to make the approximation (2.53) meaningful. We introduce what we will refer to as the natural conditions (m being an arbitrary integer)

$$\mathcal{D}^k f(mh) = p^k f(mh) + \mathcal{O}(h), \qquad k = 0, \ldots, n \qquad (2.55)$$

Assume that $f(t)$ is $(k+1)$-times differentiable and consider the case of a short sampling period h. By series expansion, we then have

$$\mathcal{D}^k f(mh) = \frac{1}{h^k} \sum_j \beta_{k,j} f(mh + jh)$$

$$= \frac{1}{h^k} \sum_j \beta_{k,j} \Big(\sum_{\nu=0}^{k} \frac{1}{\nu!} j^\nu h^\nu p^\nu f(mh) + \mathcal{O}(h^{k+1}) \Big)$$

$$= \sum_{\nu=0}^{k} \frac{1}{\nu!} p^\nu f(mh) \Big(\sum_j \beta_{k,j} j^\nu \Big) h^{\nu-k} + \mathcal{O}(h) \qquad (2.56)$$

The natural conditions (2.55) can now be expressed as

$$\sum_j \beta_{k,j} j^\nu = \begin{cases} 0, & \nu = 0, \ldots, k-1, \\ k!, & \nu = k \end{cases} \qquad (2.57)$$

The frequency function of the filter will be

$$\mathcal{D}^k(e^{i\omega h}) = \frac{1}{h^k} \sum_j \beta_{k,j} e^{ij\omega h} = \frac{1}{h^k} \sum_j \beta_{k,j} \sum_{\nu=0}^{\infty} \frac{(i\omega h)^\nu}{\nu!} j^\nu$$

$$= (i\omega)^k + \mathcal{O}(|\omega|^{k+1}) \qquad (2.58)$$

The low-frequency asymptote of the filter frequency function is hence $(i\omega)^k$.

The minimal number of terms in the sum (2.57) is apparently $k+1$. In such a case $\mathcal{D}^k$ will be a high-pass filter. If the number of $\beta_{k,j}$ coefficients is increased, the gained degrees of freedom can be used to decrease the high-frequency gain of the filter $\mathcal{D}^k$, for example by imposing

$$\mathcal{D}^k(e^{i\omega h})_{|\omega=\pi/h} = 0 \tag{2.59}$$

In addition to (2.59), it is possible to require some further high derivatives $\mathcal{D}^k(e^{i\omega h})$ to vanish at $\omega = \pi/h$. Such measures would make the filter bandpass instead of high-pass and hence more robust to unmodelled wide-band noise.

In the literature on CT identification, a common approach is to rewrite the differential equation (2.49) into an equivalent integral equation and to approximate the integrals, see [46, 60]. For the approximation, the use of block-pulse functions is popular. As noted in [46] this would correspond to the substitution

$$p \to \frac{2}{h}\frac{q-1}{q+1} \tag{2.60}$$

It will hence appear as a special case of the general framework given here. In particular, for small h, it will behave as a delta forward method, cf. Example 2.4.

The processes $y(t), u(t)$ are observed at $t = h, 2h, \ldots, Nh$. The model order n is supposed to be known. It is of interest to estimate the parameter vector $\boldsymbol{\theta}$ from the available data.

After substituting the derivatives in (2.50) by approximations (2.53), we can form the following linear regression model,

$$\begin{aligned} w(kh) &= \boldsymbol{\varphi}^T(kh)\boldsymbol{\theta} + \varepsilon(kh) \\ w(kh) &= \mathcal{D}^n y(kh) \\ \boldsymbol{\varphi}^T(kh) &= \begin{bmatrix} -\mathcal{D}^{n-1}y(kh) & \ldots & -\mathcal{D}^0 y(kh) & \mathcal{D}^{n-1}u(kh) & \ldots & \mathcal{D}^0 u(kh) \end{bmatrix} \end{aligned} \tag{2.61}$$

It turns out that a standard least squares estimate of $\boldsymbol{\theta}$

$$\hat{\boldsymbol{\theta}}_N = \left(\frac{1}{N}\sum_{k=1}^{N} \boldsymbol{\varphi}(kh)\boldsymbol{\varphi}^T(kh)\right)^{-1} \left(\frac{1}{N}\sum_{k=1}^{N} \boldsymbol{\varphi}(kh)w(kh)\right) \tag{2.62}$$

and an instrumental variable estimate of $\boldsymbol{\theta}$

$$\hat{\boldsymbol{\theta}}_N = \left(\frac{1}{N}\sum_{k=1}^{N} \boldsymbol{\zeta}(kh)\boldsymbol{\varphi}^T(kh)\right)^{-1} \left(\frac{1}{N}\sum_{k=1}^{N} \boldsymbol{\zeta}(kh)w(kh)\right) \tag{2.63}$$

where $\zeta(t)$ is the vector of instruments, will in general be heavily biased, also for N large and h small. This was illustrated in Example 2.3. The reason is that $\varepsilon(t)$ in general is correlated with $\varphi(t)$.

For an analysis, similarly to (2.49) we decompose the regressor vector $\varphi(kh)$ as

$$\begin{aligned}\varphi(kh) &= \left[-\mathcal{D}^{n-1}y(kh) \ldots -\mathcal{D}^0 y(kh)\ \mathcal{D}^{n-1}u(kh) \ldots \mathcal{D}^0 u(kh)\right]^T \\ &= \left[-\mathcal{D}^{n-1}y_u(kh) \ldots -\mathcal{D}^0 y_u(kh)\ \mathcal{D}^{n-1}u(kh) \ldots \mathcal{D}^0 u(kh)\right]^T \\ &\quad + \left[-\mathcal{D}^{n-1}y_e(kh) \ldots -\mathcal{D}^0 y_e(kh)\ 0 \ldots 0\right]^T \\ &\triangleq \varphi_u(kh) + \varphi_e(kh)\end{aligned} \quad (2.64)$$

Assuming that $u(t)$ is sufficiently differentiable, the regressor $\varphi_u(t)$ has a limit as $h \to 0$,

$$\lim_{h \to 0} \varphi_u(t) = \left[-p^{n-1}y_u(t) \ldots -y_u(t)\ p^{n-1}u(t) \ldots u(t)\right]^T \triangleq \tilde{\varphi}_u(t) \quad (2.65)$$

Hence, for the deterministic part we have

$$\begin{aligned}\lim_{h \to 0} \lim_{N \to \infty} \hat{\boldsymbol{\theta}}_N &= \lim_{h \to 0} \left(\mathsf{E}\varphi_u(t)\varphi_u^T(t)\right)^{-1}\left(\mathsf{E}\varphi_u(t)w(t)\right) \quad (2.66)\\ &= \left(\mathsf{E}\tilde{\varphi}_u(t)\tilde{\varphi}_u^T(t)\right)^{-1}\left(\mathsf{E}\tilde{\varphi}_u(t)p^n y_u(t)\right) \\ &= \left(\mathsf{E}\tilde{\varphi}_u(t)\tilde{\varphi}_u^T(t)\right)^{-1}\left(\mathsf{E}\tilde{\varphi}_u(t)(\tilde{\varphi}_u^T(t)\boldsymbol{\theta})\right) \\ &= \boldsymbol{\theta} \quad (2.67)\end{aligned}$$

This means that the least squares estimate is close to the true value of the parameter vector for large data sets and small sampling periods.

The stochastic part is more intricate to analyse. The main reason is that the derivative $p^n y_e(t)$ does not exist in a mean square sense (it will not have a finite variance). As will be shown below, by a careful choice of the weights $\{\beta_{k,j}\}$ it is though still possible to estimate $\boldsymbol{\theta}$ without any significant bias.

In order to satisfy the desired condition (2.67), it is sufficient that this condition applies to the stochastic part $y_e(t)$. In order to simplify the treatment we therefore restrict the analysis to the stochastic part. Note that such a case is of interest by itself in CAR models of time-series analysis.

Consider thus a CAR model, see Section 2.2. As shown in Section 2.5, a straightforward application of the least squares method can lead to a very large bias in the parameter estimates. There are, however, several simple means to modify the least squares estimates (2.62) and the instrumental variable estimates (2.63) to obtain consistent estimates. With consistent estimates we here mean, cf. (2.67),

$$\lim_{h \to 0} \lim_{N \to \infty} \hat{\boldsymbol{\theta}}_N = \boldsymbol{\theta} \tag{2.68}$$

The approaches we consider all fit into the same general framework,

$$\hat{\boldsymbol{\theta}}_N = \left(\frac{1}{N}\sum_{t=1}^{N} \boldsymbol{\zeta}(kh)\boldsymbol{\varphi}^T(kh)\right)^{-1} \boldsymbol{F} \left(\frac{1}{N}\sum_{t=1}^{N} \boldsymbol{\zeta}(kh)w(kh)\right) \tag{2.69}$$

The methods are (the β-weights have to fulfil the natural conditions, (2.57))

1. A least squares scheme where

$$\boldsymbol{\zeta}(kh) = \boldsymbol{\varphi}(kh), \qquad \boldsymbol{F} = \boldsymbol{I} \tag{2.70}$$

and the β-weights fulfil (in addition to the natural conditions)

$$\sum_j \sum_k \beta_{n,j}\beta_{n-1,k}\bigl(|j-k|^{2n-1} - (j-k)^{2n-1}\bigr) = 0 \tag{2.71}$$

A simple mean to satisfy the condition (2.71) is to require $j \geq k$. In some sense, this is the same as imposing a shifted structure for estimating the nth-order derivative of $y(t)$, i.e., $w(kh) = \boldsymbol{\mathcal{D}}^n y(kh+jh)$. For example, when the delta forward approximation is used, this means that $j > (n-1)$, see [54]. Consequently, this approach is often referred to as the shifted least squares method (SLS). A way to avoid the shift is presented in [14].

Note that the standard choices of integral filters and block-pulse functions will not satisfy the condition (2.71). These methods will hence suffer from the same bias problem as methods based on the standard derivative approximation. However, by using the condition (2.71), the user choices in these methods may be modified appropriately.

2. A least squares scheme, with a bias compensation feature, see [14, 55]. In this case we take

$$\boldsymbol{\zeta}(kh) = \boldsymbol{\varphi}(kh), \qquad \boldsymbol{F} = \mathrm{diag}\begin{bmatrix} 1/\xi_n & 1 & \ldots & 1 \end{bmatrix} \tag{2.72}$$

where

$$\xi_n = \frac{(-1)^{n-1}}{(2n-1)!} \sum_l \sum_m \beta_{n,l}\beta_{n-1,m}|l-m|^{2n-1} \tag{2.73}$$

This approach is commonly referred to as the bias-compensated least squares method (BCLS).

3. An instrumental variable scheme, with delayed values of the output signal as instruments, see [7]. Here

$$\boldsymbol{\zeta}(kh) = \begin{bmatrix} y(kh-lh-h) & \ldots & y(kh-lh-nh) \end{bmatrix}^T, \qquad \boldsymbol{F} = \boldsymbol{I} \tag{2.74}$$

where l is chosen appropriately. For example, for the delta forward operator we need to impose the constraint $l \geq -1$.

These three approaches all lead to an estimate $\hat{\boldsymbol{\theta}}$ with a small bias of order $\mathcal{O}(h)$,

$$\hat{\boldsymbol{\theta}} \triangleq \lim_{N \to \infty} \hat{\boldsymbol{\theta}}_N = \boldsymbol{\theta} + \tilde{\boldsymbol{\theta}}h + \mathcal{O}(h^2) \tag{2.75}$$

where explicit expressions for the dominating bias $\tilde{\boldsymbol{\theta}}$ are available, see [38, 56]. Furthermore, the statistical properties (in terms of the covariance matrix of the asymptotic parameter estimates) have been clarified. Define

$$\boldsymbol{R}_c \triangleq \operatorname{Cov} \left\{ \begin{bmatrix} p^{n-1}y(t) \\ \vdots \\ y(t) \end{bmatrix} \right\} \tag{2.76}$$

Then it has been shown, see [38, 56], that the parameter estimates are asymptotically Gaussian distributed

$$\sqrt{N}(\hat{\boldsymbol{\theta}}_N - \hat{\boldsymbol{\theta}}) \to \mathcal{N}(\boldsymbol{0}, \boldsymbol{C}) \tag{2.77}$$

where the covariance matrix $\boldsymbol{C}$ satisfies

$$\lim_{h \to 0} h\boldsymbol{C} = \sigma^2 \boldsymbol{R}_c^{-1} \tag{2.78}$$

Consequently, for large values of N and small values of h, the covariance matrix of the estimate $\hat{\boldsymbol{\theta}}_N$ may be approximated as

$$\operatorname{Cov}\{\hat{\boldsymbol{\theta}}_N\} \approx \frac{\sigma^2}{Nh} \boldsymbol{R}_c^{-1} \tag{2.79}$$

Some possible advantages of taking the described approach are given next:

- it is a numerically sound approach compared to, for instance, conventional methods using the q-formalism, which suffers from ill-conditioning problems. Moreover, it is well known that the least squares method is robust with respect to numerical problems;
- the approach is well suited for non-uniformly sampled data. The form (2.53) of derivative approximation can be extended to also handle the case of irregular sampling, where the sampling interval varies in an arbitrary fashion, see Section 2.6.1;
- it is computationally very efficient, and does not suffer from the problem of possible local minima.

The basic methods, described above, have been extended in various ways:

- by further restrictions on the derivative approximations, the bias can be reduced from $\mathcal{O}(h)$ to $\mathcal{O}(h^2)$, see [14, 54];
- an order recursive algorithm with lattice structure has been developed, see, for instance [12].
- the sensitivity to additional measurement noise has been treated in [15]. The measurement noise is modelled as CT white noise;

- various performance measures for the aforementioned methods are reported in [56]. In particular, formulas for quantifying the bias and formulas for the estimation error variance are presented. Note also that it has been shown in [52] that the estimation error variance reaches the CRB as the sampling interval tends to zero. In other words, the methods are in this sense asymptotically (statistically) efficient;
- the case of integrated sampling is treated in [13];
- a further problem concerns the possible estimation of σ^2. If this can be done a parametric estimate of the (CT) spectral density can be found, see (2.1). For details of estimating σ^2, see [54];
- it would be of interest to have an order estimation scheme, assuming that the true value of n is not known. Needless to say, in practice n is seldom *a priori* known. Some details of such an order estimation algorithm are given in [54].

There exist several other related methods that are worth mentioning. The approaches taken in [10, 11, 40] are closely related to the framework described above. Identification of CAR systems when the observations contain DT measurement noise using Bayesian inference is the topic of [16]. Parameter estimation of CAR systems from randomly sampled observations using the so-called pseudo-correlation vector concept is treated in [45]. Further analysis of the effects of fast sampling, also for output error models are given in [17, 18], where most of the analysis is carried out in the frequency domain.

2.6.1 Non-uniformly Sampled Data

Here, we will see how the form (2.53) of a derivative approximation can be extended to alsohandle the case of irregular sampling. Assume that for $h_k = t_{k+1} - t_k$, it holds that $\underline{h} \leqslant h_k \leqslant \bar{h}$, $\forall k$, where $\underline{h} > 0$ and $\bar{h}$ are two finite constants. This assumption is essential in order to prove ergodicity of the process $\{y(t_k)\}$, and to ensure that the estimate has a well-defined limit as the number of data tends to infinity. Consider the following linear approximation of the differentiation operator p^j,

$$p^j f(t_k) \approx \boldsymbol{D}_k^j f(t_k) \triangleq \sum_{\mu=0}^{j} \beta_k(j,\mu) f(t_{k+\mu}) \qquad (2.80)$$

where $f(t)$ is a smooth enough function. In order to make the approximation (2.80) meaningful, some conditions on the β_k-weights must be imposed. A natural request is then that, cf. (2.55),

$$p^j f(t_k) = \boldsymbol{D}_k^j f(t_k) + \mathcal{O}(\bar{h}), \quad j = 0,\ldots,n \qquad (2.81)$$

holds, which, by using (2.80) and a Taylor series expansion, can be reformulated as

$$\sum_{\mu=0}^{j} \beta_k(j,\mu) \lambda_k^\nu(\mu) = \begin{cases} 0, & \nu = 0,\ldots,j-1, \\ j!, & \nu = j \end{cases} \quad (2.82)$$

where we have introduced $\lambda_k(\mu)$ as

$$\lambda_k(\mu) = (t_{k+\mu} - t_k) = \begin{cases} 0, & \mu = 0, \\ \sum_{s=0}^{\mu-1} h_{k+s}, & \text{otherwise} \end{cases} \quad (2.83)$$

Under the above assumption regarding the irregular sampling interval, the solution to (2.82) exists and is unique. Furthermore, the solution is given by

$$\beta_k(j,\mu) = \frac{j!}{\prod_{\substack{s=0 \\ s \neq \mu}}^{j} (\lambda_k(\mu) - \lambda_k(s))}, \quad \mu = 0,\ldots,j \quad (2.84)$$

and the β_k-weights that fulfil (2.81) are thereby found. A proof is given in [32]. To prove (2.84) is basically the same as finding the solution of a system of equations with a Vandermonde structure [8, 19]. Furthermore, there is a connection between this setup and the method of undetermined coefficients [21].

Due to the nice structure of the solution (2.84), it is fairly easy to develop some recursion formulas for the β-weights. The result turns out to be

$$\beta_k(j+1,\mu) = \begin{cases} \dfrac{(j+1)\beta_k(j,\mu)}{\lambda_k(\mu) - \lambda_k(j+1)}, & \mu = 0,\ldots,j, \\ \dfrac{(j+1)\beta_{k+1}(j,j)}{\lambda_k(j+1)}, & \mu = j+1, \end{cases} \quad (2.85)$$

$$\beta_{k+1}(j,\mu) = \frac{\lambda_k(\mu+1)\beta_k(j,\mu+1)}{\lambda_k(\mu+1) - \lambda_k(j+1)}, \quad \mu = 0,\ldots,j-1, \quad (2.86)$$

$$\beta_k(0,0) = 1 \quad (2.87)$$

The derivation of (2.85)–(2.87) follows by direct use of (2.84). The derivatives that fulfil (2.81) can be generated by the recursion formula

$$\mathcal{D}_k^j f(t_k) = \frac{j}{\lambda_k(j)} (\mathcal{D}_{k+1}^{j-1} f(t_{k+1}) - \mathcal{D}_k^{j-1} f(t_k)), \quad j = 1,\ldots,n \quad (2.88)$$

which is given from (2.85)–(2.87), see [32] for a proof.

Next, we describe how to obtain an estimate $\hat{\boldsymbol{\theta}}$ satisfying

$$\hat{\boldsymbol{\theta}} \triangleq \lim_{N \to \infty} \hat{\boldsymbol{\theta}}_N = \boldsymbol{\theta} + \mathcal{O}(\bar{h}) \quad (2.89)$$

- A sufficient condition for (2.89) is that (2.81) and

$$w(t_k) = \mathcal{D}_{k+n-1}^n y(t_{k+n-1}) \tag{2.90}$$

are fulfilled. This gives us the SLS method for irregularly sampled data, see [32]. Note that (2.90) means that a shift is introduced in the data when forming the nth-order derivative approximation.

- A sufficient condition for (2.89) is that (2.81) is fulfilled and that ξ_n in (2.72) is taken as

$$\xi_n = \frac{(-1)^{n-1}}{(2n-1)!} \sum_{l=0}^{n} \sum_{m=0}^{n-1} \mathsf{E}_h \{\beta_k(n,l)\beta_k(n-1,m)|\lambda_k(l) - \lambda_k(m)|^{2n-1}\} \tag{2.91}$$

where E_h means expectation with respect to the sampling process, see [32]. This gives us the BCLS method for irregularly sampled data.

We conclude this section by pointing out that the computational complexity of this approach for identifying CARX models is modest due to the recursion (2.88). In contrast, use of an 'exact' method such as PEM, (2.28), or ML, (2.29), will require a several magnitudes larger computational time. The reason is that for such methods, the system needs to be sampled repeatedly for each new measurement time t_k. The difference in computational load is illustrated in Section 2.8.

2.7 The Cramér–Rao Bound

In any parameter estimation problem a common way of assessing the performance of the estimator is to derive the estimation error covariance matrix. However, this accuracy measure may be of limited interest unless one can compare it with the best possible accuracy. The by far most used bound on the estimation error covariance matrix is the Cramér–Rao lower bound (CRB), see, e.g. [23, 58]. The reason for its popularity is basically twofold. Firstly, under a Gaussian noise assumption, the CRB is often relatively simple to compute. Secondly, the bound is often tight, i.e., there exists an estimator that asymptotically achieves the CRB.

Assume that $\hat{\boldsymbol{\theta}}$ is an unbiased estimate of $\boldsymbol{\theta}$ determined from the data vector $\boldsymbol{y}$, and let $\boldsymbol{P}$ denote the covariance matrix of $\hat{\boldsymbol{\theta}}$. Then, the following relation holds (see, e.g. [58])

$$\boldsymbol{P} \geq \boldsymbol{J}^{-1} \tag{2.92}$$

where the matrix inequality $\boldsymbol{A} \geq \boldsymbol{B}$ means that $\boldsymbol{A} - \boldsymbol{B}$ is non-negative-definite, and the matrix $\boldsymbol{J}$ is the Fisher information matrix. The relation (2.92) is the celebrated CRB result, and $\boldsymbol{J}^{-1}$ stands for the CRB. In general, it holds that

$$\boldsymbol{J} = \mathsf{E}\left(\frac{\partial \ln p(\boldsymbol{y};\boldsymbol{\theta})}{\partial \boldsymbol{\theta}}\right)\left(\frac{\partial \ln p(\boldsymbol{y};\boldsymbol{\theta})}{\partial \boldsymbol{\theta}}\right)^T = -\mathsf{E}\left(\frac{\partial^2 \ln p(\boldsymbol{y};\boldsymbol{\theta})}{\partial \boldsymbol{\theta}\, \partial \boldsymbol{\theta}^T}\right) \tag{2.93}$$

where $p(\boldsymbol{y};\boldsymbol{\theta})$ denotes the likelihood function of $\boldsymbol{y}$. To obtain explicit expressions for $\boldsymbol{J}$ is, in general, a tedious procedure. However, if the data is assumed to be Gaussian distributed, the calculations are simplified. For Gaussian data, and a finite sample size N, the result turns out to be given by the Slepian–Bang formula, see, e.g. [58]. Furthermore, it is known that the normalised Fisher information matrix of a zero-mean stationary Gaussian process tends, with the sample size N, to Whittle's formula, see, e.g. [57]. Another convenient methodology for computing the CRB follows by noting that the ML estimate is in general asymptotically efficient, i.e., the covariance matrix of the ML estimate tends to the CRB as the number of data tends to infinity.

2.7.1 The Cramér–Rao Bound for Irregularly Sampled CARMA Models

Let us consider a CARMA process $y(t)$ (see Section 2.2) represented in state-space form as

$$\begin{aligned} \mathrm{d}\boldsymbol{x}(t) &= \boldsymbol{A}\boldsymbol{x}(t)\,\mathrm{d}t + \boldsymbol{w}(t), \\ y(t) &= \boldsymbol{C}\boldsymbol{x}(t) \end{aligned} \quad (2.94)$$

where $\boldsymbol{x}(t)$ and $y(t)$ are the state vector and the output signal, respectively, and where the disturbance $\boldsymbol{w}(t)$ is a Wiener process with incremental covariance matrix $\boldsymbol{\Sigma}\,\mathrm{d}t$. The model (2.94) is parameterised by $\boldsymbol{\theta}$, which may appear in the matrices $\boldsymbol{A} \in \mathbb{R}^{n \times n}$, $\boldsymbol{C} \in \mathbb{R}^{1 \times n}$, and $\boldsymbol{\Sigma} \in \mathbb{R}^{n \times n}$. The matrix $\boldsymbol{A}$ is asymptotically stable, i.e., all eigenvalues of $\boldsymbol{A}$ have a strictly negative real part. Furthermore, for simplicity we will assume that $\boldsymbol{A}$ is diagonalisable. Also, we would like to point out that the framework presented below for computing the CRB for CARMA models is also applicable for CARMAX models. Essentially this can be seen from Example 2.1 in Section 2.3.2. However, there are some technical details that need to be properly handled; for details, see [31].

The CRB for estimating $\boldsymbol{\theta}$ given the samples $\{y(t_1), y(t_2), \ldots, y(t_N)\}$ is given by $\mathrm{CRB} = \boldsymbol{J}^{-1}$, where the (k,l)th element of $\boldsymbol{J}$ is given by the Slepian–Bang formula [58]

$$[\boldsymbol{J}]_{k,l} = \frac{1}{2}\mathrm{tr}\{\boldsymbol{R}^{-1}\boldsymbol{R}'_k\boldsymbol{R}^{-1}\boldsymbol{R}'_l\} \quad (2.95)$$

where

$$\boldsymbol{R} = \mathsf{E}\begin{bmatrix} y(t_1) \\ \vdots \\ y(t_N) \end{bmatrix}\begin{bmatrix} y(t_1) & \ldots & y(t_N) \end{bmatrix} \quad (2.96)$$

is the covariance matrix of the sampled data, and where $\boldsymbol{R}'_k = \partial \boldsymbol{R}/\partial \theta_k$, with θ_k denoting the kth element of $\boldsymbol{\theta}$. Next, compact closed-form expressions for $\boldsymbol{R}$ and its derivatives $\boldsymbol{R}'_k$ are derived.

To find an expression for $\boldsymbol{R}$, it is first noted that the covariance function, $r(\tau) = \mathsf{E} y(t) y(t-\tau)$, of $y(t)$ can be written as [4,53]

$$r(\tau) = \boldsymbol{C} \mathrm{e}^{\boldsymbol{A}\tau} \boldsymbol{P} \boldsymbol{C}^T \quad (2.97)$$

for $\tau \geq 0$. In (2.97), $\mathrm{e}^{\boldsymbol{A}\tau}$ is the standard matrix exponential, which has the spectral representation

$$\mathrm{e}^{\boldsymbol{A}\tau} = \sum_{k=1}^{n} \boldsymbol{\rho}_k \boldsymbol{\xi}_k^H \mathrm{e}^{\lambda_k \tau} \quad (2.98)$$

where $\boldsymbol{\rho}_k$ and $\boldsymbol{\xi}_k^H$ are the right and left eigenvectors of $\boldsymbol{A}$, respectively (normalised such that $\boldsymbol{\rho}_k^H \boldsymbol{\xi}_k = 1$), and λ_k are the eigenvalues of $\boldsymbol{A}$. Also, in (2.97), $\boldsymbol{P}$ is the unique and non-negative-definite solution to the CT Lyapunov equation

$$\boldsymbol{A}\boldsymbol{P} + \boldsymbol{P}\boldsymbol{A}^T + \boldsymbol{\Sigma} = \boldsymbol{0} \quad (2.99)$$

Equation (2.99) can be written as a linear system of equations for the entries of $\boldsymbol{P}$, and can thereby be solved by standard methods. From (2.97), (2.98), and the solution to (2.99), it follows that $\boldsymbol{R}$ can be computed as

$$\boldsymbol{R} = \sum_{k=1}^{n} \gamma_k \alpha_k \boldsymbol{\Gamma}(\lambda_k) \quad (2.100)$$

where γ_k and α_k are defined as

$$\gamma_k \triangleq \boldsymbol{C}\boldsymbol{\rho}_k, \qquad \alpha_k \triangleq \boldsymbol{\xi}_k^H \boldsymbol{P} \boldsymbol{C}^T$$

and $\boldsymbol{\Gamma}(s)$ is the matrix whose (k,l)th element equals

$$[\boldsymbol{\Gamma}(s)]_{k,l} = \mathrm{e}^{|t_k - t_l| s}$$

To obtain convenient formulas for $\boldsymbol{R}_i'$, it is first noted that differentiation of (2.99) with respect to θ_i yields

$$\boldsymbol{A}\boldsymbol{P}_i + \boldsymbol{P}_i\boldsymbol{A}^T + \boldsymbol{A}_i\boldsymbol{P} + \boldsymbol{P}\boldsymbol{A}_i^T + \boldsymbol{\Sigma}_i = \boldsymbol{0} \quad (2.101)$$

where the derivatives $\boldsymbol{P}_i = \partial \boldsymbol{P}/\partial \theta_i$, $\boldsymbol{A}_i = \partial \boldsymbol{A}/\partial \theta_i$, and $\boldsymbol{\Sigma}_i = \partial \boldsymbol{\Sigma}/\partial \theta_i$. The Lyapunov equation (2.101) is straightforward to solve with respect to $\boldsymbol{P}_i$. Next, by differentiation of $r(\tau)$ in (2.97) with respect to θ_i, it is readily shown (for details, see [27, 29]) that

$$\boldsymbol{R}_i' = \sum_{k=1}^{n} (\gamma_{k,i} \alpha_k + \gamma_k \alpha_{k,i}) \boldsymbol{\Gamma}(\lambda_k) + \sum_{k=1}^{n} \sum_{l=1}^{n} (\gamma_k \alpha_l \beta_{k,l}^{(i)}) \boldsymbol{G}_{k,l} \quad (2.102)$$

where

2 Estimation of Continuous-time Stochastic System Parameters 57

$$\gamma_{k,i} \triangleq \boldsymbol{C}_i \boldsymbol{\rho}_k, \quad \alpha_{k,i} \triangleq \boldsymbol{\xi}_k^H \boldsymbol{P}_i \boldsymbol{C}^T + \boldsymbol{\xi}_k^H \boldsymbol{P} \boldsymbol{C}_i^T, \quad \beta_{k,l}^{(i)} \triangleq \boldsymbol{\xi}_k^H \boldsymbol{A}_i \boldsymbol{\rho}_l,$$

$$\boldsymbol{G}_{k,l} = \begin{cases} \boldsymbol{\Omega} \odot \boldsymbol{\Gamma}(\lambda_k), & \lambda_k = \lambda_l, \\ \frac{1}{\lambda_l - \lambda_k}(\boldsymbol{\Gamma}(\lambda_l) - \boldsymbol{\Gamma}(\lambda_k)), & \text{otherwise} \end{cases}$$

Here, $\odot$ denotes element-wise multiplication and $\boldsymbol{\Omega}$ is a matrix whose (k,l)th element is equal to

$$[\boldsymbol{\Omega}]_{k,l} = |t_k - t_l|$$

We summarise our algorithm for computing the CRB as follows:
Step 1. Compute $\boldsymbol{P}$ and $\boldsymbol{P}_i$ by solving the Lyapunov equations (2.99) and (2.101).
Step 2. Compute $\boldsymbol{R}$ by using (2.100).
Step 3. Compute $\boldsymbol{R}'_i$ via (2.102).
Step 4. Compute $\boldsymbol{J}$ via (2.95) and obtain CRB = $\boldsymbol{J}^{-1}$.

One important observation is that the evaluation of the CRB can become impractical if the number of samples N is large. Let CRB_N denote the CRB given N samples of $y(t)$. Then, Theorem 2 in [26] provides some remedy to this problem by showing that under certain circumstances, the CRB becomes inversely proportional to N, and hence it can be extrapolated as

$$\text{CRB}_N \stackrel{as.}{=} \frac{N_0}{N} \text{CRB}_{N_0} \tag{2.103}$$

where $N_0 < N$. Let us conclude this section about the CRB by considering an example.

Example 2.6. We consider a CAR process with poles at $\bar{p}_1 \pm i\tilde{p}_1$, for $\tilde{p}_1 = 1$ and $\bar{p}_1 = -0.01$ and -0.05, respectively. We fix the number of samples to $N = 100$ and vary the mean sample interval $T = \mathsf{E}\{t_{k+1} - t_k\}$. The following two sampling strategies are considered

(a) *(Deterministic) uniform*: Here $t_n = nT$, $n = 1, \ldots, N$.
(b) *Uniformly distributed*: Here $t_n = nT + \sum_{k=1}^n \delta_k$, $n = 1, \ldots, N$, where δ_k is uniformly distributed between $-\delta_0$ and δ_0; δ_k are independent of $w(t)$ (see (2.94)) for all t and k, and δ_k are independent of δ_j for all $j \neq k$. We choose $\delta_0 = T/5$ in our example.

Figure 2.2 shows the CRB for estimating $\tilde{p}_1$ for different values of T. Note that $\tilde{p}_1$ can be interpreted as the location of a 'spectral line', so estimating $\tilde{p}_1$ is of particular practical relevance. Note also that the sampling corresponds to taking $2\pi/T$ samples per cycle of a sinusoid with angular frequency $\omega = 1$ (which is the peak location of the spectrum). We observe a number of things from Figure 2.2. First, in general, the CRB decreases for increasing T. To understand why this is so, note that the length of the time interval during which the process is observed grows linearly with T (since N is fixed) and that the estimation problem is related to that of identifying the frequency of a single sinusoid in noise, where it is known that the CRB depends primarily on the length of the observation interval.

Second, for most values of T, there is no big difference between the CRB for different sampling schemes. Note that the length of the interval during which the process is observed is approximately NT, which is independent of the sampling scheme used. On the other hand, for values of T close to π and 2π, the CRB associated with the sampling pattern (a) appears to grow without bound. This is not hard to understand since the corresponding estimation problem becomes ill-conditioned in this case. The CRB associated with sampling scheme (b) has, on the other hand, a CRB that almost behaves as a straight line in this plot (at least for $\bar{p}_1 = -0.01$).

Finally, as already mentioned the estimation of $\tilde{p}_1$ can be interpreted as locating the corresponding spectral line. We can see from Figure 2.2 that the CRB for this problem decreases with decreasing $|\bar{p}_1|$. This can be understood intuitively since the accuracy for estimating $\tilde{p}_1$ is related to the sharpness of the corresponding spectral line, and this spectral line becomes sharper and shaper as $|\bar{p}_1|$ decreases. ∎

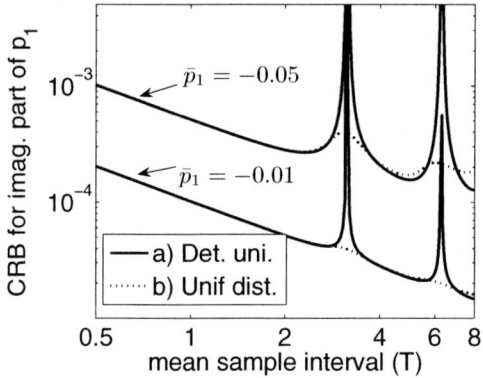

Fig. 2.2. The CRB for $\tilde{p}_1$ as a function of the mean sample interval T for two different sampling schemes and for two different values of $\bar{p}_1$. The process is described in Example 2.6. The number of samples is $N = 100$.

2.8 Numerical Studies of Direct Methods

In this section, we investigate the properties of direct methods for identification of CT stochastic systems in some simulation studies. Special attention is given to the accuracy and computational complexity of the methods. We consider CAR processes in Example 2.7 and CARX processes in Example 2.8. For CARMA processes, we refer to the methods and examples presented in [30].

Example 2.7. This example is about identification of irregularly observed CAR processes. We consider unevenly sampled data, and the methods SLS and BCLS described in Section 2.6.1 are compared with the PEM described in Section 2.4.

The second-order system

$$(p^2 + a_1 p + a_2)y(t) = e_c(t) \tag{2.104}$$

where $a_1 = a_2 = 2$ and the noise intensity equal to one is considered. In order to generate DT data, the system is sampled as described in Section 2.3, with the sampling interval h_k uniformly distributed as $h_k \sim \mathcal{U}(\underline{h}, \bar{h})$, where $\underline{h} = 0.01$ and $\bar{h}$ is varied. The time-varying DT system is then simulated to generate $N = 10,000$ samples. A Monte Carlo simulation with 100 realisations is carried out and the mean values and empirical standard deviations of the estimates are presented as functions of $\bar{h}$ in Figure 2.3. The optimisation routine used for the PEM is a simplex method with initial parameters equal to one. The theoretical estimates are found from (2.62) for $u(t) = 0$ and $N \to \infty$. The estimates of a_1 are comparable for all three methods, whereas the PEM gives the best estimate of a_2, followed by the BCLS method and the SLS method. It is also clear that there is a good match between the experimental and theoretical values for the least squares methods.

When comparing the methods for irregularly sampled data, it is important to investigate some efficiency properties of the different methods. To be more precise, we compare the computational time, as well as the computational load requested for the different methods. The computational time is the time needed to run the algorithm, while the computational load is given as the number of flops required by the algorithm in order to produce the estimates. The results for $\bar{h} = 0.05$ are found in Table 2.2. The main conclusion is that there is a significant difference between the least squares methods and the PEM concerning computational efficiency. The PEM requires a considerable execution time to produce the estimates. Furthermore, the two least squares methods require approximately the same amounts of time and flops to produce the estimates. As a final remark we would like to emphasise that the initialisation of the optimisation routine is crucial for the behaviour (*e.g.*, convergence rate) of the PEM, see [34, 42]. ∎

Example 2.8. In this example, CARX models are identified from unevenly sampled data. Consider the CARX process

$$(p^2 + a_1 p + a_2)y(t) = (b_1 p + b_2)u(t) + e_c(t) \tag{2.105}$$

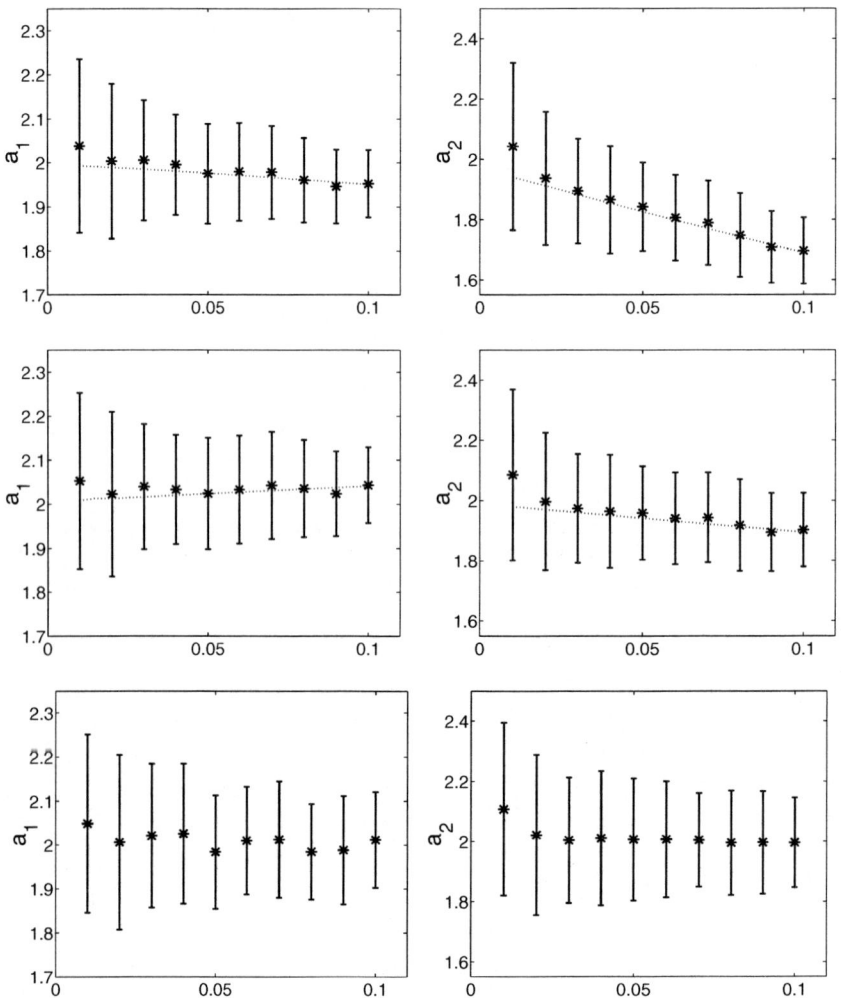

Fig. 2.3. The mean values ($*$) and empirical standard deviations for the estimates of a_1 (*left*) and a_2 (*right*) as functions of $\bar{h}$ for the SLS method (*upper*), the BCLS method (*middle*), and the PEM (*lower*). The theoretical estimates for the least squares methods are given by the dotted lines, the true values are $a_1 = a_2 = 2$, and the sampling scheme is uniformly distributed.

where $e_c(t)$ is CT white noise of intensity σ_e^2 and $u(t)$ is given by the CAR process

$$(p^2 + \bar{a}_1 p + \bar{a}_2)u(t) = v(t) \tag{2.106}$$

where $v(t)$ is CT white noise, independent of $e_c(t)$, with intensity σ_v^2. We consider the uniformly distributed sampling scheme described in Section 2.7.1 for different T and generate N data points from (2.105), with $a_1 = a_2 = 2$,

Table 2.2. Evaluation time and number of flops used for the different methods. Second-order system with $a_1 = a_2 = 2$

Method	Time [s]	Load [Megaflops]
SLS	0.118	0.29
BCLS	0.112	0.21
PEM	1120	740

$b_1 = 3$, $b_2 = 1$, and $\sigma_e^2 = 1$, and with $\bar{a}_1 = \bar{a}_2 = 3$ and $\sigma_v^2 = 1$ in (2.106). The SLS method described in Section 2.6.1 is considered together with the ML method described in Section 2.4. For the ML method, it is assumed that the input signal is constant during the sampling instants. This gives an error of order $\mathcal{O}(\bar{h}^2)$, where $\bar{h}$ is the upper bound on the sampling interval, see [51].

A Monte Carlo study with 50 realisations is carried out in which the parameters a_1, a_2, b_1, and b_2 are estimated from N data points (N is specified later). Data are affected by three random mechanisms; the two noises $e(t)$ and $v(t)$, and the stochastic sampling scheme. Therefore, it is of interest to investigate what happens when they are varying at the same time. The expected CRB is computed by using the results in Section 2.7, including the extrapolation described in (2.103), for $N_0 = 100$ data points and averaging over 50 realisations by means of a Monte Carlo simulation.

The mean values and the empirical variances for the estimates of a_1 as functions of T, given by the SLS method and the ML method, are shown in Figure 2.4, together with the CRB, for $N = 10,000$ data. The mean values and the empirical standard deviations for the estimates of a_1, a_2, b_1, and b_2 as functions of T, given by the direct method and the ML method are shown in Figure 2.5, respectively, for the case with $N = 1000$ data points. The bias is in general larger for the direct approach than for the ML method, especially for larger values of T, whereas the variance is slightly smaller for the direct approach. The CRB is reached by both the SLS method and the ML method. Note that a biased estimate may have variance lower than the CRB. An important observation is that the average computational times are considerably shorter for the direct approach, as seen in Table 2.8 for the case with $N = 10,000$. It is, however, more difficult to accurately estimate the b parameters than the a parameters. In general, it is easier to get good estimates of the a parameters since it can be said that they are excited by $u(t)$ as well as by $e(t)$. The b parameters, on the other hand, are only excited by $u(t)$. ∎

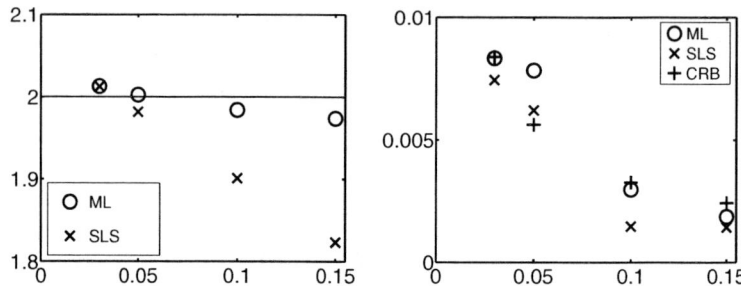

Fig. 2.4. The mean values (*left*) and the empirical variances for the estimates of a_1 for the SLS method and the ML method together with the CRB (*right*) as functions of T for $N = 10\,000$

Table 2.3. The average computational times, in seconds, for the SLS method and the ML method for different T, $N = 10,000$

T	SLS	ML
0.03	0.11	255
0.05	0.11	255
0.10	0.10	274
0.15	0.10	298

2.9 Conclusions

The important problem of identifying CT stochastic systems from DT data has been studied. Applications can be found in many different areas of science and technology since most physical systems and phenomena are of CT nature. We have described basic models of CT stochastic systems. More exactly, we have considered processes with rational spectra and their corresponding time domain representations that are formally given in terms of SDEs. As the estimation methods are based on DT data, we have described how the CT models are transformed into DT models. We have described two general approaches, the direct and indirect approaches, to estimation of CT stochastic models. Thereafter, we introduced an estimation technique where the differentiation operator is replaced by some approximation, a linear regression model is formed, and the parameters are estimated. To get consistent estimates (as the number of data tends to infinity and the sampling interval tends to zero), it was shown that the approximation of the differentiation operator must be chosen carefully. The method can also be applied to irregularly sampled data, where the sampling interval varies in an arbitrary fashion. This is advantageous, since the computational complexity of conventional methods, such as the prediction error method, is very high for irregularly sampled data. It has been described how to compute the CRB for estimation problems concern-

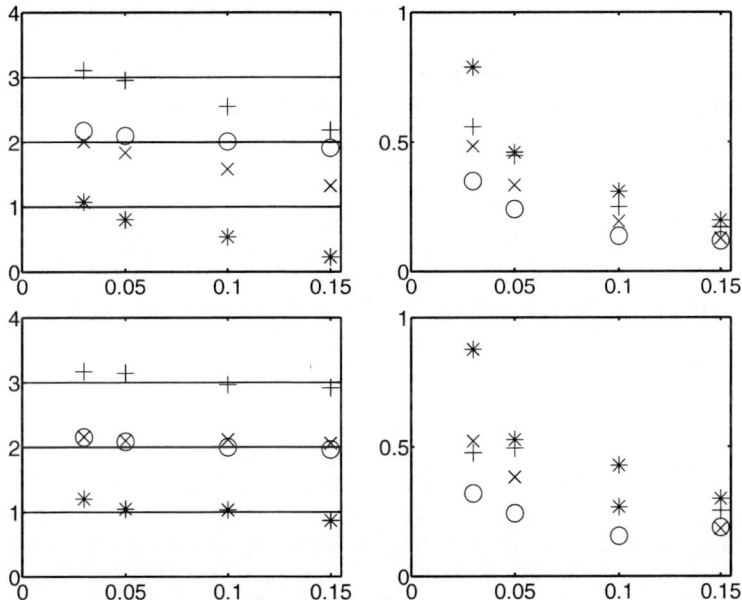

Fig. 2.5. The mean values (*left*) and the empirical standard deviations (*right*) for the estimates of a_1 (o), a_2 (×), b_1 (+), and b_2 (∗) as functions of T, for the SLS method (*upper*) and the ML method (*lower*) with $N = 1000$ data points. The true parameter values (*left*) are indicated with horizontal lines. Note that some symbols overlap.

ing CT stochastic systems. Finally, the properties of the proposed estimation technique and alternative approaches have been studied numerically.

References

1. H.M. Adorf. Interpolation of irregularly sampled data series – a survey. *Astronomical Data Analysis Software and Systems IV*, 77:460–463, 1995.
2. T.G. Andersen and J. Lund. Estimating continuous-time stochastic volatility models of the short-term interest rate. *Journal of Econometrics*, 77:343–377, 1997.
3. K.J. Åström. Maximum likelihood and prediction error methods. *Automatica*, 16:551–574, 1980.
4. K.J. Åström. *Introduction to Stochastic Control Theory*. Dover Publications, New York, NY, 2006. First published by Academic Press, New York, NY, 1970.
5. K.J. Åström, P. Hagander, and J. Sternby. Zeros of sampled systems. *Automatica*, 20:31–38, 1984.
6. K.J. Åström and C.G. Källström. Identification of ship steering dynamics. *Automatica*, 12:9–22, 1976.

7. S. Bigi, T. Söderström, and B. Carlsson. An IV-scheme for estimating continuous-time stochastic models from discrete-time data. *10th IFAC Symposium on System Identification*, Copenhagen, Denmark, July 1994.
8. Å. Björck and V. Pereyra. Solutions of Vandermonde systems of equations. *Mathematics of Computation*, 24:893–903, 1970.
9. C.W. Clark. Mathematical models in the economics of renewable resources. *SIAM Review*, 21:81–99, 1979.
10. T.E. Duncan, P. Mandl, and B. Pasik-Duncan. Numerical differentiation and parameter estimation in higher-order linear stochastic systems. *IEEE Transactions on Automatic Control*, 41(4):522–532, 1996.
11. T.E. Duncan, P. Mandl, and B. Pasik-Duncan. A note on sampling and parameter estimation in linear stochastic systems. *IEEE Transactions on Automatic Control*, 44(11):2120–2125, 1999.
12. H. Fan. An efficient order recursive algorithm with a lattice structure for estimating continuous-time AR process parameters. *Automatica*, 33:305–317, 1997.
13. H. Fan and T. Söderström. Parameter estimation of continuous-time AR processes using integrated sampling. *36th IEEE Conference on Decision and Control*, San Diego, CA, USA, December 1997.
14. H. Fan, T. Söderström, M. Mossberg, B. Carlsson, and Y. Zou. Estimation of continuous-time AR process parameters from discrete-time data. *IEEE Transactions on Signal Processing*, 47(5):1232–1244, 1999.
15. H. Fan, T. Söderström, and Y. Zou. Continuous-time AR process parameter estimation in presence of additive white noise. *IEEE Transactions on Signal Processing*, 47(12):3392–3398, 1999.
16. P. Giannopoulos and S.J. Godsill. Estimation of CAR processes observed in noise using Bayesian inference. *IEEE International Conference on Acoustics, Speech and Signal Processing*, Salt Lake City, UT, USA, May 2001.
17. J. Gillberg. *Frequency domain identification of continuous-time systems*. PhD thesis, Linköping University, Sweden, 2006.
18. J. Gillberg and L. Ljung. Frequency domain identification of continuous-time output error models from sampled data. *16th IFAC World Congress*, Prague, Czech Republic, July 2005.
19. G.H. Golub and C.F. Van Loan. *Matrix Computations*. Johns Hopkins University Press, Baltimore, MD, 3rd edition, 1996.
20. J.J. Heuring and D.W. Murray. Modeling and copying human head movements. *IEEE Transactions on Robotics and Automation*, 15(6):1095–1108, 1999.
21. E. Isaacson and H.B. Keller. *Analysis of Numerical Methods*. John Wiley and Sons, New York, NY, 1966.
22. C.V. Jakowatz (Jr.), D.E. Wahl, P.H. Eichel, D.C. Ghiglia, and P.A. Thompson. *Spotlight-mode Synthetic Aperture Radar: A Signal Processing Approach*. Kluwer Academic Publishers, Norwell, MA, 1996.
23. S.M. Kay. *Fundamentals of Statistical Signal Processing: Estimation Theory*. Prentice-Hall, Upper Saddle River, NJ, 1993.
24. P. Kinahan, J. Fessler, and J. Karp. Statistical image reconstruction in PET with compensation for missing data. *IEEE Transactions on Nuclear Science*, 44(4, part 1):1552–1557, 1997.
25. P.E. Kloeden and E. Platen. *Numerical Solution of Stochastic Differential Equations*. Springer-Verlag, Berlin/Heidelberg, Germany, 1992.

26. E.G. Larsson and E.K. Larsson. The Cramér-Rao bound for continuous-time AR parameter estimation with irregular sampling. *Circuits, Systems & Signal Processing*, 21:581–601, 2002.
27. E.K. Larsson. *Identification of Stochastic Continuous-Time Systems*. PhD thesis, Uppsala University, Sweden, 2004.
28. E.K. Larsson. Limiting sampling results for continuous-time ARMA systems. *International Journal of Control*, 78:461–473, 2005.
29. E.K. Larsson and E.G. Larsson. The CRB for parameter estimation in irregularly sampled continuous-time ARMA systems. *IEEE Signal Processing Letters*, 11(2):197–200, 2004.
30. E.K. Larsson, M. Mossberg, and T. Söderström. An overview of important practical aspects of continuous-time ARMA system identification. *Circuits, Systems & Signal Processing*, 25:17–46, 2006.
31. E.K. Larsson, M. Mossberg, and T. Söderström. Identification of continuous-time ARX models from irregularly sampled data. *IEEE Transactions on Automatic Control*, 52(3):417–427, 2007.
32. E.K. Larsson and T. Söderström. Identification of continuous-time AR processes from unevenly sampled data. *Automatica*, 38:709–718, 2002.
33. L. Ljung. *System Identification. Theory for the User*. Prentice-Hall, Upper Saddle River, NJ, 2nd edition, 1999.
34. L. Ljung. Initialisation aspects for subspace and output-error identification methods. *5th European Control Conference*, Cambridge, UK, September 2003.
35. H. Madsen and J. Holst. Estimation of continuous-time models for the heat dynamics of a building. *Energy and Buildings*, 22:67–79, 1995.
36. A.G. Malliaris. Ito's calculus in financial decision making. *SIAM Review*, 25:481–496, 1983.
37. J. Mateo and P. Laguna. Improved heart rate variability signal analysis from the beat occurrence times according to the IPFM model. *IEEE Transactions on Biomedical Engineering*, 47(8):985–996, 2000.
38. M. Mossberg. *On identification of continuous-time systems using a direct approach*. Licentiate thesis, Uppsala University, Sweden, 1998.
39. M.S. Phadke and S.M. Wu. Modeling of continuous stochastic processes from discrete observations with application to sunspots data. *Journal of the American Statistical Association*, 69(346):325–329, 1974.
40. D.T. Pham. Estimation of continuous-time autoregressive model from finely sampled data. *IEEE Transactions on Signal Processing*, 48(9):2576–2584, 2000.
41. R. Pintelon and J. Schoukens. *System Identification: a Frequency Domain Approach*. IEEE Press, Piscataway, USA, 2001.
42. G.P. Rao and H. Garnier. Numerical illustrations of the relevance of direct continuous-time model identification. *15th IFAC World Congress*, Barcelona, Spain, July 2002.
43. G.P. Rao and H. Unbehauen. Identification of continuous-time systems. *IEE Proceedings – Control Theory and Applications*, 153(2):185–220, 2006.
44. T.S. Rao, M.B. Pristley, and O. Lessi (eds). *Applications of Time Series Analysis in Astronomy and Meteorology*. Chapman & Hall/CRC Press, Boca Raton, FL, 1997.
45. A. Rivoira, Y. Moudden, and G. Fleury. Real time continuous AR parameter estimation from randomly sampled observations. *IEEE International Conference on Acoustics Speech and Signal Processing*, Orlando, FL, May 2002.

46. S. Sagara, Z.J. Yang, K. Wada, and T. Tsuji. Parameter identification and adaptive control of continuous systems with zero order hold. *12th IFAC World Congress*, Sydney, Australia, July 1993.
47. J. Schoukens, R. Pintelon, and H. Van Hamme. Identification of linear dynamic systems using piecewise constant excitations: Use, misuse and alternatives. *Automatica*, 30:1153–1169, 1994.
48. L. Shepp. A model for stock price fluctuations based on information. *IEEE Transactions on Information Theory*, 48(6):1372–1378, 2002.
49. N.K. Sinha and G.P. Rao (eds). *Identification of Continuous-time Systems. Methodology and Computer Implementation*. Kluwer Academic Publishers, Dordrecht, 1991.
50. K. Sobczyk. Modelling of random fatigue crack growth. *Engineering Fracture Mechanics*, 24:609–623, 1986.
51. T. Söderström. Algorithms for computing stochastic continuous time models from ARMA models. Technical Report UPTEC 89030 R, Uppsala University, Sweden, 1989.
52. T. Söderström. On the Cramér-Rao lower bound for estimating continuous-time autoregressive parameters. *14th IFAC World Congress*, Beijing, P.R. China, July 1999.
53. T. Söderström. *Discrete-time Stochastic Systems*. Springer-Verlag, London, UK, 2nd edition, 2002.
54. T. Söderström, H. Fan, B. Carlsson, and S. Bigi. Least squares parameter estimation of continuous-time ARX models from discrete-time data. *IEEE Transactions on Automatic Control*, 42(5):659–673, 1997.
55. T. Söderström, H. Fan, M. Mossberg, and B. Carlsson. Bias-compensating schemes for estimating continuous-time AR process parameters. *11th IFAC Symposium on System Identification*, Kitakyushu, Japan, July 1997.
56. T. Söderström and M. Mossberg. Performance evaluation of methods for identifying continuous-time autoregressive processes. *Automatica*, 36:53–59, 2000.
57. T. Söderström and P. Stoica. *System Identification*. Prentice-Hall, Hemel Hempstead, U.K., 1989.
58. P. Stoica and R. Moses. *Spectral Analysis of Signals*. Pearson Prentice-Hall, Upper Saddle River, NJ, 2005.
59. S.J. Turnovsky. Applications of continuous-time stochastic methods to models of endogenous economic growth. *Annual Reviews in Control*, 20:155–166, 1996.
60. H. Unbehauen and G.P. Rao. *Identification of Continuous Systems*. North-Holland, Amsterdam, The Netherlands, 1987.
61. B. Wahlberg. Limit results for sampled systems. *International Journal of Control*, 48:1267–1283, 1988.
62. B. Wahlberg. System identification using high-order models, revisited. *28th IEEE Conference on Decision and Control*, Tampa, Florida, USA, December 1989.

3

Robust Identification of Continuous-time Systems from Sampled Data

Juan I. Yuz[1] and Graham C. Goodwin[2]

[1] Universidad Técnica Federico Santa María, Chile
[2] University of Newcastle, Australia

Whilst most physical systems occur naturally in continuous time, it is necessary to deal with sampled data for identification purposes. In principle, one can derive an exact sampled data model for any given linear system by integration. However, conversion to sampled data form implicitly involves folding of high-frequency system characteristics back into the lower-frequency range. This means that there is an inherent loss of information. The sampling process is reversible provided one has detailed knowledge of the relationship between the low-frequency and folded components so that they can be *untangled* from the sampled model. However, it is clear from the above argument that one has an inherent sensitivity to the assumptions that one makes about the folded components. The factors that contribute to the folded components include

- the sampling rate,
- the nature of the input between samples (*i.e.*, is it generated by a first-order hold or not, or is it continuous-time white noise or not),
- the nature of the sampling process (*i.e.*, has an *anti-aliasing* filter been used and, if so, what are its frequency domain characteristics),
- the system relative degree (*i.e.*, the high-frequency *roll-off* characteristics of the system beyond the base band),
- high-frequency poles and or zeros that lie outside the base band interval.

In a recent paper [10], we have shown that the above issues lead to non-trivial robustness issues. For example, we have shown that, in the identification of continuous-time autoregressive (CAR) models from sampled data, the resultant model depends critically on the assumptions one makes about the issues outlined above. In this chapter, we will extend these ideas to general linear models. In particular, we will argue that one always needs to define a *bandwidth of validity* relative to the above assumptions and ensure that the analysis is restricted to that bandwidth. We will describe time- and frequency-domain methods for ensuring insensitivity to high-frequency folded artifacts in the identification of continuous-time systems from sampled data. We call these methods *Robust continuous-time system identification using sampled data*.

3.1 Overview

In recent years, there has been an increasing interest in the problem of identifying continuous-time models [7–9, 11, 12, 14, 16, 22, 25, 27]. This kind of model has several advantages compared to discrete-time models:

- the parameters obtained are physically meaningful, and they can be related to properties of the real system;
- the continuous-time model obtained is independent of the sampling period; and
- these models may be more suitable for *fast sampling rate* applications since a continuous-time model is the (theoretical) limit when the sampling period is infinitesimally small.

Even though it is theoretically possible to carry out system identification using continuous-time data [26, 29], this will generally involve the use of analogue operations to emulate time derivatives. Thus, in practice, one is inevitably forced to work with sampled data [21, 23].

In this chapter we explore the issues that are associated with the use of sampled-data models in continuous-time system identification. Specifically, we use sampled-data models expressed using the δ operator, to estimate the parameters of the underlying continuous-time system. In this context, one might hope that, if one samples *quickly enough*, the difference between discrete and continuous processing would become vanishing small. Thus, say we are given a set of data $\{u_k = u(t_k), y_k = y(t_k)\}$, where $t_k = kT_s$ and T_s is the sampling period. We identify a sampled-data model

$$\mathcal{M}_d : \quad y_k = G_\delta(\delta, \hat{\boldsymbol{\theta}})u_k + H_\delta(\delta, \hat{\boldsymbol{\theta}})v_k \tag{3.1}$$

where $\hat{\boldsymbol{\theta}}$ is a vector with the parameters to be estimated, then, we might hope that $\hat{\boldsymbol{\theta}}$ will converge to the corresponding continuous-time parameters, as T_s goes to zero, *i.e.*,

$$\hat{\boldsymbol{\theta}} \xrightarrow{T_s \to 0} \boldsymbol{\theta} \tag{3.2}$$

where $\boldsymbol{\theta}$ represents the *true* parameter vector of the continuous-time model

$$\mathcal{M} : \quad y(t) = G(p, \boldsymbol{\theta})u(t) + H(p, \boldsymbol{\theta})\dot{v} \tag{3.3}$$

where p denotes the differential operator.

Indeed, there are many cases that support this hypothesis. Moreover, the delta operator has been a key tool to highlight the connections between the discrete and the continuous-time domains [6, 20].

The above discussion can, however, lead to a false sense of security when using sampled data. A sampled-data model asymptotically *converges* to the continuous-time representation of a given system. However, there is an inherent loss of information when using discrete-time model representations. In the time domain, the use of sampled data implies that we do not know the intersample behaviour of the system. In the frequency domain, this fact translates to the well-known aliasing effect: high-frequency components fold back to low frequencies, in such a way that it is not possible to distinguish between them. To fill the gap between systems evolving in continuous time and their sampled-data representations, we need to make extra assumptions on the continuous-time model and signals. This is a particularly sensitive point when trying to perform system identification using sampled data. In Section 3.2, we pay particular attention to the impact of high-frequency modelling errors. This kind of errors may arise both in the discrete- and continuous-time domains. For discrete-time models, the *sampling zeros* go to infinity (in the γ-domain corresponding to the δ operator) as the sampling period is reduced, however, their effect cannot, in all cases, be neglected especially at high frequencies. For continuous-time systems, undermodelling errors may arise due to the presence of high-frequency poles and/or zeros not included in the nominal model.

Based on the above remarks, we argue here that one should always define a *bandwidth of fidelity* of a model and ensure that the model errors outside that bandwidth do not have a major impact on the identification results. In Section 3.2, we propose the use of a maximum likelihood identification procedure in the frequency domain, using a restricted bandwidth. We show that the proposed identification method is insensitive to both high-frequency undermodelling errors (in the continuous-time model), and to sampling zeros (in the sampled-data model).

A well-known instance where *naive* use of sampled data can lead to erroneous results is in the identification of continuous-time stochastic systems where the noise model has relative degree greater than zero. In this case, the sampled-data model will always have *sampling zeros* [28]. These are the stochastic equivalent of the well-known sampling zeros that occur in deterministic systems [2]. We will see in Section 3.3 that these sampling zeros play a crucial role in obtaining unbiased parameter estimates in the identification of such systems from sampled data. We show that high-frequency modelling errors can be equally as catastrophic as ignoring sampling zeros. These problems can be overcome by using the proposed frequency-domain identification procedure, restricting the estimation to a limited bandwidth.

3.2 Limited-bandwidth Estimation

In this section we discuss the issues that arise when using sampled-data models to identify the underlying continuous-time system. The discrete-time description, when expressed using the δ operator, converges to the continuous-time model as the sampling period goes to zero [20]. However, for any non-zero sampling period, there will always be a difference between the discrete- and continuous-time descriptions, due to the presence of sampling zeros. To overcome this inherent difficulty, we propose the use of maximum likelihood estimation in the frequency domain, using a restricted bandwidth.

To illustrate the differences between discrete-time models and the underlying continuous-time systems we present the following example.

Example 3.1. Consider a second-order deterministic system, described by

$$\frac{d^2}{dt^2}y(t) + \alpha_1 \frac{d}{dt}y(t) + \alpha_o y(t) = \beta_o u(t) \tag{3.4}$$

If we *naively* replace the derivatives in this continuous-time model by divided differences, we obtain the following approximate discrete-time model described in terms of the δ operator

$$\delta^2 y_k + a_1 \delta y_k + a_0 y_k = b_0 u_k \tag{3.5}$$

We see that this simple derivative replacement model has no extra zeros. However, the exact discrete-time model can be obtained, assuming a zero-order hold (ZOH) input. This model can be expressed in terms of the delta operator as follows

$$\delta^2 y_k + a_1 \delta y_k + a_0 y_k = b_0 u_k + b_1 \delta u_k \tag{3.6}$$

This model generically has a *sampling zero*. Moreover, as the sampling period T_s goes to zero, the continuous-time coefficients are recovered, and the sampling zero can be readily characterised [2,6]. Thus, for $T_s \approx 0$, the discrete-time model can be considered to be

$$\delta^2 y_k + \alpha_1 \delta y_k + \alpha_o y_k = \beta_o (1 + \tfrac{T_s}{2}\delta) u_k \tag{3.7}$$

Figure 3.1 shows a comparison of the Bode magnitude diagrams corresponding to a second-order system as (3.4) (on the left-hand side) and the exact sampled-data model (3.6), obtained for different sampling frequencies (on the right)

$$G(s) = \frac{\beta_o}{s^2 + \alpha_1 s + \alpha_o} \qquad G_\delta(\gamma) = \frac{b_1 \gamma + b_o}{\gamma^2 + a_1 \gamma + a_o} \tag{3.8}$$

The figure clearly illustrates the fact that, no matter how fast we sample, there is always a difference (near the folding frequency) between the continuous-time model and the discretised models.

The difference between discrete and continuous-time models highlighted by the previous example, in fact, corresponds to an illustration of the aliasing effect. If we assume that the continuous-time system frequency response $G(i\omega)$ goes to zero as $|\omega| \to \infty$, then the corresponding discrete-time model frequency response converges as follows

$$\lim_{T_s \to 0} G_q(e^{i\omega T_s}) = \lim_{T_s \to 0} \sum_{\ell=-\infty}^{\infty} \left[\frac{(1-e^{-sT_s})}{sT_s} G(s) \right]_{s=i\omega+i\frac{2\pi}{T_s}\ell} = G(i\omega) \tag{3.9}$$

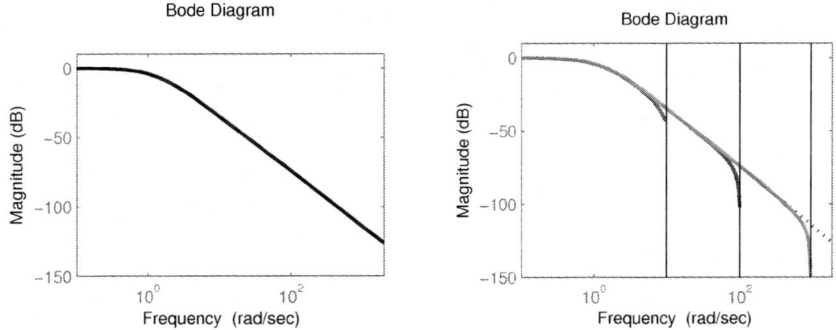

Fig. 3.1. Frequency response for continuous and discrete-time models

Remark 3.1. Equation (3.9) establishes that the frequency response of a sampled-data model converges to the continuous-time frequency response, as the sampling period goes to zero. However, for any finite sampling frequency, there is a difference between the continuous and discrete time cases, in particular, near the Nyquist frequency ($\omega_N = \frac{\omega_s}{2} = \frac{\pi}{T_s}$). Indeed, this is a direct consequence of the presence of the asymptotic sampling zeros.

A different kind of problem may arise when the *true* system contains high-frequency dynamics that are not included in the continuous-time model. We illustrate this by the following example.

Example 3.2. Consider again the continuous-time system in Example 3.1. We will consider (3.4) as the nominal model of the system. We are interested in analysing the effect of an unmodelled fast pole. Thus, let the true system be given by

$$G_o(s) = \frac{\beta_o}{(s^2 + \alpha_1 s + \alpha_o)\left(\frac{1}{\omega_u}s + 1\right)} = \frac{G_n(s)}{\left(\frac{1}{\omega_u}s + 1\right)} \tag{3.10}$$

Figure 3.2 shows the comparison of nominal and true models, both for the continuous-time system and the sampled-data models. The nominal poles of the system are at $s = -1$ and $s = -2$, the sampling frequency is $\omega_s = 250$ [rad/s], and the unmodelled fast pole is at $s = -50$.

Note that the true system has relative degree 3, and, thus, the corresponding discrete-time model will have 2 sampling zeros. As a consequence, while the asymptotic sampled-data model for the nominal system is given by (3.8), the true model will yield different asymptotic sampling zeros as T_s goes to zero. Thus, the nominal model satisfies

$$G_{n,\delta}(\gamma) \to \frac{b_o\left(1 + \frac{1}{2}T_s\gamma\right)}{\gamma^2 + a_1\gamma + a_o} \tag{3.11}$$

whereas the true model satisfies

$$G_\delta(\gamma) \to \frac{b_o \left(1 + T_s\gamma + \frac{1}{6}(T_s\gamma)^2\right)}{(\gamma^2 + a_1\gamma + a_o)\left(\frac{\gamma}{\omega_u} + 1\right)} \qquad (3.12)$$

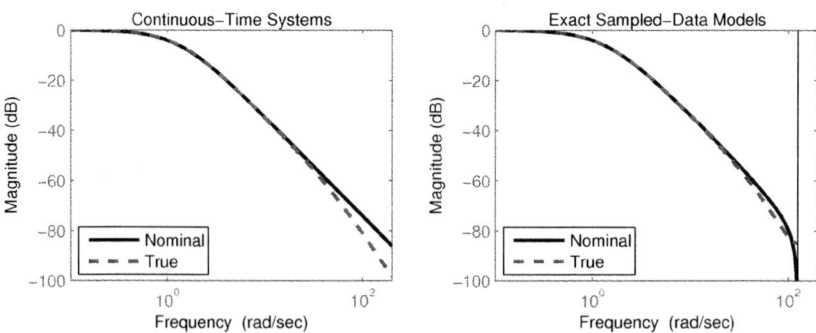

Fig. 3.2. Frequency response for nominal and true models

The previous example illustrates the problems that may arise when using *fast sampling rates*. The sampling frequency was chosen well above the nominal poles of the system, in fact, two decades. In theory, this allows one to use the asymptotic characterisation of the sampled-data model. However, we can see that, if there are any unmodelled dynamics not included in the continuous-time model (in this case, one decade above the nominal fastest pole), then there will also be undermodelling in the sampled-data description. Moreover, even though the sampling zeros go to infinity for the nominal and true models, their precise characterisation depends significantly on high-frequency aspects of the model, as shown in (3.11) and (3.12).

Remark 3.2. The above discussion highlights the issues that have to be taken into account when using sampled-data models to identify continuous-time systems. Specifically:

- any method that relies on high-frequency system characteristics will be inherently non-robust, and, as a consequence,
- models should be considered within a *bandwidth of validity*, to avoid high-frequency modelling errors — see the shaded area in Figure 3.3.

In Section 3.3, we will see how frequency-domain maximum likelihood estimation, over a *restricted bandwidth*, can be used to address these issues.

3.2.1 Frequency-domain Maximum Likelihood

In this section we describe a frequency-domain maximum likelihood (FDML) estimation procedure. Specifically, if one converts the data to the frequency

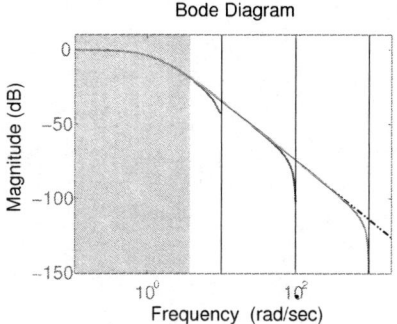

Fig. 3.3. Representation of the bandwidth of validity

domain, then one can carry out the identification over a limited range of frequencies. Note, however, that one needs to carefully define the likelihood function in this case. We use the following result (for the scalar case, the result has been derived in [17], while the multi-variable case is considered in [19]).

Lemma 3.1. *Assume a given set of input–output data* $\{u_k = u(t_k), y_k = y(t_k)\}$, *where* $t_k = kT_s$, $k = 0\ldots N$, *is generated by the exact discrete-time model*

$$y_k = G_q(q, \boldsymbol{\theta})u_k + H_q(q, \boldsymbol{\theta})v_k \tag{3.13}$$

where v_k *is Gaussian discrete-time white noise (DTWN) sequence,* $v_k \sim \mathcal{N}(0, \sigma_v^2)$.

The data is transformed to the frequency domain yielding the discrete Fourier transforms U_ℓ *and* Y_ℓ *of the input and output sequences, respectively.*

Then, the maximum likelihood estimate of $\boldsymbol{\theta}$, *when considering frequency components up to* $\omega_{\max} \leq \frac{\omega_s}{2}$, *is given by:*

$$\hat{\boldsymbol{\theta}}_{ML} = \arg\min_{\boldsymbol{\theta}} L(\boldsymbol{\theta}) \tag{3.14}$$

where $L(\boldsymbol{\theta})$ *is the negative logarithm of the likelihood function of the data given* $\boldsymbol{\theta}$, *i.e.,*

$$L(\boldsymbol{\theta}) = -\log p(Y_0, \ldots, Y_{n_{\max}} | \boldsymbol{\theta})$$
$$= \sum_{\ell=0}^{n_{\max}} \frac{|Y_\ell - G_q(e^{i\omega_\ell T_s}, \boldsymbol{\theta})U_\ell|^2}{\lambda_v^2 |H_q(e^{i\omega_\ell T_s}, \boldsymbol{\theta})|^2} + \log(\pi \lambda_v^2 |H_q(e^{i\omega_\ell T_s}, \boldsymbol{\theta})|^2) \tag{3.15}$$

where $\lambda_v^2 = T_s N \sigma_v^2$, *and* $n_{\max}$ *is the index associated with* $\omega_{\max}$.

Proof. Equation (3.13) can be expressed in the frequency domain as

$$Y_\ell = G_q(e^{i\omega_\ell T_s}, \boldsymbol{\theta})U_\ell + H_q(e^{i\omega_\ell T_s}, \boldsymbol{\theta})V_\ell \tag{3.16}$$

where Y_ℓ, U_ℓ, and V_ℓ are *scaled* discrete Fourier transforms (DFT) [6], *e.g.,*

$$Y_\ell = Y(e^{i\omega_\ell T_s}) = T_s \sum_{k=0}^{N-1} y_k e^{-i\omega_\ell k T_s} \quad , \omega_\ell = \frac{2\pi}{T_s} \frac{\ell}{N} \quad (3.17)$$

Assuming that the DTWN sequence $v_k \sim \mathcal{N}(0, \sigma_v^2)$, then V_ℓ are (asymptotically) independent and have a circular complex Gaussian distribution [4, 5]. Thus, the frequency-domain noise sequence V_ℓ has zero mean and variance $\lambda_v^2 = T_s N \sigma_v^2$. We therefore see that Y_ℓ is also complex Gaussian and satisfies

$$Y_\ell \sim \mathcal{N}(G_q(e^{i\omega_\ell T_s}, \boldsymbol{\theta}) U_\ell, \lambda_v^2 |H_q(e^{i\omega_\ell T_s}, \boldsymbol{\theta})|^2) \quad (3.18)$$

The corresponding probability density function is given by

$$p(Y_\ell) = \frac{1}{\pi \lambda_v^2 |H_q(e^{i\omega_\ell T_s}, \boldsymbol{\theta})|^2} \exp\left\{-\frac{|Y_\ell - G_q(e^{i\omega_\ell T_s}, \boldsymbol{\theta}) U_\ell|^2}{\lambda_v^2 |H_q(e^{i\omega_\ell T_s}, \boldsymbol{\theta})|^2}\right\} \quad (3.19)$$

If we consider the elements Y_ℓ within a limited bandwidth, i.e., up to some maximum frequency $\omega_{\max}$ indexed by $n_{\max}$ with $\omega_{\max} = \omega_s \frac{n_{\max}}{N} \leq \frac{\omega_s}{2}$, the appropriate log-likelihood function is given by

$$L(\boldsymbol{\theta}) = -\log p(Y_0, \ldots, Y_{n_{\max}}) = -\log \prod_{\ell=0}^{n_{\max}} p(Y_\ell)$$

$$= \sum_{\ell=0}^{n_{\max}} \frac{|Y_\ell - G_q(e^{i\omega_\ell T_s}, \boldsymbol{\theta}) U_\ell|^2}{\lambda_v^2 |H_q(e^{i\omega_\ell T_s}, \boldsymbol{\theta})|^2} + \log(\pi \lambda_v^2 |H_q(e^{i\omega_\ell T_s}, \boldsymbol{\theta})|^2) \quad (3.20)$$

Remark 3.3. The logarithmic term in the log-likelihood function (3.15) plays a key role in obtaining consistent estimates of the true system. This term can be neglected if [17]:

- the noise model is assumed to be known. In this case, H_q does not depend on $\boldsymbol{\theta}$ and, thus, plays no role in the minimisation (3.14); or
- the frequencies ω_ℓ are equidistantly distributed over the full frequency range $[0, \frac{2\pi}{T_s})$. This is equivalent to considering the full-bandwidth case in (3.15), i.e., $n_{\max} = \frac{N}{2}$ (or N, because of periodicity). This yields

$$\frac{2\pi}{N} \sum_{\ell=0}^{N-1} \log |H_q(e^{i\omega_\ell T_s}, \boldsymbol{\theta})|^2 \xrightarrow{N \to \infty} \int_0^{2\pi} \log |H_q(e^{i\omega}, \boldsymbol{\theta})|^2 d\omega \quad (3.21)$$

The last integral is equal to zero for any monic, stable and inversely stable transfer function $H_q(e^{i\omega}, \boldsymbol{\theta})$ [17].

Remark 3.4. In the previous lemma the discrete-time model (3.13) has been expressed in terms of the shift operator q. The results apply *mutatis mutandis* when the model is reparameterised using the δ operator

$$G_q(e^{i\omega_\ell T_s}) = G_\delta(\gamma_\omega) = G_\delta\left(\frac{e^{i\omega_\ell T_s} - 1}{T_s}\right) \quad (3.22)$$

3.3 Robust Continuous-time Model Identification

In this section we illustrate the problems that may arise when sampled-data models are used for continuous-time system identification. In particular, we illustrate the consequences of both types of undermodelling errors discussed earlier

- sampling zeros are not included in the sampled-data model, and
- the continuous-time system contains unmodelled high-frequency dynamics.

We show that these kinds of errors can have severe consequences in the estimation results for deterministic and stochastic systems. We show that the frequency-domain maximum likelihood (FDML) procedure, using restricted bandwidth, allows one to overcome these difficulties.

3.3.1 Effect of Sampling Zeros in Deterministic Systems

We first explore the consequences of neglecting the presence of sampling zeros in deterministic models used for identification. Specifically, the following example considers a deterministic second-order system with known input. The parameters of the system are estimated using different sampled-data model structures.

Example 3.3. Consider again the linear system in (3.4). Assume that the continuous-time parameters are $\alpha_1 = 3$, $\alpha_0 = 2$, $\beta_0 = 2$. We performed system identification assuming three different model structures:

SDRM: simple derivative replacement model. This corresponds to the structure given in (3.5), where continuous-time derivatives have been replaced by divided differences.

MIFZ: model including fixed zero. This model considers the presence of the asymptotic zero, assuming a structure as in (3.7).

MIPZ: model including parameterised zero. This model also includes a sampling zero, whose location has to be estimated, *i.e.*, we use the structure given by (3.6).

The three discrete-time models can be represented in terms of the δ operator as

$$G_\delta(\gamma) = \frac{B_\delta(\gamma)}{\gamma^2 + \hat{\alpha}_1 \gamma + \hat{\alpha}_0} \qquad (3.23)$$

where

$$B_\delta(\gamma) = \begin{cases} \hat{\beta}_0 & \text{(SDRM)} \\ \hat{\beta}_0(1 + \frac{T_s}{2}\gamma) & \text{(MIFZ)} \\ \hat{\beta}_0 + \hat{\beta}_1 \gamma & \text{(MIPZ)} \end{cases} \qquad (3.24)$$

We use a sampling period $T_s = \pi/100$ [s] and choose the input u_k to be a random Gaussian sequence of unit variance. Note that the output sequence

$y_k = y(kT_s)$ can be obtained by either simulating the continuous-time system and sampling its output, or, alternatively, by simulating the exact sampled-data model in discrete time. Also note that the data is free of any measurement noise.

The parameters are estimated in such a way as to minimise the equation error cost function

$$J(\hat{\boldsymbol{\theta}}) = \frac{1}{N} \sum_{k=0}^{N-1} e_k(\hat{\boldsymbol{\theta}})^2 = \frac{1}{N} \sum_{k=0}^{N-1} (\delta^2 y_k - \boldsymbol{\varphi}_k^T \boldsymbol{\theta})^2 \qquad (3.25)$$

where

$$\boldsymbol{\varphi}_k = \begin{cases} [-\delta y_k, -y_k, u_k]^T \\ [-\delta y_k, -y_k, (1+\frac{T_s}{2}\delta)u_k]^T \\ [-\delta y_k, -y_k, \delta u_k, u_k]^T \end{cases} \quad \text{and} \quad \hat{\boldsymbol{\theta}} = \begin{cases} [\hat{\alpha}_1, \hat{\alpha}_0, \hat{\beta}_0]^T & \text{(SDRM)} \\ [\hat{\alpha}_1, \hat{\alpha}_0, \hat{\beta}_0]^T & \text{(MIFZ)} \\ [\hat{\alpha}_1, \hat{\alpha}_0, \hat{\beta}_1, \hat{\beta}_0]^T & \text{(MIPZ)} \end{cases}$$
$$(3.26)$$

Table 3.1 shows the estimation results. Note that the system considered is linear, thus, the exact discrete-time parameters can be computed for the given sampling period. These are also given in Table 3.1.

We can see that, while both models incorporating a sampling zero (MIFZ and MIPZ) are able to recover the continuous-time parameters, when using SDRM the estimate $\hat{\beta}_0$ is clearly biased.

Table 3.1. Parameter estimates for a linear system

Parameters			Estimates		
	CT	Exact DT	SDRM	MIFZ	MIPZ
α_1	3	2.923	2.8804	2.9471	2.9229
α_0	2	1.908	1.9420	1.9090	1.9083
β_1	–	0.0305	–	$\frac{\hat{\beta}_0 T_s}{2} = 0.03$	0.0304
β_0	2	1.908	0.9777	1.9090	1.9083

The result in the previous example may be surprising since, even though the SDRM in (3.27) converges to the continuous-time system as the sampling period goes to zero, the estimate $\hat{\beta}_0$ does not converge to the underlying continuous-time parameter. This estimate is asymptotically biased. Specifically, we see that β_0 is incorrectly estimated by a factor of 2 by the SDRM. This illustrates the impact of not considering sampling effects on the sampled-data models used for continuous-time system identification.

Indeed, the following result formally establishes the asymptotic bias that was observed experimentally for the SDRM structure in the previous example. In particular, we show that β_0 is indeed underestimated by a factor of 2.

Lemma 3.2. *Consider the general second-order deterministic system given in (3.4). Assume that sampled data is collected from the system using a ZOH input generated from a DTWN sequence u_k, and sampling the output $y_k = y(kT_s)$.*
If an equation error identification procedure is utilised to estimate the parameters of the simple derivative replacement model

$$\delta^2 y + \hat{\alpha}_1 \delta y + \hat{\alpha}_0 y = \hat{\beta}_0 u \qquad (3.27)$$

then the parameter estimates asymptotically converge, as the sampling period T_s goes to zero, to

$$\hat{\alpha}_1 \to \alpha_1, \qquad \hat{\alpha}_0 \to \alpha_0, \qquad \text{and} \quad \hat{\beta}_0 \to \tfrac{1}{2}\beta_0 \qquad (3.28)$$

Proof. The parameters of the approximate SDRM (3.27) model can be obtained by simple least squares, minimising the equation error cost function

$$J(\hat{\boldsymbol{\theta}}) = \lim_{N \to \infty} \frac{1}{N} \sum_{k=0}^{N-1} e_k(\hat{\boldsymbol{\theta}})^2 = E\{e_k(\hat{\boldsymbol{\theta}})^2\} \qquad (3.29)$$

where $e_k = \delta^2 y + \hat{\alpha}_1 \delta y + \hat{\alpha}_0 y - \hat{\beta}_0 u$. The parameter estimates are given by the solution of $\frac{dJ(\hat{\boldsymbol{\theta}})}{d\hat{\boldsymbol{\theta}}} = 0$. Thus, differentiating the cost function with respect to each of the parameter estimates, we obtain

$$\begin{bmatrix} E\{(\delta y)^2\} & E\{(\delta y)y\} & -E\{(\delta y)u\} \\ E\{(\delta y)y\} & E\{y^2\} & -E\{yu\} \\ -E\{y^2\} & -E\{yu\} & E\{u^2\} \end{bmatrix} \begin{bmatrix} \hat{\alpha}_1 \\ \hat{\alpha}_0 \\ \hat{\beta}_0 \end{bmatrix} = \begin{bmatrix} -E\{(\delta y)(\delta^2 y)\} \\ -E\{y \delta^2 y\} \\ E\{u \delta^2 y\} \end{bmatrix} \qquad (3.30)$$

This equation can be rewritten in terms of (discrete-time) correlations as

$$\begin{bmatrix} \frac{2r_y(0)-2r_y(1)}{T_s^2} & \frac{r_y(1)-r_y(0)}{T_s} & \frac{r_{yu}(0)-r_{yu}(1)}{T_s} \\ \frac{r_y(1)-r_y(0)}{T_s} & r_y(0) & -r_{yu}(0) \\ \frac{r_{yu}(0)-r_{yu}(1)}{T_s} & -r_{yu}(0) & r_u(0) \end{bmatrix} \begin{bmatrix} \hat{\alpha}_1 \\ \hat{\alpha}_0 \\ \hat{\beta}_0 \end{bmatrix} = \begin{bmatrix} \frac{3r_y(0)-4r_y(1)+r_y(2)}{T_s^3} \\ \frac{-r_y(0)+2r_y(1)-r_y(2)}{T_s^2} \\ \frac{r_{yu}(0)-2r_{yu}(1)+r_{yu}(2)}{T_s^2} \end{bmatrix} \qquad (3.31)$$

To continue with the proof we need to obtain expressions for the correlations involved in the last equation. If we assume that the input sequence is a DTWN process, with unit variance then we have that

$$r_u(k) = \delta_K[k] \quad \Longleftrightarrow \quad \Phi_u^q(e^{i\omega T_s}) = 1 \qquad (3.32)$$

Then, the other correlation functions can be obtained from the relations

$$\begin{aligned} r_{yu}(k) &= \mathcal{F}_d^{-1}\{\Phi_{yu}^q(e^{i\omega T_s})\} \\ &= \mathcal{F}_d^{-1}\{G_q(e^{i\omega T_s})\Phi_u^q(e^{i\omega T_s})\} = \mathcal{F}_d^{-1}\{G_q(e^{i\omega T_s})\} \\ r_y(k) &= \mathcal{F}_d^{-1}\{\Phi_y^q(e^{i\omega T_s})\} \\ &= \mathcal{F}_d^{-1}\{G_q(e^{-i\omega T_s})\Phi_{yu}^q(e^{i\omega T_s})\} = \mathcal{F}_d^{-1}\{|G_q(e^{i\omega T_s})|^2\} \end{aligned} \qquad (3.34)$$

where $G_q(e^{i\omega T_s})$ is the exact sampled-data model corresponding to the continuous-time system (3.4). Given a sampling period T_s, the exact discrete-time model is given by

$$G_q(z) = \frac{\beta_0(c_1 z + c_0)}{(z - e^{\lambda_1 T_s})(z - e^{\lambda_2 T_s})} \qquad (3.35)$$

where

$$c_1 = \frac{(e^{\lambda_1 T_s}-1)\lambda_2 - (e^{\lambda_2 T_s}-1)\lambda_1}{(\lambda_1-\lambda_2)\lambda_1\lambda_2} = \frac{T_s^2}{2} + \frac{T_s^3}{6}(\lambda_1+\lambda_2) + \ldots \qquad (3.36)$$

$$c_0 = \frac{e^{\lambda_1 T_s}(e^{\lambda_2 T_s}-1)\lambda_1 - e^{\lambda_2 T_s}(e^{\lambda_1 T_s}-1)\lambda_2}{(\lambda_1-\lambda_2)\lambda_1\lambda_2} = \frac{T_s^2}{2} + \frac{T_s^3}{3}(\lambda_1+\lambda_2) + \ldots \qquad (3.37)$$

and λ_1 and λ_2 are the continuous-time system (stable) poles of system (3.4), i.e., $\alpha_1 = -(\lambda_1 + \lambda_2)$ and $\alpha_0 = \lambda_1\lambda_2$.

The exact discrete-time model (3.35) can be rewritten as

$$G_q(z) = \frac{C_1}{z - e^{\lambda_1 T_s}} + \frac{C_2}{z - e^{\lambda_2 T_s}} \qquad (3.38)$$

where $C_1 = \frac{\beta_0(c_1 e^{\lambda_1 T_s} + c_0)}{(e^{\lambda_1 T_s} - e^{\lambda_2 T_s})}$ and $C_2 = \frac{\beta_0(c_1 e^{\lambda_2 T_s} + c_0)}{(e^{\lambda_2 T_s} - e^{\lambda_1 T_s})}$. Substituting in (3.34), we obtain

$$r_{yu}(k) = \mathcal{F}_d^{-1}\{G_q(e^{i\omega T_s})\} = \left(C_1 e^{\lambda_1 T_s(k-1)} + C_2 e^{\lambda_2 T_s(k-1)}\right)\mu[k-1] \qquad (3.39)$$

where $\mu[k]$ is the discrete-time unitary step function. From (3.35), we have that

$$G_q(z)G_q(z^{-1}) = K_1\left(\frac{e^{\lambda_1 T_s}}{z - e^{\lambda_1 T_s}} + \frac{e^{-\lambda_1 T_s}}{z - e^{-\lambda_1 T_s}}\right)$$

$$+ K_2\left(\frac{e^{\lambda_2 T_s}}{z - e^{\lambda_2 T_s}} + \frac{e^{-\lambda_2 T_s}}{z - e^{-\lambda_2 T_s}}\right) \qquad (3.40)$$

$$K_1 = \frac{\beta_0^2(c_1^2 e^{\lambda_1 T_s} + c_0 c_1 + c_0 c_1 e^{2\lambda_1 T_s} + c_0^2 e^{\lambda_1 T_s})}{(e^{2\lambda_1 T_s} - 1)(e^{\lambda_1 T_s} e^{\lambda_2 T_s} - 1)(e^{\lambda_1 T_s} - e^{\lambda_2 T_s})} \qquad (3.41)$$

$$K_2 = \frac{\beta_0^2(c_1^2 e^{\lambda_2 T_s} + c_0 c_1 + c_0 c_1 e^{2\lambda_2 T_s} + c_0^2 e^{\lambda_2 T_s})}{(e^{2\lambda_2 T_s} - 1)(e^{\lambda_2 T_s} e^{\lambda_1 T_s} - 1)(e^{\lambda_2 T_s} - e^{\lambda_1 T_s})} \qquad (3.42)$$

Substituting in (3.34), we obtain

$$r_y(k) = \mathcal{F}_d^{-1}\{|G_q(z = e^{i\omega T_s})|^2\} = K_1 e^{\lambda_1 T_s |k|} + K_2 e^{\lambda_2 T_s |k|} \quad , \forall k \in \mathbb{Z} \qquad (3.43)$$

The correlations (3.32), (3.39), and (3.43) can be used in the normal equation (3.30) to obtain

$$\begin{bmatrix} \frac{-\beta_0^2}{2(\lambda_1+\lambda_2)}T_s & \mathcal{O}(T_s^2) & -\frac{\beta_0}{2}T_s \\ \mathcal{O}(T_s^2) & \frac{-\beta_0^2}{2(\lambda_1+\lambda_2)\lambda_1\lambda_2}T_s & 0 \\ -\frac{\beta_0}{2}T_s & 0 & 1 \end{bmatrix} \begin{bmatrix} \hat{\alpha}_1 \\ \hat{\alpha}_0 \\ \hat{\beta}_0 \end{bmatrix} = \begin{bmatrix} T_s \frac{\beta_0^2}{4} \\ T_s \frac{-\beta_0^2}{2(\lambda_1+\lambda_2)} \\ \frac{\beta_0}{2} + \mathcal{O}(T_s) \end{bmatrix} \qquad (3.44)$$

If we consider only terms up to of order T_s we obtain

$$\begin{bmatrix} \hat{\alpha}_1 \\ \hat{\alpha}_0 \\ \hat{\beta}_0 \end{bmatrix} = \begin{bmatrix} \frac{-2(\lambda_1+\lambda_2)}{2+(\lambda_1+\lambda_2)T_s} \\ \lambda_1\lambda_2 \\ \frac{\beta_0(2-(\lambda_1+\lambda_2)T_s)}{2(2+(\lambda_1+\lambda_2)T_s)} \end{bmatrix} \xrightarrow{T_s \to 0} \begin{bmatrix} -(\lambda_1+\lambda_2) \\ \lambda_1\lambda_2 \\ \beta_0/2 \end{bmatrix} \quad (3.45)$$

which corresponds to the result in (3.28).

The above results show that sampling zeros must be considered to obtain a sampled-data model accurate enough for estimation. Even though the sampling zero for the exact discrete-time model (3.7) goes asymptotically to infinity (in the γ-domain), if it is not considered, then the parameter estimates will be generically biased (for equation error structures).

3.3.2 Effect of Sampling Zeros in Stochastic Systems

A particular case of the above problem for stochastic systems has been studied in detail in the following references [15, 16, 25]. These papers deal with continuous-time autoregressive (CAR) system identification from sampled data. Such systems have relative degree n, where n is the order of the autoregressive process. Thus, consider a system described by

$$E(p)y(t) = \dot{v}(t) \quad (3.46)$$

where $\dot{v}(t)$ represents a continuous-time white noise (CTWN) process, and $E(p)$ is a polynomial in the differential operator $p = \frac{d}{dt}$, i.e.,

$$E(p) = p^n + a_{n-1}p^{n-1} + \ldots + a_0 \quad (3.47)$$

For these systems, it has been shown that one cannot ignore the presence of stochastic sampling zeros. Specifically, if derivatives are *naively* replaced by divided differences and the parameters are estimated using ordinary least squares, then the results are asymptotically biased, even when using fast sampling rates [25]. Note, however, that the exact discrete-time model that describes the continuous-time system (3.46) takes the following generic form

$$E_q(q^{-1})y(kT_s) = F_q(q^{-1})w_k \quad (3.48)$$

where w_k is a DTWN process, and E_q and F_q are polynomials in the backward shift operator q^{-1}.

The polynomial $E_q(q^{-1})$ in (3.48) is *well behaved* in the sense that it converges naturally to its continuous-time counterpart. This relationship is most readily portrayed using the delta form

$$E_\delta(\delta) = \delta^n + \bar{a}_{n-1}\delta^{n-1} + \ldots + \bar{a}_0 \quad (3.49)$$

Using (3.49), it can be shown that, as the sampling period T_s goes to zero

80   J.I. Yuz and G.C. Goodwin

$$\lim_{T_s \to 0} \bar{a}_i = a_i \quad , i = n-1, \ldots, 0 \qquad (3.50)$$

However, the polynomial $F_q(q^{-1})$ contains the stochastic sampling zeros, with no continuous-time counterpart [28]. Thus, to obtain the correct estimates– say via the prediction error method [18]—then one needs to minimise the cost function

$$J_{\text{PEM}} = \sum_{k=1}^{N} \left[\frac{E_q(q^{-1})y(kT_s)}{F_q(q^{-1})} \right]^2 \qquad (3.51)$$

Notice the key role played by the sampling zeros in the above expression. A simplification can be applied, when using high sampling frequencies, by replacing the polynomial $F_q(q^{-1})$ by its asymptotic expression. However, this polynomial has to be taken into account when estimating over the full bandwidth. Hence it is not surprising that the use of ordinary least squares, *i.e.*, a cost function of the form

$$J_{LS} = \sum_{k=1}^{N} \left[E_q(q^{-1})y(kT_s) \right]^2 \qquad (3.52)$$

leads to (asymptotically) biased results, even when using (3.49). We illustrate these ideas by the following example.

Example 3.4. Consider the continuous-time system defined by the *nominal model*

$$E(p)y(t) = \dot{v}(t) \qquad (3.53)$$

where $\dot{v}(t)$ is a CTWN process with (constant) spectral density equal to 1, and

$$E(p) = p^2 + 3p + 2 \qquad (3.54)$$

We know that the equivalent sampled-data model has the form

$$Y(z) = \frac{F_q(z)}{E_q(z)} W(z) = \frac{K(z - z_1)}{(z - e^{-T_s})(z - e^{-2T_s})} W(z) \qquad (3.55)$$

Moreover, as the sampling rate increases, the sampled model converges to

$$\frac{F_q(z)}{E_q(z)} \xrightarrow{T_s \approx 0} \frac{T_s^2}{3!} \frac{(z - z_1^*)}{(z - e^{-T_s})(z - e^{-2T_s})} \qquad (3.56)$$

where $z_1^* = -2 + \sqrt{3}$ is the asymptotic stochastic sampling zero, which corresponds to the stable root of the sampling zero polynomial $B_3(z) = z^2 + 4z + 1$ [28].

For simulation purposes we used a sampling frequency $\omega_s = 250$ [rad/s]. Note that this frequency is two decades above the fastest system pole, located at $s = -2$. We performed a Monte Carlo simulation of $N_{\text{sim}} = 250$ runs, using $N = 10,000$ data points in each run.

Test 1: If one uses ordinary least squares as in (3.52), then one finds that the parameters are (asymptotically) biased, as discussed in detail in [25]. The continuous-time parameters are extracted by converting to the delta form and then using (3.50). We obtain the following (mean) parameter estimates

$$\begin{bmatrix} \hat{a}_1 \\ \hat{a}_0 \end{bmatrix} = \begin{bmatrix} 1.9834 \\ 1.9238 \end{bmatrix} \qquad (3.57)$$

In particular, we observe that the estimate $\hat{a}_1$ is clearly biased with respect to the continuous-time value $a_1 = 3$.

Test 2: We next perform least squares estimation of the parameters, but with prefiltering of the data by the asymptotic sampling zero polynomial, *i.e.*, we use the sequence of filtered output samples given by

$$y_F(kT_s) = \frac{1}{1 + (2 - \sqrt{3})q^{-1}} y(kT_s") \qquad (3.58)$$

Note that this strategy is essentially as in [15, 16].

Again, we extract the continuous-time parameters by converting to the delta form and using (3.50). We obtain the following estimates for the coefficients of the polynomial (3.54)

$$\begin{bmatrix} \hat{a}_1 \\ \hat{a}_0 \end{bmatrix} = \begin{bmatrix} 2.9297 \\ 1.9520 \end{bmatrix} \qquad (3.59)$$

The residual small bias in this case can be explained by the use of the asymptotic sampling zero in (3.56), which is not strictly correct whenever the sampling period T_s is finite.

In the previous example we obtained an asymptotically biased estimation of the parameter $\hat{a}_1$ when the sampling zeros are ignored. In fact, the estimates obtained in (3.57) are predicted by the following lemma. This lemma is the stochastic counterpart of Lemma 3.2, namely, the asymptotic parameter estimates obtained when using the simple derivative replacement approach for second-order CAR systems.

Lemma 3.3. *Consider the second-order continuous-time autoregressive system*

$$\frac{d^2}{dt^2} y(t) + \alpha_1 \frac{d}{dt} y(t) + \alpha_0 y(t) = \dot{v}(t) \qquad (3.60)$$

where $\dot{v}(t)$ is a CTWN process. Assume that a sequence $\{y_k = y(kT_s)\}$ is obtained by sampling instantaneously the system output. If an equation error procedure is used to estimate the parameters of (3.60) using the model

$$\delta y^2 + \hat{\alpha}_1 \delta y + \hat{\alpha}_0 y = e \qquad (3.61)$$

Then, as the sampling period T_s goes to zero, the parameters go to

$$\hat{\alpha}_1 \to \tfrac{2}{3}\alpha_1 \qquad \hat{\alpha}_0 \to \alpha_0 \qquad (3.62)$$

Proof. The proof follows similar lines to the proof of Lemma 3.2. The details can be found in [24].

Up to this point, we have considered undermodelling errors that arise when *sampling zeros* (stochastic or deterministic) are not considered in the discrete-time model. In the next subsection, we show that high-frequency modelling errors in continuous-time can have an equally catastrophic effect on parameter estimation as does neglected sampling zero.

3.3.3 Continuous-time Undermodelling

In this section, we illustrate the consequences of unmodelled dynamics in the continuous-time model, when using estimation procedures based on sampled data. Our focus will be on the case of stochastic systems, however, similar issues arise for deterministic system.

The input of a stochastic system is assumed to be a CTWN process. However, such a process is only a mathematical abstraction and does not physically exist [1, 13]. In practice, we will have *wide-band* noise processes as a disturbance. This is equivalent to a form of high-frequency undermodelling.

The solution of the CAR identification problem for sampled data would seem to be straightforward given the discussion in the previous subsection. Apparently, one only needs to include the *sampling zeros* to get asymptotically unbiased parameter estimates using least squares. However, this ignores the issue of fidelity of the high-frequency components of the model. Indeed, the system relative degree cannot be robustly defined for continuous-time models due to the presence of (possibly time-varying and ill-defined) high-frequency poles or zeros. If one accepts this claim, then one cannot rely upon the integrity of the extra polynomial $F_q(q^{-1})$. In particular, the error caused by ignoring this polynomial (as suggested by the cost function (3.52)) might be as catastrophic as using a sampling zero polynomial arising from some hypothetical assumption about the relative degree. Thus, this class of identification procedures are inherently non-robust. We illustrate this by continuing Example 3.4.

Example 3.5 (Example 3.4 continued). Let us assume that the *true model* for the system (3.53) is given by the polynomial

$$E(p) = E^o(p)(0.02p + 1) \tag{3.63}$$

where we have renamed the polynomial (3.54) in the original model as $E^o(p)$. The *true* system has an unmodelled pole at $s = -50$, which is more than one decade above the fastest nominal pole in (3.53) and (3.54), but almost one decade below the sampling frequency, $\omega_s = 250$ [rad/s].

We repeat the estimation procedure described in *Test 2*, in Example 3.4, using the filtered least squares procedure. We obtain the following estimates

$$\begin{bmatrix} \hat{a}_1 \\ \hat{a}_0 \end{bmatrix} = \begin{bmatrix} 1.4238 \\ 1.8914 \end{bmatrix} \qquad (3.64)$$

These are clearly biased, even though the *nominal* sampling zero has been included in the model.

To analyse the effect of different types of undermodelling, we consider the *true* denominator polynomial (3.63) to be

$$E(p) = E^o(p)\left(\frac{1}{\omega_u}p + 1\right) \qquad (3.65)$$

We consider different values of the parameter ω_u in (3.65), using the same simulation conditions as in the previous examples (*i.e.*, 250 Monte Carlo runs using 10,000 data points each). The results are presented in Figure 3.4. The figure clearly shows the effect of the unmodelled dynamics on the parameter estimates. We see that the undermodelling has an impact even beyond the sampling frequency, which can be explained in terms of the inherent folding effect of the sampling process.

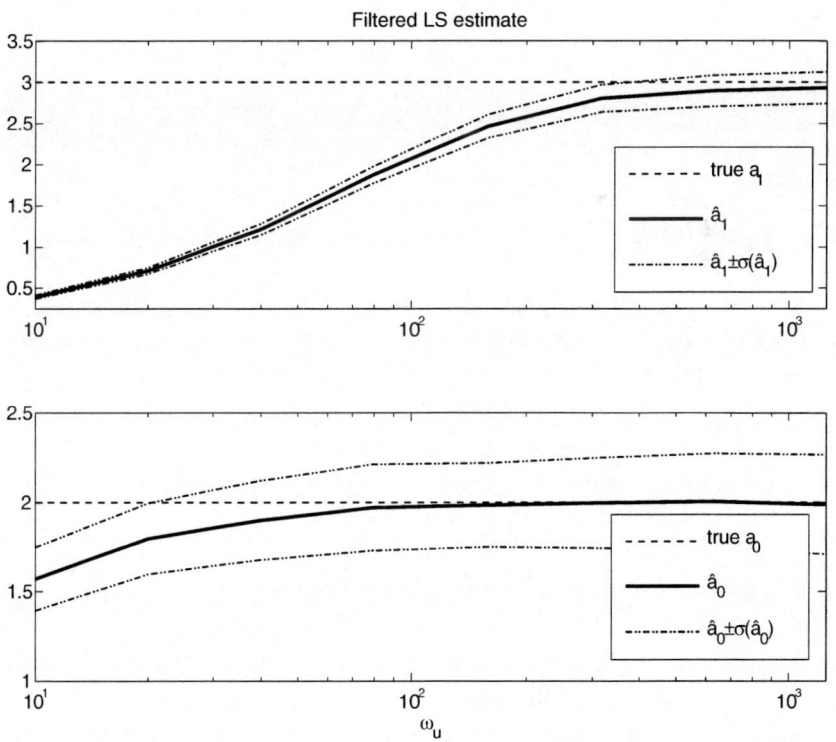

Fig. 3.4. Mean of the parameter estimates as a function of the unmodelled dynamics, using filtered LS

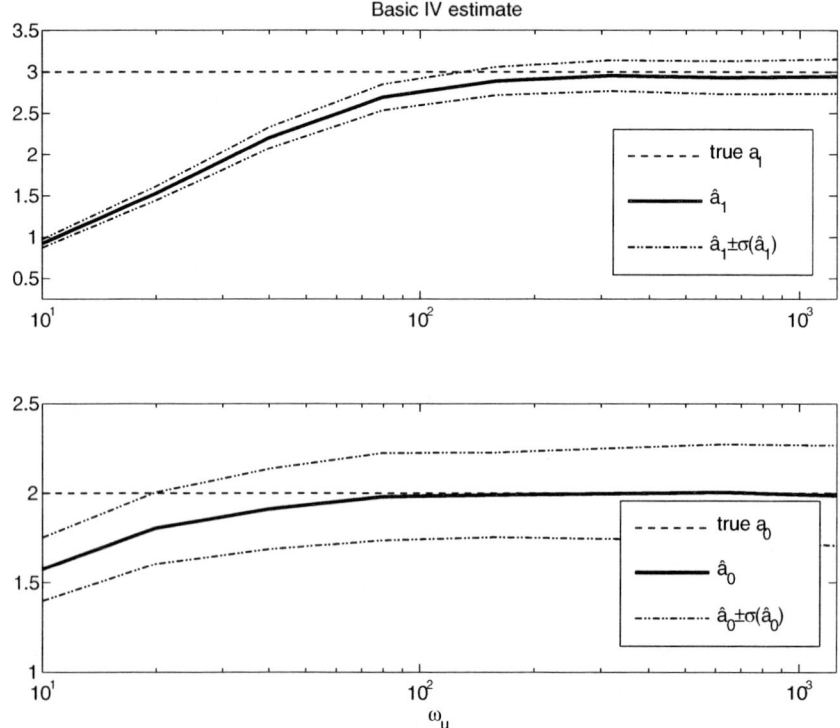

Fig. 3.5. Mean of the parameter estimates as a function of the unmodelled dynamics, using simple delayed IV

Figure 3.5 shows similar simulation results using a basic delayed instrumental variable estimator, where the IV vector consists of observations of $y(t)$ delayed one sampling period [3].

The difficulties discussed above arise due to the fact that the true high-frequency characteristics are not exactly as hypothesised in the algorithm. Thus, the folding that occurs is not governed by the anticipated sampling zero polynomial that is used to prefilter the data.

3.3.4 Restricted-bandwidth FDML Estimation

The examples presented in the previous subsections raise the question as to how these problems might be avoided or, at least, reduced, by using an identification procedure more robust to high-frequency undermodelling. Our proposal to deal with this problem is to designate a bandwidth of validity for the model and, then, to develop an algorithm that is insensitive to errors outside that range. This is most easily done in the frequency domain.

In the following example we will use the FDML procedure presented in Lemma 3.1 to estimate the parameters of CAR systems as (3.46).

Remark 3.5. For a CAR system as in (3.46), let us consider the (approximate) derivative replacement discrete-time model

$$E_q(q)y_k = w_k \qquad (3.66)$$

where y_k is the sequence of instantaneous output samples of (3.46), and w_k is a discrete-time stationary Gaussian white noise sequence with variance σ_w^2. Given N data points of the output sequence $y(kT_s)$ sampled at ω_s [rad/s], the appropriate likelihood function, in the frequency domain, takes the form

$$L = \sum_{\ell=0}^{n_{max}} \frac{|E_q(e^{i\omega_\ell T_s})Y(e^{i\omega_\ell T_s})|^2}{N\sigma_w^2} - \log \frac{|E_q(e^{i\omega_\ell T_s})|^2}{\sigma_w^2} \qquad (3.67)$$

where $\omega_\ell = \frac{\omega_s \ell}{N}$ and n_{max} corresponds to the bandwidth to be considered, i.e., $\omega_{max} = \frac{\omega_s n_{max}}{N}$.

Example 3.6. We consider again the CAR system presented in Example 3.4. If we use the result in Lemma 3.1, using the full bandwidth $[0, \pi/T_s]$ (or, equivalently, up to 125 [rad/s]) we obtain the following (mean) value for the parameter estimates

$$\begin{bmatrix} \hat{a}_1 \\ \hat{a}_0 \end{bmatrix} = \begin{bmatrix} 4.5584 \\ 1.9655 \end{bmatrix} \qquad (3.68)$$

As expected, these parameters are clearly biased because we are not taking into account the presence of the sampling zero polynomial in the true model. Next, we consider an estimation procedure restricted to a certain bandwidth of validity. For example, the usual rule of thumb is to consider up to one decade above the fastest nominal system pole, in this case, 20 [rad/s]. The resultant (mean of the) parameter estimates are then given by

$$\begin{bmatrix} \hat{a}_1 \\ \hat{a}_0 \end{bmatrix} = \begin{bmatrix} 3.0143 \\ 1.9701 \end{bmatrix} \qquad (3.69)$$

Note that these estimates are essentially equal to the (continuous-time) true values. Moreover, no prefiltering as in (3.51) or (3.58) has been used! Thus, one has achieved robustness to the relative degree at high frequencies since it plays no role in the suggested procedure. Moreover, the sampling zeros can be ignored since their impact is felt only at high frequencies.

Finally, we show that the frequency-domain procedure is also robust to the presence of unmodelled fast poles. We consider again the true system to be as in (3.63). We restrict the estimation bandwidth up to 20 [rad/s]. In this case, the mean of the parameter estimates is again very close to the nominal system coefficients, i.e., we obtain

$$\begin{bmatrix} \hat{a}_1 \\ \hat{a}_0 \end{bmatrix} = \begin{bmatrix} 2.9285 \\ 1.9409 \end{bmatrix} \qquad (3.70)$$

A more general situation is shown in Figure 3.6. The figure shows the parameter estimates obtained using the proposed FDML procedure, with the same restricted bandwidth used before $\omega_{\max} = 20$ [rad/s], for different locations of the unmodelled fast pole ω_u, as in (3.65).

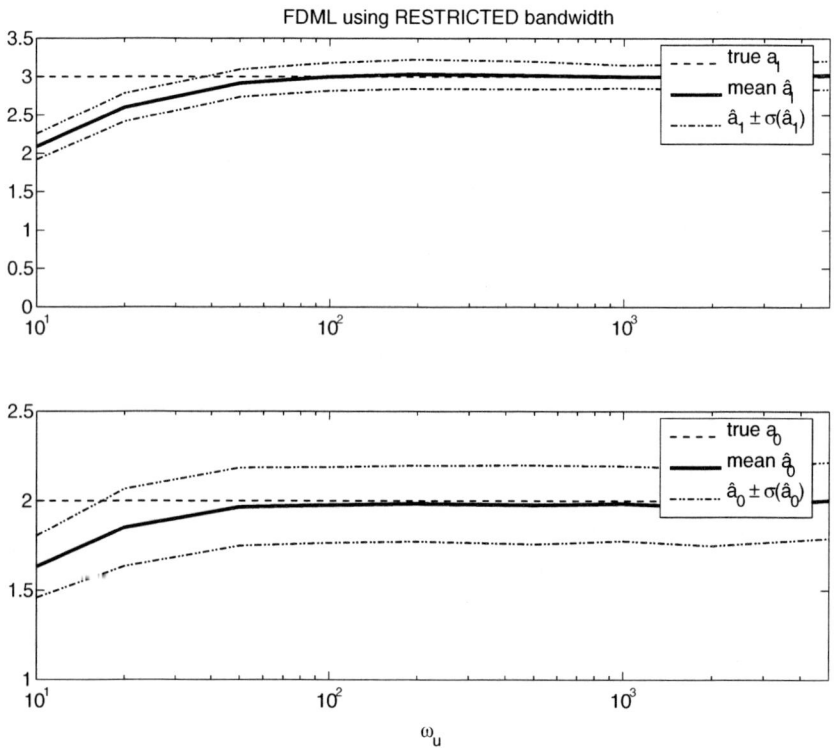

Fig. 3.6. Parameter estimates using FDML as a function of unmodelled pole

Remark 3.6. Note that the likelihood function (3.67) is not scalable by σ_w^2 and hence one needs to also include this parameter in the set to be estimated. This is an important departure from the simple least squares case.

3.4 Conclusions

In this chapter we have explored the robustness issues that arise in the identification of continuous-time systems from sampled data. A key observation is that the *fidelity* of the models at high frequencies generally plays an important role in obtaining models suitable for continuous-time system identification. In particular, we have shown that

- sampling zeros may have to be included in the discrete-time models to obtain accurate sampled-data descriptions.
- unmodelled high-frequency dynamics in the continuous-time model can have a critical impact on the quality of the estimation process when using sampled data.

This implies that any result that implicitly or explicitly depends upon the folding of high-frequency components down to lower frequencies will be inherently *non-robust*. As a consequence, we argue that models have to be considered within a *bandwidth of validity*.

To address these issues, we have proposed the use of frequency-domain maximum likelihood estimation, using a restricted bandwidth. We have shown that this approach is robust to both the presence of sampling zeros and to high-frequency modelling errors in continuous time.

The problems discussed above have been illustrated for both, deterministic and stochastic systems. Special attention was given to the identification of continuous-time autoregressive stochastic models from sampled data. We have argued that traditional approaches to this problem are inherently sensitive to high-frequency modelling errors. We have also argued that these difficulties can be mitigated by using the proposed FDML with restricted bandwidth.

Acknowledgements

The authors would like to acknowledge the support of the Centre of Complex Dynamics Systems and Control, the University of Newcastle, Australia, and National Fund for Scientific and Technological Development (FONDECYT), Chile, through the PostDoctoral Grant 3060013.

References

1. K.J. Åström. *Introduction to Stochastic Control Theory*. Academic Press, New York, 1970.
2. K.J. Åström, P. Hagander, and J. Sternby. Zeros of sampled systems. *Automatica*, 20(1):31–38, 1984.
3. S. Bigi, T. Söderström, and B. Carlsson. An IV scheme for estimating continuous-time models from sampled data. *10th IFAC Symposium on System Identification*, Copenhagen, Denmark, 1994.
4. D.R. Brillinger. Fourier analysis of stationary processes. *Proceedings of the IEEE*, 62(12):1628–1643, December 1974.
5. D.R. Brillinger. *Time Series: Data Analysis and Theory*. McGraw-Hill, New York, 1981.
6. A. Feuer and G.C. Goodwin. *Sampling in Digital Signal Processing and Control*. Birkhäuser, Boston, 1996.

7. H. Garnier, M. Mensler, and A. Richard. Continuous-time model identification from sampled data. Implementation issues and performance evaluation. *International Journal of Control*, 76(13):1337–1357, 2003.
8. J. Gillberg and L. Ljung. Frequency-domain identification of continuous-time ARMA models from sampled data. *16th IFAC World Congress*, Prague, Czech Republic, July 2005.
9. J. Gillberg and L. Ljung. Frequency-domain identification of continuous-time OE models from sampled data. *16th IFAC World Congress*, Prague, Czech Republic, July 2005.
10. G.C. Goodwin, J.I. Yuz, and H. Garnier. Robustness issues in continuous-time system identification from sampled data. *16th IFAC World Congress*, Prague, Czech Republic, July 2005.
11. R. Johansson. Identification of continuous-time models. *IEEE Transactions on Signal Processing*, 42(4):887–897, 1994.
12. R. Johansson, M. Verhaegen, and C.T. Chou. Stochastic theory of continuous-time state-space identification. *IEEE Transactions on Signal Processing*, 47(1):41–51, 1999.
13. P.E. Kloeden and E. Platen. *Numerical Solution of Stochastic Differential Equations*. Springer-Verlag, New York, 1992.
14. N.R. Kristensen, H. Madsen, and S.B. Jørgensen. Parameter estimation in stochastic grey-box models. *Automatica*, 40:225–237, 2004.
15. E.K. Larsson. Limiting properties of sampled stochastic systems. Technical Report 2003-028, Department of Information Technology, Uppsala University, Sweden, 2003.
16. E.K. Larsson and T. Söderström. Continuous-time AR parameter estimation by using properties of sampled systems. *15th IFAC World Congress*, Barcelona, Spain, July 2002.
17. L. Ljung. Some results on identifying linear systems using frequency domain data. *32nd IEEE Conference on Decision and Control*, San Antonio, Texas, USA, December 1993.
18. L. Ljung. *System Identification. Theory for the User*. Prentice Hall, Englewood Cliffs, New Jersey, 2nd edition, 1999.
19. T. McKelvey and L. Ljung. Frequency domain maximum likelihood identification. *11th IFAC Symposium on System Identification*, pages 1741–1746, Fukuoka, Japan, 1997.
20. R.H. Middleton and G.C. Goodwin. *Digital Control and Estimation. A Unified Approach*. Prentice Hall, Englewood Cliffs, New Jersey, 1990.
21. R. Pintelon and J. Schoukens. *System Identification. A Frequency Domain Approach*. IEEE Press, Piscataway, USA, 2001.
22. G.P. Rao and H. Garnier. Numerical illustrations of the relevance of direct continuous-time model identification. *15th IFAC World Congress*, Barcelona, Spain, 2002.
23. N.K. Sinha and G.P. Rao (eds). *Identification of Continuous-time Systems. Methodology and Computer Implementation*. Kluwer Academic Publishers, Dordrecht, 1991.
24. T. Söderström, B. Carlsson, and S. Bigi. On estimating continuous-time stochastic models from discrete-time data. Technical Report UPTEC 92104R, Dept. of Technology, Uppsala University, Sweden, July 1992.

25. T. Söderström, H. Fan, B. Carlsson, and S. Bigi. Least squares parameter estimation of continuous-time ARX models from discrete-time data. *IEEE Transactions on Automatic Control*, 42(5):659–673, 1997.
26. H. Unbehauen and G.P. Rao. Continuous-time approaches to system identification – a survey. *Automatica*, 26(1):23–35, 1990.
27. H. Unbehauen and G.P. Rao. A review of identification in continuous-time systems. *Annual Reviews in Control*, 22:145–171, 1998.
28. B. Wahlberg. Limit results for sampled systems. *International Journal of Control*, 48(3):1267–1283, 1988.
29. P. Young. Parameter estimation for continuous-time models – A survey. *Automatica*, 17(1):23–39, 1981.

4

Refined Instrumental Variable Identification of Continuous-time Hybrid Box–Jenkins Models

Peter C. Young[1], Hugues Garnier[2] and Marion Gilson[2]

[1] Lancaster University, UK & Australian National University
[2] Nancy-Université, CNRS, France

4.1 Introduction

This chapter describes and evaluates a statistically optimal method for the identification and estimation[3] of continuous-time (CT) hybrid Box–Jenkins (BJ) transfer function models from discrete-time, sampled data. Here, the model of the basic dynamic system is estimated in continuous-time, differential equation form, while the associated additive noise model is estimated as a discrete-time, autoregressive moving average (ARMA) process. This refined instrumental variable method for continuous-time systems (RIVC) was first developed in 1980 by Young and Jakeman [52] and its simplest embodiment, the simplified RIVC (SRIVC) method, has been used successfully for many years, demonstrating the advantages that this stochastic formulation of the continuous-time estimation problem provides in practical applications (see, *e.g.*, some recent such examples in [16, 34, 40, 45, 48]).

However, the 'simplification' that characterises the name of the SRIVC method is the assumption, for the purposes of simplicity and algorithmic development, that the additive noise is purely white in form. Such an approach is optimal under this assumption and the inherent instrumental variable aspects of the resulting algorithm ensure that the parameter estimates are consistent and asymptotically unbiased in statistical terms, even if the noise happens to be coloured. However, the SRIVC estimates are not, in general, statistically efficient (minimum variance) in this situation because the prefilters are not designed to account for the colour in the noise process.

The hybrid RIVC estimation procedure, described and evaluated in this chapter, follows logically from the refined instrumental variable (RIV) method for

---

[3] The statistical meaning of these terms will be used here, where 'identification' is taken to mean the specification of an identifiable model structure and 'estimation' relates to the estimation of the parameters that characterise this identified model structure.

discrete-time (DT) models, first developed within a maximum likelihood (ML) context by Young in 1976 [27] and comprehensively evaluated by Young and Jakeman [7,42,46]. Further developments of both the RIV and RIVC methods have been reported recently [3,37,38,41], including the use of these algorithms in a closed-loop context (see also [5] and Chapter 5 in the present book).

The RIV algorithm involves concurrent DT noise model estimation and uses this estimated noise model in the iterative-adaptive design of statistically optimal prefilters that effectively attenuate noise outside the passband of the system and prewhiten the noise remaining within the bandpass. Similarly motivated prefilters are utilised in the RIVC algorithm but they also provide a very convenient way of generating the prefiltered derivatives of the input and output variables, as required for CT model estimation.

If required, it is possible to design the RIVC algorithm entirely in CT terms by using a CT formulation that involves a continuous-time RIVC estimation algorithm with CT optimal prefilters (see [17], Equation (15)). An equivalent RIV solution of the optimal delta operator estimation problem has also been described and evaluated [39] and this provides an alternative to the RIVC approach that could be advantageous in some control applications [14].

The alternative hybrid form of the continuous-time transfer function model is considered here for two reasons. First, the approach is simple and straightforward: the theoretical and practical problems associated with the estimation of purely stochastic, continuous-time CAR or CARMA models are avoided by formulating the problem in this manner. Second, as pointed out above, one of the main functions of the noise estimation is to improve the statistical efficiency of the parameter estimation by introducing appropriately defined prefilters into the estimation procedure. And, as we shall see in this chapter, this can be achieved adequately on the basis of hybrid prefilters defined by reference to discrete-time AR or ARMA noise models.

Following the problem formulation in Section 4.2, the theoretical motivation for hybrid RIVC estimation is outlined in Section 4.3 and the RIVC/SRIVC algorithms are described in Section 4.4. The theoretical justification for the prefilters, which is discussed more formally in Section 4.5, is provided by both 'extended instrumental variable' and maximum likelihood analysis. The evaluation of the RIVC/SRIVC algorithms in Sections 4.7 and 4.8 of the chapter is based on comprehensive Monte Carlo simulation (MCS) analysis, as well as two practical examples. The first of these considers the use of RIVC in the 'dominant mode' simplification of a giant dynamic simulation model used in climate research; and the second is the re-investigation of a demonstration example [4] used in both the CAPTAIN and CONTSID Toolboxes for MATLAB® (see conclusions Section 4.9), which involves the analysis of data from a multiple-input winding process.

4.2 Problem Formulation

For simplicity of presentation, the formulation and solution of the CT estimation problem will be restricted to the case of a linear, single-input, single-output system. It should be noted, however, that the analysis extends straightforwardly to multiple-input systems and, in a more complex manner, to full multi-variable systems (*e.g.*, [7]). Indeed, a practical multiple-input example is considered in Section 4.8. In the single-input, single-output situation, it is assumed that the input $u(t)$ and the noise-free output $x(t)$ are related by the following constant coefficient, differential-delay equation,

$$\frac{d^n x(t)}{dt^n} + a_1^o \frac{d^{n-1} x(t)}{dt^{n-1}} + \cdots + a_n^o x(t) = b_0^o \frac{d^m u(t-\tau)}{dt^m} + \cdots + b_m^o u(t-\tau)$$

or,

$$x^{(n)}(t) + a_1^o x^{(n-1)}(t) + \cdots + a_n^o x(t) = b_0^o u^{(m)}(t-\tau) + \cdots + b_m^o u(t-\tau) \quad (4.1)$$

where $x^{(i)}(t)$ denotes the ith time derivative of the continuous-time signal $x(t)$ and τ is a pure time delay in time units. This is often assumed to be an integer number related to the sampling time: *i.e.*, $\tau = n_k T_s$ but this is not essential: in this CT environment, 'fractional' time delays can be introduced if required (*e.g.*, see [35]). For simplicity, the time delay will not be considered in the following analysis but it can be accommodated straightforwardly if identified from the data. Equation (4.1) can also be written in the following compact transfer function (TF) form,

$$x(t) = G_o(p) u(t) = \frac{B_o(p)}{A_o(p)} u(t) \quad (4.2)$$

with

$$B_o(p) = b_0^o p^m + b_1^o p^{m-1} + \cdots + b_m^o, \quad (4.2a)$$
$$A_o(p) = p^n + a_1^o p^{n-1} + \cdots + a_n^o, \quad n \geq m \quad (4.2b)$$

where $x(t)$ is the deterministic output of the system; p is the differential operator, *i.e.*, $p^i x(t) = \frac{d^i x(t)}{dt^i}$; $B_o(p)$ and $A_o(p)$ are assumed to be coprime; and the system is asymptotically stable. It is assumed that the input signal $\{u(t), t_1 < t < t_N\}$ is applied to the system and this gives rise to an output signal $\{x(t), t_1 < t < t_N\}$.

In order to obtain high-quality statistical estimation results, it is vital to consider the inevitable errors that will affect the measured output signal. It is assumed here that $x(t)$ is corrupted by an additive, coloured measurement noise $\xi(t)$, so that the complete equation for the data-generating system, denoted by $\mathcal{S}$, can be written in the form,

$$\mathcal{S}: y(t) = G_o(p) u(t) + H_o(p) e_o(t) \quad (4.3)$$

or, in the alternative decomposed form that is more appropriate in the present context

$$\mathcal{S} \begin{cases} x(t) = G_o(p)u(t) \\ \xi(t) = H_o(p)e_o(t) \\ y(t) = x(t) + \xi(t) \end{cases} \quad (4.4)$$

where $H_o(p)$ is stable and invertible, while $e_o(t)$ is a zero-mean, continuous-time white noise source, which is assumed to be uncorrelated with the input $u(t)$. Finally, if the additive coloured noise $\xi(t)$ has rational spectral density, then a suitable parametric representation is the following continuous-time, autoregressive moving average (CARMA) model

$$\xi(t) = H_o(p)e_o(t) = \frac{C_o(p)}{D_o(p)}e_o(t) \quad (4.5)$$

where $C_o(p)$ and $D_o(p)$ are suitably defined polynomials in the p operator. Of course, in most practical situations, the input and output signals $u(t)$ and $y(t)$ will be sampled in discrete time. In the case of uniform sampling, at a constant sampling interval T_s, these sampled signals will be denoted by $u(t_k)$ and $y(t_k)$ and the output observation equation then takes the form,

$$y(t_k) = x(t_k) + \xi(t_k) \quad k = 1, \cdots N \quad (4.6)$$

where $x(t_k)$ is the sampled value of the unobserved, noise-free output $x(t)$. The objective is then to identify a suitable model structure for (4.4) and estimate the parameters that characterise this structure, based on these sampled input and output data $Z^N = \{u(t_k); y(t_k)\}_{k=1}^N$.

Given the discrete-time, sampled nature of the data, an obvious assumption is that the discrete-time, coloured noise associated with the sampled output measurement $y(t_k)$ has rational spectral density and so can be represented by a discrete-time ARMA(p,q) model. The model set to be identified and estimated, as denoted by $\mathcal{M}$ with system ($\mathcal{G}$) and noise ($\mathcal{H}$) models parameterised independently, then takes the form,

$$\mathcal{M} : \{G(p, \boldsymbol{\rho}), H(q^{-1}, \boldsymbol{\eta})\} \quad (4.7)$$

where $\boldsymbol{\rho}$ and $\boldsymbol{\eta}$ are parameter vectors that characterise the system and noise models, respectively. In particular, the system model is formulated in continuous-time terms

$$\mathcal{G} : G(p, \boldsymbol{\rho}) = \frac{B(p, \boldsymbol{\rho})}{A(p, \boldsymbol{\rho})} = \frac{b_0 p^m + b_1 p^{m-1} + \cdots + b_m}{p^n + a_1 p^{n-1} \cdots + a_n} \quad (4.8)$$

and the associated model parameters are stacked columnwise in the parameter vector,

$$\boldsymbol{\rho} = \begin{bmatrix} a_1 & \cdots & a_n & b_0 & \cdots & b_m \end{bmatrix}^T \in \mathbb{R}^{n+m+1} \tag{4.9}$$

while the noise model is in discrete-time form

$$\mathcal{H} : H(q^{-1}, \boldsymbol{\eta}) = \frac{C(q^{-1}, \boldsymbol{\eta})}{D(q^{-1}, \boldsymbol{\eta})} = \frac{1 + c_1 q^{-1} + \cdots + c_q q^{-q}}{1 + d_1 q^{-1} + \cdots + d_p q^{-p}} \tag{4.10}$$

where q^{-r} is the backward shift operator, i.e., $q^{-r} y(t_k) = y(t_{k-r})$ and the associated model parameters are stacked columnwise in the parameter vector,

$$\boldsymbol{\eta} = \begin{bmatrix} c_1 & \cdots & c_q & d_1 & \cdots & d_p \end{bmatrix}^T \in \mathbb{R}^{p+q} \tag{4.11}$$

Consequently, the noise TF takes the usual ARMA model form

$$\xi(t_k) = \frac{C(q^{-1}, \boldsymbol{\eta})}{D(q^{-1}, \boldsymbol{\eta})} e(t_k) \qquad e(t_k) \sim \mathcal{N}(0, \sigma^2) \tag{4.12}$$

where, as shown, $e(t_k)$ is a zero-mean, normally distributed, discrete-time white noise sequence.

The structure $\mathcal{S}$ does not specify any common factors in the plant (G_o) and noise (H_o) components, so that these models can be parameterised independently. More formally, there exists the following decomposition of the parameter vector $\boldsymbol{\theta}$ for the whole hybrid model,

$$\boldsymbol{\theta} = \begin{pmatrix} \boldsymbol{\rho} \\ \boldsymbol{\eta} \end{pmatrix} \tag{4.13}$$

such that the model equations can be written in the form

$$\mathcal{M} \begin{cases} x(t) = G(p, \boldsymbol{\rho}) u(t) \\ \xi(t_k) = H(q^{-1}, \boldsymbol{\eta}) e(t_k) \\ y(t_k) = x(t_k) + \xi(t_k) \end{cases} \tag{4.14}$$

This model is considered as a hybrid Box–Jenkins model because of its close relationship to the DT model considered in great detail by Box and Jenkins in their seminal book on time-series analysis, forecasting and control [2] and used as the basis for the development of the original RIVC algorithm [52]. Alternatively, the model can be written in the following vector terms

$$\mathcal{M} \begin{cases} x^{(n)}(t) = \boldsymbol{\varphi}^T(t) \boldsymbol{\rho} \\ \xi(t_k) = \boldsymbol{\psi}^T(t_k) \boldsymbol{\eta} + e(t_k) \\ y(t_k) = x(t_k) + \xi(t_k) \end{cases} \tag{4.15}$$

where,

$$\boldsymbol{\varphi}^T(t) = \begin{bmatrix} -x^{(n-1)}(t) & \cdots & -x(t) & u^{(m)}(t) & \cdots & u(t) \end{bmatrix} \tag{4.15a}$$

$$\boldsymbol{\psi}^T(t_k) = \begin{bmatrix} -\xi(t_{k-1}) & \cdots & -\xi(t_{k-p}) & e(t_{k-1}) & \cdots & e(t_{k-q}) \end{bmatrix} \tag{4.15b}$$

For the purposes of identification, the order of this single-input model (with the pure time delay τ now added for completeness) is denoted by $[n \ m \ \tau \ p \ q]$ and the complete identification problem can now be stated as follows:

Based on N uniformly sampled measurements of the input and output, $Z^N = \{u(t_k); y(t_k)\}_{k=1}^{N}$, identify the orders n, m, p and q of the polynomials in the system and noise TF models, as well as any pure time delay τ, and estimate the parameter vector $\boldsymbol{\theta}$ in (4.13) whose parameters characterise these polynomials.

4.3 Optimal RIVC Estimation: Theoretical Motivation

The RIVC algorithm derives from the RIV algorithm for DT systems. This was evolved by converting the maximum likelihood (ML) estimation equations to a pseudo-linear form [20] involving optimal prefilters [27, 46, 52]. A similar analysis can be utilised in the present situation because the problem is very similar, in both algebraic and statistical terms. However, to conserve space, the discussion here will be restricted to a simpler development of the RIVC algorithm and we leave the interested reader to consult with these earlier references for details of the ML analysis.

4.3.1 The Hybrid Box–Jenkins Estimation Model

Following the usual *prediction error minimisation* (PEM) approach in the present hybrid situation (which is ML estimation because of the Gaussian assumptions on $e(t_k)$), a suitable error function $\varepsilon(t_k)$, at the kth sampling instant, is given by,

$$\varepsilon(t_k) = \frac{D(q^{-1}, \boldsymbol{\eta})}{C(q^{-1}, \boldsymbol{\eta})} \left\{ y(t_k) - \frac{B(p, \boldsymbol{\rho})}{A(p, \boldsymbol{\rho})} u(t_k) \right\}$$

which can be written as,

$$\varepsilon(t_k) = \frac{D(q^{-1}, \boldsymbol{\eta})}{C(q^{-1}, \boldsymbol{\eta})} \left\{ \frac{1}{A(p, \boldsymbol{\rho})} \left[A(p, \boldsymbol{\rho}) y(t_k) - B(p, \boldsymbol{\rho}) u(t_k) \right] \right\} \quad (4.16)$$

where the discrete-time prefilter $D(q^{-1}, \boldsymbol{\eta})/C(q^{-1}, \boldsymbol{\eta})$ will be recognised as the inverse of the ARMA(p,q) noise model. Note that in these equations, we are mixing discrete and continuous-time operators somewhat informally in order to indicate the hybrid computational nature of the estimation problem being considered here. Thus, operations such as,

$$\frac{B(p, \boldsymbol{\rho})}{A(p, \boldsymbol{\rho})} u(t_k)$$

imply that the input variable $u(t_k)$ is interpolated in some manner. This is to allow for the inter-sample behaviour that is not available from the sampled data and so has to be inferred in order to allow for the continuous-time numerical integration of the associated differential equations. For such integration, the discretisation interval will be varied, dependent on the numerical method employed, but it will usually be much smaller than the sampling interval T_s. Minimisation of a least squares criterion function in $\varepsilon(t_k)$, measured at the sampling instants, provides the basis for stochastic estimation. However, since the polynomial operators commute in this linear case, (4.16) can be considered in the alternative form,

$$\varepsilon(t_k) = A(p, \boldsymbol{\rho})y_\text{f}(t_k) - B(p, \boldsymbol{\rho})u_\text{f}(t_k) \tag{4.17}$$

where $y_f(t_k)$ and $u_f(t_k)$ represent the *sampled* outputs of the complete hybrid prefiltering operation involving the *continuous-time* filtering operations using the filter

$$f_\text{c}(p, \boldsymbol{\rho}) = \frac{1}{A(p, \boldsymbol{\rho})} \tag{4.18}$$

as shown in Figures 4.1 and 4.2, as well as *discrete-time* filtering operations, using the inverse noise model filter

$$f_\text{d}(q^{-1}, \boldsymbol{\eta}) = \frac{D(q^{-1}, \boldsymbol{\eta})}{C(q^{-1}, \boldsymbol{\eta})} \tag{4.19}$$

as shown in Figure 4.2. The associated, linear-in-the-parameters estimation model then takes the form

$$y_\text{f}^{(n)}(t_k) = \boldsymbol{\varphi}_\text{f}^T(t_k)\boldsymbol{\rho} + \eta(t_k) \tag{4.20}$$

where,

$$\boldsymbol{\varphi}_\text{f}^T(t_k) = \left[-y_\text{f}^{(n-1)}(t_k) \; -y_\text{f}^{(n-2)}(t_k) \cdots \; -y_\text{f}(t) \; u^{(m)}(t_k) \cdots u(t_k)\right] \tag{4.21}$$

and $\eta(t_k)$ is the continuous-time noise signal $\eta(t) = A(p, \boldsymbol{\rho})\xi(t)$ sampled at the kth sampling instant.

4.3.2 RIVC Estimation

Optimal methods of IV estimation (see, *e.g.*, [19, 27]) normally involve an iterative (or relaxation) algorithm in which, at each iteration, the 'auxiliary model' used to generate the instrumental variables, as well as the associated prefilters, are updated, based on the parameter estimates obtained at the previous iteration. Let us consider, therefore, the jth iteration where we have access to the estimate,

$$\hat{\boldsymbol{\theta}}^{j-1} = \begin{pmatrix} \hat{\boldsymbol{\rho}}^{j-1} \\ \hat{\boldsymbol{\eta}}^{j-1} \end{pmatrix} \tag{4.22}$$

obtained previously at iteration $j-1$. The most important aspect of optimal IV estimation is the definition of an optimal instrumental variable. In the present context, this is generated from the output of the continuous-time auxiliary model,

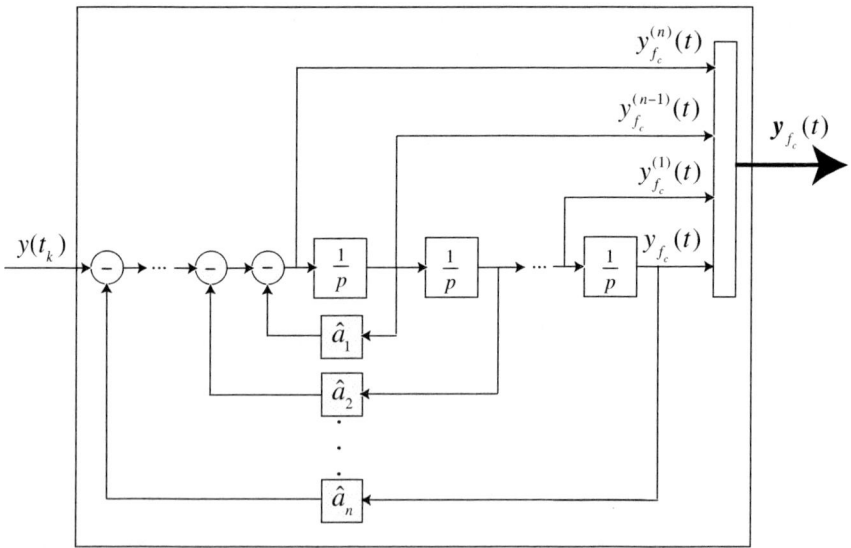

Fig. 4.1. Generation of filtered derivatives for the output $y(t)$ by the prefilter $f_c(p) = 1/A(p, \boldsymbol{\rho}^j)$, where $\hat{a}_i, i = 1, 2, \cdots, n$ are the iteratively updated parameters of the model polynomial $A(p, \boldsymbol{\rho}^j)$ at the jth iteration. This is the inside of the block marked **A** in Figure 4.2 (based on Figure 2(b) in [52]).

$$\hat{x}(t, \hat{\boldsymbol{\rho}}^{j-1}) = G(p, \hat{\boldsymbol{\rho}}^{j-1}) u(t) \quad (4.23)$$

which is prefiltered in the same hybrid manner as the other variables. The associated optimal IV vector $\hat{\boldsymbol{\varphi}}_f(t_k)$, is then an estimate of the noise-free version of the vector $\boldsymbol{\varphi}_f(t_k)$ in (4.21) and is defined as follows

$$\hat{\boldsymbol{\varphi}}_f(t_k) = \left[-\hat{x}_f^{(n-1)}(t_k) \cdots -\hat{x}_f(t_k) \; u_f^{(m)}(t_k) \cdots u_f(t_k) \right]^T \quad (4.24)$$

where it should be noted that

$$\hat{\boldsymbol{\varphi}}_f(t_k) = \hat{\boldsymbol{\varphi}}_f(t_k, \hat{\boldsymbol{\rho}}^{j-1}, \hat{\boldsymbol{\eta}}^j) \quad (4.25)$$

because the instrumental variables are now prefiltered and so are a function of both the system parameter estimates at the previous iteration and the most recent noise model parameter estimates (see later). For simplicity, however, these additional arguments will be omitted in the subsequent analysis. Note also that the noise-free version of the vector $\boldsymbol{\varphi}_f(t_k)$ in (4.21), which we will define as follows,

$$\overset{\circ}{\boldsymbol{\varphi}}_f^T(t_k) = \left[-x_f^{(n-1)}(t_k) \cdots -x_f(t_k) \; u_f^{(m)}(t_k) \cdots u_f(t_k) \right] \quad (4.26)$$

where $x(t) = G_o(p)u(t)$, is referred to in Section 4.5 when considering the statistical properties of the optimal IV parameter estimates.

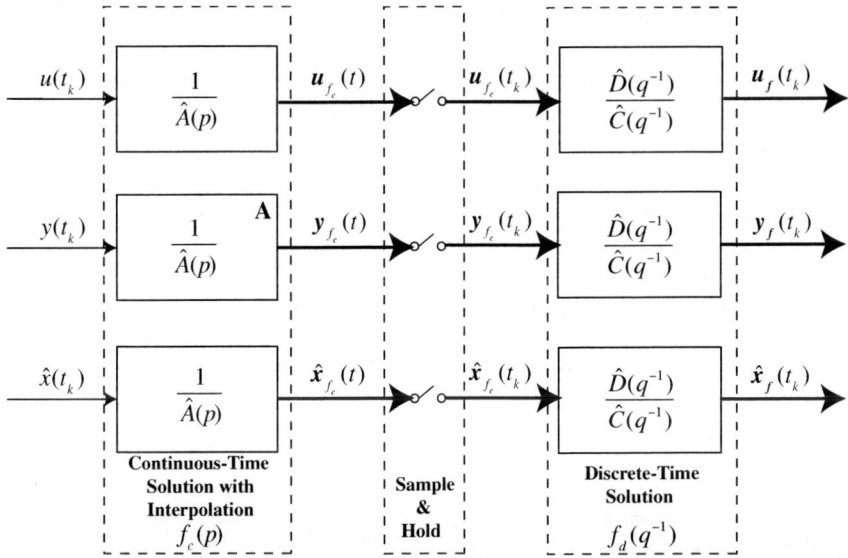

Fig. 4.2. The hybrid analogue–digital prefiltering operations used in RIVC estimation: the bold arrows and associated bold-face variables denote vector quantities with elements defined as the appropriate prefiltered derivatives of $u(t_k)$, $y(t_k)$ and $\hat{x}(t_k)$, while $\hat{A}(p) = A(p, \hat{\rho}^{j-1})$, $\hat{C}(q^{-1}) = C(q^{-1}, \hat{\eta}^j)$ and $\hat{D}(q^{-1}) = D(q^{-1}, \hat{\eta}^j)$ are the iteratively updated model polynomials that define the prefilters (based on Figure 2 in [52]). The detail of the block marked **A** is shown in Figure 4.1.

The IV optimisation problem can now be stated in the form

$$\hat{\boldsymbol{\rho}}^j(N) = \arg\min_{\boldsymbol{\rho}} \left\| \left[\frac{1}{N} \sum_{k=1}^{N} \hat{\boldsymbol{\varphi}}_{\mathrm{f}}(t_k) \boldsymbol{\varphi}_{\mathrm{f}}(t_k)^T \right] \boldsymbol{\rho} - \left[\frac{1}{N} \sum_{k=1}^{N} \hat{\boldsymbol{\varphi}}_{\mathrm{f}}(t_k) y_{\mathrm{f}}^{(n)}(t_k) \right] \right\|_{\mathbf{Q}}^2$$
(4.27)

where $\|\mathbf{x}\|^2 = \mathbf{x}^T \mathbf{Q} \mathbf{x}$ and $\mathbf{Q} = \mathbf{I}$. This results in the solution of the IV estimation (IV normal) equations

$$\hat{\boldsymbol{\rho}}^j(N) = \left[\sum_{k=1}^{N} \hat{\boldsymbol{\varphi}}_{\mathrm{f}}(t_k) \boldsymbol{\varphi}_{\mathrm{f}}(t_k)^T \right]^{-1} \sum_{k=1}^{N} \hat{\boldsymbol{\varphi}}_{\mathrm{f}}(t_k) y_{\mathrm{f}}^{(n)}(t_k)$$
(4.28)

where the $\hat{\boldsymbol{\rho}}^j(N)$ is the IV estimate of the system model parameter vector at the jth iteration based on the appropriately prefiltered input/output data $Z^N = \{u(t_k); y(t_k)\}_{k=1}^{N}$.

As regards the hybrid prefiltering, it will be noted from (4.25) that this involves the inverse noise model parameters $\hat{\boldsymbol{\eta}}^j$ obtained at the current jth iteration. This is because, given $\hat{\boldsymbol{\rho}}^{j-1}$, an estimate of the sampled noise signal $\xi(t_k)$, at the jth iteration, is obtained by subtracting the sampled output of the auxiliary model equation (4.23) from the measured output $y(t_k)$, *i.e.*,

$$\hat{\xi}(t_k) = y(t_k) - \hat{x}(t, \hat{\boldsymbol{\rho}}^{j-1}) \qquad (4.29)$$

This estimate provides the basis for the estimation of the noise model parameter vector $\boldsymbol{\eta}^j$, using whatever ARMA model estimation algorithm is selected for this task (see Section 4.4). This is then available to design and implement the latest discrete-time, inverse noise model part of the hybrid prefiltering operations shown in the right hand-side of Figure 4.2.

4.4 The RIVC and SRIVC Algorithms

The iterative RIVC and SRIVC algorithms follow directly from the RIV and SRIV algorithms for DT systems (*e.g.*, [46]). This section summarises both of the new hybrid algorithms.

4.4.1 The RIVC Algorithm

Bearing the analysis of the previous subsection 4.3.2 in mind, the main steps in the RIVC algorithm are as follows:

Step 1. Initialisation: generate an initial estimate of the TF model parameter vector $\hat{\boldsymbol{\rho}}^o$ using the simplified RIVC (SRIVC) algorithm (see subsection 4.4.2) and use this to define the initial continuous-time profilter $f_c(p, \hat{\boldsymbol{\rho}}^o)$.

Step 2. Iterative estimation.
    for $j = 1 : convergence$
     (i) *Generate the instrumental variable series $\hat{x}(t, \hat{\boldsymbol{\rho}}^{j-1})$ using the auxiliary model built up from the estimated polynomials $A(p, \hat{\boldsymbol{\rho}}^{j-1})$ and $B(p, \hat{\boldsymbol{\rho}}^{j-1})$ based on $\hat{\boldsymbol{\rho}}^{j-1}$ at the previous $(j-1)$th iteration.*
     (ii) *Prefilter the input $u(t_k)$, output $y(t_k)$ and instrumental variable $\hat{x}(t, \hat{\boldsymbol{\rho}}^{j-1})$ by the continuous-time filter $f_c(p, \hat{\boldsymbol{\rho}}^{j-1})$ in order to generate the filtered derivatives of these variables (see Figure 4.1).*
     (iii) *Obtain an optimal estimate of the noise model parameter vector $\hat{\boldsymbol{\eta}}^j$ based on the estimated noise sequence $\hat{\xi}(t_k)$ from (4.29), using a selected ARMA estimation algorithm.*
     (iv) *Sample the filtered derivative signals at the discrete-time sampling interval T_s and prefilter these by the discrete-time filter $f_d(q^{-1}, \hat{\boldsymbol{\eta}}^j)$, in order to define all the required elements in the data vector $\boldsymbol{\varphi}_f(t_k)$, the IV vector $\hat{\boldsymbol{\varphi}}_f(t_k)$ and the nth-order filtered derivative $y_f^{(n)}(t_k)$ (see Figure 4.2).*
     (v) *Based on these prefiltered data, generate the latest estimate $\hat{\boldsymbol{\rho}}^j$ of the system model parameter vector using the en bloc IV solution (4.28), or its recursive equivalent. Together with the estimate $\hat{\boldsymbol{\eta}}^j$ of the noise model parameter estimate from (iii), this provides the estimate $\hat{\boldsymbol{\theta}}^j$ of the composite parameter vector at the jth iteration.*
    end

Step 3. After the convergence of the iterations is complete, compute the estimated parametric error covariance matrix $\hat{\mathbf{P}}_\rho$, associated with the converged estimate $\hat{\rho}$ of the system model parameter vector, from the expression (see Section 4.5),

$$\hat{\mathbf{P}}_\rho = \hat{\sigma}^2 \left[\sum_{k=1}^{N} \hat{\boldsymbol{\varphi}}_f(t_k) \hat{\boldsymbol{\varphi}}_f^T(t_k) \right]^{-1} \qquad (4.30)$$

where $\hat{\boldsymbol{\varphi}}_f(t_k)$ is the IV vector obtained at convergence and $\hat{\sigma}^2$ is the estimated residual variance.

Comments

1. The fact that the ARMA noise model estimation is carried out separately on the basis of the estimated noise $\hat{\xi}(t_k)$ signal obtained from the IV part of estimation algorithm in (4.29), implies that the system and noise model parameters are statistically independent in some sense. As we shall see in Section 4.5, this is indeed the case: the parameters of the two models are asymptotically independent and so the associated parametric error covariance matrix is block diagonal.
2. As pointed out previously in Section 4.3.1, the generation of the instrumental variable $\hat{x}(t)$ in (4.23), as well as the CT part of the hybrid prefiltering operations, are solved by numerical integration at an appropriate discretisation interval that is normally much smaller than the sampling interval T_s of the data. This requires interpolation of the sampled input signal $u(t_k)$ and is at the discretion of the user. For instance, the implementations of the RIVC algorithm in the CAPTAIN and CONTSID toolboxes utilise the `lsim` routine in MATLAB®. Although a simple approach, this yields good practical results (see comments in Section 4.7.2). However, more sophisticated, or even optimal, interpolation could be used if this was thought necessary.
3. The method of discrete-time ARMA model estimation is also at the discretion of the user. However, for the examples described in Section 4.7, the IVARMA algorithm [37, 38] is used because it is an iterative IV algorithm that fits well within the present context and yields optimal estimates of the ARMA model parameters. For the implementation of the RIVC algorithm in the CAPTAIN Toolbox, however, IVARMA and PEM options are both available to perform this function.

4.4.2 The SRIVC Algorithm

It will be noted that the above formulation of the RIVC estimation problem is considerably simplified if it is assumed that the additive noise is white, *i.e.*, $C(q^{-1}, \boldsymbol{\eta}) = D(q^{-1}, \boldsymbol{\eta}) = 1$. In this case, simplified RIVC (SRIVC) estimation involves only the parameters in the $A(p,\boldsymbol{\rho})$ and $B(p,\boldsymbol{\rho})$ polynomials and the prefiltering only involves the continuous-time prefilter $f_c(p,\boldsymbol{\rho}) = 1/A(p,\boldsymbol{\rho})$. Consequently, the main steps in the SRIVC algorithm are the same as those in the RIVC algorithm, except that the noise model estimation and subsequent discrete-time prefiltering in steps (ii) and (iii) of the iterative procedure are no longer required and are omitted.

Comments

1. The fact that the SRIVC algorithm yields consistent estimates of the system model parameters, even in the coloured noise situation, means that the SRIVC estimates can provide the information required for the initiation of the full RIVC algorithm. In addition, although noise model identification and estimation is not essential in the SRIVC algorithm, separate identification and estimation of an ARMA(p,q) model for the estimated noise sequence $\hat{\xi}(t_k)$ from (4.29) can be included for diagnostic purposes. Moreover, if the estimated noise sequence is sufficiently coloured to require full RIVC estimation, then the identified ARMA(p,q) model structure obtained in this manner can be used to prime the RIVC algorithm in this regard.
2. The initial selection of $A(p,\hat{\boldsymbol{\rho}}^o)$ does not have to be particularly accurate provided the prefilter $f_c(p,\hat{\boldsymbol{\rho}}^o)$ based on it does not seriously attenuate any signals within the passband of the system being modelled. It can be based on various approaches

a) The selection of the single breakpoint parameter λ (breakpoint frequency in radians/time unit) of the filter,

$$f_c(p) = \frac{1}{E(p)} = \frac{1}{(p+\lambda)^n} \qquad (4.31)$$

which is chosen so that it is equal to, or larger than, the bandwidth of the system to be identified. This filter form was suggested long ago (*e.g.*, [15, 26]) but has proven popular ever since.

b) The incorporation of an algorithm for discrete-time model estimation, such as RIV or PEM, if necessary using a coarser sampling interval, from which the CT model polynomial can be inferred using the d2cm tool in MATLAB®.

c) The specification of an *a priori* polynomial based on prior studies.

Of these, a) is simple and, based on extensive practical experience, seems very robust in practical terms; while b) is more automatic but not so robust because of the problems that can arise in estimating a discrete-time model from rapidly sampled data. For this reason, in the CAPTAIN and CONTSID Toolbox implementations of the algorithm, the user can specify a coarser sampling interval for the DT model estimation in order to allow for better identifiability.

3. The SRIVC algorithm performs optimally if the additive noise is white: *i.e.*, estimates are consistent and asymptotically efficient. If the noise is coloured, with rational spectral density, then the estimates remain consistent, because of the inherent instrumental variable mechanism, but they are not minimum variance. However, experience has shown that they are often relatively efficient (*i.e.*, low but not minimum variance). If the noise is 'difficult' (*e.g.*, heteroscedastic) and does not conform to the usual stationary time-series assumptions, then the estimates remain consistent, again because of the instrumental variable implementation (whereas alternative approaches, such as PEM, may suffer problems in such circumstances: see *e.g.*, [37, 38]).

4. As the SRIVC method is an iterative procedure, it is stopped either when a specified number of iterations has elapsed or when the relative percentage error in the parameter estimate vector is sufficiently small. Normally, this SRIVC algorithm is rapidly convergent and has been available for many years in the CAPTAIN Toolbox, where it has proven to be very robust in practical applications, converging in all cases where the model is identifiable from the data. The convergence of the algorithm has not been considered theoretically but the algorithm is quite similar to the iterative least squares algorithm, for discrete-time systems, of Steiglitz and McBride [21], the convergence of which has been established by Stoica and Söderström [22] in the case of white additive noise. Moreover, the inherent optimal instrumental variable nature (see Section 4.5) of the SRIVC algorithm removes the limitations of the Steiglitz and McBride algorithm [22] in the coloured-noise situation.

4.4.3 Multiple-input Systems

It is clearly straightforward to extend the RIVC/SRIVC methods to the multiple-input situation if the TF denominator is common to all input channels and a three-input practical example of this is considered in Section 4.8. The situation is not so straightforward in the case where there are different denominator polynomials for each input channel. However, following the RIV approach for DT systems [6], the algorithms can be extended to handle this situation [3]: indeed, the current version of RIVC in the CONTSID Toolbox provides this option and it is referred to in Section 4.8. In multiple-input examples, the order of a general r-input model is denoted by $[n_1\ n_2\ \cdots\ n_r\ m_1\ m_2\ \cdots\ m_r\ \tau_1\ \tau_2\ \cdots\ \tau_r\ p\ q]$.

4.4.4 Non-uniformly Sampled Data

One advantage of the SRIVC approach to continuous-time modelling is that it can be based on irregularly sampled data (see also Chapter 11) and can handle 'fractional' pure time delays (see *e.g.*, Chapters 11 and 12 in the present book). The current implementation of the SRIVC algorithm in the CONTSID

Toolbox can handle irregularly sampled data. However, the RIVC algorithm has not yet been upgraded in this regard because it requires additional interpolation and re-sampling in order to generate a regularly sampled series for the ARMA noise model estimation parts of the algorithm.

4.5 Theoretical Background and Statistical Properties of the RIVC Estimates

The motivational arguments presented in Section 4.3 suggest that, upon convergence, the RIVC parameter estimates will possess the optimal statistical properties of consistency and asymptotic efficiency when the additive noise has a Gaussian normal probability distribution and rational spectral density. This section presents more formal analysis to verify further the optimality of the estimates and confirm the asymptotic independence of the system and noise model parameter estimates.

4.5.1 Optimality of RIVC Estimation

In the control and systems literature, optimal IV estimation is usually considered in relation to the so-called 'extended IV' approach to estimation, as developed for the DT case [19]. A similar approach can be applied in the present CT case by re-writing the IV optimisation equation (4.27) in the following alternative form that explicitly reveals a continuous-time prefilter $f(p)$

$$\hat{\rho}(N) = \arg\min_{\rho} \left\| \left[\frac{1}{N} \sum_{k=1}^{N} \zeta_f(t_k) f(p) \varphi^T(t_k) \right] \hat{\rho} - \left[\frac{1}{N} \sum_{k=1}^{N} \zeta_f(t_k) f(p) y^{(n)}(t_k) \right] \right\|_Q^2 \tag{4.32}$$

where $f(p)$ is the stable prefilter, $\zeta_f(t_k)$ is the prefiltered instrumental vector $\zeta_f(t_k) = f(p)\zeta(t_k)$ and $\mathbf{Q}$ is a positive-definite matrix. By definition, when $G_o \in \mathcal{G}$, the extended IV estimate provides a consistent estimate under the following two conditions

$$\begin{cases} \bar{\mathsf{E}}\{\zeta_f(t_k) f(p) \varphi^T(t_k)\} \text{ is non-singular,} \\ \bar{\mathsf{E}}\{\zeta_f(t_k) f(p) \xi(t_k)\} = 0 \end{cases} \tag{4.33}$$

Clearly, the selection of the instrumental variable vector $\zeta_f(t_k)$, the weighting matrix $\mathbf{Q}$ and the prefilter $f(p)$ may have a considerable effect on the covariance matrix $\mathbf{P}_\theta$ produced by the IV estimation algorithm.

In the open-loop situation, the Cramér–Rao lower bound on P_θ for any unbiased identification method (*e.g.*, [19, 24]) defines the optimal solution. In this regard, it has been shown that the minimum value of the covariance matrix $\mathbf{P}_\theta$, as a function of the design variables $\zeta_f(t_k)$, $f(p)$ and $\mathbf{Q}$, exists and is given by

$$\mathbf{P}_\theta \geq \mathbf{P}_\theta^{opt}$$

with
$$\mathbf{P}_\theta^{opt} = [\bar{\mathsf{E}}\{\mathring{\boldsymbol{\zeta}}_f(t_k)\mathring{\boldsymbol{\zeta}}_f^T(t_k)\}]^{-1} \qquad (4.34)$$

where $\mathring{\boldsymbol{\zeta}}_f(t_k)$ is the optimally prefiltered IV vector, with the associated design variables defined as

$$\mathbf{Q} = \mathbf{I}, \qquad (4.35a)$$

$$f(p) = \frac{1}{H_o(p)A_o(p)} = \frac{D_o(p)}{C_o(p)A_o(p)}, \qquad (4.35b)$$

$$\mathring{\boldsymbol{\zeta}}(t_k) = \begin{bmatrix} -x^{(n-1)}(t_k) & -x^{(n-2)}(t_k) \cdots & -x(t_k) & u^{(m)}(t_k) & \cdots & u(t_k) \end{bmatrix}^T \qquad (4.35c)$$

so that,

$$\mathring{\boldsymbol{\zeta}}_f(t_k) = f(p)\begin{bmatrix} -x^{(n-1)}(t_k) & -x^{(n-2)}(t_k) \cdots & -x(t_k) & u^{(m)}(t_k) & \cdots & u(t_k) \end{bmatrix}^T \qquad (4.36)$$

which will be recognised as the noise-free, prefiltered vector $\mathring{\boldsymbol{\varphi}}_f^T(t_k)$ defined earlier in (4.26).

Comments

1. Not surprisingly, the above analysis justifies the RIVC algorithmic design that iteratively updates those aspects of the theoretical solution that are not known *a priori*: in this case, the unknown model polynomials and the noise-free output of the system that is, of course, the source of the instrumental variables. If it is assumed that, in all identifiable situations, the RIVC algorithm converges in the sense that $\hat{\boldsymbol{\rho}} \Rightarrow \boldsymbol{\rho}$ and $\hat{\boldsymbol{\eta}} \Rightarrow \boldsymbol{\eta}$, then the RIVC estimates will be consistent and asymptotically efficient.
2. The optimal filter $f(p)$ in (4.35b) is formulated in CT terms. In the proposed RIVC algorithm, this filter takes a hybrid form, as discussed in the previous sections. As mentioned in the introductory Section 4.1, however, it is possible to design the prefiltering operations entirely in CT terms, if this is desired, in which case the optimal prefilters take this CT form.

4.5.2 The Asymptotic Independence of the System and Noise Model Parameter Estimates

The maximum likelihood motivation of refined IV estimation [27, 46, 52] is based on the decomposition of the estimation problem into two separate but inter-linked subproblems: first the estimation of the system transfer function model parameters under the assumption that the noise model parameters are known; and second, the estimation of the ARMA noise model parameters under the assumption that TF model parameters are known. This approach is then carried over to the formulation of both the discrete-time RIV and continuous-time RIVC algorithms, as described in previous sections of this chapter. The justification for such a simplifying approach is given by the following modified version of the theorem due originally to Pierce [15] and formulated in the present control theoretic form by Young and Jakeman [42, 46, 52].

Theorem 4.1. *If in the model* (4.15)

(i) *the* $e(t_k)$ *in the ARMA noise model equation* (4.12) *are independent and identically distributed with zero mean, variance* σ^2, *skewness* κ_1 *and kurtosis* κ_2;
(ii) *the model is stable and identifiable;*
(iii) *the* $u(t_k)$ *are persistently exciting;*

then the ML estimates $\hat{\rho}$, $\hat{\eta}$ *and* $\hat{\sigma}^2$ *obtained from a data set of* N *samples, possess a limiting normal distribution, such that the following results hold*

(a) *the asymptotic covariance matrix of the estimation errors associated with the estimate* $\hat{\rho}$ *is of the form*

$$\mathbf{P}_\rho = \frac{\sigma^2}{N} \left[plim \left\{ \frac{1}{N} \sum_{k=1}^{N} \mathring{\boldsymbol{\varphi}}_f(t_k) \mathring{\boldsymbol{\varphi}}_f^T(t_k) \right\} \right]^{-1} \quad (4.37)$$

(b) *the estimate* $\hat{\eta}$ *is asymptotically independent of* $\hat{\rho}$ *and has an error covariance matrix of the form*

$$\mathbf{P}_\eta = \frac{\sigma^2}{N} [\bar{\mathsf{E}}\{\boldsymbol{\psi}_{f_{d1}}(t_k) \boldsymbol{\psi}_{f_{d1}}^T(t_k)\}]^{-1} \quad (4.38)$$

and

(c) *the estimate* $\hat{\sigma}^2$ *has asymptotic variance* $(2\sigma^4/N)(1 + 0.5\kappa_2)$ *and, if* $\kappa = 0$, *is independent of the above estimates.*

Proof. This follows straightforwardly from Pierce [15] modified to the present hybrid BJ setting. Moreover, the main results (covariance matrices and asymptotic independence of the system and noise model parameter estimates) are obvious from the ML formulation of the RIV/RIVC algorithms [42, 46, 52].

Comments

1. The definition of $\mathbf{P}_\rho$ in (4.37) clearly confirms the analysis in Section 4.3 and the result (4.34) in the extended IV analysis. Also, we can note that, because the optimal IV vector is generated, via the auxiliary model (4.23) and the prefilters, from the noise-free input $u(t)$, which is independent of the noise $\xi(t)$ then,

$$plim \left\{ \frac{1}{N} \sum_{k=1}^{N} \hat{\boldsymbol{\varphi}}_f(t_k) \boldsymbol{\varphi}_f^T(t_k) \right\} = plim \left\{ \frac{1}{N} \sum_{k=1}^{N} \hat{\boldsymbol{\varphi}}_f(t_k) \mathring{\boldsymbol{\varphi}}_f^T(t_k) \right\} \quad (4.39)$$

In (4.37) and (4.38), $\boldsymbol{\varphi}_f(t_k)$ and $\boldsymbol{\psi}_{f_{d1}}(t_k)$ are defined as the values of these variables that would result if the auxiliary model, inverse noise model and prefilter parameters were all set at values based on the true model parameter vectors $\boldsymbol{\rho}$ and $\boldsymbol{\eta}$. The relevance of these results is obvious: if it is assumed that, in all identifiable situations, the RIVC algorithm converges in the sense that $\hat{\boldsymbol{\rho}} \Rightarrow \boldsymbol{\rho}$ and $\hat{\boldsymbol{\eta}} \Rightarrow \boldsymbol{\eta}$, then

$$\hat{\mathbf{P}}_\rho = \hat{\sigma}^2 \left[\sum_{k=1}^{N} \hat{\boldsymbol{\varphi}}_f(t_k) \hat{\boldsymbol{\varphi}}_f^T(t_k) \right]^{-1} \tag{4.40}$$

will tend in probability to $\mathbf{P}_\rho$. In the implementation of the RIVC algorithm, therefore, $\hat{\mathbf{P}}_\rho$ in (4.40) is provided as the estimate of the parametric error covariance matrix and will provide good empirical estimates of the uncertainty in the parameter estimates in $\hat{\boldsymbol{\rho}}$. And, noting the Monte Carlo results of both Pierce [15] and Young and Jakeman [42] in the DT situation, we might also assume that they will provide a good indication of the error covariance properties, even for small sample size N. This is confirmed in the later simulation examples.

2. In this theorem, $\boldsymbol{\psi}_{f_{d1}}(t_k) = f_{d1}(q^{-1})\boldsymbol{\psi}(t_k)$ is the filtered noise vector defined as

$$\boldsymbol{\psi}_{f_{d1}}^T(t_k) = [-\xi_{f_{d1}}(t_{k-1}) \cdots - \xi_{f_{d1}}(t_{k-p}) \; e_{f_{d1}}(t_{k-1}) \cdots e_{f_{d1}}(t_{k-q})] \tag{4.41}$$

where $\xi_{f_{d1}}(t_k)$ and $e_{f_{d1}}(t_k)$ are obtained, respectively, by filtering the coloured noise variable $\xi(t_k)$ and the white noise variable $e(t_k)$ by the prefilter

$$f_{d1}(q^{-1}) = \frac{1}{C_o(q^{-1})} \tag{4.42}$$

If this is computed in the RIVC algorithm, the true $\xi(t_k)$ and $e(t_k)$ are replaced by their estimates: *i.e.*, the estimate $\hat{\xi}(t_k)$ of $\xi(t_k)$ is generated by (4.29) and the $\hat{e}(t_k)$ are the residuals of the ARMA noise model[4]. The justification for this prefiltering comes from both the Pierce Theorem and the related ML analysis of Young and Jakeman cited previously (see also [37, 38], where these results are used as the basis for the design of the recently proposed IVARMA algorithm). In practice, of course, $\mathbf{P}_\eta$ would be provided by the algorithm used to estimate the ARMA noise model (IVARMA, PEM, *etc.*).

3. The asymptotic independence of the ML estimates $\hat{\boldsymbol{\rho}}$ and $\hat{\boldsymbol{\eta}}$ presented in item (c) of this theorem provides the theoretical justification for the estimation mechanism in the RIVC algorithm, where the ARMA noise model parameter estimation is carried out separately, based on the coloured noise variable estimate obtained from the system model parameter estimation step. It also justifies the assumed block-diagonal structure of the parametric error covariance matrix.

---

[4] Note that, in the case of an AR model, $C_o(q^{-1}) = 1$, so that no prefiltering is required, as would be expected in this case where simple linear least squares can be used for AR model estimation.

4. Note that this useful asymptotic independence arises from the selection of the BJ model form and is not a property of other time series models, such as the ARMAX model. Other advantages of the BJ model form, in relation to the ARMAX model, are discussed at length by Jakeman and Young [8, 9] in the case of discrete-time TF model estimation.

4.6 Model Order Identification

One very important aspect of TF modelling is the identification of the model structure: *i.e.*, the degrees n, m, p, and q of the model polynomials and any associated pure time delay τ or τ_d. One statistical measure that is useful in this regard is the simulation coefficient of determination R_T^2, defined as follows

$$R_T^2 = 1 - \frac{\sigma_{\hat{\xi}}^2}{\sigma_y^2} \qquad (4.43)$$

where $\sigma_{\hat{\xi}}^2$ is the variance of the estimated noise $\hat{\xi}(t_k)$ and σ_y^2 is the variance of the measured output $y(t_k)$. This should be differentiated from the standard coefficient of determination R^2, where the $\sigma_{\hat{\xi}}^2$ in (4.43) is replaced by the variance of the final ARMA model residuals $\hat{\sigma}^2$. R_T^2 is clearly a normalised measure of how much of the output variance is explained by the deterministic system part of the estimated model. However, it is well known that this measure, on its own, is not sufficient to avoid overparametrisation and identify a parsimonious model, so that other model order identification statistics are required. In this regard, because the SRIVC and RIVC methods exploit optimal instrumental variable methodology, they are able to utilise the special properties of the instrumental product matrix (IPM) [45, 53]; in particular, the YIC statistic [47], which is used in the numerical examples discussed in subsequent sections of this chapter. The YIC is defined as follows

$$\text{YIC} = \log_e \frac{\hat{\sigma}^2}{\sigma_y^2} + \log_e\{\text{NEVN}\}; \quad \text{NEVN} = \frac{1}{n_\theta}\sum_{i=1}^{n_\theta} \frac{\hat{p}_{ii}}{\hat{\theta}_i^2} \qquad (4.44)$$

Here, $n_\theta = n + m + p + q + 1$ is the number of estimated parameters; $\hat{p}_{ii}$ is the ith diagonal element of the block-diagonal covariance matrix $\mathbf{P}_\theta$, where,

$$\mathbf{P}_\theta = \begin{pmatrix} \mathbf{P}_\rho & \mathbf{0} \\ \mathbf{0} & \mathbf{P}_\eta \end{pmatrix} \qquad (4.45)$$

and so is an estimate of the variance of the estimated uncertainty on the ith parameter estimate. $\hat{\theta}_i^2$ is the square of the ith parameter estimate in the $\boldsymbol{\theta}$ vector, so that ratio $\hat{p}_{ii}/\hat{\theta}_i^2$ is a normalised measure of the uncertainty on the ith parameter estimate.

From the definition of R_T^2, we see that the first term in the YIC is simply a relative measure of how well the model explains the data: the smaller the model residuals the more negative the term becomes. The normalised error variance norm (NEVN) term, on the other hand, provides a measure of the conditioning of the IPM, which needs to be inverted when the IV normal equations are solved (see, *e.g.*, [46]): if the model is overparameterised, then it can be shown that the IPM will tend to singularity and, because of its ill-conditioning, the elements of its inverse (in the form here of the covariance matrix $\mathbf{P}_\theta$) will increase in value, often by several orders of magnitude. When this happens, the second term in the YIC tends to dominate the criterion function, indicating overparameterisation.

It is important to note that, based on practical experience, the YIC is normally best considered during SRIVC identification, which is much less computationally intensive than RIVC identification, so allowing for much faster investigation of the model order range selected by the user. In this situation, n_θ is replaced by $n_\rho = n + m + 1$ the $\hat{p}_{ii}$ are obtained by reference to the covariance matrix $\mathbf{P}_\rho$.

Although heuristic, the YIC has proven very useful in practical identification terms over the past ten years. It should not, however, be used as a sole arbiter of model order: rather the combination of R_T^2 and YIC provides an indication of the best parsimonious models that can be evaluated by other standard statistical measures (*e.g.*, the auto and partial autocorrelation of the model residuals, the cross-correlation of the residuals with the input signal $u(t_k)$, *etc.*). Also, within a data-based mechanistic model setting (see, *e.g.*, [32]), the physical interpretation of the model can often provide valuable information on the model adequacy: for instance, a model with complex eigenvalues caused by overparameterisation may prove incompatible with the non-oscillatory nature of the physical system under study[5].

4.7 Simulation Examples

This section considers Monte Carlo simulation examples that illustrate the performance of the SRIVC algorithm in the noise-free case and the full hybrid RIVC method in the case of coloured output noise. The simulation model used in this analysis provides a very good test for CT and DT estimation methods: it was first suggested by Rao and Garnier [18] (see also [4, 17]); and has been commented upon by Ljung [25].

4.7.1 The Rao–Garnier Test System

The test model is a linear [4 2 0], non-minimum phase system with complex poles. Its operator model form is given by [18]

---

[5] Of course, this could be due to the uncertainty in the parameters of an otherwise parsimonious model, in which case constrained estimation needs to be used [36].

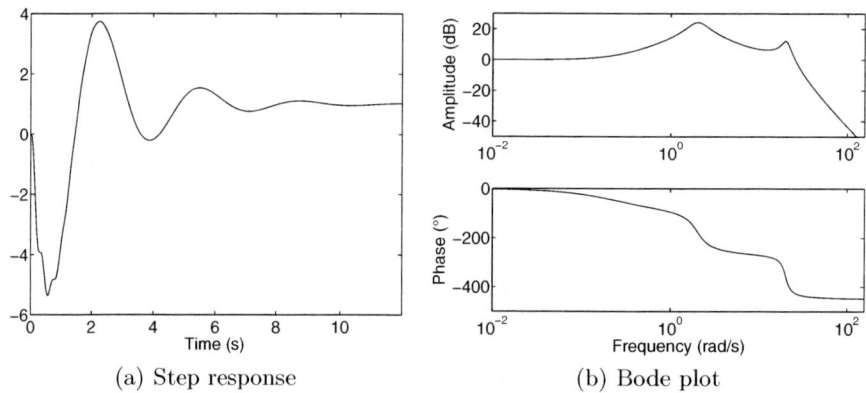

Fig. 4.3. Step and frequency response of the fourth-order Rao–Garnier system

$$G_o(p) = \frac{-6400p + 1600}{p^4 + 5p^3 + 408p^2 + 416p + 1600} \tag{4.46}$$

This is an interesting system from two points of view. First, it has one fast oscillatory mode with relative damping 0.1 and one slow oscillatory mode with relative damping 0.25. Secondly, the system is non-minimum phase, with a zero in the right half-plane. The step response and Bode plots of the system are shown in Figure 4.3. The settling time of the system is about 10 s.

4.7.2 Noise-free Case

The noise-free case is considered first in order to investigate the effect of systematic interpolation errors that are introduced into the computation of the prefiltered time derivatives (see Figure 4.2 and the earlier discussion on this topic in Sections 4.3.1 and 4.4). This is due to the fact that, in general, the inter-sample behaviour of the prefilter inputs is not known, so that time-domain approaches to CT model identification often require a high sampling frequency, compared with the bandwidth of the system to be identified, in order to minimise these systematic interpolation errors. As we shall see, this is no longer the case with the SRIVC method, which is not sensitive to the sampling interval.

The observation time is fixed at $T_f = 20$s and the number of data points available, N, is inversely proportional to the sampling interval T_s. The input $u(t)$ is chosen as a binary ± 1 signal. Note that in the case of the chosen piecewise constant excitation signal, the system response can be calculated exactly at the sampling instances via appropriate zero-order hold (ZOH) discretisation of the CT system. The estimation methods considered here are the simple least squares-based state variable filter (LSSVF) method (see, *e.g.*, [4]); the iterative SRIVC method; and, finally, an indirect method based on the estimation

4 Refined IV Identification of CT Hybrid Box–Jenkins Models 111

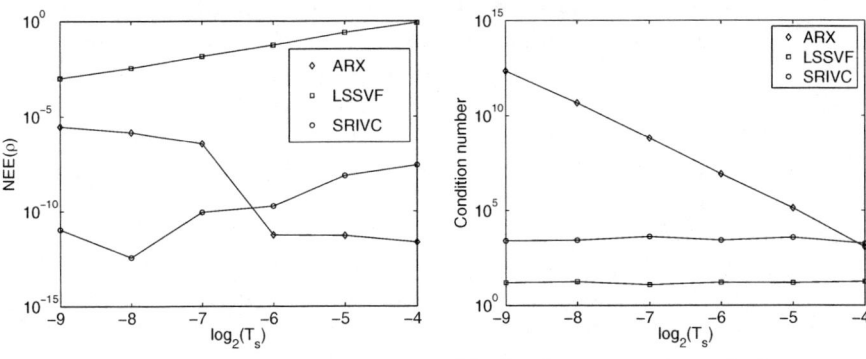

(a) NEE versus the sampling interval

(b) Condition number of the normal equations

Fig. 4.4. Results in the noise-free case for the three estimation algorithms

of a DT ARX model converted to a CT model by using the appropriate d2cm conversion tool in MATLAB® with the ZOH option. Note that the SRIVC routine is initialised here with the estimates obtained from the LSSVF method. The cut-off frequency of the SVF filter is set to 10 rad/s.

In order to compare the accuracy of the different approaches, the sum of the normalised estimation error (NEE) of the parameters is considered

$$\text{NEE}(\boldsymbol{\rho}) = \left\| \frac{\boldsymbol{\rho}_o - \hat{\boldsymbol{\rho}}}{\boldsymbol{\rho}_o} \right\|_2 = \sqrt{\sum_{i=1}^{n_\rho} \left(\frac{\rho_i^o - \hat{\rho}_i}{\rho_i^o} \right)^2} \qquad (4.47)$$

where $\hat{\rho}_i$ represents the ith element of $\hat{\boldsymbol{\rho}}$, while the superscript 'o' is used to denote the true value.

The normalised estimation error versus the sampling period for the three methods is plotted in Figure 4(a). The LSSVF method yields relatively poor results. This is caused by the numerical errors introduced into the simulation of the CT filtering operations required to generate the time derivatives of the filtered output. The use of a more sophisticated interpolation technique than the lsim routine used here would make it possible to obtain better results. However, this is not necessary in the case of the SRIVC method because its iterative IV-based procedure is able to counteract the presence of the simulation-induced errors, which are effectively treated as additive noise on the filtered output time derivatives.

Although the SRIVC and the indirect ARX-based estimation approaches perform similarly overall in this noise-free situation, it is clear that, at small sampling intervals, the SRIVC estimation results are better; while at the coarser sampling intervals, the position is reversed. This is an expected result, since it is well known that DT estimation schemes suffer from numerical issues at small sampling intervals because the eigenvalues of the DT model migrate ever closer to the unit circle in the complex domain as the sampling frequency increases. Figure 4(b) shows how the condition number of the normal equations for the three methods varies with the sampling interval. This reveals a significant numerical advantage of the direct approach for all chosen sampling periods. The indirect approach also suffers from additional numerical problems that are encountered in the conversion from the DT to the CT domain as the sampling interval decreases.

The SRIVC algorithm needs between 2 to 9 iterations to converge from the LSSVF parameter estimates. In the case of coarse sampling periods, an alternative to LSSVF initiation are the estimates obtained from the discrete-time RIV algorithm: this is an option available in both the CAPTAIN and CONTSID Toolboxes[6].

4.7.3 Noisy-output Case

Of course, model order selection and model parameter estimation are easy in the case of noise-free data and it is necessary to consider the performance of estimation algorithms in the presence of noise if we are to get some idea of their practical utility. In this subsection, the MCS analysis is applied in the case of additive noise at the output of the system, under the usual assumption that the input signal is noise free. The case of observational errors on the input signal is a problem of errors-in-variables [13, 23] or closed-loop control (see Chapter 5 in the present book) and the parameter estimates from all the algorithms considered here will be biased unless the algorithm is modified to account for this. As in the previous section, the underlying deterministic system is the model (4.46) and two noisy situations are considered, both with additive ARMA(2,1) noise models. The denominator polynomial is the same in each case but the numerator is changed in order to investigate the effect, at different sampling intervals, of numerator zeros near the unit circle in the complex z domain. Two estimation methods are considered, in addition to the SRIVC and RIVC methods, both based on gradient optimisation: direct CT estimation by the continuous-time output error (COE) algorithm in the CONTSID Toolbox; and indirect estimation using the prediction error minimisation (PEM) algorithm in the MATLAB® system identification (SID) Toolbox. The SRIVC and RIVC algorithms used in the analysis are those available in the CAPTAIN and CONTSID Toolboxes mentioned previously.

Sampling Frequency

The usual rule of the thumb for data acquisition is that the sampling frequency is to be about ten times the bandwidth of the system under study. In the present example, the highest resonant frequency of the model (4.46) is $\omega_{n,1} = 20$ rad/s which suggests, therefore, that

---

[6] See Section 4.9.

$$\omega_s = \frac{2\pi}{T_s} \approx 10\omega_{n,1} \tag{4.48}$$

$$T_s \approx \frac{2\pi}{10\omega_{n,1}} = 0.0314\,\text{s} \tag{4.49}$$

It is well known that the DT model estimation techniques suffer from ill-conditioned numerical problems at high sampling frequencies because the poles lie close to the unit circle in the complex z domain. In the simulation experiments, therefore, two sampling intervals were considered: $T_s = 50$ ms and $T_s = 10$ ms. As we shall see, indirect identification and estimation of the model (4.46) is possible (although with undesirable features) at $T_s = 50$ ms but proves problematical at $T_s = 10$ ms.

Input Signals

In these experiments, a PRBS (pseudo-random binary signal) of maximum length is used to excite the system over its whole bandwidth. The characteristics of the signal, whose amplitude switches between ± 1, are the following: in the case of $T_s = 50$ ms, the number of stages of the shift register is set to 9, the clock period is set to 3, which results in the number of samples $N = 1533$; in the case of $T_s = 10$ ms, the number of stages of the shift register is set to 10, the clock period is set to 7, which results in the number of samples $N = 7161$. The duration of the experiment is about seven times the settling time for both sampling periods.

Type of Measurement Noise

Based on the CT model (4.46), the following hybrid system is considered

$$\begin{cases} x(t) = G_o(p)u(t) \\ y(t_k) = x(t_k) + \xi(t_k) \end{cases} \tag{4.50}$$

Here, $\xi(t_k)$ is coloured noise derived from an ARMA(2,1) process with a denominator polynomial,

$$D(q^{-1}) = 1 + 1.9628q^{-1} + 0.9851q^{-2} \tag{4.51}$$

but with different MA polynomials at the two sampling intervals: in the case of $T_s = 50$ ms,

$$C(q^{-1}) = 1 + 0.98q^{-1} \tag{4.52}$$

and with $T_s = 10$ ms,

$$C(q^{-1}) = 1 - 0.99q^{-1} \tag{4.53}$$

In both cases, the variance of $\xi(t_k)$ is adjusted to obtain a signal-to-noise ratio of 10 dB (defined in terms of the variances).

Note that this choice of MA polynomials is meant to test the algorithmic performance because, in both cases, it means that the inverse noise model used in the prefiltering has a pole very close to unity in the complex z plane, so that the prefilter dynamics are close to either exponential ($C(q^{-1}) = 1 + 0.98q^{-1}$) or oscillatory ($C(q^{-1}) = 1 - 0.99q^{-1}$) instability. As we shall see, the latter case is particularly challenging in the case of indirect estimation, since both the stability margin and the sampling interval are very small.

Monte Carlo Simulation Results

The MCS results obtained using different algorithms are presented in Tables 4.1 and 4.2. Here, the algorithms considered are: the COE, SRIVC, RIVC based on an ARMA(2,1) noise model, and RIVC based on an AR(23) noise model, where the AR(23) model is the AR model identified by the Akaike Information Criterion (AIC) when applied to the noise signal. The latter results allow us to compare the results obtained using the actual ARMA(2,1) model structure with those obtained using this high-order AR(23) approximation. In addition, indirect PEM estimation is evaluated, with the estimated DT model converted to CT using the MATLAB® d2cm function with zero-order hold. All the MCS results are based on $N_{\text{exp}} = 100$ random realisations, with the Gaussian white noise input to the ARMA noise model being selected randomly for each realisation. In order to compare the statistical performance of the different approaches, the computed mean and standard deviation of the estimated parameter ensemble are presented, as well as the results obtained from typical single runs in the case of the RIVC algorithms, where the standard error (SE) estimates obtained from the RIVC algorithm can be compared with the computed MCS standard deviations.

Direct Estimation

As expected, both tables show that the COE and SRIVC methods provide similar, unbiased estimates of the model parameters with reasonable standard deviations (although it should be noted that, on average, the COE requires more iterations to converge). However, the estimates are clearly not statistically efficient if they are compared with the estimates produced by the RIVC algorithm, where the standard deviations are always smaller and in some cases, such as the estimate of b_0 in Table 4.1, the standard deviation is some 50 times less. It is clear from the two sets of results obtained using the RIVC algorithm, with different assumed noise model structures, that the mean estimates and standard deviations are quite similar although, of course, the AR(23) noise model is much less parsimonious. On the other hand, this suggests that the simpler implementation of the RIVC algorithm using an AR noise model (as used for many years in the discrete-time RIV algorithm implemented in the CAPTAIN Toolbox) is a reasonable alternative to the full RIVC algorithm based on an ARMA noise model, even in cases where the root of the MA is

4 Refined IV Identification of CT Hybrid Box–Jenkins Models

Table 4.1. Monte Carlo simulation results with additive coloured noise numerator $C(q^{-1}) = 1+0.98q^{-1}$ and $T_s = 50$ ms: SR denotes single-run results and MC denotes results from MCS analysis based on 100 random realisations (note that no standard errors on the single-run PEM estimates is given because the CT model parameters have to be derived from the DT estimates)

METHOD	Parameter	$\hat{b}_0$	$\hat{b}_1$	$\hat{a}_1$	$\hat{a}_2$	$\hat{a}_3$	$\hat{a}_4$	$\hat{d}_1$	$\hat{d}_2$	$\hat{c}_1$
	True value	−6400	1600	5	408	416	1600	−1.9628	0.9851	0.9800
COE (MC)	Mean	−6433.1	1555.2	5.023	408.26	416.96	1609.3			
	St. Dev.	267.0	366.3	0.2171	1.337	22.52	35.62			
SRIVC (MC)	Mean	−6433.1	1555.2	5.023	408.26	416.97	1609.3			
	St. Dev.	267.0	366.3	0.2172	1.338	22.52	35.63			
RIVC ARMA (MC)	Mean	−6400.0	1605.2	5.002	408.01	416.74	1602.1	−1.9613	0.9836	0.9768
	St. Dev.	5.618	144.3	0.0229	0.146	8.589	23.25	0.006	0.006	0.008
RIVC ARMA (SR)	Mean	−6402.6	1650.8	5.016	408.0	422.6	1603.6	−1.9644	0.9863	0.9739
	SE	4.887	116.9	0.0188	0.127	7.149	18.49	0.004	0.004	0.006
RIVC AR (MC)	Mean	−6399.9	1605.5	5.002	408.01	416.64	1602.1			
	St. Dev.	7.396	145.1	0.0230	0.146	8.503	23.22			
RIVC AR (SR)	Mean	−6404.9	1664.8	5.022	407.97	424.03	1602.4			
	SE	5.802	110.4	0.0176	0.123	6.532	16.64			
PEM ARMA (MC)	Mean	−6398.9	1600.1	5.002	408.0	416.78	1601.2	−1.9614	0.9836	0.9773
	St. Dev.	10.556	191.6	0.0274	0.167	9.536	24.50	0.006	0.006	0.008
PEM ARMA(SR)	Mean	−6408.8	1777.8	5.024	408.1	425.1	1600.8	−1.9647	0.9868	0.9732

quite close to unity. In fact, the comparative performance remains reasonably good even if the MA polynomial is set to $C(q^{-1}) = 1 - 0.99q^{-1}$, as shown in Table 4.2.

The typical single-run results in Table 4.1 show that the RIVC algorithm, using either the correct ARMA(2,1) or the approximate AR(23) models, predicts the mean and standard deviation obtained in the MCS results well, fully justifying the algorithm in theoretical and practical terms. The associated parameter estimates produce models that are close to actual simulated system, with coefficients of determination R_T^2, based on the error between the simulated output of the estimated model and the true deterministic output, very close to unity. Not surprisingly, as shown in Figure 4.5, the Bode plots for the RIVC (SR) estimated model (with correctly specified ARMA(2,1) order) and the actual Bode plots are virtually identical.

The performance of the various algorithms shown in Table 4.2 is broadly similar to that in Table 4.1. In particular, despite the close proximity of the MA zero to the unit circle in the complex z plane, the inverse noise prefiltering operations are not adversely affected and so not sensitive under these con-

Table 4.2. Monte Carlo simulation results with additive coloured noise numerator $C(q^{-1}) = 1 - 0.99q^{-1}$ and $T_s = 10$ ms: SR denotes single-run results; MC denotes results from MCS analysis based on 100 random realisations (except OE that had 17 failures); and **IND** denotes indirect estimation using DT algorithms OE and SRIV, with conversion to CT. In these indirect estimation cases, the SR entries are those used later for Figure 4.7

METHOD	Parameter	$\hat{b}_0$	$\hat{b}_1$	$\hat{a}_1$	$\hat{a}_2$	$\hat{a}_3$	$\hat{a}_4$	$\hat{d}_1$	$\hat{d}_2$	$\hat{c}_1$
	True value	−6400	1600	5	408	416	1600	−1.9628	0.9851	0.9800
COE (MC)	Mean	−6403.1	1600.1	5.00	408.0	416.3	1599.8			
	St. Dev.	61.70	25.52	0.1565	2.589	3.976	10.00			
SRIVC (MC)	Mean	−6403.1	1600.1	5.01	408.0	416.3	1599.8			
	St. Dev.	61.70	25.52	0.1565	2.590	3.976	10.00			
RIVC ARMA	Mean	−6400	1599.9	4.996	408.1	416.0	1600.2	−1.9622	0.9846	0.9903
(MC)	St. Dev.	32.22	12.22	0.0973	1.916	2.128	7.562	0.002	0.002	0.002
RIVC ARMA	Mean	−6360.3	1597.2	5.0819	406.49	414.51	1594.3	−1.9627	0.9847	0.9912
(SR)	SE	34.51	11.79	0.1029	2.098	2.322	8.380	0.002	0.002	0.0015
RIVC AR	Mean	−6398.8	1599.7	4.991	408.1	415.9	1600.2			
(MC)	St. Dev.	33.44	15.97	0.1040	1.957	2.215	7.723			
RIVC AR	Mean	−6357.0	1601.9	5.109	406.1	414.6	1592.3			
(SR)	SE	36.19	40.48	0.0913	1.732	2.716	7.735			
IND:OE(MC)	Mean	−6404.1	1598.3	4.9747	408.32	416.65	1600.3			
(only 83 runs)	St. Dev.	129.8	89.153	0.2312	5.776	14.768	29.523			
IND:OE(SR)	Mean	−7746.1	−1692.1	8.5269	435.63	1819.5	4333.3			
IND:SRIV(MC)	Mean	−6417.3	1617.1	4.9769	407.95	415.51	1599.6			
	St. Dev.	105.58	80.063	0.2204	5.680	7.531	23.953			
IND:SRIV(SR)	Mean	−6438.9	1612.1	4.692	410.14	415.12	1607.9			

ditions, suggesting that the RIVC algorithm should be robust in practical applications.

Indirect Estimation

Table 4.1 shows that, at the coarser sampling interval of $T_s = 50$ ms, the PEM algorithm in the MATLAB® SID toolbox yields indirect estimates of the CT model parameters that are reasonably comparable with those of the RIVC algorithm. The discrete-time RIV algorithm in the CAPTAIN Toolbox gives very similar results so these are not shown. However, in both cases, the standard deviations are always higher than those of RIVC and, in the single-run case, the standard errors are not available directly because of the need

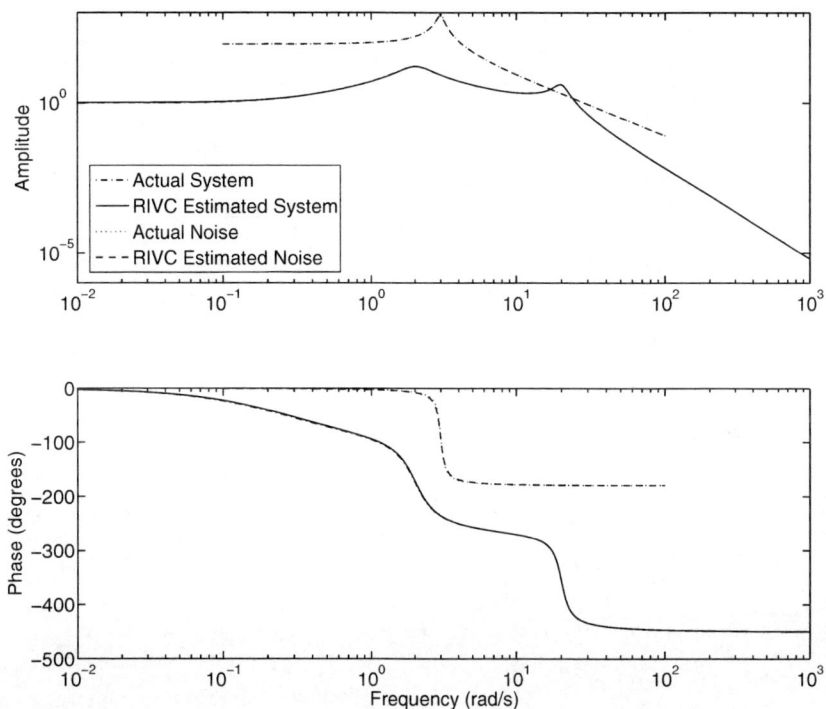

Fig. 4.5. Actual and RIVC estimated Bode plots for the case of $T_s = 50$ ms and the noise MA polynomial $C(q^{-1}) = 1 + 0.98q^{-1}$: other model parameters are given in Table 4.1 (RIVC ARMA (SR))

to convert the model to continuous time. So, although this indirect method could be used at this coarser sampling interval, it is obviously better to use RIVC. As shown in Figure 4.6, the Bode plots for the RIVC (SR) estimated model (with correctly specified ARMA(2,1) order) and the actual Bode plots are virtually identical.

At the finer sampling interval of $T_s = 10$ ms, the indirect approach using PEM or RIV breaks down, with unreliable results: 86% of PEM estimation runs and 44% of the RIV estimation runs failing to provide satisfactory convergence (the results are not shown in Table 4.2). Consequently, the main difference in the results of Table 4.2 is that the PEM results have been replaced by output error (OE) results, using the OE routine in the MATLAB® SID Toolbox, where the performance is much better now that the noise model parameters do not have to be estimated concurrently. Even here, however, 17 of the OE ensemble fail to converge and the means/standard deviations are computed from only the 83 successful outcomes. In contrast to this, the SRIV algorithm in the CAPTAIN Toolbox (which is, effectively, the iterative IV equivalent

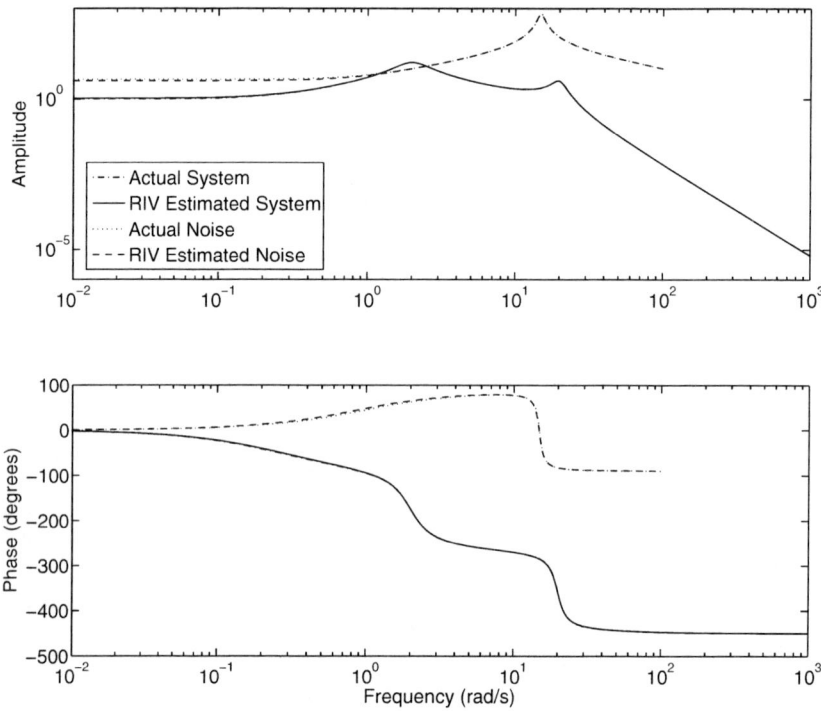

Fig. 4.6. Actual and RIVC estimated Bode plots for the case of $T_s = 10$ ms and the noise MA polynomial $C(q^{-1}) = 1 - 0.99q^{-1}$: the estimated model parameters are given in Table 4.2. Note that, for clarity, the amplitude plots for the noise model have been suitably scaled (10 times estimated model gain) to separate them from the system plots.

of the OE gradient optimisation algorithm) converged in all 100 realisations and so the MCS results obtained in this case are also given in Table 4.2 for comparison.

For interest, two of the single runs obtained using the indirect OE and SRIV approaches are reported in Table 4.2; and the associated Bode plots, based on the estimated parameters, are shown in Figure 4.7. Here, the results relate to one of the MCS realisations where the OE algorithm fails to converge satisfactorily and is only able to identify the higher-frequency spectral peak characteristics. However, using the same data set, the SRIV algorithm provides good estimates and an almost perfect match to the actual Bode plots (the SRIV lines mask the actual lines almost completely). Similar comparative performance has been reported in another situation [37] where the model has two oscillatory modes although, in this other case, the resonant frequencies were close together. The problem with OE and PEM when applied to these

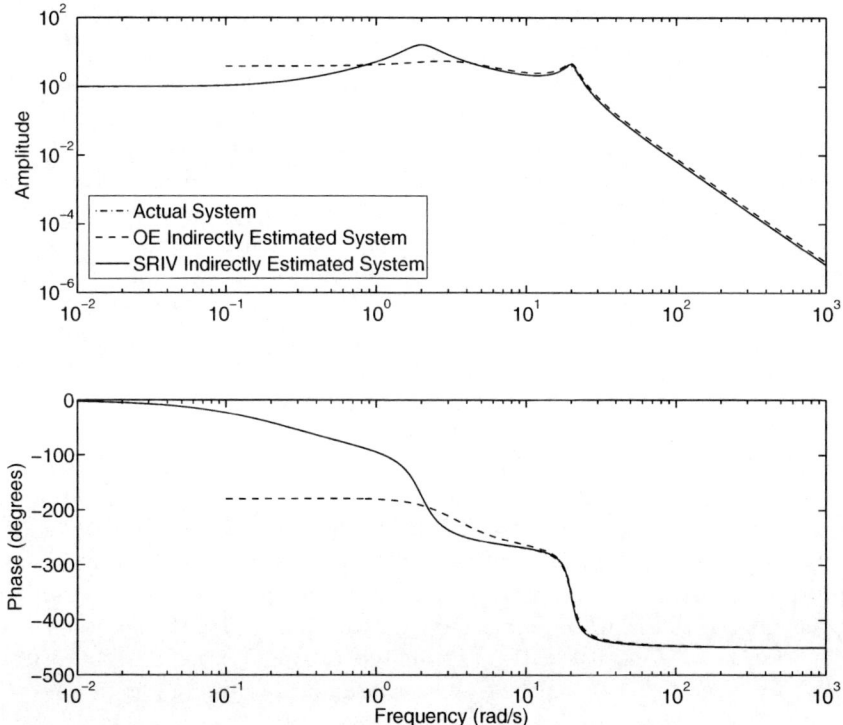

Fig. 4.7. Actual system Bode plots compared with those of the models indirectly estimated by the OE and SRIV algorithms for the case $T_s = 10$ ms and the noise MA polynomial $C(q^{-1}) = 1 - 0.99q^{-1}$. The estimated model parameters are given in Table 4.2 (**IND**: OE (SR) and **IND**: SRIV (SR)).

kind of multi-mode systems appears, at least in part, to be related to the specification of starting values for the estimates in the gradient optimisation [37]. The iterative IV optimisation used in the SRIV and RIV algorithms does not appear to suffer from these problems.

4.8 Practical Examples

In this section, we consider two practical examples of very different types. The first is concerned with the topical subject of 'global warming' and considers the simple time-series model simplification of the giant dynamic simulation models used in climate research. The second is a demonstration example [4] used in both the CAPTAIN and CONTSID Toolboxes, which involves the analysis of data from a multiple-input winding process.

4.8.1 Hadley Centre Global Circulation Model (GCM) Data

In the study of global climatic change, the global circulation model (GCM) is one of the principal tools used for evaluating possible future scenarios of climate variation. These models are very large indeed: probably the largest computer simulation models ever constructed. This example is concerned with modelling the dynamic relationship, over a period of 1100 years, between the variables shown in Figure 4.8. Here, the annual sampled input $u(t_k)$ is the radiative forcing (W/m^2) and the associated output $y(t_k)$ is the global mean temperature (deg °C) obtained from an experimental simulation of the UK Hadley Centre GCM. The TF modelling in this case can be seen as an exercise in dominant-mode analysis [33] or model simplification (order reduction) [46]. The first 1000 annual samples of the data, up to the vertical dash-dot line in Figure 4.8, are used for estimation and the last 100 years are set aside for validation. When the SRIVC algorithm is used for initial model order identification, it clearly identifies a second-order $[n\, m\, \tau\, p\, q] = [2\, 2\, 0\, 0\, 0]$ model, with a high $R_T^2 = 0.9953$ and a low YIC$= -9.3$. Although, the third-order $[3\, 3\, 0\, 0\, 0]$ model is identified as best by the Akaike information criterion (AIC), it has only a marginally better $R_T^2 = 0.9955$ and a significantly less negative YIC$= -3.6$, suggesting that it is probably overparameterised. All higher-order models greater than three are rejected as being poorly identifiable because of overparameterisation. The SRIVC estimated parameters for the $[2\, 2\, 0\, 0\, 0]$ model are given below, with the estimated standard errors in parentheses

$$\begin{cases} \hat{x}(t_k) = \frac{0.3516(\pm 0.0206)p + 0.001625(\pm 0.000128)}{p^2 + 0.1625(\pm 0.0101)p + 0.0004365(\pm 0.0000355)} u(t_k) \\ \hat{\xi}(t_k) = \frac{1}{1 - 0.4793(\pm 0.031)q^{-1} + 0.111(\pm 0.035)q^{-2} - 0.128(\pm 0.031)q^{-3}} e(t_k) \end{cases} \quad (4.54)$$

Also shown here are the estimates of the parameters in an AIC identified AR(3) model of the additive noise, based on the estimated noise signal $\hat{\xi}(t_k)$ obtained from (4.29), suggesting that RIVC estimation should consider the $[2\, 2\, 0\, 3\, 0]$ model.

The equivalent estimates obtained by RIVC estimation for the $[2\, 2\, 0\, 3\, 0]$ model are as follows

$$\begin{cases} \hat{x}(t_k) = \frac{0.3514(\pm 0.0361)p + 0.001624(\pm 0.000224)}{p^2 + 0.1624(\pm 0.0177)p + 0.0004361(\pm 0.0000621)} u(t_k) \\ \hat{\xi}(t_k) = \frac{1}{1 - 0.4802(\pm 0.031)q^{-1} + 0.111(\pm 0.035)q^{-2} - 0.128(\pm 0.031)q^{-3}} e(t_k) \end{cases} \quad (4.55)$$

It is clear that the estimated parameters are very similar to those in (4.54) and that the main difference lies in the estimated standard errors, where the RIVC estimates are significantly larger and indicative of greater uncertainty. In other words, the SRIVC estimation provides excellent unbiased estimates of the parameters but indicates more confidence in these estimates than is justified. In the subsequent validation analysis, however, both models explain the last 100 years of data (see Figure 4.8) very well, with identical coefficients determination of $R_T^2 = 0.9952$, only marginally less than that obtained in

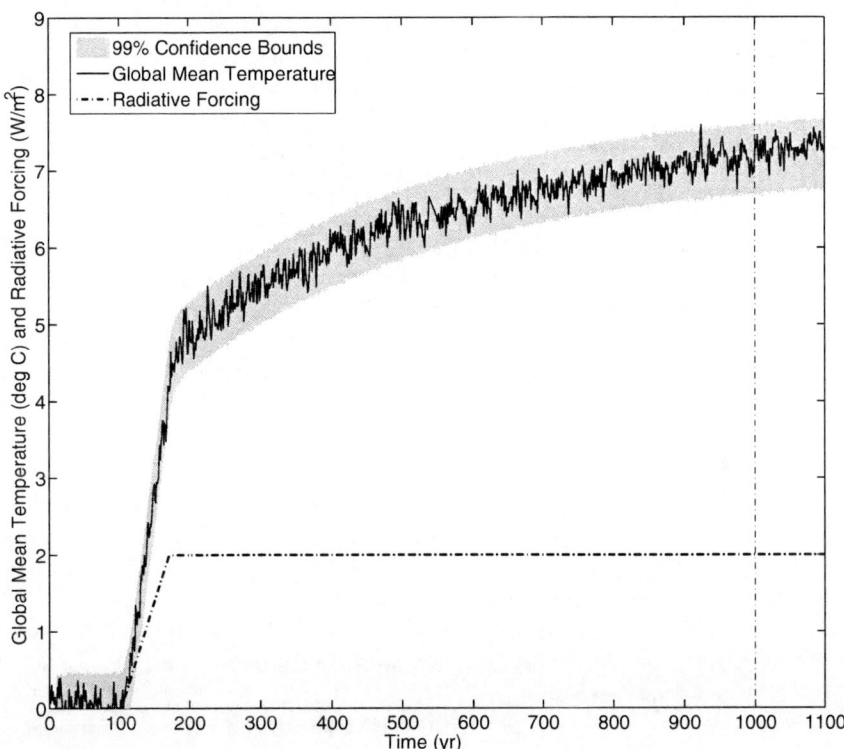

Fig. 4.8. Estimation and validation results for an RIVC model of the Hadley Global Circulation Model (GCM) experimental data: estimation/validation boundary shown as dash-dot line. Also shown are the 99 percentile bounds obtained in Monte Carlo simulation analysis of the RIVC model.

estimation, although the estimated confidence bounds in the SRIV case are, of course, a little smaller because of its (incorrect) lower level of uncertainty. One of the main reasons for requiring estimates of parametric uncertainty in models is to evaluate how much confidence we can have in the model in any application for which it is intended. In the present example, it is important that the model can be interpreted in physically meaningful terms, as required in DBM modelling (see earlier, Section 4.6). In this regard, it is instructive, to decompose the deterministic system model by partial fraction expansion into either the following parallel pathway form

$$\hat{x}(t_k) = \left(\frac{2.174}{1 + 6.263p} + \frac{1.549}{1 + 366.2p} \right) u(t_k) \tag{4.56}$$

or the alternative feedback form

$$\hat{x}(t_k) = \frac{2.227}{1 + 6.338p} \left(u(t_k) - \frac{0.1804}{1 + 216.5p} \hat{x}(t_k) \right) \tag{4.57}$$

These decompositions show that the model can be considered in either of these forms, both of which have physical significance. The parallel pathway form (4.56) suggests that the forcing signal affects two 'compartments' in parallel, one with a short residence time (time constant) of 6.263 years and the other with a long residence time of 366.2 years; while the feedback form (4.57) suggests that there is a forward path characterised by a short residence time of 6.338 years and a negative-feedback path compartment with a long residence time of 216.5 years. The physical interpretation of these compartments is outside the scope of the present chapter but the interested reader can consult related publications (*e.g.*, [10, 40]).

Considering the parallel pathway form (4.56), Figure 4.9 shows the results of MCS analysis in which the effects of model uncertainty, as defined by the SRIVC and RIVC estimated parametric covariance matrices, is mapped into its effect on the derived residence times in (4.56) and (4.57). Comparing the top histograms (SRIVC) with the bottom ones (RIVC), we see that, although the estimated residence times are identical, the SRIVC-derived distributions incorrectly suggest more confidence than is justified when they are compared with the RIVC-derived distributions. These latter distributions are wider and more skewed towards larger values. In this MCS analysis, the effect of both the parametric uncertainty and the uncertainty injected by the additive noise is also used to derive 99% uncertainty bounds on the model output, as shown by the grey band in Figure 4.8.

It is interesting to observe that if the AIC identified [3 3 0 0 0] model is considered in the same way as that described above, the overparameterisation becomes obvious. In the MCS analysis, 30% of the SRIVC realisations and 42% of the RIVC realisations are unstable because the additional, very long residence time is so ill-defined and the MCS realisations often produce a model with an eigenvalue in the right half of the s-plane. In the experience of the first author, the AIC often indicates such overparameterised models when applied in TF identification, rather than its originally intended application to fully stochastic AR models.

Finally, it should be noted that the data used in this example are all that could be obtained and provide some difficulties as regards statistical inference. In particular, the input signal is not persistently exciting and the system response has not reached a steady state by the end of the data set (Figure 4.8). Consequently, the results obtained here should be treated as preliminary ones that need to be confirmed by subsequent modelling based on a better planned experiment and a larger data set. Nevertheless, it is an interesting example to illustrate the wide application potential of the RIVC/SRIVC methodology.

4.8.2 A Multiple-input Winding Process

Model identification in this winding process is based on the experimental data set plotted in Figure 4.10, where the measured output is shown in the top panel and the $r = 3$ input signals in the panels below this. Since the data

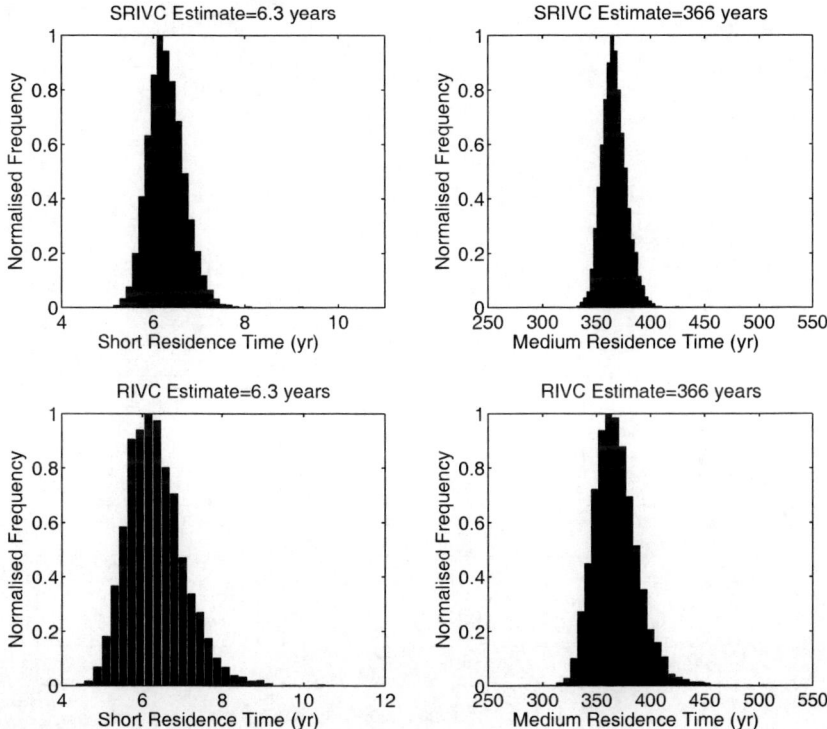

Fig. 4.9. Monte Carlo simulation of the models estimated from the Hadley Global Circulation Model experimental data: normalised histograms of the two residence times that characterise the parallel pathways, as estimated by the SRIVC (*top panels*) and RIVC (*bottom panels*) algorithms.

are plentiful, the identification and estimation analysis is applied to the first half of the data and validation analysis to the latter half.

When the SRIVC algorithm is used for model order identification over a range of models from first to third order, it shows that many models have similar explanatory ability, with R_T^2 values around $0.98 - 0.99$. However, the $[n\ m_1\ m_2\ m_3\ \tau_1\ \tau_2\ \tau_3\ p\ q] = [3\ 3\ 3\ 1\ 0\ 0\ 0\ 0\ 0]$ model, which looks well identified with an $R_T^2 = 0.990$ and YIC=−7.1661 and is favoured by the AIC, performs badly in validation, with $R_T^2 = 0.954$ for SRIVC and 0.917 for RIVC. The best validation performance is obtained with the $[2\ 1\ 1\ 1\ 0\ 0\ 0\ 0\ 0]$ model, although the $[1\ 1\ 1\ 1\ 0\ 0\ 0\ 0\ 0]$ model also does quite well. This SRIVC estimated second-order model has an $R_T^2 = 0.986$ and a YIC=−7.6830. The SRIVC model has the following form

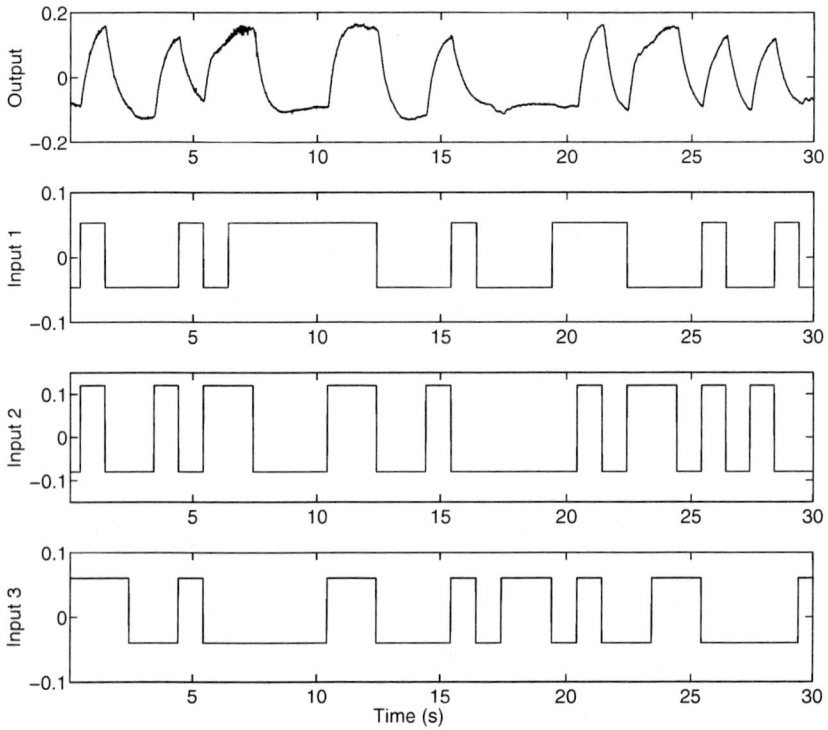

Fig. 4.10. Winding process data

$$\begin{cases} \hat{x}(t_k) = \dfrac{\hat{B}_1(p)}{\hat{A}(p)} u_1(t_k) + \dfrac{\hat{B}_2(p)}{\hat{A}(p)} u_2(t_k) + \dfrac{\hat{B}_3(p)}{\hat{A}(p)} u_3(t_k) \\ y(t_k) = \hat{x}(t_k) + e(t_k) \end{cases} \quad (4.58)$$

with the following estimated polynomials

$$\hat{B}_1(p) = 23.341(\pm 1.996); \hat{B}_2(p) = 134.4(\pm 10.632); \hat{B}_3(p) = 21.425(\pm 1.808) \quad (4.59)$$

$$\hat{A}(p) = p^2 + 36.609(\pm 2.911)p + 108.02(\pm 8.280) \quad (4.60)$$

In validation, it achieves an $R_T^2 = 0.976$. The AIC identifies an AR(14) noise model from the SRIVC model residuals and the RIVC estimated model based on this AR model order has an estimation $R_T^2 = 0.983$, a little less than the SRIVC estimated model, and a YIC=-7.1234. The RIVC AR(14) model, based on the estimation data set, has the following form

$$\begin{cases} \hat{x}(t_k) = \dfrac{\hat{B}_1(p)}{\hat{A}(p)} u_1(t_k) + \dfrac{\hat{B}_2(p)}{\hat{A}(p)} u_2(t_k) + \dfrac{\hat{B}_3(p)}{\hat{A}(p)} u_3(t_k) \\ y(t_k) = \hat{x}(t_k) + \hat{\xi}(t_k) \end{cases} \quad (4.61)$$

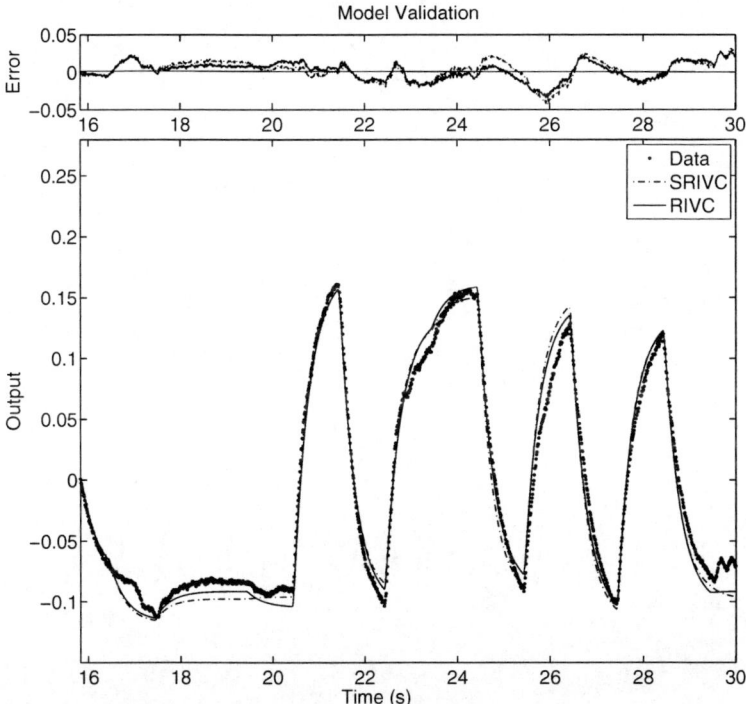

Fig. 4.11. Validation of the SRIVC and RIVC models on the second half of the data set

with the following estimated polynomials and AR noise model (for convenience the estimated AR(14) parameters are not cited)

$$\begin{cases} \hat{B}_1(p) = 14.587(\pm 9.114); \hat{B}_2(p) = 157.03(\pm 25.65); \hat{B}_3(p) = 30.843(\pm 10.363) \\ \hat{A}(p) = p^2 + 45.057(\pm 7.531)p + 124.54(\pm 19.556) \\ \hat{\xi}(t_k) = \frac{1}{1+\sum_{i=1}^{14} \hat{d}_i q^{-1}} e(t_k) \end{cases}$$

(4.62)

However, in validation, it achieves a somewhat higher $R_T^2 = 0.982$ than the SRIVC estimated model and this is only marginally less that its estimation performance. The comparative validation performance is shown graphically in Figure 4.11, where the error series in the upper panel reveal the small advantage of the RIVC estimated model.

The RIVC AR(14) estimated model of the form of (4.61), based on all of the data, has the following estimated parameters

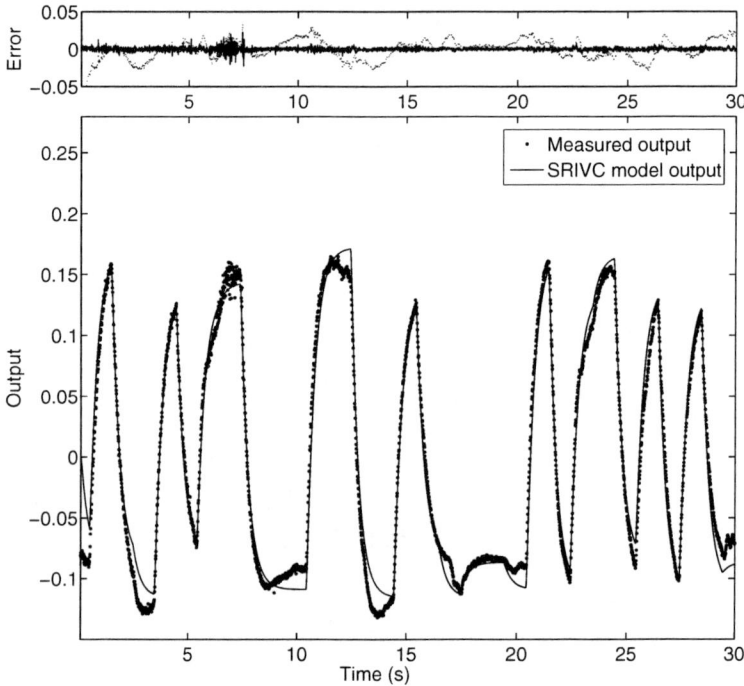

Fig. 4.12. The output of final RIVC estimated model, based on the complete data set, compared with the measured output. The error plot at the top shows both the estimated noise signal $\hat{\xi}(t_k)$, as a light dotted line, and the estimated final residuals $\hat{e}(t_k)$ as a dark full line.

$$\begin{cases} \hat{B}_1(p) = 7.923(\pm 5.032); \ \hat{B}_2(p) = 144.61(\pm 13.608); \ \hat{B}_3(p) = 33.131(\pm 5.785) \\ \hat{A}(p) = p^2 + 43.108(\pm 4.167)p + 115.70(\pm 10.446) \\ \hat{\xi}(t_k) = \frac{1}{1+\sum_{i=1}^{14} \hat{d}_i q^{-1}} e(t_k) \end{cases}$$

(4.63)

and has an $R_T^2 = 0.980$. The final residuals of the model pass all the diagnostic tests, with no significant autocorrelation or cross-correlation with the three input signals.

An alternative model can be obtained by identifying a more parsimonious ARMA model for the noise. Identification analysis using the IVARMA algorithm yields an ARMA(2,2) model which produces a quite similar model to (4.63), with an $R_T^2 = 0.979$. The RIVC ARMA(2,2) model of the form of (4.61), based on all of the data, has the following parameter ARMA noise model estimates

$$\begin{cases} \hat{B}_1(p) = 7.672(\pm 4.139);\ \hat{B}_2(p) = 148.54(\pm 12.180);\ \hat{B}_3(p) = 33.116(\pm 4.834) \\ \hat{A}(p) = p^2 + 44.335(\pm 3.711)p + 117.72(\pm 9.169) \\ \hat{\xi}(t_k) = \dfrac{1 - 0.1098(\pm 0.063)q^{-1} - 0.0959(\pm 0.041)q^{-2}}{1 - 0.4245(\pm 0.058)q^{-1} + 0.5622(\pm 0.058)q^{-2}} e(t_k) \end{cases}$$

(4.64)

The main advantage of this model is its parsimony, which reduces the standard errors on the parameters with minimal loss in explanatory ability. Its main disadvantage is that, although the final residuals are not significantly cross-correlated with the input signals, there is some autocorrelation. However, the variance is very small: $\hat{\sigma}^2 = 0.00000965$ compared with $\sigma_y^2 = 0.0091$ for the measured output signal, giving a conventional coefficient of determination based on these final residuals of $R^2 = 0.999$. There is some heteroscedasticity in the final residuals that can be seen in Figure 4.12, which compares the model output and the measured output, with both the estimate of the noise $\hat{\xi}(t_k)$ (light dotted line) and these final residuals $\hat{e}(t_k)$ (black full line) plotted in the upper panel. Some minor improvements in the model might be obtained by taking the heteroscedasticity into account and also allowing for initial condition effects. Nevertheless, it should be satisfactory for most practical purposes, although this would normally involve additional experiments on the winding process in order to further validate the model.

The model (4.64) makes reasonable physical sense. It is non-oscillatory, characterised by eigenvalues at -41.449 and -2.8368 (*i.e.*, time constants of 0.0241 and 0.3525 s). The output response is dominated by the effects of the second input $u_2(t)$ whose output accounts for about 97% of the measured output variance, with the $u_1(t)$ and $u_3(t)$ contributing very little (particularly $u_1(t)$). Finally, it is worth considering how the multiple-input TF model with different denominator polynomials in each input path [3] compares with the common-denominator model evaluated above. Here, convergence problems are encountered if only the estimation data set is used but reasonable SRIVC estimates are obtained with the full data set ($R_T^2 = 0.979$), although at the cost of quite a large computation time (7 times that of the common denominator model). When RIVC is applied to the full data set with AR(14) and ARMA(2,2) noise models, the results are not so good, with $R_T^2 = 0.972$ and $R_T^2 = 0.973$, respectively. Overall, there seems little advantage in moving to this more complicated model that, with its four additional parameters, shows some signs of overparameterisation.

4.9 Conclusions

This chapter has described how the full RIVC algorithm for identifying and estimating hybrid Box–Jenkins transfer function models for linear, continuous-time systems from discrete-time, sampled data has been developed from the earlier SRIVC algorithm.

Based on simulation results obtained in the noise-free case, it is clear that the performance of the SRIVC method is not sensitive to the sampling frequency, in contrast to earlier continuous-time estimation methods based on simple state-variable filtering methods. Also, as expected, it performs better than alternative indirect approaches, based on DT estimation followed by conversion to CT, particularly at higher sampling frequencies. In the case of noisy data, the results obtained in the Monte Carlo simulation studies, using a difficult fourth-order model, demonstrate the improvement in statistical efficiency that accrues from the explicit inclusion of noise model effects in the RIVC algorithm. In particular, the empirical parametric error variance is reduced and agrees with that obtained from the algorithm's estimate of the parametric error covariance matrix, as required. The MCS and practical example results also suggest that, in many applications, the use of a higher-order AR noise model, rather than an ARMA model in RIVC estimation, can provide similar results while, at the same time, providing a simpler algorithm that only requires least squares estimation of the AR model parameters.

The results of the MCS simulation analysis also show that the estimates obtained from the simple SRIVC algorithm are very good. Although not as statistically efficient as the RIVC estimates in conditions where the noise is highly coloured (*i.e.*, eigenvalues near the unit circle), the SRIVC estimates are always consistent and asymptotically unbiased. Moreover, they have relatively good statistical efficiency, which is quite close to that of the RIVC estimates when the noise is not highly coloured. These results are consistent with previous practical experience of the SRIVC algorithm over many years and demonstrate its utility as a simple, fast and robust, day-to-day tool in continuous-time modelling.

It is felt that the results presented in this chapter, as well as other associated papers that have appeared recently, show that continuous-time identification and estimation, based on a stochastic formulation of the transfer function estimation problem, provides a theoretically elegant and practically useful approach to the modelling of stochastic linear systems from discrete-time sampled data. It is an approach that has many advantages in scientific terms since it provides a differential equation model that conforms with models used in most scientific research, where conservation equations (mass, energy, *etc.*) are normally formulated in terms of differential equations. It is also a model defined by a unique set of parameter values that are not dependent on the sampling interval, so eliminating the need for conversion from discrete to continuous time that is an essential element of indirect approaches to estimation based on discrete-time estimation.

Finally, note that most of the continuous-time model analysis reported in this chapter was carried out using both the CONTSID[7] and CAPTAIN[8] Toolboxes for MATLAB®. The CONTSID toolbox contains the major identification

---

[7] http://www.cran.uhp-nancy.fr/contsid/
[8] http://www.es.lancs.ac.uk/cres/captain/

and estimation tools for continuous-time systems, including SRIVC/RIVC (see Chapter 9 in the present book); the CAPTAIN toolbox has the same SRIVC/RIVC tools but also has additional optimal algorithms (SRIV/RIV) for discrete-time systems (used in the present chapter for comparative, indirect discrete-time modelling), as well as other time-series algorithms, including: time-variable and state-dependent parameter (non-linear) model estimation; non-stationary signal extraction; and adaptive forecasting.

References

1. T. Bastogne, H. Garnier, and P. Sibille. A PMF-based subspace method for continuous-time model identification. Application to a multivariable winding process. *International Journal of Control*, 74(2):118–132, 2001.
2. G.E.P. Box and G.M. Jenkins. *Time Series Analysis Forecasting and Control.* Holden-Day: San Francisco, 1970.
3. H. Garnier, M. Gilson, P.C. Young, and E. Huselstein. An optimal IV technique for identifying continuous-time transfer function model of multiple input systems. *Control Engineering Practice*, 15(4):471–486, 2007.
4. H. Garnier, M. Mensler, and A. Richard. Continuous-time model identification from sampled data. Implementation issues and performance evaluation. *International Journal of Control*, 76(13):1337–1357, 2003.
5. M. Gilson, H. Garnier, P.C. Young, and P. Van den Hof. A refined IV method for closed-loop system identification. *14th IFAC Symposium on System Identification*, Newcastle, Australia, pages 903–908, March 2006.
6. A.J. Jakeman, L.P. Steele, and P.C. Young. Instrumental variable algorithms for multiple input systems described by multiple transfer functions. *IEEE Transactions on Systems, Man, and Cybernetics*, SMC-10:593–602, 1980.
7. A.J. Jakeman and P.C. Young. Refined instrumental variable methods of time-series analysis: Part II, multivariable systems. *International Journal of Control*, 29:621–644, 1979.
8. A.J. Jakeman and P.C. Young. On the decoupling of system and noise model parameter estimation in time-series analysis. *International Journal of Control*, 34:423–431, 1981.
9. A.J. Jakeman and P.C. Young. Advanced methods of recursive time-series analysis. *International Journal of Control*, 37:1291–1310, 1983.
10. A.J. Jarvis, P.C. Young, D.T Leedal, and A. Chotai. An incremental carbon emissions strategy for the global carbon cycle using state variable feedback control design. *Avoiding Dangerous Climate Change: International Symposium on Stabilisation of Greenhouse Gases*, Exeter, UK, pages 45–49, Meteorological Office, 2005.
11. L. Ljung. *System Identification. Theory for the User.* Prentice Hall, Upper Saddle River, 2nd edition, 1999.
12. L. Ljung. Initialisation aspects for subspace and output-error identification methods. *European Control Conference*, Cambridge, UK, 2003.
13. K. Mahata and H. Garnier. Identification of continuous-time errors-in-variables models. *Automatica*, 46(9):1477–1490, 2006.
14. R.H. Middleton and G.C. Goodwin. *Digital Control and Estimation: a Unified Approach.* Prentice Hall, Englewood Cliffs, N.J., 1990.

15. D.A. Pierce. Least squares estimation in dynamic disturbance time-series models. *Biometrika*, 5:73–78, 1972.
16. L. Price, P.C. Young, D. Berckmans, K. Janssens, and J. Taylor. Data-based mechanistic modelling and control of mass and energy transfer in agricultural buildings. *Annual Reviews in Control*, 23:71–82, 1999.
17. G.P. Rao and H. Garnier. Identification of continuous-time systems: direct or indirect? *Systems Science*, 30(3):25–50, 2004.
18. G.P. Rao and H. Garnier. Numerical illustrations of the relevance of direct continuous-time model identification. *15th IFAC World Congress*, Barcelona, Spain, July 2002.
19. T. Söderström and P. Stoica. *Instrumental Variable Methods for System Identification*. Springer Verlag, New York, 1983.
20. V. Solo. *Time Series Recursions and Stochastic Approximation*. PhD thesis, Australian National University, Canberra, Australia, 1978.
21. K. Steiglitz and L.E. McBride. A technique for the identification of linear systems. *IEEE Transactions on Automatic Control*, 10:461–464, October 1965.
22. P. Stoica and T. Söderström. The Steiglitz-McBride identification algorithms revisited. Convergence analysis and accuracy aspects. *IEEE Transactions on Automatic Control*, AC-26:712–717, 1981.
23. S. Thil and H. Garnier and M. Gilson Third-order cumulants based methods for continuous-time errors-in-variables model identification. *Automatica*, 44(3), 2008.
24. P.E. Wellstead. An instrumental product moment test for model order estimation. *Automatica*, 14:89–91, 1978.
25. P.C. Young. In flight dynamic checkout - a discussion. *IEEE Transactions on Aerospace*, AS2(3):1100–1111, 1964.
26. P.C. Young. The determination of the parameters of a dynamic process. *Radio and Electronic Engineering (Journal of IERE)*, 29:345–361, 1965.
27. P.C. Young. Some observations on instrumental variable methods of time-series analysis. *International Journal of Control*, 23:593–612, 1976.
28. P.C. Young. Parameter estimation for continuous-time models - a survey. *Automatica*, 17(1):23–39, 1981.
29. P.C. Young. *Recursive Estimation and Time-Series Analysis*. Springer-Verlag, Berlin, 1984.
30. P.C. Young. Recursive estimation, forecasting and adaptive control. In C.T. Leondes (ed), *Control and Dynamic Systems*, pages 119–166. Academic Press: San Diego, USA, 1989.
31. P.C. Young. Data-based mechanistic modeling of engineering systems. *Journal of Vibration and Control*, 4:5–28, 1998.
32. P.C. Young. Data-based mechanistic modeling of environmental, ecological, economic and engineering systems. *Journal of Environmental Modelling and Software*, 13:105–122, 1998.
33. P.C. Young. Data-based mechanistic modelling, generalised sensitivity and dominant mode analysis. *Computer Physics Communications*, 117:113–129, 1999.
34. P.C. Young. Identification and estimation of continuous-time hydrological models from discrete-time data. In B. Webb, N. Arnell, C. Onf, N. MacIntyre, R. Gurney, and C. Kirby, (eds), *Hydrology: Science and Practice for the 21st Century, Vol. 1*, pages 406–413. British Hydrological Society: London, 2004.

35. P.C. Young. The data-based mechanistic approach to the modelling, forecasting and control of environmental systems. *Annual Reviews in Control*, 30:169–182, 2006.
36. P.C. Young. Data-based mechanistic modelling and river flow forecasting. *14th IFAC Symposium on System Identification*, Newcastle, Australia, pages 756–761, March 2006.
37. P.C. Young. An instrumental variable approach to ARMA model identification and estimation. *14th IFAC Symposium on System Identification*, Newcastle, Australia, pages 410–415, March 2006.
38. P.C. Young. The refined instrumental variable method: unified estimation of discrete and continuous-time transfer function models. *Journal Européen des Systèmes Automatisés*, 2008.
39. P.C. Young, A. Chotai, and W. Tych. Identification, estimation and control of continuous-time systems described by delta operator models. In *Identification of Continuous-Time Systems. Methodology and computer implementation*, Kluwers Academic Publishers: Dordrecht, N.K. Sinha and G.P. Rao (eds), pages 363–418, 1991.
40. P.C. Young and H. Garnier. Identification and estimation of continuous-time, data-based mechanistic models for environmental systems. *Environmental Modelling & Software*, 21:1055–1072, 2006.
41. P.C. Young, H. Garnier, and M. Gilson. An optimal instrumental variable approach for identifying hybrid continuous-time Box-Jenkins models. *14th IFAC Symposium on System Identification*, Newcastle, Australia, pages 225–230, March 2006.
42. P.C. Young and A.J. Jakeman. Refined instrumental variable methods of time-series analysis: Part I, SISO systems. *International Journal of Control*, 29:1–30, 1979.
43. P.C. Young and A.J. Jakeman. Refined instrumental variable methods of time-series analysis: Part III, extensions. *International Journal of Control*, 31:741–764, 1980.
44. P.C. Young, A.J. Jakeman, and R. McMurtrie. An instrumental variable method for model order identification. *Automatica*, 16:281–296, 1980.
45. P.C. Young and S. Parkinson. Simplicity out of complexity. In M.B. Beck (ed), *Environmental Foresight and Models: A Manifesto*, pages 251–294. Elsevier: Oxford, UK, 2002.
46. P.C. Young, S. Parkinson, and M.J. Lees. Simplicity out of complexity: Occam's razor revisited. *Journal of Applied Statistics*, 23:165–210, 1996.

5

Instrumental Variable Methods for Closed-loop Continuous-time Model Identification

Marion Gilson[1], Hugues Garnier[1], Peter C. Young[2] and Paul Van den Hof[3]

[1] Nancy-Université, CNRS, France
[2] Lancaster University, UK & Australian National University
[3] Delft University of Technology, Delft, The Netherlands

5.1 Introduction

For many industrial production processes, safety and production restrictions are often strong reasons for not allowing identification experiments in open loop. In such situations, experimental data can only be obtained under closed-loop conditions. The main difficulty in closed-loop identification is due to the correlation between the disturbances and the control signal, induced by the loop. Several alternatives are available to cope with this problem, broadly classified into three main approaches: direct, indirect and joint input–output [9, 14]. Some particular versions of these methods have been developed more recently in the area of control-relevant identification as, *e.g.*, the two-stage, the coprime factor and the dual-Youla methods. An overview of these recent developments can be found in [2] and [19].

When considering methods that can be used to identify models of systems operating in closed loop, instrumental variable (IV) techniques are rather attractive because they are normally simple or iterative modifications of the linear regression algorithm. For instance, when dealing with highly complex processes that are high-dimensional in terms of inputs and outputs, it can be attractive to rely on methods, such as these, that do not require non-convex optimisation algorithms. In addition to this computationally attractive property, IV methods also have the potential advantage that they can yield consistent and asymptotically unbiased estimates of the plant model parameters if the noise does not have rational spectral density or the noise model is misspecified; or even if the control system is non-linear and/or time varying.

For closed-loop identification, a basic IV estimator was first suggested assuming knowledge of the controller [20]; and the topic was later discussed in more detail in [15]. More recently, a so-called 'tailor-made' IV algorithm has been suggested [6], where the closed-loop plant is parameterised using (open-loop) plant parameters. The class of algorithms denoted by BELS (bias-eliminated least squares), *e.g.*, [27], is also directed towards the use of only simple linear regression-like algorithms. Recently, it has been shown that these algorithms are, in fact, particular forms of IV estimation [6, 16].

When comparing the various available IV algorithms that all lead to consistent plant estimates, it is important to ask how the algorithm can be made statistically efficient: *i.e.*, how is it possible to achieve the smallest variance of the parameter estimates. Concerning extended IV methods, an optimal variance result has been developed in the open-loop identification case, showing consequences for the choice of weights, filters, and instruments [9, 14, 17]. Similar enhancements of the basic IV approach are also the basis of the optimal refined instrumental variable (RIV) method [8, 21, 26] that is designed specifically for the Box-Jenkins transfer function model in discrete (RIV) or continuous (RIVC) time. For the closed-loop case, a statistical analysis has been provided [13, 15]; and this analysis has been used to compare several closed-loop identification methods [1]. More recently, some attention has been given to a characterisation of the properties of the several (extended) IV methods [7].

All of the above methods, except RIVC, focus on the identification of discrete-time (DT) models. Recently, however, there has been renewed interest in the relevance of continuous-time (CT) model identification methods: see, *e.g.*, papers at the 15th IFAC World Congress in 2002, such as [23], which compares RIVC with another optimal approach; and [11, 12], where the advantages of direct CT approaches are illustrated by extensive simulation examples. Unfortunately, however, closed-loop model CT identification is still an issue that has not so far received adequate attention. Indeed, there are only a few recent publications that deal with closed-loop CT identification: amongst these, a bias-eliminated least squares method [3], some basic instrumental variable methods [5], and a two-step algorithm using the RIVC algorithm [24], appear to be the most successful (see also Chapter 13 in the present book).

This chapter aims at filling this gap: it describes and evaluates statistically more efficient IV methods for the closed-loop identification of 'hybrid' CT transfer function models from discrete-time, sampled data (based, in part, on the analysis of the optimal open-loop approach in Chapter 4 of this book). Here, the model of the basic dynamic system is estimated in continuous-time, differential equation form, while the associated additive-noise model is estimated as a discrete-time process. Several IV and IV-related methods are presented and they are unified in an extended IV framework. Then, the minimum-variance closed-loop IV estimation approach is introduced, with the consequences of this formulation on the several design variables. Since such an optimal statistical approach requires the concurrent estimation of a model for the noise process, several bootstrap methods are proposed for accomplishing this. A comparison between these different proposed methods is made with the help of simulation examples, showing that more statistically efficient estimation can be achieved by an appropriate choice of the design parameters.

5.2 Problem Formulation

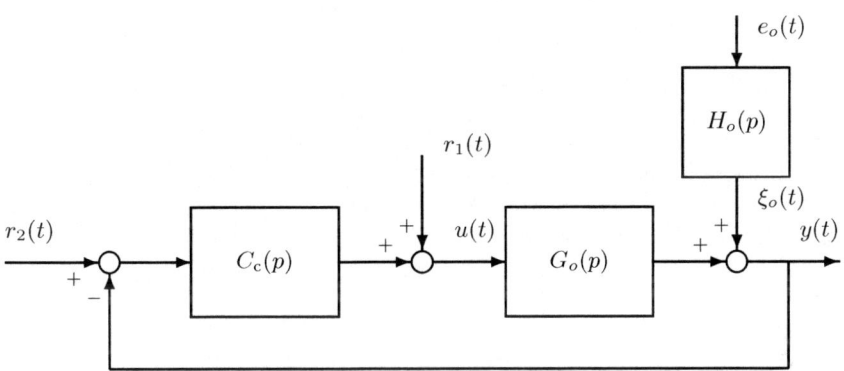

Fig. 5.1. Closed-loop configuration

Consider a stable, linear, SISO, closed-loop system of the form shown in Figure 5.1. The data-generating system is assumed to be given by the relations

$$\mathcal{S}: \begin{cases} y(t) = G_o(p)u(t) + H_o(p)e_o(t) \\ u(t) = r(t) - C_c(p)y(t) \end{cases} \quad (5.1)$$

The process is denoted by $G_o(p) = B_o(p)/A_o(p)$ and the controller by $C_c(p)$ where p is the differential operator ($p = d/dt$). $u(t)$ describes the process input signal, $y(t)$ the process output signal. For ease of notation we introduce an external signal $r(t) = r_1(t) + C_c(p)r_2(t)$. Moreover, it is also assumed that the CT signals $u(t)$ and $y(t)$ are uniformly sampled at T_s. A coloured disturbance is assumed to affect the closed loop: bearing in mind the spectral factorisation theorem, this noise term, $\xi_o(t) = H_o(p)e_o(t)$, is modelled as linearly filtered white noise. The external signal $r(t)$ is assumed to be uncorrelated with the noise disturbance $\xi_o(t)$.

The CT model identification problem is to find estimates of $G_o(p)$ from finite sequences $\{r(t_k)\}_{k=1}^N$, $\{u(t_k)\}_{k=1}^N$, $\{y(t_k)\}_{k=1}^N$ of, respectively, the external signal, the process input and output DT data. The model is then described by the following hybrid equation,

$$\mathcal{M}: y(t_k) = G(p, \boldsymbol{\theta})u(t_k) + H(q^{-1}, \boldsymbol{\theta})e(t_k) \quad (5.2)$$

where q^{-i} is the backward shift operator, i.e., $q^{-i}y(t_k) = y(t_{k-i})$; $e(t_k)$ is a discrete-time white noise, with zero mean and variance σ_e^2. The conventional notation $w(t_k)$ is used here to denote the sampled value of $w(t)$ at time instant t_k.

The hybrid form (5.2) of the continuous-time transfer function model is considered here for two reasons. First, the approach is simple and straightforward: the theoretical and practical problems associated with the estimation of purely stochastic, continuous-time noise models are avoided by formulating the problem in this manner. Second, one of the main functions of the noise estimation is to improve the statistical efficiency of the parameter estimation by introducing appropriately defined prefilters into the estimation procedure. And, as we shall see in this chapter, this can be achieved adequately by assuming that the additive coloured noise $\xi(t_k)$ has rational spectral density, so that it can be represented in the form of a discrete-time, autoregressive moving average (ARMA) model (see below).

With the above assumptions, the parameterised CT hybrid process model takes the form,

$$\mathcal{G}: G(p, \boldsymbol{\rho}) = \frac{B(p, \boldsymbol{\rho})}{A(p, \boldsymbol{\rho})} = \frac{b_0 p^{n_b} + b_1 p^{n_b-1} + \cdots + b_{n_b}}{p^{n_a} + a_1 p^{n_a-1} + \cdots + a_{n_a}} \quad (5.3)$$

where n_b, n_a denote the degrees of the process numerator and denominator polynomials, respectively, with the pair (A, B) assumed to be coprime. The process model parameters are stacked columnwise in the parameter vector

$$\boldsymbol{\rho} = [a_1 \cdots a_{n_a} \; b_0 \cdots b_{n_b}]^T \in \mathbb{R}^{n_a+n_b+1} \quad (5.4)$$

The numerator and denominator orders n_b and n_a are to be identified from the data and the parameterised DT noise model is assumed to be in the form of the following discrete-time ARMA process,

$$\xi(t_k) = H(q^{-1}, \boldsymbol{\eta}) e(t_k) \quad (5.5a)$$

$$\mathcal{H}: H(q^{-1}, \boldsymbol{\eta}) = \frac{C(q^{-1}, \boldsymbol{\eta})}{D(q^{-1}, \boldsymbol{\eta})} = \frac{1 + c_1 q^{-1} + \cdots + c_{n_c} q^{-n_c}}{1 + d_1 q^{-1} + \cdots + d_{n_d} q^{-n_d}} \quad (5.5b)$$

$$e(t_k) \sim \mathcal{N}(0, \sigma_e^2) \quad (5.5c)$$

where the associated noise model parameters are stacked columnwise in the parameter vector,

$$\boldsymbol{\eta} = [d_1 \cdots d_{n_d} \; c_1 \cdots c_{n_c}]^T \in \mathbb{R}^{n_c+n_d} \quad (5.6)$$

where, as shown, $e(t_k)$ is a zero-mean, normally distributed, discrete-time white noise sequence.

The model structure $\mathcal{M}$ (5.2) is chosen so that the process and noise models do not have common factors; these models can therefore be parameterised independently. More formally, there exists the following decomposition of the parameter vector $\boldsymbol{\theta}$ for the whole hybrid model,

$$\boldsymbol{\theta} = \begin{pmatrix} \boldsymbol{\rho} \\ \boldsymbol{\eta} \end{pmatrix} \quad (5.7)$$

Additionally, the controller $C_c(p)$ is given by

$$C_c(p) = \frac{Q(p)}{P(p)} = \frac{q_0 p^{n_q} + q_1 p^{n_q-1} + \cdots + q_{n_q}}{p^{n_p} + p_1 p^{n_p-1} + \cdots + p_{n_p}} \quad (5.8)$$

with the pair (P, Q) assumed to be coprime. Of course, in this hybrid context, the continuous-time controller could be replaced by a DT alternative if this is required (see, e.g., [24] where the continuous-time process is estimated within a DT, non-minimal state-space control loop [18]). In the following, the closed-loop system is assumed to be asymptotically stable and $r(t)$ is an external signal that is persistently exciting of sufficient high order.

With these notations, the closed-loop system can be described as

$$\begin{cases} y(t_k) = \dfrac{G_o(p)}{1 + C_c(p)G_o(p)} r(t_k) + \dfrac{1}{1 + C_c(p)G_o(p)} \xi_o(t_k) \\ u(t_k) = \dfrac{1}{1 + C_c(p)G_o(p)} r(t_k) - \dfrac{C_c(p)}{1 + C_c(p)G_o(p)} \xi_o(t_k) \end{cases} \quad (5.9)$$

In the following instrumental variable algorithms, use is made of the noise-free input–output signals deduced from (5.9) and denoted from hereon as

$$\begin{cases} x(t_k) = \dfrac{G_o(p)}{1 + C_c(p)G_o(p)} r(t_k) \\ \nu(t_k) = \dfrac{1}{1 + C_c(p)G_o(p)} r(t_k) \end{cases} \quad (5.10)$$

Now consider the relationship between the process input and output signals in (5.1),

$$y(t) = G_o(p)u(t) + H_o(p)e_o(t) \quad (5.11)$$

This latter can also be written in the vector form at time instant $t = t_k$

$$y^{(n_a)}(t_k) = \boldsymbol{\varphi}^T(t_k)\boldsymbol{\rho}_o + v_o(t_k) \quad (5.12)$$

where $\boldsymbol{\rho}_o$ denotes the true process parameter vector,

$$\boldsymbol{\varphi}^T(t_k) = [-y^{(n_a-1)}(t_k) \cdots - y(t_k) \; u^{(n_b)}(t_k) \cdots u(t_k)] \quad (5.13)$$

$w^{(i)}(t_k)$ denotes the ith time derivative of the CT signal $w(t)$ at time instant t_k and

$$v_o(t_k) = A_o(p)H_o(p)e_o(t_k) \quad (5.14)$$

Note that the noise-free signals $x(t_k)$ and $\nu(t_k)$ in (5.10) are not available from measurements, therefore the several closed-loop methods presented in this chapter make use of this noisy regressor $\boldsymbol{\varphi}(t_k)$.

There are two main time-domain approaches to estimate a CT model in this form. The first, indirect approach, is to estimate an initial DT model from the sampled data and then convert this into a CT model. The second, direct approach, that we consider in the present chapter, is to identify a CT model directly from the DT data.

5.3 Basic Instrumental Variable Estimators

The process model parameters $\boldsymbol{\rho}$ can be estimated using a basic instrumental variable (IV) estimator. By assuming that the time derivatives of the input, output and external signals are available (see Section 5.5.2), the CT version of the basic IV estimate of $\boldsymbol{\rho}$ is given by

$$\hat{\boldsymbol{\rho}}_{iv} = sol\left\{\frac{1}{N}\sum_{k=1}^{N}\boldsymbol{\zeta}(t_k)[y^{(n_a)}(t_k) - \boldsymbol{\varphi}^T(t_k)\boldsymbol{\rho}] = 0\right\} \quad (5.15)$$

where N denotes the number of data and $\boldsymbol{\zeta}(t_k)$ is a vector of instrumental variables.

There is a considerable amount of freedom in the choice of the instruments. A first solution is to adapt the closed-loop IV method developed for DT models in [15] to the CT model identification case. This method is referred to as CLIVC and was first presented in [5]. It involves using the external signal time derivatives as instruments. The so-called basic IV estimate for closed-loop CT models is then given by

$$\hat{\boldsymbol{\rho}}_{\text{clivc}} = \left[\sum_{k=1}^{N}\boldsymbol{\zeta}(t_k)\boldsymbol{\varphi}^T(t_k)\right]^{-1}\sum_{k=1}^{N}\boldsymbol{\zeta}(t_k)y^{(n_a)}(t_k) \quad (5.16)$$

$$\text{with } \boldsymbol{\zeta}^T(t_k) = \left[r^{(n_a+n_b)}(t_k) \cdots r(t_k)\right] \in \mathbb{R}^{n_a+n_b+1} \quad (5.17)$$

In contrast with the basic IV for DT model identification that uses a difference equation model, the CT version makes use of an instrument built up from the time-derivatives of the external signals.

5.3.1 Consistency Properties

By inserting (5.12) into (5.15), the following equation is obtained

$$\hat{\boldsymbol{\rho}}_{iv} = \boldsymbol{\rho}_o + \left[\sum_{k=1}^{N}\boldsymbol{\zeta}(t_k)\boldsymbol{\varphi}^T(t_k)\right]^{-1}\left[\sum_{k=1}^{N}\boldsymbol{\zeta}(t_k)v_o(t_k)\right] \quad (5.18)$$

where $\boldsymbol{\varphi}^T(t_k)$ and $v_o(t_k)$ are given by (5.13) and (5.14), respectively. It can be deduced from (5.18) that $\hat{\boldsymbol{\rho}}_{iv}$ is a consistent estimate of $\boldsymbol{\rho}$ if[4]

$$\begin{cases} \bar{\mathsf{E}}[\boldsymbol{\zeta}(t_k)\boldsymbol{\varphi}^T(t_k)] \text{ is non-singular} \\ \bar{\mathsf{E}}[\boldsymbol{\zeta}(t_k)v_o(t_k)] = 0 \end{cases} \quad (5.19)$$

Several IV variants can be obtained by different choices of the instruments $\boldsymbol{\zeta}(t_k)$ in (5.15), respecting the conditions given by (5.19).

---

[4] The notation $\bar{\mathsf{E}}[.] = \lim_{N\to\infty}\frac{1}{N}\sum_{k=1}^{N}\mathsf{E}[.]$ is adopted from the prediction error framework of [9].

5.3.2 Accuracy Analysis

The asymptotic distribution of the parameter estimate $\hat{\boldsymbol{\rho}}_{iv}$ in (5.15) has been investigated extensively in the open-loop DT context (e.g., [9, 13, 14]). More recently, this work also has been extended to the closed-loop DT model identification framework [7]. By considering (5.15), these previous results can be applied to the case of the CT hybrid model given by (5.2). As a result, under the assumptions formulated in Section 5.2 and $G_o \in \mathcal{G}$, $\hat{\boldsymbol{\rho}}_{iv}$ is asymptotically Gaussian distributed

$$\sqrt{N}(\hat{\boldsymbol{\rho}}_{iv} - \boldsymbol{\rho}^*) \stackrel{\text{dist}}{\to} \mathcal{N}(0, \mathbf{P}_{iv}) \tag{5.20}$$

where $\boldsymbol{\rho}^*$ represents the limit of $\hat{\boldsymbol{\rho}}_{iv}$ when $N \to \infty$ and where the covariance matrix is given by

$$\mathbf{P}_{iv} = \sigma_{e_o}^2 \left[\bar{\mathbb{E}}\boldsymbol{\zeta}(t_k)\boldsymbol{\varphi}^T(t_k)\right]^{-1} \left[\bar{\mathbb{E}}\tilde{\boldsymbol{\zeta}}(t_k)\tilde{\boldsymbol{\zeta}}^T(t_k)\right] \left[\left(\bar{\mathbb{E}}\boldsymbol{\zeta}(t_k)\boldsymbol{\varphi}^T(t_k)\right)^{-1}\right]^T \tag{5.21}$$

with $\tilde{\boldsymbol{\zeta}}(t_k) = H_o(p)A_o(p)\boldsymbol{\zeta}(t_k)$ and $\sigma_{e_o}^2$ denotes the intensity of $\{e_o(t_k)\}$.

5.4 Extended Instrumental Variable Estimators

There are various ways of considering IV estimation from an optimal standpoint. One such approach is to consider an extended IV solution (see introduction section). In CT model identification, if the time-derivative signals are assumed to be known, the extended IV estimate of $\boldsymbol{\rho}$ is obtained by prefiltering the input–output data appearing in (5.15) and by generalising the basic IV estimates $\hat{\boldsymbol{\rho}}_{iv}$ using an augmented instrument $\boldsymbol{\zeta}(t_k) \in \mathbb{R}^{n_\zeta}$ ($n_\zeta \geq n_a + n_b + 1$) so that an overdetermined system of equations is obtained in the form,

$$\hat{\boldsymbol{\rho}}_{xiv} = \arg\min_{\boldsymbol{\rho}} \left\| \left[\sum_{k=1}^{N} f(p)\boldsymbol{\zeta}(t_k)f(p)\boldsymbol{\varphi}^T(t_k)\right]\boldsymbol{\rho} - \left[\sum_{k=1}^{N} f(p)\boldsymbol{\zeta}(t_k)f(p)y^{(n_a)}(t_k)\right] \right\|_W^2 \tag{5.22}$$

where $f(p)$ is a stable prefilter, and $\|x\|_W^2 = x^T W x$, with W a positive-definite weighting matrix. This extended IV gives a parameter estimator that requires more computations than the basic IV. However, the enlargement of the IV vector can be used for improving the accuracy of the parameter estimates [14]. Note that, when $f(p) = 1$ and $n_\zeta = n_a + n_b + 1$ ($W = I$), the basic IV estimate (5.15) is obtained.

5.4.1 Consistency Properties

The consistency conditions are easily obtained by generalising (5.19) to the estimator (5.22)

$$\begin{cases} \bar{\mathsf{E}}[f(p)\zeta(t_k)f(p)\varphi^T(t_k)] \text{ is full-column rank} \\ \bar{\mathsf{E}}[f(p)\zeta(t_k)f(p)v_o(t_k)] = 0 \end{cases} \tag{5.23}$$

5.4.2 Accuracy Analysis

The asymptotic distribution of parameter vector (5.22) is obtained by following the same reasoning as in Section 5.3.2. Therefore, by considering the results given in Section 5.2, under the assumption that $G_o \in \mathcal{G}$, $\hat{\boldsymbol{\rho}}_{xiv}$ is asymptotically Gaussian distributed,

$$\sqrt{N}(\hat{\boldsymbol{\rho}}_{xiv} - \boldsymbol{\rho}^*) \stackrel{\text{dist}}{\to} \mathcal{N}(0, \mathbf{P}_{xiv}) \tag{5.24}$$

where the covariance matrix is given by

$$\mathbf{P}_{xiv} = \sigma_{e_{of}}^2 \left[\mathbf{R}^T\mathbf{W}\mathbf{R}\right]^{-1} \mathbf{R}^T\mathbf{W} \left[\bar{\mathsf{E}}\tilde{\boldsymbol{\zeta}}(t_k)\tilde{\boldsymbol{\zeta}}^T(t_k)\right] \mathbf{W}\mathbf{R} \left[\mathbf{R}^T\mathbf{W}\mathbf{R}\right]^{-1}$$

with

$$\tilde{\boldsymbol{\zeta}}(t_k) = f(p)H_o(p)\Lambda_o(p)\boldsymbol{\zeta}(t_k) \text{ and } \mathbf{R} = \bar{\mathsf{E}}f(p)\boldsymbol{\zeta}(t_k)f(p)\mathring{\boldsymbol{\varphi}}^T(t_k)$$

where $\mathring{\varphi}(t_k)$ is the noise-free part of the regressor $\varphi(t_k)$ (5.13), built up from the noise-free input–output signals $\nu(t_k)$ and $x(t_k)$ (5.10) as

$$\mathring{\varphi}^T(t_k) = [-x^{(n_a-1)}(t_k) \cdots - x(t_k) \; \nu^{(n_b)}(t_k) \cdots \nu(t_k)] \tag{5.25}$$

Note that the noise-free part of the regressor is partly defined by the noise-free output variable $x(t_k)$ in (5.10) and its derivatives. It is well known in open-loop estimation that an estimate of this variable, generated as the output of an 'auxiliary model', is normally used as the prime source of the instrumental variable for the output variable. In the closed-loop context, however, the measured regression vector also contains the filtered process input and its derivatives, it is clear, therefore, that a suitable estimate of the noise-free process input $\nu(t_k)$ will also be required for accurate IV estimation.

5.5 Optimal Instrumental Variable Estimators

5.5.1 Main Results

The choice of the instruments $\boldsymbol{\zeta}(t)$, the number of IV components n_ζ, the weighting matrix W and the prefilter $f(p)$ may have a considerable effect on the covariance matrix $\mathbf{P}_{xiv}$. In the open-loop DT situation, the lower bound of $\mathbf{P}_{xiv}$ for any unbiased identification method is given by the Cramér–Rao lower bound [9,13]. Optimal choices of the above-mentioned design variables exist so that $\mathbf{P}_{xiv}$ reaches the Cramér–Rao lower bound. These results cannot be applied to the closed-loop IV case because of the correlation between the process input signal $u(t_k)$ and the noise. In this regard, it has been shown in [15] that, for a model given by (5.12), there exists a minimum value of the covariance matrix $\mathbf{P}_{xiv}$ as a function of the design variables $\zeta(t_k)$, $f(p)$ and W, under the restriction that $\zeta(t_k)$ is a function of the external signal $r(t_k)$ only. Although these results have been obtained for the case of DT models, a similar analysis applies in the CT case and the covariance matrix can be optimised with respect to the design variables. The optimal covariance matrix (different from the Cramér–Rao lower bound) for a data-generating closed-loop system given by (5.2), where $u(t_k)$ and $y(t_k)$ are correlated by noise, is then

$$\mathbf{P}_{xiv} \geq \mathbf{P}_{xiv}^{\text{opt}} \text{ and}$$

$$\mathbf{P}_{xiv}^{\text{opt}} = \sigma_{e_{o^{\text{opt}}}}^2 \left\{ \bar{\mathsf{E}} \left[[A_o(p)H_o(p)]^{-1} \mathring{\varphi}^T(t_k) \right]^T \left[[A_o(p)H_o(p)]^{-1} \mathring{\varphi}^T(t_k) \right] \right\}^{-1}$$
(5.26)

$\mathbf{P}_{xiv}^{\text{opt}}$ is then obtained by taking,

$$\begin{aligned}
\zeta(t_k) &= f^{\text{opt}}(p)\mathring{\varphi}(t_k), \\
f^{\text{opt}}(p) &= [A_o(p)H_o(p)]^{-1}, \\
n_\zeta &= n_a + n_b + 1, \\
W &= I
\end{aligned}$$
(5.27)

Therefore, the only difference between open-loop and closed-loop cases is that in the latter, the input process signal is correlated with the noise, so that the instruments must be correlated with the noise-free part of $u(t)$ but uncorrelated with the noisy part of $u(t)$ (due to the feedback loop).

Moreover, when defined in this manner, it would appear that the optimal IV estimator can only be obtained if, first, the true noise (and process) model is exactly known and secondly the noise-free part of the regressor is available. However, this is a probabilistic estimation problem and therefore the statistically optimal estimates can be obtained if these TF model polynomials are replaced by their optimal estimates. Moreover, practically useful suboptimal solutions can be obtained by utilising good, if not optimal, estimates. This is discussed in the next subsections.

5.5.2 Implementation Issues

Handling of the Unmeasurable Time-derivative Signals

In comparison with the DT counterpart, direct CT model identification raises several technical issues. The first is related to implementation. Unlike the difference equation model, the differential equation model is not a linear combination of the sampled process input and output signals but contains time-derivative signals. The theoretical study presented in the previous section assumes that these time-derivative signals are available, and therefore, the parameter estimation procedure can be directly applied on them. However, these input, output and external time-derivative signals are not available as measurement data in most practical cases. A standard approach used in CT model identification is to introduce a low-pass stable filter $f_c(p)$, *i.e.*, define

$$y_{f_c}(t_k) = f_c(p)y(t_k), \quad u_{f_c}(t_k) = f_c(p)u(t_k) \quad (5.28)$$

where the subscript f_c is used to denote the prefiltered forms of the associated variables. The filtered time derivatives can then be obtained by sending both input–output signals to a bench of filters of the form $f_c(p)p^i$

$$y_{f_c}^{(i)}(t_k) = f_c(p)p^i y(t_k), \quad i \leq n_a \quad (5.29)$$

$$u_{f_c}^{(i)}(t_k) = f_c(p)p^i u(t_k), \quad i \leq n_b \quad (5.30)$$

The motivation is that the filtered signals $u_{f_c}(t)$ and y_{f_c} satisfy

$$y_{f_c}(t_k) = G_o(p)u_{f_c}(t_k) + f_c(p)H_o(p)e_o(t_k) \quad (5.31)$$

i.e., the process transfer function is not changed but the noise transfer function is modified by the introduction of the filter. Equation (5.31) can be rewritten under the following linear regression form

$$y_{f_c}^{(n_a)}(t_k) = \varphi_{f_c}^T(t_k)\rho_o + v_{of_c}(t_k) \quad (5.32)$$

with

$$\varphi_{f_c}^T(t_k) = [-y_{f_c}^{(n_a-1)}(t_k) \cdots - y_{f_c}(t_k) \; u_{f_c}^{(n_b)}(t_k) \cdots u_{f_c f_c}(t_k)] \quad (5.33)$$

$$v_{of_c}(t_k) = f_c(p)v_o(t_k) \quad (5.34)$$

Various types of CT filters have been devised to deal with the need to reconstruct the time derivatives [4] and the continuous-time system identification (CONTSID) toolbox has been developed on the basis of these methods (see Chapter 9 in the present book). A usual filter that has been used in simple IV methods is as follows

$$f_c(p) = \left(\frac{\beta}{p+\lambda}\right)^{n_a} \quad (5.35)$$

which is the filter used in the case of the minimum-order multiple filter (also referred to as state-variable filter) method. Note however that other filters can be used [4]. Moreover, it is clearly possible to select the prefilter $f_c(p)$ in order to achieve some form of optimal IV estimation and this is considered later in Section 5.5.

For simplicity, it has been assumed above that the differential equation model (5.2) is initially at rest. However, note that, in the general case, the initial condition terms do not vanish in (5.32). Whether they require estimation or they can be neglected depends upon the selected signal-prefiltering method.

Noise Modelling and Hybrid Filtering

The choice of the instruments and prefilter in the IV method affects the asymptotic variance, while consistency properties are generically secured by the IV implementation. It has been found that minor deviations from the optimal estimates of the polynomials required for the implementation of the auxiliary model and prefilters will normally only cause second-order effects in the resulting accuracy. Therefore, a reasonable IV estimator can be obtained if consistent, but not necessarily efficient estimates of the polynomials are utilised (see [9] for a discussion in the DT case). In addition, the computational procedures can be kept simple and tractable if linear regression estimates are used in the preliminary stages of the estimation.

Several bootstrap IV methods have been proposed, in an attempt to approximate the optimal IV method (see *e.g.*, [9, 13, 21] for the open-loop situation and [7] for the closed-loop one). As explained in Section 5.5, the difference between open-loop and closed-loop cases lies in the input process signal that is correlated with the noise in the latter. Therefore, the instrumental variable vector must include IVs associated with the input as well as the output signal, and these must be correlated with the noise-free part of $u(t)$ but uncorrelated with the noise on $u(t)$ arising from the feedback loop.

Following the discussion in Section 5.2, CT models are estimated to represent the transfer between the external signal and the output, as well as for the transfer between the external signal and the input. And according to the hybrid model (5.2) we are using here, DT models are used to estimate the noise contribution.

From (5.27) and Section 5.5.2, the IV filter involves a filter $f(p)$ required for handling the time derivatives along with the CT process TF denominator polynomial and noise model contributions. As a result, the IV estimation will require hybrid filtering operations involving:

- a CT filter $f(p) = f_c(p)$ needed to compute the time derivatives (see (5.28) and also Chapter 4 in the present book);
- a DT filtering operation needed to approximate the inverse of the CT process TF denominator polynomial and noise model contributions (see (5.27)), denoted from hereafter as $f_d(q^{-1}, \boldsymbol{\eta})$.

To realise the optimal choices for the instruments, two alternatives are developed in the following sections: the first relies on multi-step algorithms while the second is based on iterative (adaptive) solutions. As we will see, the form of the CT and DT filters will differ according to the assumed true CT system model structures.

5.5.3 Multi-step Approximate Implementations of the Optimal IV Estimate

Two-step CLIVC2 algorithm

The two-step IV algorithm, denoted as CLIVC2, is based on the following CT ARX model

$$\begin{cases} A_o(p)y(t_k) = B_o(p)u(t_k) + e_o(t_k) \\ \text{with } u(t_k) = r(t_k) - C_c(p)y(t_k) \end{cases} \qquad (5.36)$$

or its filtered version

$$\begin{cases} y_{f_c}(t_k) = \frac{B_o(p)}{A_o(p)} u_{f_c}(t_k) + f_c(p) \frac{1}{A_o(p)} e(t_k) \\ \text{with } u_{f_c}(t_k) = r_{f_c}(t_k) - C_c(p) y_{f_c}(t_k) \end{cases} \qquad (5.37)$$

where we see that the noise model is constrained to include the process TF denominator polynomial $A_o(p)$.
In this particular case, the approximate optimal filter f_{clivc2} is composed of:

- the CT filter $f_c(p)$ (see (5.35)). It could also be chosen amongst several options [17];
- the DT filter $f_d(q^{-1}) = 1$ since the noise model of the assumed CT ARX data-generating system is $H_o(p) = 1/A_o(p)$.

Since the CT filter $f_c(p)$ can be chosen amongst several non-optimal filters, the resulting CLIVC2 algorithm is an approximate implementation of the optimal IV solution presented in Section 5.5.1.

The outline of the CLIVC2 algorithm is then the following

1. Choose a CT prefilter $f_c(p)$ to compute $y_{f_c}^{(i)}(t_k)$, $u_{f_c}^{(i)}(t_k)$ and $r_{f_c}^{(i)}(t_k)$, for $i \leq n_a$.
   Write the filtered CT ARX model structure as a linear regression

   $$y_{f_c}^{(n_a)}(t_k) = \varphi_{f_c}^T(t_k)\rho \qquad (5.38)$$

   and obtain an estimate $\hat{\rho}_1$ of ρ by the least squares method.
2. Use this estimate $\hat{\rho}_1$ along with the process model, as defined by,

   $$G(p, \hat{\rho}_1) = \frac{B(p, \hat{\rho}_1)}{A(p, \hat{\rho}_1)}$$

   to generate the instruments $\zeta_{f_c}(t_k, \hat{\rho}_1)$ using the following closed-loop auxiliary models

5 Optimal IV Methods for Closed-loop CT Model Identification   145

$$\hat{x}_{f_c}(t_k, \hat{\boldsymbol{\rho}}_1) = \frac{G(p, \hat{\boldsymbol{\rho}}_1)}{1 + C_c(p)G(p, \hat{\boldsymbol{\rho}}_1)} r_{f_c}(t_k) \tag{5.39}$$

$$\hat{\nu}_{f_c}(t_k, \hat{\boldsymbol{\rho}}_1) = \frac{1}{1 + C_c(p)G(p, \hat{\boldsymbol{\rho}}_1)} r_{f_c}(t_k) \tag{5.40}$$

$$\boldsymbol{\zeta}_{f_c}(t_k, \hat{\boldsymbol{\rho}}_1) = [-\hat{x}_{f_c}^{(n_a-1)}(t_k, \hat{\boldsymbol{\rho}}_1) \cdots - \hat{x}_{f_c}(t_k, \hat{\boldsymbol{\rho}}_1)$$
$$\hat{\nu}_{f_c}^{(n_b)}(t_k, \hat{\boldsymbol{\rho}}_1) \cdots \hat{\nu}_{f_c}(t_k, \hat{\boldsymbol{\rho}}_1)]^T \tag{5.41}$$

$\boldsymbol{\zeta}_{f_c}(t_k, \hat{\boldsymbol{\rho}}_1)$ represents an estimate of the noise-free part of the regressor $\boldsymbol{\varphi}_{f_c}(t_k)$ and according to the notations used in Chapter 4, it will be denoted from hereafter as $\boldsymbol{\zeta}_{f_c}(t_k, \hat{\boldsymbol{\rho}}_1) = \hat{\boldsymbol{\varphi}}_{f_c}(t_k, \hat{\boldsymbol{\rho}}_1)$.
Using the instrument $\hat{\boldsymbol{\varphi}}_{f_c}(t_k, \hat{\boldsymbol{\rho}}_1)$ and the prefilter $f_d(q^{-1}, \boldsymbol{\eta}) = 1$, determine the IV estimate in (5.38) as

$$\hat{\boldsymbol{\rho}}_{\text{clivc2}} = \left[\sum_{k=1}^{N} \hat{\boldsymbol{\varphi}}_{f_{\text{clivc2}}}(t_k, \hat{\boldsymbol{\rho}}_1) \boldsymbol{\varphi}_{f_{\text{clivc2}}}^T(t_k)\right]^{-1} \left[\sum_{k=1}^{N} \hat{\boldsymbol{\varphi}}_{f_{\text{clivc2}}}(t_k, \hat{\boldsymbol{\rho}}_1) y_{f_{\text{clivc2}}}^{(n_a)}(t_k)\right] \tag{5.42}$$

where $f_{\text{clivc2}}(p) = f_c(p)$ here.

Remark 5.1. In contrast to the discrete-time case, a high-order least squares estimator should not be used in the first step of the continuous-time system identification procedure because of the numerical errors induced by the simulation method required for the generation of the filtered variables in (5.33) and (5.41).

Four-step CLIVC4 algorithm

Although the process parameter estimates from the CLIVC2 algorithm are consistent, it is worthwhile considering improved noise model estimation in order to construct an estimator with a smaller variance (closer to the optimal solution). One improvement is to assume the following CT ARARX model structure

$$\begin{cases} A_o(p)y(t_k) = B_o(p)u(t_k) + \frac{1}{D_o(p)} e_o(t_k) \\ \text{with } u(t_k) = r(t_k) - C_c(p)y(t_k) \end{cases} \tag{5.43}$$

or its filtered version

$$\begin{cases} y_{f_c}(t_k) = \frac{B_o(p)}{A_o(p)} u_{f_c}(t_k) + f_c(p) \frac{1}{A_o(p)D_o(p)} e_o(t_k) \\ \text{with } u_{f_c}(t_k) = r_{f_c}(t_k) - C_c(p)y_{f_c}(t_k) \end{cases} \tag{5.44}$$

where we see that the noise model is also constrained to include the TF denominator polynomial $A_o(p)$.
In this particular case, the approximate optimal filter f_{clivc4} is composed of:

- the CT filter $f_c(p)$ (see (5.35)). It could also be chosen amongst several options [17]; as, e.g., (5.35);
- the DT filter $f_d(q^{-1}, \boldsymbol{\eta}) = 1/D(q^{-1}, \boldsymbol{\eta})$ (AR model of order to be chosen or identified) since the noise model of the assumed CT ARARX data-generating CT system is $H_o(p) = 1/A_o(p)D_o(p)$.

As a result, the proposed CLIVC4 algorithm is then based on the following CT hybrid ARARX model structure [7]

$$\begin{cases} A(p,\boldsymbol{\rho})y(t_k) = B(p,\boldsymbol{\rho})u(t_k) + \dfrac{1}{D(q^{-1},\boldsymbol{\eta})}e(t_k) \\ \text{with } u(t_k) = r(t_k) - C_c(p)y(t_k) \end{cases} \quad (5.45)$$

Note that in the above equation, we are mixing discrete and continuous-time operators somewhat informally in order to indicate the hybrid computational nature of the estimation problem being considered here. Thus, operations such as,

$$\frac{B(p,\boldsymbol{\rho})}{A(p,\boldsymbol{\rho})}u(t_k)$$

imply that the input variable $u(t_k)$ is interpolated in some manner. This is to allow for the inter-sample behaviour that is not available from the sampled data and so has to be inferred in order to allow for the continuous-time numerical integration of the associated differential equations. For such integration, the discretisation interval will be varied, dependent on the numerical method employed, but it will usually be much smaller than the sampling interval T_s (see Chapter 4 in the present book).

This proposed solution may be seen as an extension of the four-step IV technique for open-loop DT model identification (IV4) [9] to the CT hybrid closed-loop framework. The difference between the two algorithms is that in the CT version, a filter is needed to handle the time-derivatives problem. As previously, since it is carried out by a CT filter $f_c(p)$ chosen amongst several non-optimal filters, the resulting CLIVC4 algorithm is an approximate implementation of the optimal IV solution presented in Section 5.5.1.

The outline of the CLIVC4 algorithm is as follows:

1. Choose a CT prefilter $f_c(p)$ to compute $y_{f_c}^{(i)}(t_k)$, $u_{f_c}^{(i)}(t_k)$ and $r_{f_c}^{(i)}(t_k)$, for $i \leq n_a$.
   Write the filtered model structure as a linear regression

   $$y_{f_c}^{(n_a)}(t_k) = \boldsymbol{\varphi}_{f_c}^T(t_k)\boldsymbol{\rho} \quad (5.46)$$

   Obtain an estimate $\hat{\boldsymbol{\rho}}_1$ of $\boldsymbol{\rho}$ by the least squares method and use this to define the corresponding CT transfer function $G(p, \hat{\boldsymbol{\rho}}_1)$.

2. Generate the instruments $\zeta_{f_c}(t_k, \hat{\rho}_1) = \hat{\varphi}_{f_c}(t_k, \hat{\rho}_1)$ using the closed-loop auxiliary models as in (5.41). $\hat{\varphi}_{f_c}(t_k, \hat{\rho}_1)$ represents an estimate of the noise-free part of the filtered regressor $\varphi_{f_c}(t_k)$. Determine the IV estimate of ρ in (5.46) as

$$\hat{\rho}_2 = \left[\sum_{k=1}^{N} \hat{\varphi}_{f_c}(t_k, \hat{\rho}_1)\varphi_{f_c}^T(t_k)\right]^{-1} \left[\sum_{k=1}^{N} \hat{\varphi}_{f_c}(t_k, \hat{\rho}_1)y_{f_c}^{(n_a)}(t_k)\right] \quad (5.47)$$

and use this to define the corresponding CT transfer function $G(p, \hat{\rho}_2)$.

3. Let $\hat{w}(t_k) = y_{f_c}^{(n_a)}(t_k) - \varphi_{f_c}^T(t_k)\hat{\rho}_2$. Now, an AR model[5] of order $2n_a$ can be postulated for $\hat{w}(t_k)$:

$$f_d(q^{-1}, \hat{\eta})\hat{w}(t_k) = e(t_k)$$

and then $f_d(q^{-1}, \hat{\eta})$ can be estimated using the least squares method.

4. Generate the instruments $\zeta_{f_c}(t_k, \hat{\rho}_2) = \hat{\varphi}_{f_c}(t_k, \hat{\rho}_2)$ as

$$\hat{\varphi}_{f_c}(t_k, \hat{\rho}_2) = [-\hat{x}_{f_c}^{(n_a-1)}(t_k, \hat{\rho}_2) \cdots - \hat{x}_{f_c}(t_k, \hat{\rho}_2)$$
$$\hat{\nu}_{f_c f_c}^{(n_b)}(t_k, \hat{\rho}_2) \cdots \hat{\nu}_{f_c}(t_k, \hat{\rho}_2)]^T \quad (5.48)$$

where $\hat{x}_{f_c}(t_k, \hat{\rho}_2)$ and $\hat{\nu}_{f_c}(t_k, \hat{\rho}_2)$ are the estimated noise-free output of the closed-loop auxiliary models computed as in (5.39) and (5.40) on the basis of $G(p, \hat{\rho}_2)$.

Using these instruments $\hat{\varphi}_{f_c}(t_k, \hat{\rho}_2)$ and the prefilter $f_d(q^{-1}, \hat{\eta})$, determine the IV estimate of ρ in (5.46) as

$$\hat{\rho}_{\text{clivc4}} = \left[\sum_{k=1}^{N} \hat{\varphi}_{f_{\text{clivc4}}}(t_k, \hat{\rho}_2)\varphi_{f_{\text{clivc4}}}^T(t_k)\right]^{-1} \left[\sum_{k=1}^{N} \hat{\varphi}_{f_{\text{clivc4}}}(t_k, \hat{\rho}_2)y_{f_{\text{clivc4}}}^{(n_a)}(t_k)\right]$$
(5.49)

where

$$\hat{\varphi}_{f_{\text{clivc4}}}(t_k, \hat{\rho}_2) = f_d(q^{-1}, \hat{\eta})\hat{\varphi}_{f_c}(t_k, \hat{\rho}_2), \quad (5.50)$$

$$\varphi_{f_{\text{clivc4}}}(t_k) = f_d(q^{-1}, \hat{\eta})\varphi_{f_c}(t_k), \quad (5.51)$$

and $\quad y_{f_{\text{clivc4}}}^{(n_a)}(t_k) = f_d(q^{-1}, \hat{\eta})y_{f_c}^{(n_a)}(t_k) \quad (5.52)$

5.5.4 Iterative Implementations of the Optimal IV Estimate

In the previous algorithms, the filter $f_c(p)$ used to compute the time-derivative signals, is fixed *a priori* by the user and is not included into the design variables of the method. Furthermore, the CLIVC2 and CLIVC4 approaches rely

---

[5] Or the AR order can be identified using a model order identification method, such as the Akaike information criterion (AIC).

on a noise model that is constrained to include the process TF denominator polynomial (see (5.43)).

An alternative approach is to consider instead, a CT Box–Jenkins (BJ) transfer function (TF) model defined as follows

$$\begin{cases} y(t_k) = \dfrac{B_o(p)}{A_o(p)} u(t_k) + \dfrac{C_o(p)}{D_o(p)} e_o(t_k) \\ \text{with } u(t_k) = r(t_k) - C_c(p) y(t_k) \end{cases} \quad (5.53)$$

For most practical purposes, this model is the most natural one to use since it does not constrain the process and the noise models to have common denominator polynomials. It also has the advantage that the maximum likelihood estimates of the process model parameters are asymptotically independent of the noise model parameter estimates (see Chapter 4 in this book and [10]). The problem introduced by considering (5.53), however, is that the model is non-linear-in-the-parameters so that simple IV estimation cannot be directly applied.

Fortunately, this problem of non-linear estimation can be overcome by designing an iterative estimation algorithm on the basis of the procedures used in the refined instrumental variable (RIV) algorithm [8, 21, 22, 25] and its CT equivalent, the refined instrumental variable for a continuous system (RIVC) algorithm [26], as discussed fully in Chapter 4, suitably extended to handle the closed-loop identification case.

Following the usual *prediction error minimisation* (PEM) approach in the present hybrid situation (which is ML estimation because of the Gaussian assumptions on $e(t_k)$), a suitable error function $\varepsilon(t_k)$, at the kth sampling instant, is given by

$$\varepsilon(t_k) = \dfrac{D_o(p)}{C_o(p)} \left\{ y(t_k) - \dfrac{B_o(p)}{A_o(p)} u(t_k) \right\}$$

which can be written as

$$\varepsilon(t_k) = \dfrac{D_o(p)}{C_o(p)} \left\{ \dfrac{1}{A_o(p)} [A_o(p) y(t_k) - B_o(p) u(t_k)] \right\} \quad (5.54)$$

where the CT prefilter $D_o(p)/C_o(p)$ will be recognised as the inverse of the continuous-time autoregressive moving average (CARMA) noise model in (5.53).

Minimisation of a least squares criterion function in $\varepsilon(t_k)$, measured at the sampling instants, provides the basis for stochastic estimation. However, since the polynomial operators commute in this linear case, (5.54) can be considered in the alternative form,

$$\varepsilon(t_k) = A_o(p) y_\mathrm{f}(t_k) - B_o(p) u_\mathrm{f}(t_k) \quad (5.55)$$

where $y_f(t_k)$ and $u_f(t_k)$ represent the *sampled* outputs of the complete CT prefiltering operation

$$y_f(t_k) = \frac{1}{A_o(p)} \frac{D_o(p)}{C_o(p)} y(t_k), \quad (5.56)$$

$$u_f(t_k) = \frac{1}{A_o(p)} \frac{D_o(p)}{C_o(p)} u(t_k) \quad (5.57)$$

In this particular case, the optimal filter f_{clrivc} is composed of:

- $f_c(p, \boldsymbol{\rho}) = 1/A(p, \boldsymbol{\rho})$ that is used to generate the time derivatives;
- $f_d(q^{-1}, \boldsymbol{\eta}) = D(q^{-1}, \boldsymbol{\eta})/C(q^{-1}, \boldsymbol{\eta})$ since the noise model of the assumed CT BJ data-generating system is $H_o(p) = C_o(p)/D_o(p)$.

As a result, the proposed CLRIVC algorithm is then based on the following CT hybrid Box–Jenkins model structure

$$\begin{cases} y(t_k) = \dfrac{B(p, \boldsymbol{\rho})}{A(p, \boldsymbol{\rho})} u(t_k) + \dfrac{C(q^{-1}, \boldsymbol{\eta})}{D(q^{-1}, \boldsymbol{\eta})} e(t_k) \\ \text{with } u(t_k) = r(t_k) - C_c(p) y(t_k) \end{cases} \quad (5.58)$$

It involves an iterative (or relaxation) algorithm in which, at each iteration, the auxiliary model (see Section 5.4.2) used to generate the instrumental variables, as well as the associated prefilters, are updated, based on the parameter estimates obtained at the previous iteration.

Iterative CLRIVC Algorithm

The outline of the CLRIVC algorithm is as follows:

1. Set $C(q^{-1}, \hat{\boldsymbol{\eta}}^0) = D(q^{-1}, \hat{\boldsymbol{\eta}}^0) = 1$. Choose an initial CT prefilter $f_c(p)$ to compute $y_{f_c}^{(i)}(t_k)$, $u_{f_c}^{(i)}(t_k)$ and $r_{f_c}^{(i)}(t_k)$, for $i \leq n_a$.
   From the linear model structure (5.55), generate an initial estimate $\hat{\boldsymbol{\rho}}^0$ of $\boldsymbol{\rho}$ using, e.g., the CLSRIVC algorithm (see next section): the corresponding TF is denoted by $G(p, \hat{\boldsymbol{\rho}}^0)$. Use this initial estimate to define the CT pre-filter $f_c(p, \hat{\boldsymbol{\rho}}^0) = 1/A(p, \hat{\boldsymbol{\rho}}^0)$, and set $j = 1$.
2. Iterative estimation.
   for $j = 1 : convergence$
   a) Generate the filtered instrumental variables $\boldsymbol{\zeta}_{f_c}(t_k, \hat{\boldsymbol{\rho}}^{j-1}) = \hat{\boldsymbol{\varphi}}_{f_c}(t_k, \hat{\boldsymbol{\rho}}^{j-1})$ from the estimates of the noise-free input and output variables using the following closed-loop auxiliary models

$$\hat{x}_{f_c}(t_k, \hat{\boldsymbol{\rho}}^{j-1}) = \frac{G(p, \hat{\boldsymbol{\rho}}^{j-1})}{1 + C_c(p)G(p, \hat{\boldsymbol{\rho}}^{j-1})} r_{f_c}(t_k) \qquad (5.59)$$

$$\hat{\nu}_{f_c}(t_k, \hat{\boldsymbol{\rho}}^{j-1}) = \frac{1}{1 + C_c(p)G(p, \hat{\boldsymbol{\rho}}^{j-1})} r_{f_c}(t_k) \qquad (5.60)$$

$$\hat{\varphi}_{f_c}(t_k, \hat{\boldsymbol{\rho}}^{j-1}) = [-\hat{x}_{f_c}^{(n_a-1)}(t_k, \hat{\boldsymbol{\rho}}^{j-1}) \cdots -\hat{x}_{f_c}(t_k, \hat{\boldsymbol{\rho}}^{j-1})$$
$$\hat{\nu}_{f_c}^{(n_b)}(t_k, \hat{\boldsymbol{\rho}}^{j-1}) \cdots \hat{\nu}_{f_c}(t_k, \hat{\boldsymbol{\rho}}^{j-1})]^T \qquad (5.61)$$

where the CT filter is given as

$$f_c(p, \hat{\boldsymbol{\rho}}^{j-1}) = \frac{1}{A(p, \hat{\boldsymbol{\rho}}^{j-1})}.$$

Use this filter to compute $y_{f_c}^{(i)}(t_k, \hat{\boldsymbol{\rho}}^{j-1})$ and $u_{f_c}^{(i)}(t_k, \hat{\boldsymbol{\rho}}^{j-1})$, for $i \leq n_a$ and update the filtered regression filter

$$\varphi_{f_c}(t_k, \hat{\boldsymbol{\rho}}^{j-1}) = [-y_{f_c}^{(n_a-1)}(t_k, \hat{\boldsymbol{\rho}}^{j-1}) \cdots -y_{f_c}(t_k, \hat{\boldsymbol{\rho}}^{j-1})$$
$$u_{f_c}^{(n_b)}(t_k, \hat{\boldsymbol{\rho}}^{j-1}) \cdots u_{f_c}(t_k, \hat{\boldsymbol{\rho}}^{j-1})]^T \qquad (5.62)$$

b) Obtain an optimal estimate of the noise model parameter vector $\boldsymbol{\eta}^j$ based on the estimated noise sequence

$$\hat{\xi}(t_k) = y(t_k) - \hat{x}(t_k, \hat{\boldsymbol{\rho}}^{j-1}) \qquad (5.63)$$

using a selected ARMA estimation algorithm and use this to define the corresponding TF: $H(q^{-1}, \hat{\boldsymbol{\eta}}^j)$.

c) Use the estimated noise model parameters in $\hat{\boldsymbol{\eta}}^j$ to define the DT filter $f_d(q^{-1}, \hat{\boldsymbol{\eta}}^j)$, which takes the form

$$f_d(q^{-1}, \hat{\boldsymbol{\eta}}^j) = \frac{D(q^{-1}, \hat{\boldsymbol{\eta}}^j)}{C(q^{-1}, \hat{\boldsymbol{\eta}}^j)}$$

Then, sample the filtered derivative signals at the discrete-time sampling interval T_s and prefilter these by the discrete-time filter $f_d(q^{-1}, \hat{\boldsymbol{\eta}}^j)$.

d) Based on these prefiltered data, generate an updated estimate $\hat{\boldsymbol{\rho}}^j$ of the process model parameter vector as

$$\hat{\boldsymbol{\rho}}^j = \left[\sum_{k=1}^{N} \hat{\varphi}_{f_{clrivc}}(t_k, \hat{\boldsymbol{\rho}}^{j-1}) \varphi_{f_{clrivc}}^T(t_k, \hat{\boldsymbol{\rho}}^{j-1}) \right]^{-1}$$
$$\left[\sum_{k=1}^{N} \hat{\varphi}_{f_{clrivc}}(t_k, \hat{\boldsymbol{\rho}}^{j-1}) y_{f_{clrivc}}^{(n_a)}(t_k, \hat{\boldsymbol{\rho}}^{j-1}) \right] \qquad (5.64)$$

where

$$\hat{\varphi}_{\text{f}_{\text{clrive}}}(t_k, \hat{\boldsymbol{\rho}}^{j-1}) = f_{\text{d}}(q^{-1}, \hat{\boldsymbol{\eta}}^j)\hat{\varphi}_{\text{f}_{\text{c}}}(t_k, \hat{\boldsymbol{\rho}}^{j-1}), \tag{5.65}$$

$$\varphi_{\text{f}_{\text{clrive}}}(t_k, \hat{\boldsymbol{\rho}}^{j-1}) = f_{\text{d}}(q^{-1}, \hat{\boldsymbol{\eta}}^j)\varphi_{\text{f}_{\text{c}}}(t_k, \hat{\boldsymbol{\rho}}^{j-1}), \tag{5.66}$$

$$y_{\text{f}_{\text{clrive}}}^{(n_a)}(t_k, \hat{\boldsymbol{\rho}}^{j-1}) = f_{\text{d}}(q^{-1}, \hat{\boldsymbol{\eta}}^j)y_{\text{f}_{\text{c}}}^{(n_a)}(t_k, \hat{\boldsymbol{\rho}}^{j-1}) \tag{5.67}$$

Together with the estimate $\hat{\boldsymbol{\eta}}^j$ of the noise model parameter estimate from step (2b), this provides the estimate $\hat{\boldsymbol{\theta}}^j$ of the composite parameter vector at the jth iteration.

3. After the convergence of the iterations is complete, compute the estimated parametric error covariance matrix $\hat{\mathbf{P}}_\rho$, associated with the converged estimate $\hat{\boldsymbol{\rho}}$ of the system model parameter vector, from the expression (see Chapter 4 in this book),

$$\hat{\mathbf{P}}_\rho = \hat{\sigma}_e^2 \left[\sum_{k=1}^N \hat{\varphi}_{\text{f}_{\text{clrive}}}(t_k, \hat{\boldsymbol{\rho}})\hat{\varphi}_{\text{f}_{\text{clrive}}}^T(t_k, \hat{\boldsymbol{\rho}}) \right]^{-1} \tag{5.68}$$

where $\hat{\varphi}_{\text{f}_{\text{clrive}}}(t_k, \hat{\boldsymbol{\rho}})$ is the IV vector obtained at convergence and $\hat{\sigma}_e^2$ is the estimated residual variance.

Simplified Iterative CLSRIVC Algorithm

It will be noted that the above formulation of the CLRIVC estimation problem is considerably simplified if it is assumed in the CT BJ model structure that the additive noise is white, i.e., $C_o(p) = D_o(p) = 1$. In this case, the assumed model structure is a CT hybrid OE model given as

$$\begin{cases} y(t_k) = \dfrac{B(p, \boldsymbol{\rho})}{A(p, \boldsymbol{\rho})} u(t_k) + e(t_k) \\ \text{with } u(t_k) = r(t_k) - C_{\text{c}}(p) y(t_k) \end{cases} \tag{5.69}$$

The simplified CLRIVC (denoted as CLSRIVC) algorithm may be used here; the estimation only involves the parameters in the $A(p, \boldsymbol{\rho})$ and $B(p, \boldsymbol{\rho})$ polynomials and the optimal filter f_{clsrivc} involves:

- the CT filter $f_{\text{c}}(p, \boldsymbol{\rho}) = 1/A(p, \boldsymbol{\rho})$;
- the DT filter $f_{\text{d}}(q^{-1}, \boldsymbol{\eta}) = 1$ since the noise model of the associated CT OE data-generating system is $H_o(p) = 1$.

Consequently, the main steps in the CLSRIVC algorithm are the same as those in the CLRIVC algorithm, except that the noise model estimation and subsequent discrete-time prefiltering in steps (2b) and (2c) of the iterative procedure are no longer required and are omitted.

Comments

1. Note that the IV vector used in (5.64) should be written as

$$\hat{\varphi}_{f_{\text{clrivc}}}(t_k, \hat{\boldsymbol{\rho}}^{j-1}) = \hat{\varphi}_{f_{\text{clrivc}}}(t_k, \hat{\boldsymbol{\rho}}^{j-1}, \hat{\boldsymbol{\eta}}^j) \tag{5.70}$$

   because the instrumental variables are prefiltered and therefore are a function of both the system parameter estimates at the previous iteration and the most recent noise model parameter estimates (see algorithm). For simplicity, however, these additional arguments are omitted in the algorithm.

2. The fact that the ARMA noise model estimation is carried out separately on the basis of the estimated noise signal $\hat{\xi}(t_k)$ obtained from the IV part of the estimation algorithm in (5.63), implies that the system and noise model parameters are statistically independent (see Chapter 4 for a thorough analysis).

3. The initial selection of $A(p, \hat{\boldsymbol{\rho}}^0)$ does not have to be particularly accurate provided the prefilter $f_c(p, \hat{\boldsymbol{\rho}}^0)$ based on it does not seriously attenuate any signals within the passband of the system being modelled (see Chapter 4).

4. These bootstrap algorithms (CLIVC2, CLIVC4, CLRIVC, CLSRIVC) require knowledge of the controller. However, when it is unknown, another solution may be used to build up the instrumental vector that satisfies the optimal conditions (5.27). Indeed, the noise-free estimation of this instrumental vector can be achieved by using the two closed-loop transfers between $r(t_k), u(t_k)$ and between $r(t_k), y(t_k)$ instead of the open-loop one (between $u(t_k)$ and $y(t_k)$). The second step consists then in identifying the two closed-loop transfers $G_{yr}(p, \boldsymbol{\rho})$ and $G_{ur}(p, \boldsymbol{\rho})$ to compute the instruments as

$$\hat{x}(t_k, \hat{\boldsymbol{\rho}}) = G_{yr}(p, \hat{\boldsymbol{\rho}}) r(t_k)$$
$$\hat{\nu}(t_k, \hat{\boldsymbol{\rho}}) = G_{ur}(p, \hat{\boldsymbol{\rho}}) r(t_k) \tag{5.71}$$

5. Another solution is to estimate the closed-loop TF $G_{ur}(p, \hat{\boldsymbol{\rho}})$ by SRIVC or RIVC and then this can be used to obtain an estimate of the noise-free input for use as the input in the direct RIVC estimation of the process TF. This solution is close to the two-step method [19]; it is not optimal but yields good results with reasonable, albeit not minimum, variance parameter estimates [24].

5.6 Summary

The theoretical optimal choices for the design variables of the two multi-step and two iterative algorithms for complete CT modelling are summarised in Table 5.1, while the CT and DT filter forms required, for implementation, in each optimal IV version for CT hybrid modelling are given in Table 5.2.

Table 5.1. Optimal choices for the design variables of the proposed IV methods for complete CT modelling

Method	Assumed filtered data generating CT system (5.72)	Model structure	Shaping noise model $f(p)H_o(p)$	$f^{\mathrm{opt}}(p) = f_c(p)f_\mathrm{d}(p)$ see (5.27)	
				$f_c(p)$	$f_\mathrm{d}(p)$
CLIVC2	$y_\mathrm{f}(t_k) = \frac{B_o(p)}{A_o(p)}u_\mathrm{f}(t_k) + \frac{f(p)}{A_o(p)}e_o(t_k)$	CT ARX	$\frac{f(p)}{A_o(p)}$	$f(p)$	1
CLIVC4	$y_\mathrm{f}(t_k) = \frac{B_o(p)}{A_o(p)}u_\mathrm{f}(t_k) + \frac{f(p)}{A_o(p)D_o(p)}e_o(t_k)$	CT ARARX	$\frac{f(p)}{A_o(p)D_o(p)}$	$f(p)$	$D_o(p)$
CLSRIVC	$y_\mathrm{f}(t_k) = \frac{B_o(p)}{A_o(p)}u_\mathrm{f}(t_k) + e_o(t_k)$	CT OE	1	$\frac{1}{A_o(p)}$	1
CLRIVC	$y_\mathrm{f}(t_k) = \frac{B_o(p)}{A_o(p)}u_\mathrm{f}(t_k) + \frac{C_o(p)}{D_o(p)}e_o(t_k)$	CT BJ	$\frac{C_o(p)}{D_o(p)}$	$\frac{1}{A_o(p)}$	$\frac{D_o(p)}{C_o(p)}$

It will be noticed that, in Table 5.1, the assumed filtered data-generating CT system is given as

$$\begin{cases} y_\mathrm{f}(t_k) = G_o(p)u_\mathrm{f}(t_k) + f(p)H_o(p)e_o(t_k) \\ u_\mathrm{f}(t_k) = r_\mathrm{f}(t_k) - C_\mathrm{c}(p)y_\mathrm{f}(t_k) \end{cases} \quad (5.72)$$

Table 5.2. Implemented filter forms in the multi-step and iterative IV methods for CT hybrid modelling

Method	Model structure	$f_c(p)$	$f_\mathrm{d}(q^{-1})$
CLIVC2	CT ARX	$\left(\frac{\lambda}{p+\lambda}\right)^{n_a}$	1
CLIVC4	CT hybrid ARARX	$\left(\frac{\lambda}{p+\lambda}\right)^{n_a}$	$\mathrm{AR}(2n_a)$
CLSRIVC	CT hybrid OE	$\frac{1}{A(p,\hat{\rho})}$	1
CLRIVC	CT hybrid BJ	$\frac{1}{A(p,\hat{\rho})}$	$\mathrm{ARMA}(n_d, n_c)$

5.7 Numerical Examples

The following numerical example is used to illustrate and compare the performances of the proposed approaches. The process to be identified is described by (5.1), where

$$G_o(p) = \frac{p+1}{p^2 + 0.5p + 1} \quad (5.73)$$

$$C_c(p) = \frac{10p + 15}{p} \quad (5.74)$$

An external signal is added to $r_1(t_k)$ (see Figure 5.1) and chosen to be a pseudo-random binary signal of maximum length generated from a shift register with 4 stages and a clock period of 500 ($N = 7500$ data points). The sampling period T_s is chosen equal to 5 ms.

From the comparative studies presented recently in [4], the state-variable filter (SVF) approach can be considered as one of the simplest methods to handle the time-derivative problem. This latter approach has been used here with the basic (CLIVC) and multi-step estimators (CLIVC2, CLIVC4). It is not required in the case of CLRIVC because the continuous-time part of the optimal hybrid prefilter is used to generate the filtered derivatives.

Table 5.3. Mean and standard deviations of the open-loop parameter estimates for 100 Monte Carlo runs – White noise

Method	$\hat{b}_0 \pm \sigma_{\hat{b}_0}$	$\hat{b}_1 \pm \sigma_{\hat{b}_1}$	$\hat{a}_1 \pm \sigma_{\hat{a}_1}$	$\hat{a}_2 \pm \sigma_{\hat{a}_2}$	Norm
True value	1	1	0.5	1	
CLIVC	0.995 ± 0.036	1.050 ± 0.047	0.536 ± 0.021	1.015 ± 0.030	0.936
CLIVC2	0.994 ± 0.005	1.018 ± 0.046	0.520 ± 0.021	1.012 ± 0.028	0.875
CLSRIVC	0.995 ± 0.003	0.990 ± 0.050	0.518 ± 0.020	1.013 ± 0.030	0.910

5.7.1 Example 1: White Noise

First, a Gaussian white noise disturbance ($H_o(p) = 1$) is considered in order to illustrate the performance of the CLIVC, CLIVC2 and CLSRIVC algorithms. The process model parameters are estimated on the basis of closed-loop data sequences. A Monte Carlo simulation of 100 experiments is performed for a signal-to-noise (SNR) ratio given as

$$\text{SNR} = 10 \log \left(\frac{P_x}{P_e} \right) = 15 \text{ dB} \quad (5.75)$$

where P_e represents the average power of the zero-mean additive noise on the system output (*e.g.*, the variance) while P_x denotes the average power of the noise-free output fluctuations.

The Monte Carlo simulation (MCS) results are presented in Table 5.3 where the mean and standard deviation of the estimated parameters are displayed. It can be seen that the three IV methods deliver similar unbiased estimates of the model parameters with reasonable standard deviations. However, as expected, note that the basic CLIVC estimates are not as statistically efficient as the CLIVC2 and CLSRIVC estimates, where the standard deviations are smaller and, in the case of b_0, the standard deviation is some 7 times smaller.

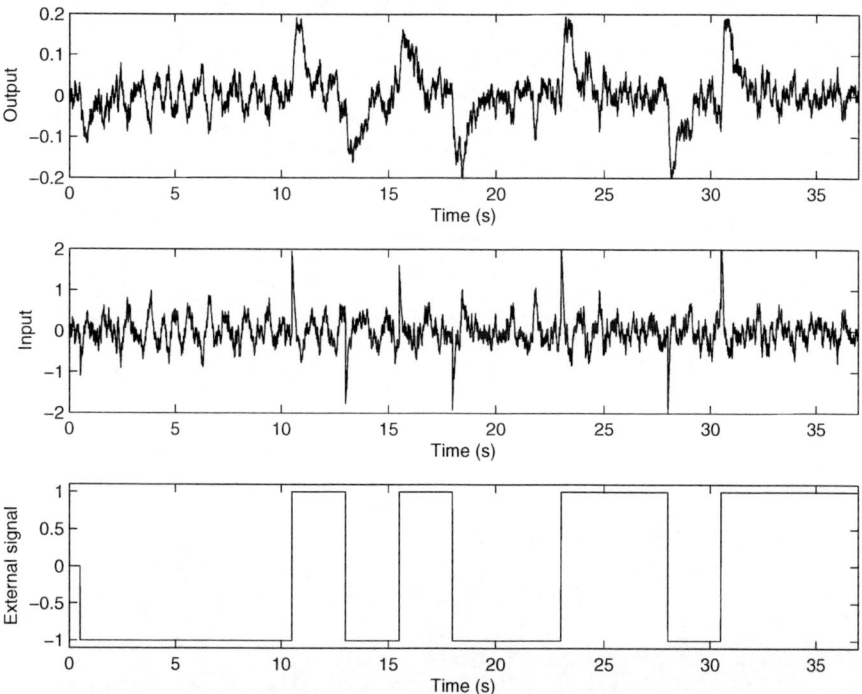

Fig. 5.2. Closed-loop data used in Example 2 – Coloured noise

Furthermore, the 2-norm of the difference between the true $(G(e^{i\omega},\boldsymbol{\rho}_o))$ and estimated $(G(e^{i\omega},\hat{\boldsymbol{\rho}}_j))$ transfer functions is also computed for each method

$$\text{Norm} = \frac{1}{N_{\text{exp}}} \sum_{j=1}^{N_{\text{exp}}} \int |G(e^{i\omega},\boldsymbol{\rho}_o) - G(e^{i\omega},\hat{\boldsymbol{\rho}}_j))|^2 d\omega \quad (5.76)$$

where N_{exp} is the number of Monte Carlo simulation runs. The results are given in Table 5.3 and confirm the previous results: the three IV methods lead to accurate results; moreover, the bootstrap methods provide slightly better results than the basic IV technique.

5.7.2 Example 2: Coloured Noise

A second example is used to analyse the performance of the proposed methods in the case of a coloured noise, with

$$H(q^{-1},\boldsymbol{\eta}_o) = \frac{1 - 0.98q^{-1}}{1 - 1.9747q^{-1} + 0.9753q^{-2}}$$

The process parameters are estimated on the basis of closed-loop data sequences described previously. A Monte Carlo simulation of 100 experiments

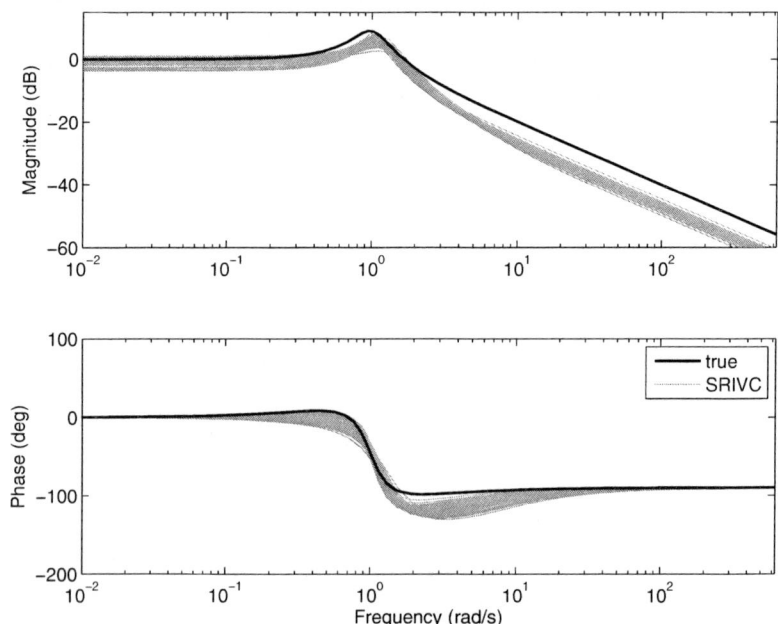

Fig. 5.3. Bode plots of the 100 identified SRIVC models – Coloured noise

is performed for a SNR = 15dB. The first 100 points of the external signal are forced to zeros in order to be free of the prefiltering initial conditions. The external signal, input and output data are plotted in Figure 5.2. The process model parameters are estimated by using methods CLIVC, CLIVC4, CLRIVC. Moreover, the direct closed-loop approach (see [19]) is also used in this example, in order to illustrate the difficulties of identifying a closed-loop model in a coloured noise situation and to see how much bias is introduced into the parameter estimates in this direct approach, when the closed-loop operation is not really taken into account. The open-loop SRIVC algorithm (see Chapter 4) is used for this purpose.

The mean and standard deviation of the 100 sets of estimated model parameters from the MCS analysis are given in Table 5.4. The Bode diagrams of the 100 identified models are displayed in Figures 5.3 to 5.6. As expected, the direct closed-loop approach using the open-loop SRIVC method clearly leads to biased results: however, it will be noticed that, although the SRIVC estimates are biased, the inherent pre-filtering introduced by CT estimation allows us to obtain better results than those obtained from indirect DT estimation. Furthermore, the three closed-loop IV methods provide similar unbiased estimates of the model parameters with reasonable standard deviations. However, again as expected, the CLIVC estimates are not as statistically efficient as the estimates produced by the multi-step CLIVC4 and iterative CLRIVC

5 Optimal IV Methods for Closed-loop CT Model Identification

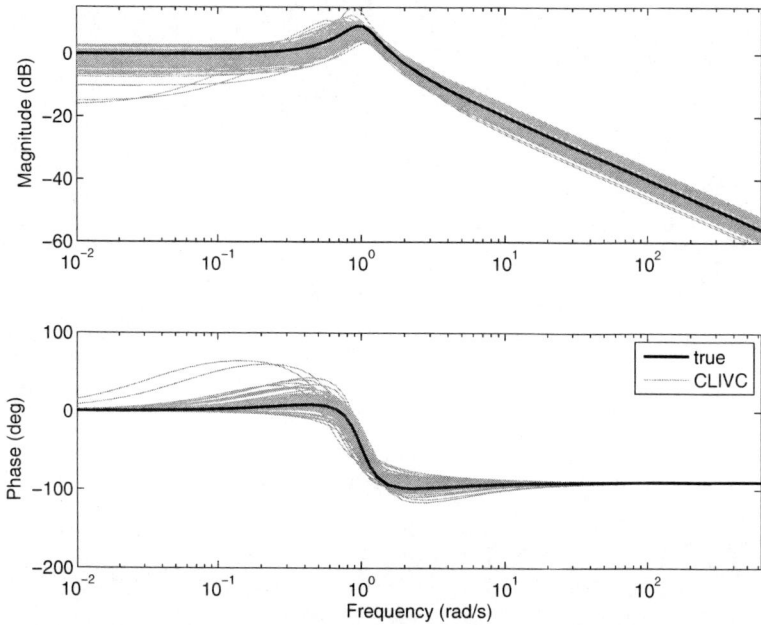

Fig. 5.4. Bode plots of the 100 identified CLIVC models – Coloured noise

algorithms, where the standard deviation are always smaller, thanks to the prefiltering and associated noise model estimation. Furthermore, thanks to its iterative structure and its prefilter updating operation, the CLRIVC algorithm leads to better results than the CLIVC4 method. Moreover, it is interesting to note that, from our experience, the basic CLIVC method provides better results than the DT version (using the sampled external signal), thanks to the inherent CT prefiltering (see [7]).

Table 5.4. Mean and standard deviations of the open-loop parameter estimates for 100 Monte Carlo runs - Colored noise

Method	$\hat{b}_0 \pm \sigma_{\hat{b}_0}$	$\hat{b}_1 \pm \sigma_{\hat{b}_1}$	$\hat{a}_1 \pm \sigma_{\hat{a}_1}$	$\hat{a}_2 \pm \sigma_{\hat{a}_2}$	Norm
True value	1	1	0.5	1	
SRIVC	0.449 ± 0.050	1.254 ± 0.247	0.636 ± 0.140	1.354 ± 0.156	0.855
CLIVC	1.011 ± 0.278	0.812 ± 0.299	0.546 ± 0.099	0.963 ± 0.231	0.784
CLIVC4	0.960 ± 0.131	0.977 ± 0.240	0.563 ± 0.104	1.015 ± 0.119	0.767
CLRIVC	0.972 ± 0.112	0.973 ± 0.191	0.557 ± 0.083	1.007 ± 0.094	0.779

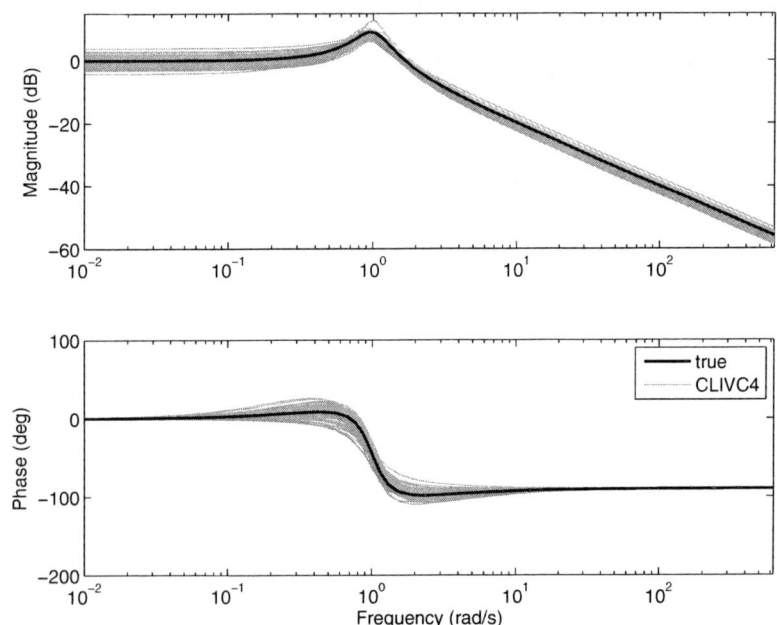

Fig. 5.5. Bode plots of the 100 identified CLIVC4 models – Coloured noise

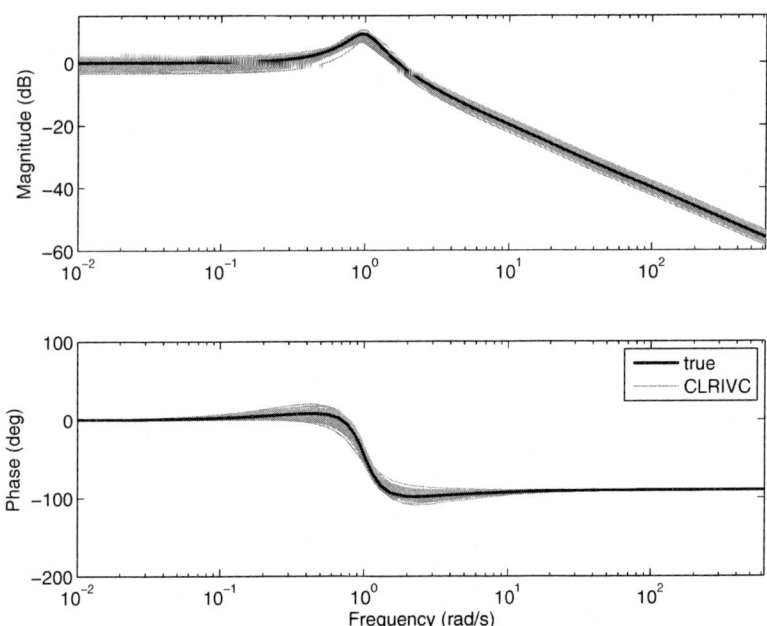

Fig. 5.6. Bode plots of the 100 identified CLRIVC models – Coloured noise

5.8 Conclusions

This chapter has addressed the problem of estimating the parameters of continuous-time transfer functions models for linear dynamic systems operating in closed loop using instrumental variable techniques. Several closed-loop IV estimators have been described, including the development of explicit expressions for the parametric error covariance matrix. In particular, the chapter has shown that reduced values of this covariance matrix can be achieved for a particular choice of instruments and prefilter; and both multi-step and iterative solutions have been developed to determine the design parameters that allow for such improved closed-loop IV estimation.

References

1. U. Forssell and C.T. Chou. Efficiency of prediction error and instrumental variable methods for closed-loop identification. In *37th IEEE Conference on Decision and Control*, Tampa, USA, pages 1287–1288, December 1998.
2. U. Forssell and L. Ljung. Closed-loop identification revisited. *Automatica*, 35(7):1215–1241, 1999.
3. H. Garnier, M. Gilson, and W.X. Zheng. A bias-eliminated least-squares method for continuous-time model identification of closed-loop systems. *International Journal of Control*, 73(1):38–48, 2000.
4. H. Garnier, M. Mensler, and A. Richard. Continuous-time model identification from sampled data. Implementation issues and performance evaluation. *International Journal of Control*, 76(13):1337–1357, 2003.
5. M. Gilson and H. Garnier. Continuous-time model identification of systems operating in closed-loop. In *13th IFAC Symposium on System Identification*, Rotterdam, Netherlands, pages 425–430, August 2003.
6. M. Gilson and P. Van den Hof. On the relation between a bias-eliminated least-squares (BELS) and an IV estimator in closed-loop identification. *Automatica*, 37(10):1593–1600, 2001.
7. M. Gilson and P. Van den Hof. Instrumental variable methods for closed-loop system identification. *Automatica*, 41(2):241–249, 2005.
8. A.J. Jakeman, L.P. Steele, and P.C. Young. Instrumental variable algorithms for multiple input systems described by multiple transfer functions. *IEEE Transactions on Systems, Man, and Cybernetics*, 10:593–602, 1980.
9. L. Ljung. *System Identification. Theory for the User*. Prentice Hall, Upper Saddle River, 2nd edition, 1999.
10. D.A. Pierce. Least squares estimation in dynamic disturbance time-series models. *Biometrika*, 5:73–78, 1972.
11. G.P. Rao and H. Garnier. Numerical illustrations of the relevance of direct continuous-time model identification. *15th IFAC World Congress*, Barcelona, Spain, July 2002.
12. G.P. Rao and H. Garnier. Identification of continuous-time models: direct or indirect? *Invited semi-plenary paper for the XV International Conference on Systems Science*, Wroclaw, Poland, September 2004.

13. T. Söderström and P. Stoica. *Instrumental Variable Methods for System Identification.* Springer-Verlag, New York, 1983.
14. T. Söderström and P. Stoica. *System Identification.* Series in Systems and Control Engineering. Prentice Hall, Englewood Cliffs, 1989.
15. T. Söderström, P. Stoica, and E. Trulsson. Instrumental variable methods for closed-loop systems. In *10th IFAC World Congress*, Munich, Germany, pages 363–368, 1987.
16. T. Söderström, W.X. Zheng, and P. Stoica. Comments on 'On a least-squares-based algorithm for identification of stochastic linear systems'. *IEEE Transactions on Signal Processing*, 47(5):1395–1396, 1999.
17. P. Stoica and T. Söderström. Optimal instrumental variable estimation and approximate implementations. *IEEE Transactions on Automatic Control*, 28(7):757–772, 1983.
18. C.J. Taylor, A. Chotai, and P.C. Young. State space control system design based on non-minimal state variable feedback: further generalization and unification results. *International Journal of Control*, 73:1329–1345, 2000.
19. P. Van den Hof. Closed-loop issues in system identification. *Annual Reviews in Control*, 22:173–186, 1998.
20. P.C. Young. An instrumental variable method for real-time identification of a noisy process. *Automatica*, 6:271–287, 1970.
21. P.C. Young. Some observations on instrumental variable methods of time-series analysis. *International Journal of Control*, 23(5):593–612, 1976.
22. P.C. Young. *Recursive Estimation and Time-Series Analysis.* Springer-Verlag, Berlin, 1984.
23. P.C. Young. Optimal IV identification and estimation of continuous-time TF models. *15th IFAC World Congress*, Barcelona, Spain, July 2002.
24. P.C. Young. The refined instrumental variable method: unified estimation of discrete and continuous-time transfer function models. *Journal Européen des Systèmes Automatisés*, 2008.
25. P.C. Young and A.J. Jakeman. Refined instrumental variable methods of time-series analysis: Part I, SISO systems. *International Journal of Control*, 29:1–30, 1979.
26. P.C. Young and A.J. Jakeman. Refined instrumental variable methods of time-series analysis: Part III, extensions. *International Journal of Control*, 31:741–764, 1980.
27. W.X. Zheng. Identification of closed-loop systems with low-order controllers. *Automatica*, 32(12):1753–1757, 1996.

6

Model Order Identification for Continuous-time Models

Liuping Wang[1] and Peter C. Young[2]

[1] RMIT University, Melbourne, Australia
[2] Lancaster University, UK & Australian National University

6.1 Introduction

For the past four decades, the principle of instrumental variables (IV) has been quite a popular approach to the identification and estimation[3] of discrete-time dynamic systems (*e.g.*, [5–8,14,16,18,20]). Amongst these references, the research monograph by Söderström and Stoica [6] in 1983 is devoted entirely to IV methods; whilst the text by Young [18] in 1984 concentrates on standard and optimal IV methods within a wider context. However, the IV method has also been used very successfully in continuous-time model estimation. For instance Young and Jakeman [52] proposed an optimal IV approach for continuous-time model estimation over a quarter of a century ago and there are many examples where IV estimation has been applied within a continuous-time context since then.

One of the well-documented issues in instrumental variable identification is the singularity, or near-singularity, of the instrumental product matrix (IPM) when the model structure is overparameterised; a useful property that has been exploited as the basis for model order estimation by Wellstead [12], Young *et al.* [22], and Wellstead and Rojas [13]. This chapter presents new results on model structure identification in IV estimation using a rather different approach. In particular, by assuming *a priori* knowledge of the relative degree and the maximum model order for the continuous-time system structure, an instrumental variable model order identification procedure is proposed, based on UDV matrix factorisation. Here, the higher-order model parameter estimates do not affect the lower order parameter estimates, so allowing for the natural truncation of model orders. In addition, the predictive errors, which are exploited in the context of cross-validation via the predicted resid-

---

[3] In this chapter, the statistical meanings of 'identification' and 'estimation' are utilised: namely, 'identification' refers to the identification of a uniquely identifiable model order and structure; while 'estimation' refers to the estimation of the parameters that characterise this identified model form.

ual sums of squares (PRESS) statistic, are calculated in a systematic manner for all candidate models.

It should be noted that, although the results presented in this chapter are derived for continuous-time systems, they can be extended to discrete-time systems in a straightforward manner.

6.2 Instrumental Variable Identification

Assume that a plant has a continuous-time operator model form

$$G(p) = \frac{B(p)}{A(p)} \tag{6.1}$$

where

$$A(p) = p^n + a_1 p^{n-1} + a_2 p^{n-2} + \ldots + a_n$$
$$B(p) = b_0 p^m + b_1 p^{m-1} + b_2 p^{m-2} + \ldots + b_m$$

are the polynomials in the differential p operator. Given the degrees of $A(p)$ and $B(p)$ as n and m, the relative degree of a continuous-time model is defined as $\gamma = n - m$. We assume that the plant input and output signals are $u(t)$ and $y(t)$ with band-limited additive white noise $\xi(t)$, so that

$$y(t) = \frac{B(p)}{A(p)} u(t) + \xi(t) \tag{6.2}$$

or, in decomposed form

$$\begin{cases} x(t) = \frac{B(p)}{A(p)} u(t) \\ y(t) = x(t) + \xi(t) \end{cases} \tag{6.3}$$

where $x(t)$ is the 'noise-free' output of the system. The optimal refined IV solution of the estimation problem (RIVC) for this continuous-time model, based on the assumption that the sampled noise $\xi(t_k)$ can be described by a discrete-time autoregressive moving average (ARMA) process (the 'hybrid Box–Jenkins' model) is considered in Chapter 4 of this book. This optimal solution involves hybrid prefilters that are updated iteratively to reflect the estimated parameters of the model polynomials in both the system and ARMA noise models. Here, however, we consider an alternative, non-iterative 'state-variable filter' (SVF) solution [1,15,17] in which the fixed prefilter is an all-pole process with a denominator $C(p) = p^n + c_1 p^{n-1} + c_2 p^{n-2} + \ldots + c_n$.

By passing both the sampled input and output measurements $u(t_k)$ and $y(t_k)$ ($k = 0, 1, 2, \ldots$) through this all-pole prefilter, we obtain the filtered input and output signals. When applied to the model (6.2), this operation yields

$$\frac{A(p)}{C(p)}y(t_k) = \frac{B(p)}{C(p)}u(t_k) + \frac{A(p)}{C(p)}\xi(t_k) \tag{6.4}$$

where the continuous-time filtering operation includes interpolation of the sampled input and output signals (see Chapter 4 of the present book). In the case where the sampled noise $\xi(t_k)$ can be considered as a zero-mean, serially uncorrelated and normally distributed white noise process, it can be shown that the prefilter $C(p)$ is optimal in statistical terms if $C(p) = A(p)$. Indeed, this is the basis for the simplified refined instrumental variable (SRIVC) approach to continuous-time model estimation that has been available for many years in the CAPTAIN toolbox[4] for MATLAB®. However, its principle use in the present context is to facilitate the estimation of the prefiltered derivatives used in the IV estimation procedure that forms part of the proposed model order identification strategy (see later example in Section 6.5).

The derivatives of the filtered input and output signals are obtained as shown in Figure 6.1. This can be considered in SVF terms by letting $y_f^{(n)}(t)$, $y_f^{(n-1)}(t)$, ..., $y_f^{(0)}(t)$ denote, respectively, $\frac{p^n}{C(p)}y(t)$, $\frac{p^{n-1}}{C(p)}y(t)$, ..., $\frac{1}{C(p)}y(t)$; and $u_f^{(n)}(t)$, $u_f^{(n-1)}(t)$, ..., $u_f^{(0)}(t)$ denote, respectively, $\frac{p^n}{C(p)}u(t)$, $\frac{p^{n-1}}{C(p)}u(t)$, ..., $\frac{1}{C(p)}u(t)$. To obtain the derivatives of the filtered output signals, we simply define a state-variable vector

$$X^y(t) = [y_f^{(n)}(t)\ y_f^{(n-1)}(t)\ \ldots\ y_f^{(0)}(t)]^T$$

Then, by choosing the state space model in a control canonical form, we have

$$\begin{bmatrix} \frac{dy_f^{(n)}(t)}{dt} \\ \frac{dy_f^{(n-1)}(t)}{dt} \\ \vdots \\ \frac{dy_f^{(0)}(t)}{dt} \end{bmatrix} = \begin{bmatrix} -c_1 & -c_2 & \ldots & -c_n \\ 1 & 0 & \ldots & 0 \\ \ldots & & & \\ 0 & \ldots & 1 & 0 \end{bmatrix} \begin{bmatrix} y_f^{(n)}(t) \\ y_f^{(n-1)}(t) \\ \vdots \\ y_f^{(0)}(t) \end{bmatrix} + \begin{bmatrix} 1 \\ 0 \\ \vdots \\ 0 \end{bmatrix} y(t)$$

$$= \Omega \begin{bmatrix} y_f^{(n)}(t) \\ y_f^{(n-1)}(t) \\ \vdots \\ y_f^{(0)}(t) \end{bmatrix} + \boldsymbol{\Gamma} y(t) \tag{6.5}$$

The solution of the state space equation (6.5), assuming zero initial conditions, generates the prefiltered derivatives of the output signal. Similarly,

$$X^u(t) = \left[u_f^{(n)}(t)\ u_f^{(n-1)}(t)\ \ldots\ u_f^{(0)}(t) \right]^T$$

defines the prefiltered derivatives of the input signal; and

---

[4] CAPTAIN can be downloaded from http://www.es.lancs.ac.uk/cres/captain/

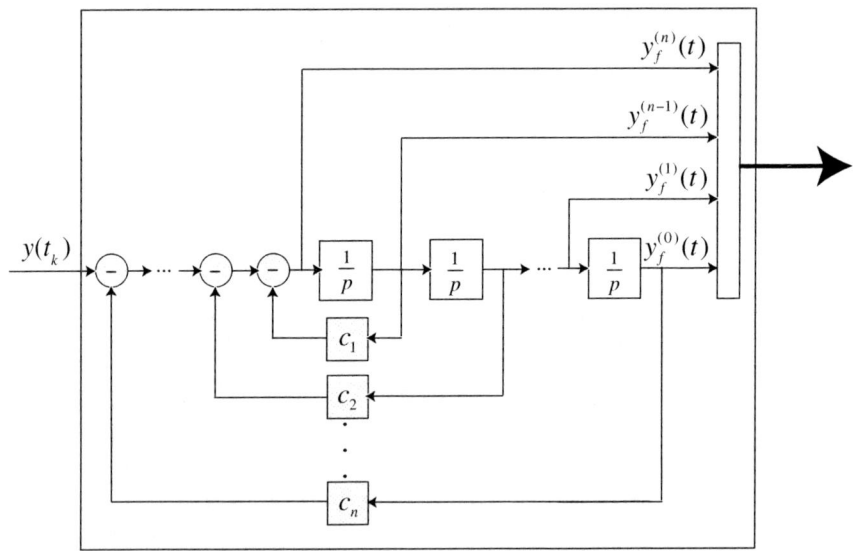

Fig. 6.1. Generation of prefiltered derivatives for the output $y(t)$ by the prefilter $f(p) = 1/C(p)$ (based on Figure 2(b) in [52])

$$X^h(t) = [h_n(t) \ h_{n-1}(t) \ldots h_0(t)]^T$$

is used to capture the initial conditions of the state variables (when necessary). These state variables satisfy the following differential equations, respectively

$$\dot{X}^u(t) = \Omega X^u(t) + \Gamma u(t)$$
$$X^u(0) = \mathbf{0}_n \qquad (6.6)$$

$$\dot{X}^h(t) = \Omega X^h(t)$$
$$X^h(0) = \mathbf{I}_n \qquad (6.7)$$

where $\mathbf{0}_n$ is a zero-column vector of length n and $\mathbf{I}_n$ is a column vector of length n with the first element unity and the rest zero.

With the state variables defined by (6.5)–(6.7), (6.4) can be written as

$$y_f^{(n)}(t) + a_1 y_f^{(n-1)}(t) + \ldots + a_n y_f^{(0)}(t) =$$
$$b_0 u_f^{(m)}(t) + b_1 u_f^{(m-1)}(t) + \ldots + b_m u_f^{(0)}(t) \qquad (6.8)$$
$$+ g_0 h_n(t) + g_1 h_{n-1}(t) + \ldots g_n h_0(t) + \xi_f(t)$$

where $\xi_f(t)$ denotes the filtered noise variable and $\{g_i\}$, $i = 0, \ldots, n$, are extra coefficients used to capture the transient response of the filter. In the following analysis, we make the normal assumption that the zeros of $C(p)$

lie strictly inside the left-half complex plane so that the state variables $h_i(t)$, $i = 0, 1, \ldots n$, decay exponentially to zero. For simplicity, their effect on the solution is neglected here: however, they can be included easily if this is desired (see, *e.g.*, [1])

To estimate the parameters a_i, $i = 1, 2, \ldots, n$, b_i, $i = 0, 1, \ldots, m$, (6.9), with the terms in $h_i(t)$, $i = 0, 1, \ldots, n$ omitted, is reformulated into a standard linear regression form as

$$y_\text{f}^{(n)}(t) = \boldsymbol{\varphi}^T(t)\boldsymbol{\rho} + \xi_\text{f}(t) \tag{6.9}$$

where

$$\boldsymbol{\varphi}^T(t) = [-y_\text{f}^{(n-1)}(t) \ \ldots \ -y_\text{f}^{(0)}(t) \ u_\text{f}^{(m)}(t) \ \ldots u_\text{f}^{(0)}(t)]$$

and

$$\boldsymbol{\rho}^T = [a_1 \ \ldots a_n \ b_0 \ \ldots b_m]$$

We now define the instrumental variable as the output of the following 'auxiliary model'

$$\hat{x}(t) = \frac{\hat{B}(p)}{\hat{A}(p)} u(t) \tag{6.10}$$

which will be recognised from (6.3) as an estimate of the noise-free output $x(t)$ of the system if $\hat{A}(p)$ and $\hat{B}(p)$ are statistically consistent estimates of the system model parameters $A(p)$ and $B(p)$. In practice, these estimates are generated using an iterative updating algorithm, such as that used for the optimal RIVC and SRIVC algorithms described in Chapter 4 of this book. The filtered nth derivative of the instrumental variable is defined as

$$\hat{x}_\text{f}^{(n)}(t) = \frac{p^n}{C(p)} \hat{x}(t) \tag{6.11}$$

and the prefiltered IV vector is then given by

$$\hat{\boldsymbol{\varphi}}(t)^T = [-\hat{x}_\text{f}^{(n-1)}(t) \ \ldots \ -\hat{x}_\text{f}^{(0)}(t) \ u_\text{f}^{(m)}(t) \ \ldots u_\text{f}^{(0)}(t)]$$

Note that, for notational simplicity, we do not add the subscript f to both regressor vectors $\boldsymbol{\varphi}(t)$ and $\hat{\boldsymbol{\varphi}}(t)$ as they are understood to contain the filtered plant input and output variables. Note also that the algorithm derived below is applicable to the general case with or without prefilters.

If T_N is the actual measurement time for the given set of data, then the completely continuous-time, IV estimate of the parameter vector $\boldsymbol{\rho}$ is then given by

$$\hat{\boldsymbol{\rho}} = \left(\int_0^{T_N} \hat{\boldsymbol{\varphi}}(t) \boldsymbol{\varphi}(t)^T dt \right)^{-1} \int_0^{T_N} \hat{\boldsymbol{\varphi}}(t) y_\text{f}^{(n)}(t) dt \tag{6.12}$$

However, we are concerned here with discrete-time, sampled input and output data $\{u(t_k)\}$ and $\{y(t_k)\}$, $k = 1, 2, \ldots, N$, with a uniform sampling interval $T_\text{s} = t_{k+1} - t_k$. As a result, (6.12) can be written as,

$$\lim_{T_s \to 0} \left[\sum_{k=1}^{N} \hat{\varphi}(t_k)\varphi^T(t_k)T_s \right]^{-1} \sum_{k=1}^{N} \hat{\varphi}(t_k)y_{\mathrm{f}}^{(n)}(t_k)T_s \qquad (6.13)$$

which can be approximated either by using the finite-sum approximation given by

$$\hat{\rho} \approx \left[\sum_{k=1}^{N} \hat{\varphi}(t_k)\varphi^T(t_k) \right]^{-1} \sum_{k=1}^{N} \hat{\varphi}(t_k)y_{\mathrm{f}}^{(n)}(t_k) \qquad (6.14)$$

or a numerical integration scheme with a higher accuracy. Normally, the solution (6.14) or its recursive equivalent is utilised because, as mentioned previously and discussed in Chapter 4 of this book, the interpolation inherent in the continuous-time prefiltering operations is sufficient to ensure good estimation performance.

6.3 Instrumental Variable Estimation using a Multiple-model Structure

Note that, in the instrumental variable solution (6.14), we have assumed known orders for the numerator and denominator polynomials. In practice, these are not known and the model structure identification scheme, such as that discussed below, will encounter situations where the model is overparameterised. In such situations, as indicated previously, it is likely that the instrumental product matrix (IPM), $\sum_{k=1}^{N} \hat{\varphi}(t_k)\varphi^T(t_k)$, will be ill conditioned. In order to avoid this, a matrix decomposition algorithm is required to generate the inverse of the IPM without actually performing the inversion. In addition, it is an advantage for model structure identification if: (i) a threshhold can be set to detect the situation when no more information can be extracted from the given data set; and (ii) the model parameters are available from the estimation for all candidate model structures, so that the best model structure can be determined for a given data set and a relative degree.

In this chapter, these algorithmic objectives are achieved using UDV matrix factorisation with a specific configuration of the input and output data that is termed a 'multiple-model structure'. The idea of exploiting a multiple model structure for IV estimation has been described previously for discrete-time systems [3, 4, 11]. Here, we extend this basic idea to continuous-time system identification using UDV factorisation, including the computation and exploitation of prediction error statistics.

6.3.1 Augmented Data Regressor

The basic idea of a multiple-model structure is to augment the original regression vector $\varphi(t_k)$ given by (6.9) with $-y_{\mathrm{f}}^{(n)}(t_k)$ and re-arrange the new regressor, which replaces this, in such a way that the variables corresponding

6 Model Order Identification for Continuous-time Models

to a lower index i of $y_\mathrm{f}^{(i)}(t_k)$ and $u_\mathrm{f}^{(i)}(t_k)$ enter the regressor first, with $y_\mathrm{f}^{(i)}(t_k)$ and $u_\mathrm{f}^{(i)}(t_k)$ appearing in pairs. This new regressor will be denoted by $\phi_i(t_k)$, where the subscript $i = 1, 2, \cdots, n$ indicates the order of the model. However, unlike the discrete-time counterpart of this procedure, the fact that the minimal order of a continuous-time model is no less than its relative degree demands special consideration. Here, we propose that the effect of relative degree is incorporated into the regressor. For example, if a continuous-time transfer function has relative degree one, then a first-order structure for the augmented regressor has the form,

$$\phi_1(t_k) = \left[-y_\mathrm{f}^{(0)}(t_k)\; u_\mathrm{f}^{(0)}(t_k)\; -y_\mathrm{f}^{(1)}(t_k) \right]^T$$

and a second-order structure follows with,

$$\phi_2(t_k) = \left[-y_\mathrm{f}^{(0)}(t_k)\; u_\mathrm{f}^{(0)}(t_k)\; -y_\mathrm{f}^{(1)}(t_k)\; u_\mathrm{f}^{(1)}(t_k)\; -y_\mathrm{f}^{(2)}(t_k) \right]^T \quad (6.15)$$

$$= \left[\phi_1^T(t_k)\; u_\mathrm{f}^{(1)}(t_k)\; -y_\mathrm{f}^{(2)}(t_k) \right]^T \quad (6.16)$$

and so on, to the nth-order structure,

$$\phi_n(t_k) = \left[\phi_{n-1}^T(t_k)\; u_\mathrm{f}^{(n-1)}(t_k)\; -y_\mathrm{f}^{(n)}(t_k) \right]^T$$

For a relative degree of γ, the minimal order of the model is γ. The augmented regressor for the γ-order model is defined as,

$$\phi_\gamma^T(t_k) = \left[-y_\mathrm{f}^{(0)}(t_k)\; -y_\mathrm{f}^{(1)}(t_k)\; \ldots\; -y_\mathrm{f}^{(\gamma-1)}(t_k)\; u_\mathrm{f}^{(0)}(t_k)\; -y_\mathrm{f}^{(\gamma)}(t_k) \right] \quad (6.17)$$

and the augmented regressor for order $\gamma + 1$ follows with,

$$\phi_{\gamma+1}^T(t_k) = \left[\phi_\gamma^T(t_k)\; u_\mathrm{f}^{(1)}(t_k)\; -y_\mathrm{f}^{(\gamma+1)}(t_k) \right] \quad (6.18)$$

For any model order $n > \gamma$, the augmented regressor is defined as,

$$\phi_n^T(t_k) = [-y_\mathrm{f}^{(0)}(t_k)\; -y_\mathrm{f}^{(1)}(t_k) \ldots\; -y_\mathrm{f}^{(\gamma-1)}(t_k)u_\mathrm{f}^{(0)}(t_k)\; -y_\mathrm{f}^{(\gamma)}(t_k)$$
$$u_\mathrm{f}^{(1)}(t_k)\; -y_\mathrm{f}^{(\gamma)}(t_k) \ldots\; -y_\mathrm{f}^{(n-1)}(t_k)u_\mathrm{f}^{(m)}(t_k)\; -y_\mathrm{f}^{(n)}(t_k)] \quad (6.19)$$

which is, equivalently,

$$\phi_n^T(t_k) = \left[\phi_{n-1}^T(t_k)\; u_\mathrm{f}^{(m)}(t_k)\; -y_\mathrm{f}^{(n)}(t_k) \right] \quad (6.20)$$

With this special arrangement of the regressor, the linear regression equation (6.9) takes the new form

$$\phi_n^T(t_k)\boldsymbol{\theta}_\mathrm{aug} = -\xi_\mathrm{f}(t_k) \quad (6.21)$$

where the new, re-ordered and augmented parameter vector $\boldsymbol{\theta}_{\text{aug}}$ is defined by
$$\boldsymbol{\theta}_{\text{aug}}^T = \begin{bmatrix} a_n \ a_{n-1} \ldots a_{n-\gamma+1} \ b_m \ a_{n-\gamma} \ b_{m-1} \ldots a_1 \ b_0 \ 1 \end{bmatrix}$$
while the associated, augmented IV vector is defined as follows
$$\hat{\boldsymbol{\phi}}_n^T(t_k) = [-\hat{x}_f^{(0)}(t_k) - \hat{x}_f^{(1)}(t_k) \ldots - \hat{x}_f^{(\gamma-1)}(t_k) u_f^{(0)}(t_k) - \hat{x}_f^{(\gamma)}(t_k)$$
$$u_f^{(1)}(t_k) - \hat{x}_f^{(\gamma)}(t_k) \ldots - \hat{x}_f^{(n-1)}(t_k) u_f^{(m)}(t_k) - \hat{x}_f^{(n)}(t_k)] \quad (6.22)$$

The subscript 'aug' indicates that the parameter vector $\boldsymbol{\theta}_{\text{aug}}$ contains unity in its last element.

Remark 6.1. The discrete version of the structure shown in (6.21) corresponds to the 'shift structure' in [2], where recursive algorithms are proposed for fast calculation of gain matrices.

Remark 6.2. The relative degree problem in a continuous-time system model is addressed here by first entering the regressor with the extra terms in $y_f^{(i)}(t_k)$, $i = 0, 1, \ldots, n - m$, which is equivalent to forcing zero into the leading coefficients of the numerator of the continuous-time transfer function.

6.3.2 Instrumental Variable Solution Using UDV Factorisation

Note that, with the special multiple-model arrangement of the data regressor explained above, the combinations of the elements in the augmented regressor, for a given model order n and relative degree γ, provide the information about the candidate models for all possible orders. The next step in the development of the order identification procedure is, therefore, to extract the information for the instrumental variable estimates of the candidate models for all possible orders that are less than n. This is achieved through UDV factorisation of the augmented IPM defined by

$$\mathbf{P} = \sum_{k=1}^{N} \hat{\boldsymbol{\phi}}_n(t_k) \boldsymbol{\phi}_n^T(t_k)$$

This computation is performed such that the estimation of lower-order models does not affect the estimation of the higher-order models in the structure. The following theorem summarises the results for IV estimation of the multiple-model structure.

Theorem 6.1. *Suppose that the IPM is decomposed into*

$$\mathbf{P} = \mathbf{UDV} \quad (6.23)$$

where $\mathbf{U}$ is a lower triangular matrix with unit diagonal elements, $\mathbf{D}$ is a diagonal matrix and $\mathbf{V}$ is an upper triangular matrix with unit diagonal elements. Then for the model order k ($k = \gamma, \gamma + 1, \gamma + 2, \ldots, n$), the following statements are true

- the IV solution $\hat{\boldsymbol{\theta}}_k$ of the parameter estimation problem is given by the jth column (above the diagonal) in $\mathbf{V}^{-1}$, where $j = \gamma + 2 * k$.
- the residual error $e(t_k)$ for the kth model order is located at the jth ($j = \gamma + 2 * k$) row of the vector

$$-\phi_n^T(t_k)\mathbf{V}^{-1}$$

The results are illustrated in the following example, which can be extended to a general case without any difficulty.

Example 6.1. In order to understand how the decomposition algorithm works, let us consider a second-order system with a relative degree of unity, where the operator model form is

$$G(p) = \frac{b_0 p + b_1}{p^2 + a_1 p + a_2} \tag{6.24}$$

The filtered augmented regressor and augmented IV vector are

$$\boldsymbol{\phi}_2(t) = \left[-y_f^0(t)\ u_f^{(0)}(t)\ -y_f^{(1)}(t)\ u_f^{(1)}(t)\ -y_f^{(2)}(t)\right]^T$$

$$\hat{\boldsymbol{\phi}}_2(t) = \left[-\hat{x}_f^{(0)}(t)\ u_f^{(0)}(t)\ -\hat{x}_f^{(1)}(t)\ u_f^{(1)}(t)\ -\hat{x}_f^{(2)}(t)\right]^T$$

The IPM defined below is arranged into block matrices

$$\mathbf{P} = \sum_{k=1}^{N} \hat{\boldsymbol{\phi}}_2(t_k)\boldsymbol{\phi}_2^T(t_k)$$

$$= \begin{bmatrix} \mathbf{A}_{44} & \mathbf{B}_{41} \\ \mathbf{C}_{14} & \mathbf{A}_{5s} \end{bmatrix} \tag{6.25}$$

where

$$\mathbf{A}_{44} = \sum_{k=1}^{N} \begin{bmatrix} -\hat{x}_f^{(0)}(t_k) \\ u_f^{(0)}(t_k) \\ -\hat{x}_f^{(1)}(t_k) \\ u_f^{(1)}(t_k) \end{bmatrix} \left[-y_f^{(0)}(t_k)\ u_f^{(0)}(t_k)\ -y_f^{(1)}(t_k)\ u_f^{(1)}(t_k)\right];$$

$$\mathbf{B}_{41} = \sum_{k=1}^{N} \begin{bmatrix} -\hat{x}_f^{(0)}(t_k) \\ u_f^{(0)}(t_k) \\ -\hat{x}_f^{(1)}(t_k) \\ u_f^{(1)}(t_k) \end{bmatrix} (-y_f^{(2)}(t_k));$$

$$\mathbf{A}_{5s} = \sum_{k=1}^{N} \hat{x}_f^{(2)}(t_k) y_f^{(2)}(t_k);$$

$$\mathbf{C}_{14} = \sum_{k=1}^{N} \left[-y_f^{(0)}(t_k)\ u_f^{(0)}(t_k)\ -y_f^{(1)}(t_k)\ u_f^{(1)}(t_k)\right](-\hat{x}_f^{(2)}(t_k))$$

Note that the matrix $\mathbf{A}_{44}$ is a square matrix with dimensions of 4×4, which is essentially the IPM for a second-order model (in the single-model structure case). $\mathbf{A}_{5s}$ is a scalar, $\mathbf{B}_{41}$ and $\mathbf{C}_{14}$ are column and row vectors with dimensions of 4×1 and 1×4, respectively. We also assume that $\mathbf{A}_{44}$ is invertible, so that it is identifiable for a second-order model. With these assumptions, the following matrix equality is true

$$\mathbf{P} = \mathbf{U}_5 \mathbf{D}_5 \mathbf{V}_5 \tag{6.26}$$

$$= \begin{bmatrix} \mathbf{I}_{44} & \mathbf{0}_{41} \\ \mathbf{C}_{14}\mathbf{A}_{44}^{-1} & 1 \end{bmatrix} \begin{bmatrix} \mathbf{A}_{44} & \mathbf{0}_{41} \\ \mathbf{0}_{14} & \mathbf{A}_{44} - \mathbf{C}_{14}\mathbf{A}_{44}^{-1}\mathbf{B}_{41} \end{bmatrix} \begin{bmatrix} \mathbf{I}_{44} & \mathbf{A}_{44}^{-1}\mathbf{B}_{41} \\ \mathbf{0}_{14} & 1 \end{bmatrix} \tag{6.27}$$

where $\mathbf{I}_{44}$ is the identity matrix with the same dimension as $\mathbf{A}_{44}$, while $\mathbf{0}_{14}$ and $\mathbf{0}_{41}$ are zero matrices with dimensions of 1×4 and 4×1, respectively. Note from the definitions of $\mathbf{A}_{44}$ and $\mathbf{B}_{41}$ that the parameter vector for the second-order model with relative degree one is described by

$$\hat{\boldsymbol{\theta}}_2 = -\mathbf{A}_{44}^{-1}\mathbf{B}_{41} = \begin{bmatrix} \hat{a}_2 & \hat{b}_1 & \hat{a}_1 & \hat{b}_0 \end{bmatrix}^T \tag{6.28}$$

which are the negative elements above the diagonal of the 5th column of the $\mathbf{V}_5$ matrix. This estimated parameter vector corresponds to the transfer function model given by (6.24). It is easy to see that

$$\mathbf{V}_5^{-1} = \begin{bmatrix} \mathbf{I}_{44} & -\mathbf{A}_{44}^{-1}\mathbf{B}_{41} \\ \mathbf{0}_{14} & 1 \end{bmatrix}$$

Therefore the last column above the diagonal on $\mathbf{V}_5^{-1}$ provides the estimates for the second-order continuous-time model. Let us now proceed to the second step of the decomposition with

$$\mathbf{A}_{44} = \begin{bmatrix} \mathbf{A}_{33} & \mathbf{B}_{31} \\ \mathbf{C}_{13} & \mathbf{A}_{4s} \end{bmatrix} \tag{6.29}$$

where

$$\mathbf{A}_{33} = \sum_{k=1}^{N} \begin{bmatrix} -\hat{x}_{\mathrm{f}}^{(0)}(t_k) \\ u_{\mathrm{f}}^{(0)}(t_k) \\ -\hat{x}_{\mathrm{f}}^{(1)}(t_k) \end{bmatrix} \begin{bmatrix} -y_{\mathrm{f}}^{(0)}(t_k) & u_{\mathrm{f}}^{(0)}(t_k) & -y_{\mathrm{f}}^{(1)}(t_k) \end{bmatrix};$$

$$\mathbf{B}_{31} = \sum_{k=1}^{N} \begin{bmatrix} -\hat{x}_{\mathrm{f}}^{(0)}(t_k) \\ u_{\mathrm{f}}^{(0)}(t_k) \\ -\hat{x}_{\mathrm{f}}^{(1)}(t_k) \end{bmatrix} u_{\mathrm{f}}^{(1)}(t_k);$$

$$\mathbf{A}_{4s} = \sum_{k=1}^{N} u_{\mathrm{f}}^{(1)}(t_k) u_{\mathrm{f}}^{(1)}(t_k);$$

$$\mathbf{C}_{13} = \sum_{k=1}^{N} \begin{bmatrix} -y_{\mathrm{f}}^{(0)}(t_k) & u_{\mathrm{f}}^{(0)}(t_k) & -y_{\mathrm{f}}^{(1)}(t_k) \end{bmatrix} u_{\mathrm{f}}^{(1)}(t_k).$$

6 Model Order Identification for Continuous-time Models

where we assume that $a_{11} \neq 0$,

$$\begin{bmatrix} u_{21} \\ u_{31} \end{bmatrix} = \begin{bmatrix} \frac{a_{21}}{a_{11}} \\ \frac{a_{31}}{a_{11}} \end{bmatrix} ; \begin{bmatrix} v_{12} & v_{13} \end{bmatrix} = \begin{bmatrix} \frac{a_{12}}{a_{11}} & \frac{a_{13}}{a_{11}} \end{bmatrix}$$

$$\begin{bmatrix} d_{11} & d_{12} \\ d_{21} & d_{22} \end{bmatrix} = \mathbf{A}_{22} - \begin{bmatrix} a_{21} \\ a_{31} \end{bmatrix} \begin{bmatrix} a_{12} & a_{13} \end{bmatrix} / a_{11}$$

Step 2. This concentrates on decomposing the matrix

$$\bar{\mathbf{X}} = \begin{bmatrix} a_{11} & 0 & 0 \\ 0 & d_{11} & d_{12} \\ 0 & d_{21} & d_{22} \end{bmatrix} \tag{6.37}$$

By letting

$$\mathbf{A}_{11} = \begin{bmatrix} a_{11} & 0 \\ 0 & d_{11} \end{bmatrix} ; \mathbf{A}_{22} = d_{22}; \mathbf{B} = \begin{bmatrix} 0 \\ d_{12} \end{bmatrix} ; \mathbf{C} = \begin{bmatrix} 0 & d_{21} \end{bmatrix}$$

we produce

$$\mathbf{X} = \begin{bmatrix} 1 & 0 & 0 \\ 0 & 1 & 0 \\ 0 & \frac{d_{21}}{d_{11}} & 1 \end{bmatrix} \begin{bmatrix} a_{11} & 0 & 0 \\ 0 & d_{11} & 0 \\ 0 & d_{21} & d_{22} - \frac{d_{21}d_{12}}{d_{11}} \end{bmatrix} \begin{bmatrix} 1 & 0 & 0 \\ 0 & 1 & \frac{d_{12}}{d_{11}} \\ 0 & 0 & 1 \end{bmatrix}$$

Note that

$$\begin{bmatrix} 1 & 0 & 0 \\ u_{21} & 1 & 0 \\ u_{31} & \frac{d_{21}}{d_{11}} & 1 \end{bmatrix} = \begin{bmatrix} 1 & 0 & 0 \\ u_{21} & 1 & 0 \\ u_{31} & 0 & 1 \end{bmatrix} \begin{bmatrix} 1 & 0 & 0 \\ 0 & 1 & 0 \\ 0 & \frac{d_{21}}{d_{11}} & 1 \end{bmatrix}$$

which is also true for the **V** matrix.

One key point should be noted here: the decomposition is constructed in such a way that the decomposed elements in the earlier columns and rows are not affected by the computation introduced in the later columns and rows. This is, of course, important in the multiple-model structure approach, where the model estimated from higher-order should not affect the estimated lower-order models.

If **X** is symmetric and positive-definite, then $\mathbf{U} = \mathbf{V}^T$, and the diagonal elements in **D** are positive. However, in the case of IV estimation, the IPM (here **X**) is not necessarily symmetric, and as a result, some diagonal elements in **D** could become negative. As the procedure demonstrates, however, the later decomposition involves only the division of a diagonal element from the previous operation. Consequently, the inaccuracy caused by the singularity or near singularity of the IPM at a higher order does not affect the lower-order estimated model, and the model order can be safely truncated without re-computing the whole set of of candidate models.

To facilitate the further application of this method, the script of the MATLAB® m-file for the UDV factorisation, based on the above decomposition procedure, is shown below.

```
function [U,D,V]=UDV_f(A)
%UDV factorisation of a matrix A
[n,m]=size(A);
D=zeros(n,n);
D(1,1)=A(1,1);
J=A;
[U,V,J]=UDV_de(J);   %function is included in the following
D(2,2)=J(1,1);
U_mat(:,1)=U;
V_mat(:,1)=V;
for kk=2:n;
    [U,V,J]=UDV_de(J);
    U_mat=[U_mat [zeros(kk-1,1);U]];
    V_mat=[V_mat [zeros(kk-1,1);V]];
    if kk<n
        D(kk+1,kk+1)=J(1,1);
    end
  end
U=U_mat;
V=V_mat';


function [U,V,J]=UDV_de(J);
%J is the submatrix to be decomposed
    [n,m]=size(J);
    U=J(:,1)/J(1,1);
    Jt=J';
    V=Jt(:,1)/J(1,1);
    J=J(2:n,2:n)-U(2:n,1)*V(2:n,1)'*J(1,1);
```

6.4 Model Structure Selection Using PRESS

Cross-validation is one way of assessing the quality of an estimated model. In the simplest situation, the N sample experimental data set is separated into an estimation set of M samples and a validation set of $N - M$ samples. The estimation data set is used for estimating the parameters of the model; while the remaining validation data set provides the basis for testing its predictive capability. In the present context, the prediction error set is generated by calculating, for $k = M + 1, M + 2, \ldots, N$,

$$\hat{y}(t_k) = \phi^T(t_k)\hat{\boldsymbol{\theta}}$$
$$e_p(t_k) = y(t_k) - \hat{y}(t_k) \qquad (6.38)$$

The error $e_p(t_k)$ can be considered as the true prediction error because the noise disturbances in the estimation set are independent of those in the validation set. Consequently, the sum of squared prediction errors $\sum_{k=M+1}^{N} e_p^2(t_k)$ can be used for determining the quality of the estimated model.

More complex cross-validation procedures involve separation of the observational data into several sets. Then, each time, one set of data is left out for model validation and the rest of the sets are used for parameter estimation, with the prediction errors computed for the omitted set. This procedure is repeated for all the validation sets in order to obtain the prediction errors. The limit of this practice is the so-called 'leave-one-out' validation procedure, where the prediction error is defined for the case of estimating a continuous-time model with prefiltering as

$$e_{-t_m}(t_m) = y_f^{(n)}(t_m) - \varphi^T(t_m)\hat{\boldsymbol{\theta}}_{-t_m} \tag{6.39}$$

in which $\varphi(t_m)$ is the data regressor and defined as

$$\varphi^T(t_m) = [-y_f^{(n-1)}(t_m) \ \ldots \ -y_f^{(0)}(t_m) \ u_f^{(m)}(t_m) \ \ldots u_f^{(0)}(t_m)]$$

and, in the present context, $\hat{\boldsymbol{\theta}}_{-t_m}$ is the IV estimate of $\boldsymbol{\theta}$ without the information $\hat{\varphi}(t_m)$, $\varphi(t_m)$ and $y_f^{(n)}(t_m)$. Mathematically, $\hat{\boldsymbol{\theta}}_{-t_m}$ is computed as

$$\hat{\boldsymbol{\theta}}_{-t_m} = [\sum_{k=1}^{N} \hat{\varphi}(t_k)\varphi^T(t_k) - \hat{\varphi}(t_m)\varphi^T(t_m)]^{-1} \sum_{k=1}^{N} \hat{\varphi}(t_k)y_f^{(n)}(t_k) - \hat{\varphi}(t_m)y_f^{(n)}(t_m) \tag{6.40}$$

The sum of the squared prediction errors $e_{-t_m}(.)$ is called the predicted residual sums of squares (PRESS) and is defined as follows

$$\text{PRESS} = \sum_{k=1}^{N} e_{-t_k}^2(t_k) \tag{6.41}$$

This is a computationally demanding task when the PRESS statistic is computed from its original form. However, it was shown in [9] that, when a least squares algorithm is used, PRESS can be computed using a rather simple formula. Fortunately, this approach can be extended to the case where the parameter vector $\hat{\boldsymbol{\theta}}$ is estimated using the IV approach [9, 10] and it can be used for both continuous and discrete-time model identification.

To avoid confusion, we will first discuss the computation of PRESS for a basic IV algorithm, and then extend the results to the multiple-model structure with UDV factorisation.

Theorem 6.2. *The prediction error is computed via the following equation*

$$e_{-t_m}(t_m) = \frac{y_f^{(n)}(t_m) - \varphi^T(t_m)\hat{\boldsymbol{\theta}}}{1 - \varphi^T(t_m)(\sum_{k=1}^{N} \hat{\varphi}(t_k)\varphi^T(t_k))^{-1}\hat{\varphi}(t_m)} \tag{6.42}$$

where,

$$\hat{\boldsymbol{\theta}} = \left[\sum_{k=1}^{N} \hat{\boldsymbol{\varphi}}(t_k)\boldsymbol{\varphi}(t_k)^T\right]^{-1} \sum_{k=1}^{N} \hat{\boldsymbol{\varphi}}(t_k)y_f^{(n)}(t_k)$$

is the IV estimated parameter vector.

Remark 6.3. The key point is that the leave-one-out prediction error is produced by the standard residual from the IV estimation, multiplied by a weighting function defined as

$$\frac{1}{1 - \boldsymbol{\varphi}^T(t_m)(\sum_{k=1}^{N} \hat{\boldsymbol{\varphi}}(t_k)\boldsymbol{\varphi}^T(t_k))^{-1}\hat{\boldsymbol{\varphi}}(t_m)}$$

This formulae makes the computation simple and straightforward.

Proof: Let $\mathbf{A}$ be a non-singular matrix. Let $\boldsymbol{\alpha}$ and $\boldsymbol{\beta}$ be column matrices, and assume that $\mathbf{A} + \boldsymbol{\alpha}\boldsymbol{\beta}^T$ is non-singular. Then, the well-known matrix inversion lemma states that

$$(\mathbf{A} + \boldsymbol{\alpha}\boldsymbol{\beta}^T)^{-1} = \mathbf{A}^{-1} - \frac{\mathbf{A}^{-1}\boldsymbol{\alpha}\boldsymbol{\beta}^T\mathbf{A}^{-1}}{1 + \boldsymbol{\beta}^T\mathbf{A}^{-1}\boldsymbol{\alpha}}$$

From this, we have

$$\left[\sum_{k=1}^{N} \hat{\boldsymbol{\varphi}}(t_k)\boldsymbol{\varphi}^T(t_k) - \hat{\boldsymbol{\varphi}}(t_m)\boldsymbol{\varphi}(t_m)^T\right]^{-1} = \sum_{k=1}^{N} \hat{\boldsymbol{\varphi}}(t_k)\boldsymbol{\varphi}^T(t_k)$$
$$+ \frac{(\sum_{k=1}^{N} \hat{\boldsymbol{\varphi}}(t_k)\boldsymbol{\varphi}^T(t_k))^{-1}\hat{\boldsymbol{\varphi}}(t_m)\boldsymbol{\varphi}^T(t_m)(\sum_{k=1}^{N} \hat{\boldsymbol{\varphi}}(t_k)\boldsymbol{\varphi}^T(t_k))^{-1}}{1 - \boldsymbol{\varphi}^T(t_m)(\sum_{k=1}^{N} \hat{\boldsymbol{\varphi}}(t_k)\boldsymbol{\varphi}^T(t_k))^{-1}\hat{\boldsymbol{\varphi}}(t_m)} \quad (6.43)$$

Then, from the definition of $\hat{\boldsymbol{\theta}}_{-t_m}$ by (6.40) in conjunction with (6.43), we have

$$\hat{\boldsymbol{\theta}}_{-t_m} = [\sum_{k=1}^{N} \hat{\boldsymbol{\varphi}}(t_k)\boldsymbol{\varphi}(t_k)^T$$
$$+ \frac{(\sum_{k=1}^{N} \hat{\boldsymbol{\varphi}}(t_k)\boldsymbol{\varphi}^T(t_k))^{-1}\hat{\boldsymbol{\varphi}}(t_m)\boldsymbol{\varphi}^T(t_m)(\sum_{k=1}^{N} \hat{\boldsymbol{\varphi}}(t_k)\boldsymbol{\varphi}^T(t_k))^{-1}}{1 - \boldsymbol{\varphi}^T(t_m)(\sum_{k=1}^{N} \hat{\boldsymbol{\varphi}}(t_k)\boldsymbol{\varphi}^T(t_k))^{-1}\hat{\boldsymbol{\varphi}}(t_m)}]$$
$$\times [\sum_{k=1}^{N} \hat{\boldsymbol{\varphi}}(t_k)y_f^{(n)}(t_k) - \hat{\boldsymbol{\varphi}}(t_m)y_f^{(n)}(t_m)] \quad (6.44)$$

Although, at first sight, (6.44) looks complicated, simple algebraic manipulation produces two basic terms

6 Model Order Identification for Continuous-time Models

$$\hat{\boldsymbol{\theta}}_{-t_m} = [\hat{\boldsymbol{\theta}} + \frac{(\sum_{k=1}^{N} \hat{\boldsymbol{\varphi}}(t_k)\boldsymbol{\varphi}^T(t_k))^{-1}\hat{\boldsymbol{\varphi}}(t_m)\boldsymbol{\varphi}^T(t_m)\hat{\boldsymbol{\theta}}}{1 - \boldsymbol{\varphi}^T(t_m)(\sum_{k=1}^{N}\hat{\boldsymbol{\varphi}}(t_k)\boldsymbol{\varphi}^T(t_k))^{-1}\hat{\boldsymbol{\varphi}}(t_m)}]$$
$$- [\mathbf{I} + \frac{(\sum_{k=1}^{N}\hat{\boldsymbol{\varphi}}(t_k)\boldsymbol{\varphi}^T(t_k))^{-1}\hat{\boldsymbol{\varphi}}(t_m)\boldsymbol{\varphi}^T(t_m)}{1 - \boldsymbol{\varphi}^T(t_m)(\sum_{k=1}^{N}\hat{\boldsymbol{\varphi}}(t_k)\boldsymbol{\varphi}^T(t_k))^{-1}\hat{\boldsymbol{\varphi}}(t_m)}]$$
$$[(\sum_{k=1}^{N}\hat{\boldsymbol{\varphi}}(t_k)\boldsymbol{\varphi}^T(t_k))^{-1}\hat{\boldsymbol{\varphi}}(t_m)y_{\mathrm{f}}^{(n)}(t_m)] \quad (6.45)$$

where $\hat{\boldsymbol{\theta}} = (\sum_{k=1}^{N}\hat{\boldsymbol{\varphi}}(t_k)\boldsymbol{\varphi}^T(t_k))^{-1}\sum_{k=1}^{N}\hat{\boldsymbol{\varphi}}(t_k)y_{\mathrm{f}}^{(n)}(t_k)$ is the IV estimated parameter vector and $\mathbf{I}$ is the identity matrix of dimension equal to that of the IPM.

On the basis of this estimated $\hat{\boldsymbol{\theta}}_{-t_m}$, the predicted output of the model at time t_m is

$$y_{-t_m}(t_m) = \boldsymbol{\varphi}^T(t_m)\hat{\boldsymbol{\theta}}_{-t_m} \quad (6.46)$$

Substituting (6.45) into (6.46) yields

$$y_{-t_m}(t_m) = \frac{\boldsymbol{\varphi}^T(t_m)\hat{\boldsymbol{\theta}} - \boldsymbol{\varphi}^T(t_m)(\sum_{k=1}^{N}\hat{\boldsymbol{\varphi}}(t_k)\boldsymbol{\varphi}^T(t_k))^{-1}\hat{\boldsymbol{\varphi}}(t_m)y_{\mathrm{f}}^n(t_m)}{1 - \boldsymbol{\varphi}^T(t_m)(\sum_{k=1}^{N}\hat{\boldsymbol{\varphi}}(t_k)\boldsymbol{\varphi}^T(t_k))^{-1}\hat{\boldsymbol{\varphi}}(t_m)} \quad (6.47)$$

Now, the prediction error at time t_m is

$$e_{-t_m}(t_m) = y_{\mathrm{f}}^{(n)}(t_m) - y_{-t_m}(t_m) \quad (6.48)$$

and the prediction error specified by (6.42) is obtained by substitution from (6.47).

The results presented in Theorem 6.2 are derived for a general IV estimation algorithm. In this general framework and given a specific model structure, we estimate $\hat{\boldsymbol{\theta}}$, and then evaluate the prediction error $e_{-t_m}(t_m)$ using (6.42) for $t_m = 0, t_1, t_2, \ldots, T_m$. However, when the multiple-model structure is used in conjunction with UDV matrix decomposition, it is possible to further simplify the computation and obtain the weights for the computation of PRESS for all candidate models.

In order to explain this approach, let us first examine the structure of the augmented data regressor $\boldsymbol{\phi}_n(t_k)$ in (6.19) for a maximum model order of n. This regressor contains all the elements required for a family of regressors up to the maximum model order. For instance, by blocking out the augmented $-y_{\mathrm{f}}^{(n)}(t_k)$, then we have the regressor for model order n. Therefore, we can construct the weight in the prediction error for all candidate models from the augmented data regressors (e.g., $\boldsymbol{\phi}_n(t_k)$ and $\hat{\boldsymbol{\phi}}_n(t_k)$).

Note that for a given maximum model order n, we can write

$$\boldsymbol{\phi}_n^T(t_m)(\sum_{k=1}^{N}\hat{\boldsymbol{\phi}}_n(t_k)\boldsymbol{\phi}_n^T(t_k))^{-1}\hat{\boldsymbol{\phi}}_n(t_m)$$
$$= \boldsymbol{\phi}_n(t_m)^T\mathbf{V}^{-1}\mathbf{D}^{-1}\mathbf{U}^{-1}\hat{\boldsymbol{\phi}}_n(t_m) \quad (6.49)$$

where $\sum_{k=1}^{N} \hat{\boldsymbol{\phi}}_n(t_k)\boldsymbol{\phi}_n(t_k)^T = \mathbf{UDV}$.
Now, if we define two transformed data vectors as

$$\boldsymbol{w}_\phi(t_m)^T = \boldsymbol{\phi}_n(t_m)^T\mathbf{V}^{-1}; \boldsymbol{w}_\psi(t_m) = \mathbf{U}^{-1}\hat{\boldsymbol{\phi}}_n(t_m)$$

then we have

$$\boldsymbol{\phi}_n(t_m)^T\mathbf{V}^{-1}\mathbf{D}^{-1}\mathbf{U}^{-1}\hat{\boldsymbol{\phi}}_n(t_m) = \boldsymbol{w}_\phi(t_m)^T\mathbf{D}^{-1}\boldsymbol{w}_\psi(t_m)$$
$$= \sum_{k=1}^{n_m} \frac{\boldsymbol{w}_\phi(t_m)^k \boldsymbol{w}_\psi(t_m)^k}{d_k} \quad (6.50)$$

where n_m is the dimension of the augmented data vector, $\boldsymbol{w}_\phi(t_m)^k$ is the kth element in the vector $\boldsymbol{w}_\phi(t_m)$, $\boldsymbol{w}_\psi(t_m)^k$ is the kth element in the vector $\boldsymbol{w}_\psi(t_m)$ and d_k is the kth diagonal element in the $\mathbf{D}$ matrix. Since $\mathbf{U}^{-1}$ and $\mathbf{V}^{-1}$ matrices are lower and upper triangular matrices and the decomposition is performed with the lower-order structure first, the elements in the transformed data vectors are preserved with respect to the orders in the original regressors. In other words, the truncations of $\boldsymbol{w}_\phi(t_m)$ and $\boldsymbol{w}_\psi(t_m)$ will lead to the transformed vectors for a lower-order structure. Therefore, for a given model order, the quantity for the weight of the error equation is a truncated sum of (6.50). For instance, given a maximum model order n and a relative model order 1, the weight for the prediction error is as follows

First order

$$\frac{\boldsymbol{w}_\phi(t_m)^1 \boldsymbol{w}_\psi(t_m)^1}{d_1} + \frac{\boldsymbol{w}_\phi(t_m)^2 \boldsymbol{w}_\psi(t_m)^2}{d_2}$$

Second order

$$\frac{\boldsymbol{w}_\phi(t_m)^1 \boldsymbol{w}_\psi(t_m)^1}{d_1} + \frac{\boldsymbol{w}_\phi(t_m)^2 \boldsymbol{w}_\psi(t_m)^2}{d_2} + \frac{\boldsymbol{w}_\phi(t_m)^3 \boldsymbol{w}_\psi(t_m)^3}{d_3} + \frac{\boldsymbol{w}_\phi(t_m)^4 \boldsymbol{w}_\psi(t_m)^4}{d_4}$$

It is quite easy to use MATLAB® to compute the PRESS statistic in conjunction with the UDV decomposition, as shown by the MATLAB® code below.

```
A=Z*Phi'; %generate the cross-correlation matrix
[U,D,V]=UDV_f(A); %UDV factorisation of A matrix;
%calculating PRESS for all possible model structures
%the residual errors are obtained Phi'*inv(V) as inv(V) is
%structured to contain estimated parameter vector with unit
%diagonal element via instrumental variable
Vinv=inv(V);
E=-Phi'*Vinv; %the error for all decomposition
W_left=Phi'*inv(V); %Phi is a flat matrix
W_right1=inv(U)*Z; %Z is a flat matrix
W_right=W_right1'; % W_right to be a tall matrix
[nr,nc]=size(W_left);
```

```
        h=zeros(nr,1);
        one=ones(nr,1);
        %calculation of PRESS begins here
        for i=2:nc
                % calculate prediction error
                % h is the inflation matrix
                % epress is the press error
                     h=h+W_left(:,i-1).*W_right(:,i-1)/D(i-1,i-1);
                     epress(:,i)=E(:,i)./(one-h);
        end

        loss=diag(epress'*epress);    %this is the PRESS
                                      %for all possible models
```

6.5 Simulation Studies

Consider the continuous-time system with Laplace transfer function

$$G(s) = \frac{0.25(s+1)}{(s^2 + 2 \times 0.1 \times 0.5s + 0.5^2)(s+0.1)(s+0.6)} \quad (6.51)$$

where s denotes the Laplace variable. For the purposes of the simulation study, the input is selected as a PRBS signal of unit amplitude with a switching period of 15 samples. The input and noise free output signals are sampled at a uniform sampling interval of 0.1 time units, over a period of 600 time units. Additive measurement noise, in the form of a normally distributed, discrete-time white noise sequence, with a standard deviation of 1.4, is added to the sampled, noise-free output signal $x(t_k)$ to yield the noisy, measured output signal $y(t_k)$.

In order to assist in the application of the multiple-model identification strategy, let us first consider the results obtained with the standard identification approach for optimal RIVC/SRIVC estimation, as described in Chapter 4 of this book. With the help of the RIVCID identification routine in the CAPTAIN Toolbox, the SRIVC algorithm (which is optimal in the above experimental situation) is used to estimate all models from first to sixth order. However, models greater than 4th order are all rejected by RIVCID as unidentifiable because of the ill-conditioning of the IPM (correctly rejected, of course: this is one of the primary objectives of the RIVCID identification routine). The identification results for the 4 best identified models are given in Table 6.1.

In Table 6.1, R_T^2 is the simulation coefficient of determination defined by,

$$R_T^2 = 1 - \frac{\hat{\sigma}^2}{\sigma_y^2} \quad (6.52)$$

Table 6.1. SRIVC identification results

Model	YIC	R_T^2
4 1 0	−12.5	0.986
4 2 0	−9.3	0.986
4 3 0	−5.9	0.986
3 4 0	−9.3	0.985

where $\hat{\sigma}^2$ is the variance of the error between the simulated model output and the measured output; while σ_y^2 is the variance of the noisy output signal. YIC is a model order information criterion, based on the inverse of the IPM (see [47] and Chapter 4 of this book), where more negative values indicate models with good conditioning of the IPM and the low average parametric estimation uncertainty. On this basis, the [4 1 0] model appears best identified; while the true [4 2 0] model is next best but it has a marginally larger R_T^2. Both of these models yield step response characteristics that are virtually indistinguishable from the actual system step response. However, while the Bode diagram for the [4 2 0] model is virtually the same as the actual system Bode diagram, that for the [4 1 0] model diverges somewhat at the highest frequencies. The SRIVC parameter estimates obtained for the [4 2 0] model are given in Table 2, where SE denotes the estimated standard errors on the parameter estimates.

In order to demonstrate the procedure of UDV factorisation and model structure selection using PRESS, we consider the case when the instrumental variable is selected as the simulated, noise-free output of the SRIVC estimated [4 2 0] model, above[5]; and the state-variable filter is the denominator of the same estimated model, with 2 added roots at $s+10$ to increase it to 6th order, as required if we wish to evaluate all models up to 6th order.

In order to demonstrate that the higher-order model estimates do not affect the lower-order model estimates, the maximum model order is selected equal to 5 and 6, respectively, assuming the correct relative degree of 3. The estimation results are illustrated below for the corresponding columns of $\mathbf{V}^{-1}$.

---

[5] Very similar results are obtained if the SRIVC estimated [4 1 0] model is used, rather than the [4 2 0] model.

Table 6.2. SRIVC estimation results

Model	a_1	a_2	a_3	a_4	b_0	b_1
True	0.8000	0.3800	0.1810	0.0150	0.2500	1.0000
Estimated	0.7890	0.3775	0.1778	0.0148	0.2668	0.9854
SE	0.0142	0.0029	0.0036	0.0003	0.0184	0.0196

$n_{\max} = 5$ (third-, fourth- and fifth-order model parameters)

```
0.0208    0.0156   -0.0153
0.2484    0.1886   -0.1524
0.1954    0.3876    0.0013
0.3601    0.2613   -0.2573
1.0000    0.8324   -0.2812
     0    0.2565    0.2817
     0    1.0000    0.5943
     0         0    0.1349
     0         0    1.0000
```

$n_{\max} = 6$ (third-, fourth-, fifth- and sixth-order model parameters)

```
0.0208    0.0156   -0.0153    0.0062
0.2484    0.1886   -0.1524    0.1975
0.1954    0.3876    0.0013    1.5731
0.3601    0.2613   -0.2573    0.1013
1.0000    0.8324   -0.2812    2.1437
     0    0.2565    0.2817    2.1826
     0    1.0000    0.5943    6.4232
     0         0    0.1349    0.2456
     0         0    1.0000    3.1008
     0         0         0    0.8283
     0         0         0    1.0000
```

The elements on the diagonal of the **D** matrix are given below for $n_{\max} = 6$.

```
1.3118e+4   631.2602    96.8180   26.3013    0.0481   0.4863
1.9957e-6   0.0081    -7.9386e-8  1.2535e-5  2.9247e-8
```

It can be seen that the 7th element on the diagonal of **D** ($= 1.9957\mathrm{e} - 6$) is small, corresponding to the correct, 4th-order model. If the estimation is continued, the 5th-order model is obtained. However, this is unstable, as indicated by the negative sign of some coefficients. In fact, further inspection shows that this estimated model exhibits a near cancellation of an unstable pole at 0.6162 with a zero at 0.6872. The good news is that the poor estimation results for the 5th and 6th model orders did not affect the estimation

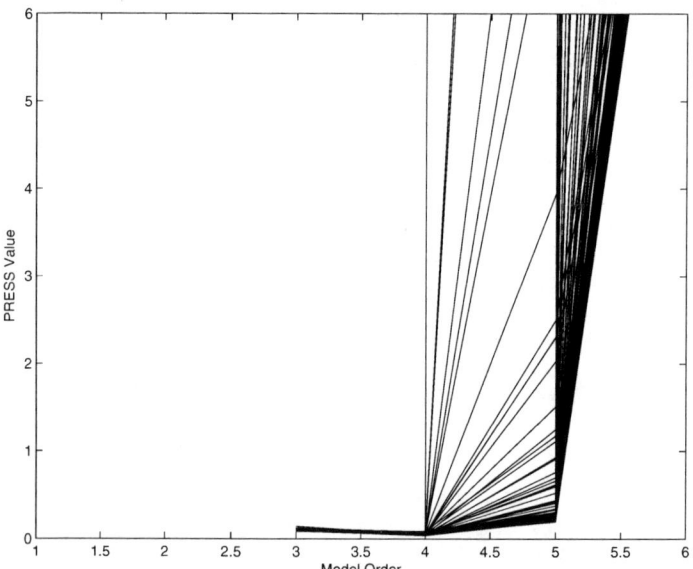

Fig. 6.2. PRESS values for the 100 noise realisations with correct relative degree

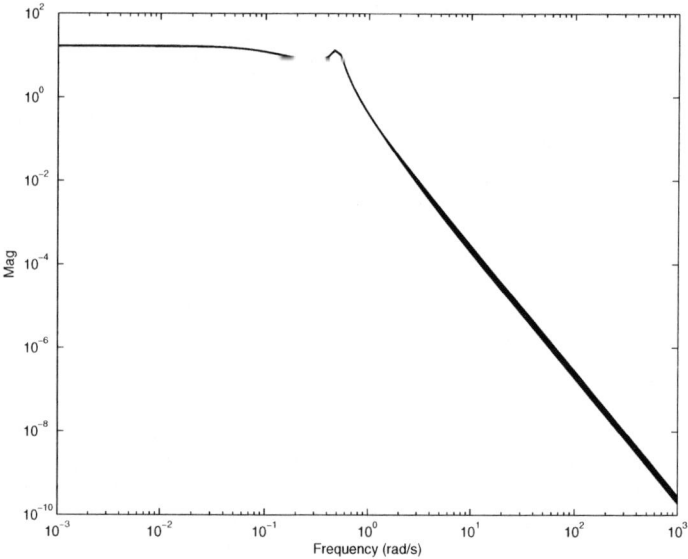

Fig. 6.3. The estimated frequency responses for the 100 noise realisations with correct relative degree

6 Model Order Identification for Continuous-time Models   183

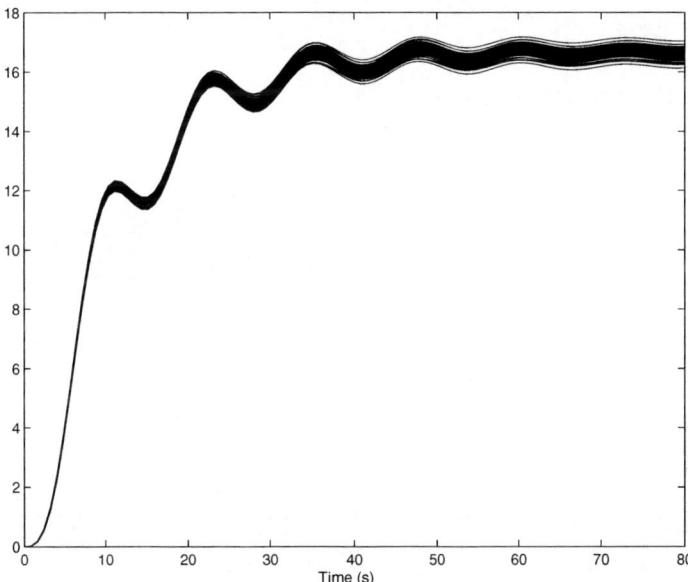

Fig. 6.4. The estimated step responses for the 100 noise realisations with correct relative degree

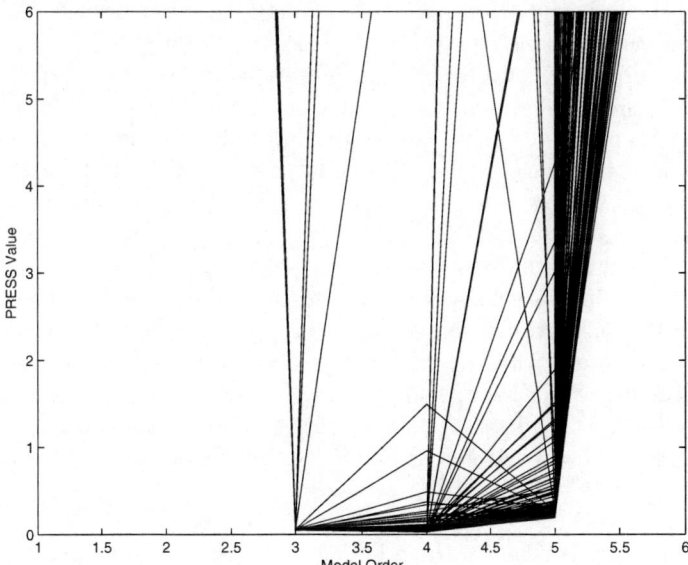

Fig. 6.5. PRESS values for the 100 noise realisations with incorrect relative degree

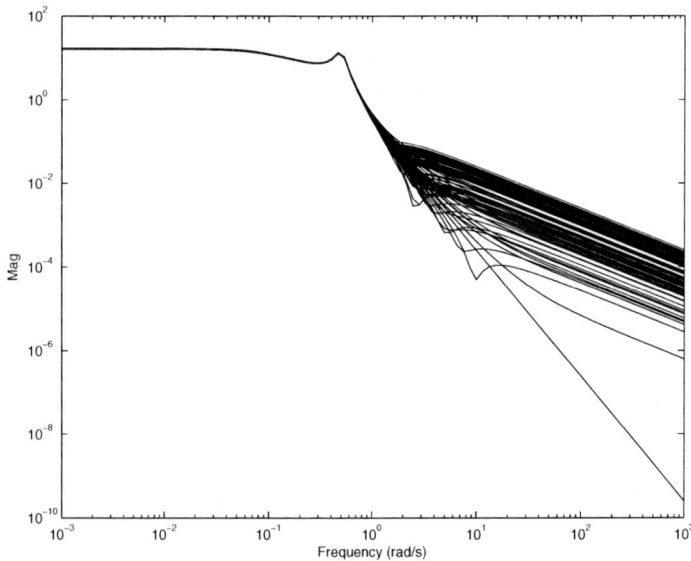

Fig. 6.6. The estimated frequency responses for the 100 noise realisations with incorrect relative degree

results for the 3rd- and 4th-order models. Thus, we can utilise the estimation results from the lower-order estimated models.

In order to illustrate how PRESS can be used to detect the model order, a Monte Carlo simulation is utilised, based on 100 independent realisations of the measurement noise, under the assumption that the maximum model order is 6, with a relative degree of 3. Figure 6.2 shows the PRESS values for all the realisations and we see that the correct order is identified in all cases. Figures 6.3 and 6.4 show the amplitude plots and step response plots for all these PRESS-selected models.

In order to examine what happens if the relative degree is selected incorrectly, we let $\gamma = 1$, which gives the model an extra degree of freedom in the numerator. The same kind of Monte Carlo simulation experiment as that used above yields Figure 6.5, which shows the resulting PRESS values. The identification is not as clear cut as when the relative degree is correctly specified, but the values differ very little between 3rd- and 4th-order models for the majority of the cases.

It is clear that, because of the mismatch in the relative degree, the best model order is selected according to the minimum PRESS, which measures the predictive capability of the model. Figures 6.6 and 6.7 show the frequency response amplitude and temporal step response plots for all the PRESS-selected models. These show that, in general, despite the mismatch of the relative degree, the estimation results are good. There is only one case where an unstable

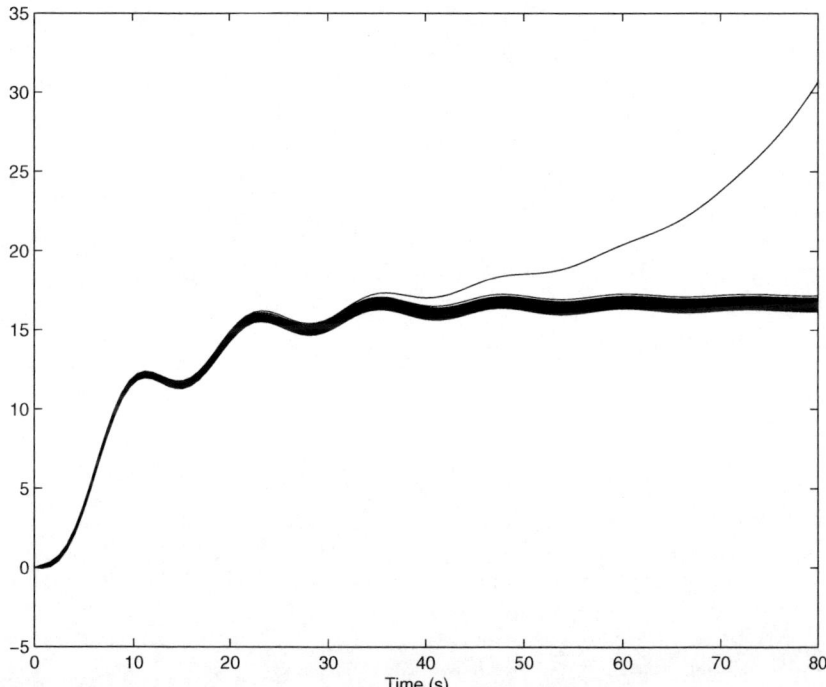

Fig. 6.7. The estimated step responses for the 100 noise realisations with incorrect relative degree

model is selected as seen from Figure 6.6. However, the estimated frequency response is worse in the high-frequency region when compared with the results obtained when the correct relative degree is selected.

It seems that, in this example, there is a compromise between relative degree and model order in this mismatch situation. If the relative degree γ is too low, which means that there are more degrees of freedom for the numerator, then the identified model order is lower than the actual model order. However, if the proposed **UDV** factorisation/PRESS approach is utilised in association with the standard RIVC/SRIVC identification procedure, as we have done in this example, this deficiency would normally be avoided. In the present example, for instance, the SRIVC identification suggests strongly that the relative degree is 3 or 4.

6.6 Conclusions

This chapter has described a new instrumental variable model structure identification procedure for continuous-time models based on the exploitation of a multiple-model structure, UDV matrix factorisation and the PRESS statistic. By assuming a relative degree and maximum model order, a set of candidate continuous-time transfer function models is estimated using UDV matrix decomposition, where the models are estimated sequentially from lower order to higher order. In this procedure, the computation of the prediction error is simplified when using the UDV decomposition algorithm, so that the computational requirements are reasonable. In addition, the leave-one-out cross-validation procedure, as used previously in simple least squares estimation of discrete-time models, is extended to the instrumental variable estimation of continuous-time models.

References

1. P.J. Gawthrop. Parametric identification of transient signals. *IMA Journal of Mathematical Control and Information*, 1:117–128, 1984.
2. L. Ljung, M. Morf, and D. Falconer. Fast calculation of gain matrices for recursive estimation. *International Journal of Control*, 1–19, 1978.
3. S. Niu. Augmented UD identification for process control. *PhD Thesis, Department of Chemical Engineering, University of Alberta, Canada*, 1993.
4. S. Niu, D.G. Fisher, and D. Xiao. An augmented UD identification algorithm. *International Journal of Control*, 193–211, 1992.
5. T. Söderström and P. Stoica. Comparison of some instrumental variable methods: Consistency and accuracy aspects. *Automatica*, 17:101–115, 1981.
6. T. Söderström and P. Stoica. *Instrumental variable methods for system identification*. Springer-Verlag, New York, Lecture Notes in Control and Information Sciences, 1983.
7. P. Stoica and T. Söderström. Instrumental variable methods for identification of Hammerstein systems. *International Journal of Control*, 35:459–476, 1982.
8. P. Stoica, T. Söderström, and B. Friedlander. Optimal instrumental estimates of the AR-parameters of an ARMA process. *IEEE Transactions on Automatic Control*, 30:1066–1074, 1985.
9. L. Wang and W.R. Cluett. Use of PRESS residuals in dynamic system identification. *Automatica*, 32:781–784, 1996.
10. L. Wang and W.R. Cluett. *From Plant Data to Process Control: Ideas for Process Identification and PID Design*. Taylor and Francis, London, 2000.
11. L. Wang and P.J. Gawthrop. On the estimation of continuous time transfer functions. *International Journal of Control*, 74:889–904, 2001.
12. P.E. Wellstead. An instrument product moment test for model order estimation. *Automatica*, 14:89–91, 1978.
13. P.E. Wellstead and R.A. Rojas. Instrumental product model order testing: extensions and applications. *International Journal of Control*, 35:1013–1027, 1982.
14. K.Y. Wong and E. Polak. Identification of linear discrete time system using the instrumental variable approach. *IEEE Transactions on Automatic Control*, 12:707–718, 1967.
15. P.C. Young. In flight dynamic checkout - a discussion. *IEEE Transactions on Aerospace*, AS-2(3):1106–1111, 1964.

16. P.C. Young. Some observations on instrumental variable methods of time series analysis. *International Journal of Control*, 23:593–612, 1976.
17. P.C. Young. Parameter estimation for continuous-time models - a survey. *Automatica*, 17(1):23–39, 1981.
18. P.C. Young. *Recursive Estimation and Time Series Analysis*. Springer-Verlag, Berlin, 1984.
19. P.C. Young. Recursive estimation, forecasting and adaptive control. In C.T. Leondes (ed), *Control and Dynamic Systems*, pages 119–166, Academic Press: San Diego, 1989.
20. P.C. Young and A.J. Jakeman. Refined instrumental variable methods of recursive time-series analysis: Part I: single input and single output systems. *International Journal of Control*, 29:1–30, 1979.
21. P.C. Young and A.J. Jakeman. Refined instrumental variable methods of time-series analysis: Part III, extensions. *International Journal of Control*, 31:741–764, 1980.
22. P.C. Young, A.J. Jakeman, and R. McMurtrie. An instrumental variable method for model order identification. *Automatica*, 16:281–296, 1980.

7

Estimation of the Parameters of Continuous-time Systems Using Data Compression

Liuping Wang[1] and Peter J. Gawthrop[2]

[1] RMIT University, Melbourne, Australia
[2] University of Glasgow, Scotland

7.1 Introduction

This chapter provides a unified introductory account of the estimation of the parameters of continuous-time systems using data compression based on a number of previous publications [17, 18, 28, 30–32].

The outline of the chapter is indicated in Figure 7.1. In particular, the core of our approach is the frequency-sampling filter (FSF) of Wang and Cluett [28, 30] where time- or frequency-domain data – within a predefined bandwidth – are represented as a set of (complex) filter coefficients; this can be viewed as a form of identification-orientated data compression.

The FSF coefficients are used to derive a system step response that is used in one of two ways:

1. to generate the parameters of a transfer function;
2. to optimise the parameters of a physical model represented by a bond graph [16].

The methods will be described, analysed and also illustrated using data from a real example.

7.2 Data Compression Using Frequency-sampling Filters

This section will describe the data-compression process using frequency-sampling filters (FSF). The original work on system identification using FSF was oriented for discrete-time system identification of step-response coefficients used for industrial process control in the 1990s [28, 30] at the University of Toronto where the first author worked during that period. Recent years have seen the FSF model being used as an effective part of the vehicle designed for continuous-time system identification [17, 18, 31, 32]. The connection between continuous-time and discrete-time identification is the fact that

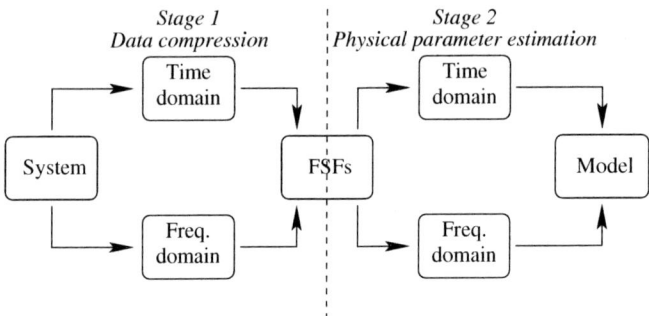

Fig. 7.1. Identification overview. Stage 1 identifies FSF parameters from either time or frequency data; stage two uses this compressed data to estimate either an empirical transfer function or the parameters of a physical model.

the estimated step-response coefficients in the discrete case correspond to the continuous-time step response at the sampling time instant, *i.e.*, the scenario of step-response invariance. In addition, the discrete frequency response can also be used to approximate the continuous-time frequency response when the sampling rate is chosen to be sufficiently fast. Based on these observations, we can estimate the discrete-time step-response coefficients or frequency response using the FSF model, which effectively compressed the original experimental data set to a set of estimated step-response coefficients. The reasons why we choose FSF model in the compression process are: (1) its simplicity and robustness in the estimation process; (2) relatively small number of parameters required to be estimated in the environment of fast sampling; (3) *a priori* knowledge can be incorporated in the compression process to improve the accuracy of the compressed model.

7.2.1 FSF Model

Let us start with a single input and single output system, and extend the model to multi-input and multi-output system. Assume a stable discrete-time system with a sampling interval of T_s. This system can be described by the finite impulse response (FIR) model

$$G(z) = \sum_{j=0}^{N-1} h_j z^{-j} \tag{7.1}$$

where N is the model order chosen such that the impulse response coefficients $h_j \approx 0$ for all $j \geq N$. Note that in the FIR model (7.1), the number of coefficients is determined by the settling time (let us call it T_m) of the underlying continuous-time system as well as the sampling interval we use. The general rule is to choose $N \approx \frac{T_m}{T_s}$. In the context of continuous-time system identification, T_s is set to be sufficiently small so that the discrete-time process closely mimics the underlying continuous-time system. In other words, the model order N linearly increases as the sampling interval T_s decreases. We typically choose N in the range of $400 - 800$ for a near-continuous-time system. There are two related issues. One is the computational load; and the other is the quality of the model. Instead, the FSF model is introduced to preserve the original advantages of an FIR model, yet overcome the difficulties encountered by an FIR in the fast sampling environment, through a linear transformation. The essence here is to capture the coefficients in the frequency domain, instead of the time domain. Let us assume that N is an odd number, and the set of the discrete-time frequency response is $G(e^{i\frac{2\pi j}{N}})$ for $j = 0, \pm 1, \pm 2, \ldots, \pm\frac{N-1}{2}$. Odd number N is selected to ensure that we include the zero frequency and the rest of the frequencies appear in complex-conjugate pairs. Then, from the inverse discrete Fourier transform (IDFT) relationship, we have the jth impulse response coefficient

$$h_j = \frac{1}{N} \sum_{l=-\frac{N-1}{2}}^{\frac{N-1}{2}} G(e^{i\frac{2\pi l}{N}}) e^{i\frac{2\pi l j}{N}} \tag{7.2}$$

Substituting (7.2) into (7.1) gives

$$G(z) = \sum_{j=0}^{N-1} \frac{1}{N} \sum_{l=-\frac{N-1}{2}}^{\frac{N-1}{2}} G(e^{i\frac{2\pi l}{N}}) e^{i\frac{2\pi l j}{N}} z^{-j} \tag{7.3}$$

$$= \sum_{l=-\frac{N-1}{2}}^{\frac{N-1}{2}} G(e^{i\frac{2\pi l}{N}}) \frac{1}{N} \frac{1-z^{-N}}{1-e^{i\frac{2\pi l}{N}} z^{-1}} \tag{7.4}$$

where (7.4) is obtained by interchanging the summations in (7.3) and using the result

$$\sum_{j=0}^{N-1} e^{i\frac{2\pi l j}{N}} z^{-j} = \frac{1-z^{-N}}{1-e^{i\frac{2\pi l}{N}} z^{-1}} \tag{7.5}$$

The z-transfer function model defined by (7.4) is called the frequency-sampling filter model. Through linear transformation from the FIR model, the coefficients of the FSF model are the set of discrete frequency responses in an interval of $\frac{2\pi}{N}$. Let us define $\Omega = \frac{2\pi}{N}$ and

$$H_l(z) = \frac{1}{N} \frac{1-z^{-N}}{1-e^{il\Omega} z^{-1}} \tag{7.6}$$

for $l = 0, \pm 1, \pm 2, \ldots, \pm\frac{N-1}{2}$, as the frequency-sampling filters. At $z = e^{il\Omega}$, $H^l(z) = 1$. The FSF filters are narrow-band limited, with their centre frequencies located at $\frac{2\pi l}{N}$ radians. All the filters have identical frequency responses except for the locations of their centre frequencies. As the parameter N increases, the bandwidth of the filters reduces and tends to behave more like a

δ function. Namely, at $\Omega l = \frac{2\pi l}{N}$, the filter magnitudes are equal to one and are small or equal to zero at all other frequencies.

7.2.2 FSF Model in Data Compression

It is easy to understand that the parameters for a model that is based on time-domain descriptions such as FIR will increase linearly as T_s reduces. With the rational z-transfer function models, it was shown in [1] that a continuous-time process with zeros in the left half-plane will often give rise to a discrete-time model having zeros outside the unit circle as the sampling period tends to zero. In addition, the poles of the discrete-time model will tend to the unit circle. For these reasons, the main approach for identification of fast sampled systemsss has been based on continuous-time transfer function models using either state-variable filter or approximation of derivatives (see for example, Chapters 1–3 in this book). It is worthwhile to point out that the estimation of a transfer function model requires *a priori* knowledge about the underlying system as well as an algorithm with a higher complexity.

In the first instance, it is beneficial for the experimental data to be treated with filtering to remove the disturbance and compressed into a set of data that contains the vital information about the dynamics of the system. This is particularly useful if a non-linear optimisation is to be used in the derivation of the final model, which is often the case where a physical model is desired. What makes the FSF model a better candidate for its use in data compression is its distinct property with respect to sampling rate and its parameter distribution over the frequency range. This property is summarised below.

Theorem 7.1. *Say that the underlying continuous-time system has a Laplace transfer function $G_c(s)$, and its continuous-time impulse response $h_c(t)$ has a finite settling time T_m such that for $t \geq T_m$, $h_c(t) \approx 0$. We set the parameter N in the FSF model given by (7.4) as $N = \frac{T_m}{T_s}$ for a given sampling interval $T_s > 0$. Then, as $T_s \to 0$, the lth parameter of the FSF model, $G(e^{i\frac{2\pi l}{N}})$, converges to $G_c(jw_l)$ at $w_l = \frac{2\pi l}{T_m}$ radians/time.*

The proof of this theorem is based on the relationship between continuous-time and discrete-time Fourier transforms of a stable dynamic system (see [30] for details). This result indicates that the parameters of the FSF model in the low- and median-frequency range are relatively invariant with respect to sampling interval T_s, when T_s is sufficiently small. The key here is to look at the set of FSF model parameters corresponding to the continuous-time system frequency response evaluated at $w = 0, \frac{2\pi}{T_m}, \ldots, \frac{\pi}{T_s}$ for a fixed value of T_m. As T_m is the settling time for the underlying continuous-time system independent of the sampling interval T_s, as T_s reduces, $N \approx \frac{T_m}{T_s}$ increases. However, with this selection of N, the additional parameters to the FSF model when reducing sampling interval T_s are all in the higher-frequency region.

What we try to do is to take advantage of the frequency distribution of the FSF parameters in the data-compression process. Assuming that the underlying continuous-time system has a relative degree greater than 1 (*i.e.*, the order of the denominator is greater than the order of the numerator), the magnitude of the continuous-time frequency response $|G_c(iw)| \to 0$ as $\omega \to \infty$. This translates to the fact that for this class of systems, there is an odd number $n > 0$, such that for all $k > n$, $|G_c(i\frac{2\pi k}{T_m})| \approx 0$. In the environment of fast sampling (or a sufficiently small T_s), the FSF model parameter $|G(e^{i\frac{2\pi k}{N}})| \approx 0$ for $\frac{n-1}{2} < k \leq \frac{N-1}{2}$ (counting on the complex conjugate pairs of frequency response). n becomes independent of the choice of sampling interval T_s for a sufficiently small T_s. Under the assumption of relative degree and stable system, the original FSF model can be reduced to

$$G(z) = \sum_{l=-\frac{n-1}{2}}^{\frac{n-1}{2}} G(e^{il\Omega}) H^l(z) \tag{7.7}$$

where n is an odd number, and $1 + \frac{n-1}{2}$ is the number of frequencies included in the frequency-sampling filter model. Note that with the assumption of n being an odd number, N can be either an odd or an even number.

Equation (7.7) can also be written in terms of real and imaginary parts of the discrete frequency response $G(e^{il\Omega})$ [4] as

$$G(z) = \frac{1}{N} \frac{1 - z^{-N}}{1 - z^{-1}} G(e^{i0}) + \sum_{l=1}^{\frac{n-1}{2}} [\text{Re}(G(e^{il\Omega})) F_R^l(z) + \text{Im}(G(e^{il\Omega})) F_I^l(z)] \tag{7.8}$$

where $F_R^l(z)$ and $F_I^l(z)$ are the lth second-order filters given by

$$F_R^l(z) = \frac{1}{N} \frac{2(1 - \cos(l\Omega)z^{-1})(1 - z^{-N})}{1 - 2\cos(l\Omega)z^{-1} + z^{-2}}$$

$$F_I^l(z) = \frac{1}{N} \frac{2\sin(l\Omega)z^{-1}(1 - z^{-N})}{1 - 2\cos(l\Omega)z^{-1} + z^{-2}}$$

The parameters in the FSF model lead to other information about the underlying system. In particular, the discrete step-response coefficient g_m at the sample instant m is in a linear relation to the discrete-frequency parameters

$$g_m = Q(m)^T \boldsymbol{\theta} \tag{7.9}$$

where

$$Q(m) = \begin{bmatrix} \frac{m+1}{N} \\ 2\text{Re}(S(1,m)) \\ 2\text{Im}(S(1,m)) \\ \vdots \\ 2\text{Re}(S(\frac{n-1}{2},m)) \\ 2\text{Im}(S(\frac{n-1}{2},m)) \end{bmatrix} ; \boldsymbol{\theta} = \begin{bmatrix} G(e^{i0}) \\ \text{Re}(G(e^{i\Omega})) \\ \text{Im}(G(e^{i\Omega})) \\ \vdots \\ \text{Re}(G(e^{i\frac{n-1}{2}\Omega})) \\ \text{Im}(G(e^{i\frac{n-1}{2}\Omega})) \end{bmatrix}$$

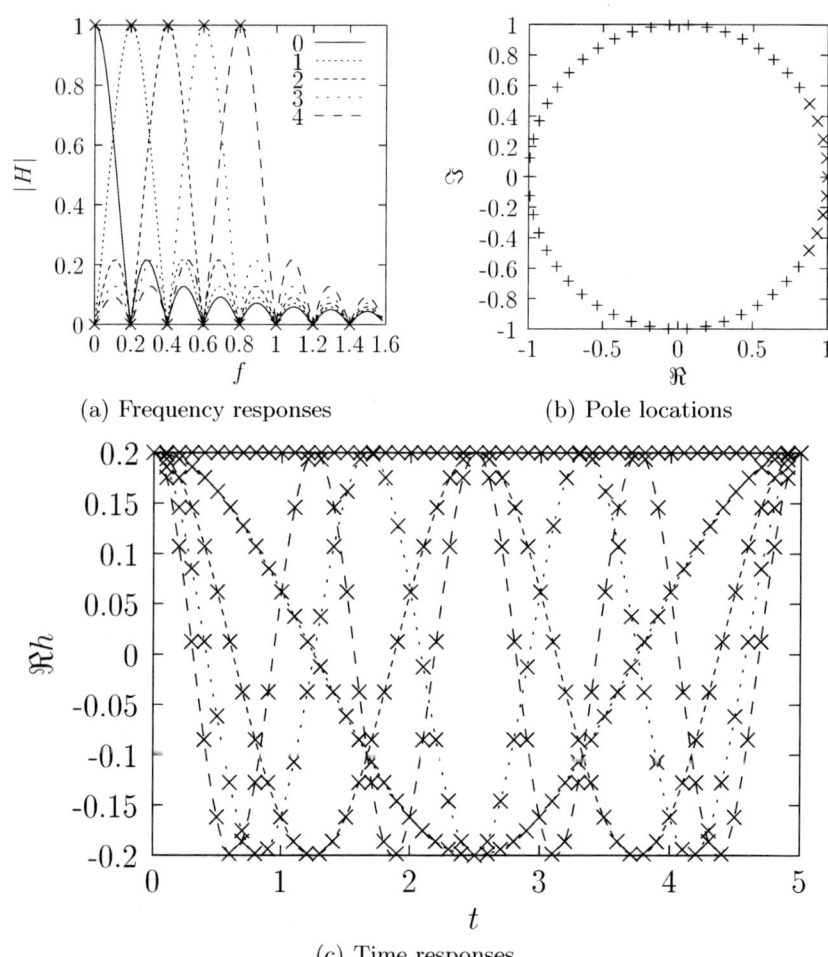

Fig. 7.2. Frequency sampling filters: $N = 50$, $n = 9$, $\frac{n-1}{2} = 4$, $T_m = 5$ s, $F_s = 0.2$ Hz and $f_c = 0.8$ Hz

and $S(l, m) = \frac{1}{N} \frac{1-e^{il\Omega(m+1)}}{1-e^{il\Omega}}$, for $l = 1, 2, \ldots, \frac{n-1}{2}$.

To further understand the idea of FSF model, we also express the frequency sample interval in Hz, which is $F_s = \frac{1}{T_m}$, and the cut-off frequency f_c in Hz, where

$$f_c = \frac{n-1}{2}F_s = \frac{n-1}{2T_m} \tag{7.10}$$

Figure 7.2 shows the graphic illustration of the frequency response, pole location and time response of a set of FSF filters. Figure 7.2(a) shows the superimposed frequency responses of $|\bar{H}_k(z)|$ for $0 \leq k \leq 4$ when $T_m = 5$ (implying $F_s = 0.2$) for a frequency range $0 \leq \omega \leq 10$. The symbol '×' marks the frequency samples that coincide with the peaks of the FSFs. Figure 7.2(b) shows $N = 50$ potential FSF poles (marked by '+') equispaced around the unit circle and the $n = 9$ actual FSF poles (marked by '×') clustered around $z = 1$ on the unit circle.

7.2.3 Estimation Using FSF Structure

With respect to the application in data compression, the central idea is to estimate a reduced-order FSF model using discrete-time system identification techniques in the environment of fast sampling. The reduced-order FSF model provides compressed information about the underlying continuous-time system, including continuous-time step response and frequency response. Suppose that $u(k)$ is the process input, $y(k)$ is the process output and $v(k)$ is the disturbance signal. The output $y(k)$ can be expressed in a linear regression form by defining the parameter vector and the regressor vector as

$$\boldsymbol{\theta} = \begin{bmatrix} G(e^{i0}) \\ \operatorname{Re}(G(e^{i\Omega})) \\ \operatorname{Im}(G(e^{i\Omega})) \\ \vdots \\ \operatorname{Re}(G(e^{i\Omega\frac{n-1}{2}})) \\ \operatorname{Im}(G(e^{i\Omega\frac{n-1}{2}})) \end{bmatrix} ; \quad \boldsymbol{\varphi}(k) = \begin{bmatrix} f(k)^0 \\ f(k)_R^1 \\ f(k)_I^1 \\ \vdots \\ f(k)_R^{\frac{n-1}{2}} \\ f(k)_I^{\frac{n-1}{2}} \end{bmatrix}$$

where

$$f(k)^0 = \frac{1}{N} \frac{1-z^{-N}}{1-z^{-1}} u(k)$$

$$f(k)_R^l = F_R^l(z)u(k); f(k)_I^l = F_I^l(z)u(k)$$

for $l = 1, 2, \ldots, \frac{n-1}{2}$. This allows us to write the linear regression with correlated residuals as

$$y(k) = \boldsymbol{\varphi}^T(k)\boldsymbol{\theta} + v(k)$$
$$v(k) = \frac{\epsilon(k)}{D(z)} \quad (7.11)$$

where $\epsilon(k)$ is a white noise sequence with zero mean and standard deviation σ and $v(k)$ is not correlated with the input signal $u(k)$. Given a set of sampled finite amount of data

$$\{y(1), y(2), y(3), \ldots, y(M)\}$$
$$\{u(1), u(2), u(3), \ldots, u(M)\}$$

we can obtain an estimate of the frequency-sampling filter model and an estimate of the noise model $\frac{1}{D(z)}$ using the generalised least squares method [6, 27]. More specifically, in the core-estimation algorithm, we let

$$y_D(k) = \hat{D}(z)y(k); \varphi_D(k) = \hat{D}(z)\varphi(k)$$

The estimation of $\hat{\boldsymbol{\theta}}$ is obtained by minimising the quadratic performance index

$$J = \sum_{k=1}^{M}[y_D(k) - \varphi_D(k)\boldsymbol{\theta}]^2$$

$$= \boldsymbol{\theta}^T \sum_{k=1}^{M}[\varphi_D(k)\varphi_D^T(k)]\boldsymbol{\theta} - 2\boldsymbol{\theta}^T \sum_{k=1}^{M}[\varphi_D(k)y_D(k)] + cons \quad (7.12)$$

$\hat{D}(z)$ is estimated from the error sequence $e(k) = y(k) - \varphi^T(k)\hat{\boldsymbol{\theta}}$, $k = 1, 2, 3, \ldots, M$. The generalised least squares method is based on an iterative procedure and the iteration stops after the estimated parameters converge.

The extension to a multi-input and single-output system is straightforward. Suppose that the inputs available are numbered as $u_1(k)$, $u_2(k), \ldots u_p(k)$, the times to steady state for the individual subsystems are chosen as N_1, $N_2, \ldots, N_p$ and the number of parameters for each FSF model is chosen as $n_1, n_2, \ldots, n_p$. In the FSF model structure for the MISO system, the first input $u_1(k)$ is passed through a set of n_1 frequency-sampling filters based on N_1 to form the first n_1 elements in the data regressor, followed by passing the second input $u_2(k)$ through a set of n_2 frequency-sampling filters based on N_2 to form the next n_2 elements in the the regressor, and so on. With this data regressor, the generalised least squares algorithm can be directly applied to estimate the frequency-response parameters associated with the MISO system, also allowing the individual selection of N_1, N_2, ..., N_p and $n_1, n_2, \ldots, n_p$.

One might ask since the model parameters in the FSF model are the discrete-frequency response, why we would not simply estimate the plant frequency response using the discrete Fourier transform, which is a well-known approach for data analysis and compression. The estimation using the discrete Fourier transform, called empirical transfer function estimation (ETFE) is known to have problems such as leakage and large variances in the estimated frequency response. Although optimisation techniques existed to improve the variance of the estimate (for example, see Chapter 8 of this book), the ETFE has estimated the number of frequencies proportional to the actual data length, as a consequence, there is a large information redundancy as well as poor quality of information provided by ETFE. In contrast, the data compression by using a reduced-order FSF model captures the dynamics of the system with a fixed settling time T_m of the impulse response of the continuous-time system, regardless of the length of the experimental data. It captures the frequency response to the Nyquist sampling frequency $\frac{\pi}{T_s}$, using the frequency parameters at $0, \frac{2\pi}{T_m}, \frac{4\pi}{T_m}, \ldots$. The rest of the frequency response is obtained through interpolation based on the FSF model. Furthermore, the data compression using FSF allows natural application of filtering, disturbance modelling and many other existing techniques in dynamic system identification.

7.3 Data Compression with Constraints

Although structural constraints such as model order and time delay have been incorporated in the continuous-time system identification since its origin, the constraints on the estimated model parameters were rarely enforced. This section shows that by incorporating physical parameter information known *a priori* as hard constraints, the traditional parameter estimation schemes are modified to minimise a quadratic cost function with linear inequality constraints. More specifically, *a priori* knowledge in both time- and frequency-domains is utilised simultaneously in the data-compression process as the constraints for the optimal parameter solution. In addition, the optimal solution for the constrained case is obtained using a quadratic programming procedure.

7.3.1 Formulation of the Constraints

Constraints on Frequency Parameters

Suppose that the continuous-time system is known at frequency $\gamma(< \frac{\pi}{T_s})$. By converting it to the discrete frequency γT_s, from (7.8) the frequency information can be expressed as

$$G(e^{i\gamma T_s}) = L(e^{i\gamma T_s})^T \boldsymbol{\theta} \tag{7.13}$$

where

$$L(e^{i\gamma T_s}) = \begin{bmatrix} F(e^{i\gamma T_s})^0 \\ F(e^{i\gamma T_s})^1_R \\ F(e^{i\gamma T_s})^1_I \\ \ldots \\ F(e^{i\gamma T_s})^{\frac{n-1}{2}}_R \\ F(e^{i\gamma T_s})^{\frac{n-1}{2}}_I \end{bmatrix}$$

This equation is then split into real and imaginary parts

$$\text{Re}(G(e^{i\gamma T_s})) = \text{Re}(L(e^{i\gamma T_s}))^T \boldsymbol{\theta} \tag{7.14}$$
$$\text{Im}(G(e^{i\gamma T_s})) = \text{Im}(L(e^{i\gamma T_s}))^T \boldsymbol{\theta} \tag{7.15}$$

If the frequency information is known quite accurately, then equality constraints based on (7.14) and (7.15) can be imposed in the solutions. This is particularly useful when the system has strong resonance, and the critical frequency information is used in the constraints to ensure good fitting. If the frequency information is known within certain bounds, then the inequality constraints can be imposed as

$$\text{Re}(G(e^{i\gamma T_s}))_{\min} \leq real(L(e^{i\gamma T_s})\boldsymbol{\theta} \leq \text{Re}(G(e^{i\gamma T_s}))_{\max}$$
$$\text{Im}(G(e^{i\gamma T_s}))_{\min} \leq \text{Im}(L(e^{i\gamma T_s})\boldsymbol{\theta} \leq \text{Im}(G(e^{i\gamma T_s}))_{\max} \tag{7.16}$$

Constraints on Step-response Parameters

Constraints on step response parameters will be based on (7.9). Given the *a priori* information about some step-response coefficients g_m, $0 \leq m \leq N-1$, the equality constraint is formulated as

$$g_m = Q(m)^T \boldsymbol{\theta} \qquad (7.17)$$

where $Q(m)$ is defined by (7.9). For inequality constraints, with specification of minimum and maximum of step responses, say $\underline{g}_m \leq g_m \leq \overline{g}_m$, then the inequality constraint on a step-response coefficient g_m is formulated as

$$\underline{g}_m \leq Q(m)^T \boldsymbol{\theta} \leq \overline{g}_m \qquad (7.18)$$

7.3.2 Solution of the Estimation Problem with Constraints

The estimation problem with constraints is essentially to minimise the quadratic cost function

$$J = \boldsymbol{\theta}^T \sum_{k=1}^{M} [\boldsymbol{\varphi}_D(k)\boldsymbol{\varphi}_D^T(k)]\boldsymbol{\theta} - 2\boldsymbol{\theta}^T \sum_{k=1}^{M} [\boldsymbol{\varphi}_D(k)y_D(k)] + constant \qquad (7.19)$$

subject to equality constraints

$$M_1 \boldsymbol{\theta} = \beta_1$$

and inequality constraints

$$M_2 \boldsymbol{\theta} \leq \beta_2$$

By defining $E = \sum_{k=1}^{M}[\boldsymbol{\varphi}_D(k)\boldsymbol{\varphi}_D^T(k)]$ and $F = -2\sum_{k=1}^{M}[\boldsymbol{\varphi}_D(k)y_D(k)]$, $M = [M_1^T \ M_2^T]^T$, $\beta = [\beta_1^T \ \beta_2^T]^T$ the necessary conditions for this optimisation problem (Kuhn–Tucker condition) are [24]

$$E\boldsymbol{\theta} + F + M^T \lambda = 0$$
$$M\boldsymbol{\theta} - \beta \leq 0$$
$$\lambda^T (M\boldsymbol{\theta} - \beta) = 0$$
$$\lambda \geq 0 \qquad (7.20)$$

where the vector λ contains the Lagrange multipliers. These conditions can be expressed in a simpler form in terms of the set of active constraints. Let S_{act} denote the index set of active constraints. Then, the necessary conditions become

$$E\boldsymbol{\theta} + F + \sum_{i \subset S_{\text{act}}} \lambda_i M_i^T = 0$$

$$\begin{aligned} M_i \boldsymbol{\theta} - \beta_i &= 0 & i \subset S_{\text{act}} \\ M_i \boldsymbol{\theta} - \beta_i &< 0 & i \not\subset S_{\text{act}} \\ \lambda_i &\geq 0 & i \subset S_{\text{act}} \\ \lambda_i &= 0 & i \not\subset S_{\text{act}} \end{aligned}$$

where M_i is the ith row of the M matrix. It is clear that if the active set were known, the original problem could be replaced by the corresponding problem having equality constraints only. Alternatively, suppose an active set is guessed and the corresponding equality constrained problem is solved. Then if the other constraints are satisfied and the Lagrange multipliers turn out to be non-negative, that solution would be correct. In the case that only equality constraints are involved, the optimal solution has a closed form as

$$\begin{bmatrix} E & M_1^T \\ M_1 & 0 \end{bmatrix} \begin{bmatrix} \boldsymbol{\theta} \\ \lambda_1 \end{bmatrix} = \begin{bmatrix} -F \\ \beta_1 \end{bmatrix} \qquad (7.21)$$

Explicitly

$$\lambda_1 = -(M_1 E^{-1} M_1^T)^{-1}(\beta_1 + M_1 E^{-1} F) \qquad (7.22)$$
$$\hat{\boldsymbol{\theta}} = -E^{-1}(F + M_1^T \lambda_1) \qquad (7.23)$$

When inequality constraints are required, an iterative algorithm is needed to solve the quadratic programming problem [24].

7.3.3 Monte Carlo Simulation Study

This section is to illustrate the data-compression process through a Monte Carlo simulation study. The relay experiment proposed by Åström and Hagglund [2] is particularly suitable for continuous-time identification as the sampling interval in the experiment can be chosen as small as desired. This experiment [29, 30] was extended to include identification of more general classes of models other than simple frequency-response points. The same set of design parameters as in [32] is used here to generate the input excitation signal. In the Monte Carlo simulation study, a white noise sequence with standard deviation of 0.8 is used to generate $e(k)$ and the disturbance $\boldsymbol{\xi}(k) = \frac{0.1}{1-0.9z^{-1}}e(k)$. 100 realisations of the white noise sequence are generated by changing the seed of the generator from 1 to 100.

The system used for simulation is given by the transfer function

$$G(s) = \frac{e^{-3s}}{(s^2 + 0.4s + 1)(s+1)^3} \qquad (7.24)$$

The sampling interval for this system is $T_s = 0.1$ s. The settling time T_m is estimated as 40 s, hence the number of samples to steady state $N = \frac{T_m}{T_s} = 400$. By using frequency-sampling filters to parameterise this system, the number of frequencies required is 63, yielding the number of parameters in the FSF model as $n = 125$. Figures 7.3(a)–7.5(a) show the magnitude and phase of the frequency responses when estimated without constraints. From the distribution of the responses, it is seen that the estimation of the non-parametric models is unbiased. However, the variances are large both for the frequency response and step response.

To introduce equality constraints on the estimation, *a priori* knowledge about the system is required. The *a priori* knowledge for this system is assumed as a time delay being approximately 1 s, the gain being 1, and the first pair of frequency responses $G_c(\frac{2\pi}{T_s}) = 0.5305 - j0.8317$. The *a priori* knowledge about the steady-state gain of the system is translated into the constraint on the first parameter of the FSF model while, the *a priori* knowledge about the frequency-response information is translated into two equality constraints on the second and third parameters of the FSF model. Note that frequency information at an arbitrary frequency can be translated into constraints in a linear combination of the parameters of the FSF model. Similarly, the *a priori* information about time delay is translated into a set of linear equality constraints in terms of the parameters of the FSF model. Four constraints have been put on the time delay at the sampling instant $k = 0, 3, 6, 9$. The reason for not using every sampling instant is because the solution is ill-conditioned when a constraint is imposed on every sampling instant. As is seen from the Monte Carlo simulation study, this approach is adequate for this purpose. With the equality constraints imposed on the estimated parameters, the generalised least squares method is modified to have the constraints on the system parameters, but not on the noise model parameters. Figure 7.3 compares the estimated frequency amplitudes with and without constraints. Figure 7.4 shows the comparison results of the estimated phase with and without constraints. Figure 7.5 compares the estimated step response with and without constraints. All results confirm that the estimation results have a smaller uncertainty bound when the constrained estimation is used with correct *a priori* knowledge.

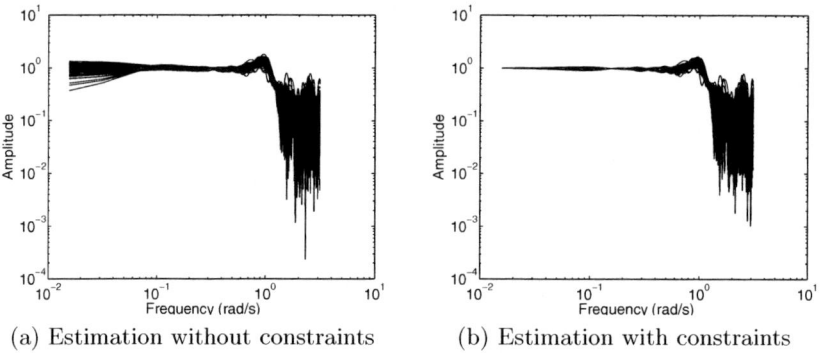

(a) Estimation without constraints   (b) Estimation with constraints

Fig. 7.3. Monte Carlo simulation results: estimated amplitudes

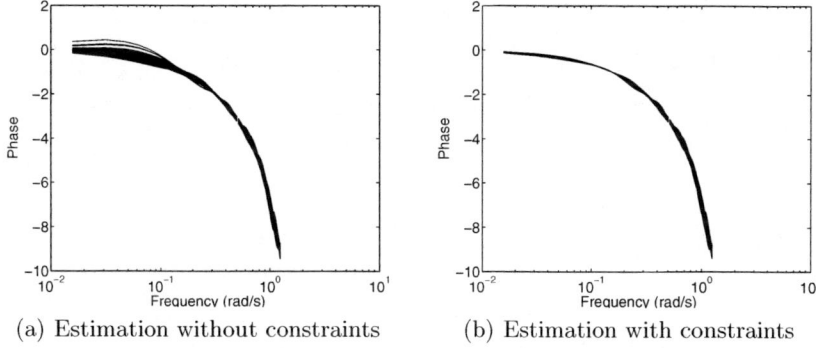

Fig. 7.4. Monte Carlo simulation results: estimated phases

7.4 Physical-model-based Estimation

Many engineering systems of interest to the control engineer are *partially known* in the sense that the system structure, together with some system parameters are known, but some system parameters are unknown. This gives rise to a problem of *parameter estimation* when values for the unknown parameters are to be determined from experimental data comprising measurements of system inputs and outputs. There is a considerable literature in the area including [3, 5, 7, 11, 12]. Although in special cases such identification may be *linear*-in-the-parameters [3] or *polynomial*-in-the parameters [11, 12] in general the problem is *non-linear*-in-the-parameters. This means that, in general, the resultant optimisation problem is not quadratic or polynomial, and may even be non-convex. In such cases, the optimisation task is eased by knowing (rather than deducing numerically) the first and second derivative of the error function with respect to the unknown system parameters.

Symbolic methods for non linear systems modelling, analysis and optimisation are currently strong research areas [25] driven by the ready availability of symbolic computational tools. In particular, the bond-graph approach [13, 16, 20] has been used to generate models applicable to control design [10].

For the purposes of this chapter, a *physically plausible* model of a physical system is defined as a model that represents a different physical system that shares key behaviours of the actual system. Typically, the physically plausible model will be simpler than the model itself and will be represented by a bond graph.

The advantages of having a simpler model are:

- it is easier to understand a simple model than a complex model;
- the computation and numerical aspects of identification and control are eased.

The advantages of a physical model are that:

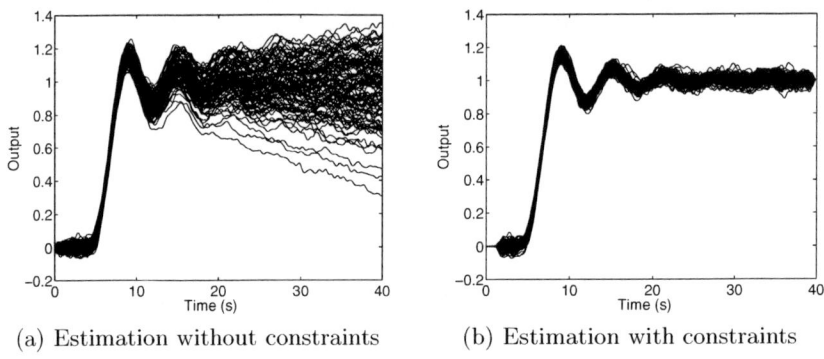

(a) Estimation without constraints   (b) Estimation with constraints

Fig. 7.5. Monte Carlo simulation results: estimated step responses

- the parameters of a physical model have a clearer interpretation than those of a purely empirical model and
- the behaviour of the model can be understood in physical terms.

The disadvantage of a physical model is that it is *not* usually linear in the physical parameters, thus leading to a non-linear optimisation problem. The *time-domain* parameter estimation problem posed in this chapter is to estimate the unknown *physical* parameters Θ from the estimated system *impulse response* $h(t_i)$ at a finite number of discrete time instants $t_i, 1 \geq i \geq N_{\text{opt}}$. The usual least squares estimation problem is posed; that is to minimise the cost function J with respect to the vector of unknown parameters Θ where

$$J(\hat{\Theta}) = \frac{1}{2N_{\text{opt}}} \sum_{i=1}^{N_{\text{opt}}} e_i^2 \qquad (7.25)$$

where the output error $e(t_i)$ is defined as

$$e_i = \hat{h}(t_i, \hat{\Theta}) - h(t_i, \Theta) \qquad (7.26)$$

In a similar fashion the *frequency-domain* parameter estimation estimates Θ from the estimated *frequency response* $G(i\omega_i)$ at a finite number of discrete frequencies $\omega_i, 1 \geq i \geq N_{\text{opt}}$ with

$$e_i = \hat{G}(i\omega_i, \hat{\Theta}) - G(i\omega_i, \Theta) \qquad (7.27)$$

These non-linear least squares problems do not admit an explicit solution in general; instead, numerical techniques must be used. Each iteration of such an algorithm requires evaluation of the function J for the current estimate $\hat{\Theta}$ and thus an evaluation of $\hat{g}(t_i, \hat{\Theta})$ or $\hat{G}(i\omega_i, \hat{\Theta})$ for that value of $\hat{\Theta}$. Thus, each iteration is computationally expensive and therefore an efficient algorithm is desirable.

A number of optimisation methods are available, the main division is between those that use gradient information and those that don't. The former have been discussed in this context previously [14, 15, 19] and include the Levenberg–Marquardt [9] and the 'projected BFGS-Armijo' algorithm of Kelley [21, Section 5.5.3]. The latter includes the Broyden–Fletcher–Goldfarb–Shanno (BFGS) method [9].

As the gradient-based approaches have been considered previously, this chapter uses the (non-gradient) BFGS method (as implemented as `bfgsmin` in Octave [8]).

7.5 Example: Inverted Pendulum

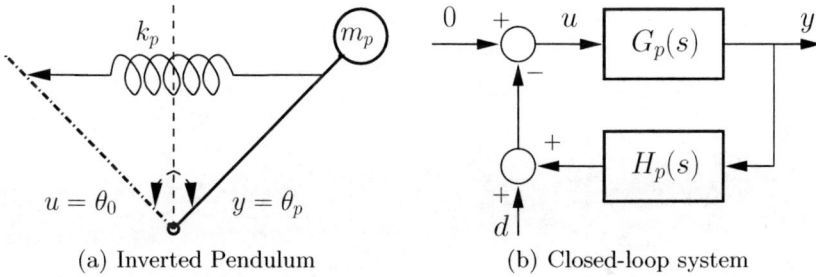

(a) Inverted Pendulum (b) Closed-loop system

Fig. 7.6. Experimental system

This section provides an illustrative example where the parameters of an inverted pendulum are identified using the two stage process of:

- identifying the FSF parameters from the closed-loop experimental data and
- identifying the physical parameters from the corresponding impulse or frequency responses.

A simple model human standing is equivalent to controlling an inverted pendulum (the body) via a spring (tendons and muscle) [23, Figure 1]. It is convenient to represent such a model by Figure 7.6 where the input u is the effective input angle θ_0 and the output y is the pendulum angle θ_p and the length of the pendulum is l. The system can be modelled with three parameters:

- the inertia about the pivot J_p
- the effective gravitational spring k_g and
- the ratio α of the effective spring constant to the gravitational spring.

Using the usual small-angle approximation, the system has the transfer function

$$G_{\rm p} = \frac{\alpha k_{\rm g}}{(1-\alpha)k_{\rm g} - J_{\rm p}s^2} \quad (7.28)$$

It is known [22] that $\alpha < 1$ (that is, the spring is not stiff enough to hold up the pendulum) so that the system of (7.28) is unstable and therefore requires regulation.

The feedback structure is given in Figure 7.6(b) where $H_{\rm p}$ is the stabilising controller and d a disturbance signal. The corresponding *closed-loop transfer function* $G(s)$ is given by

$$G(s) = \frac{y}{d} = -\frac{G_{\rm p}(s)}{1 + G_{\rm p}(s)H_{\rm p}(s)} \quad (7.29)$$

As part of a programme to investigate the dynamics of human standing, an initial experimental setup replaces both pendulum and controller by digital equivalents within separate computers connected together, and to a third data-collection computer, via analogue instrumentation[3].

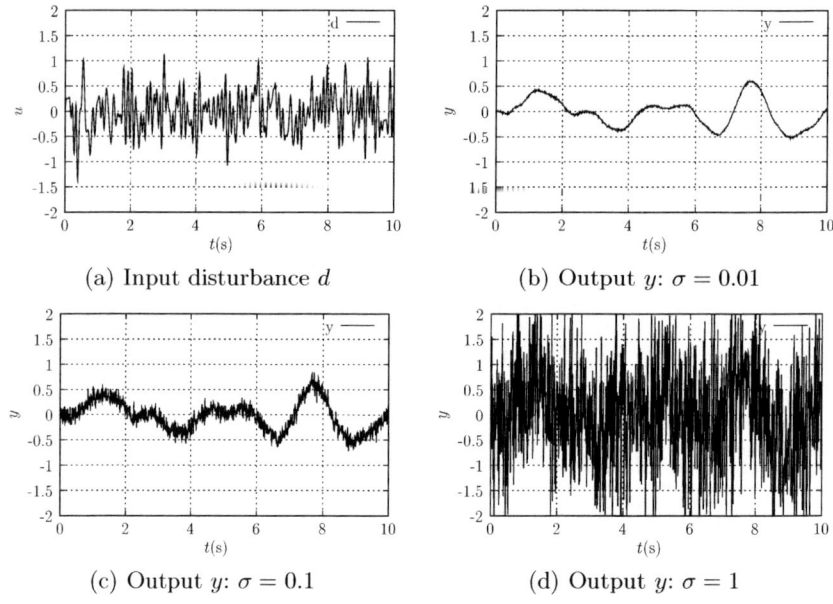

Fig. 7.7. Data sampled with $T_{\rm s} = 0.01$ s. The first 10 s of 100 s of data is shown. Additional noise with standard deviation σ is artificially added to the system output to give Figures (b)–(d).

---

[3] The data used here was collected at the Department of Sports Science at the University of Birmingham in June 2006. It is used with the kind permission of Dr Martin Lakie and Dr Ian Loram.

The data collected from this setup is used as an illustrative example in this chapter; it has the advantage that the exact model is known. For the purposes of this chapter, a data set of length 100 s is used that has been sampled with interval $T_s = 0.01$ s giving about 10000 data points for each signal. The input disturbance d is the multi-sine signal of Figure 7.7(a); it has the power spectral density shown in Figure 7.8.

To illustrate the properties of the FSF approach as noise levels increase, white noise with variance σ^2 is added to the measured output data y; the result is shown in Figures 7.7(b)–(d). Figures 7.8(b)–(d) show some standard

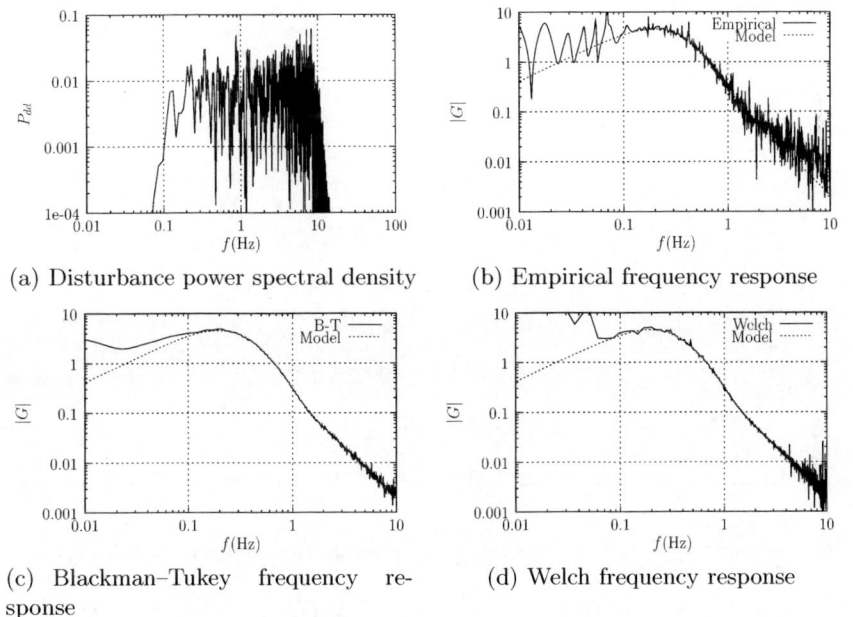

Fig. 7.8. Other non-parametric methods

non-parametric estimation results for the data without any added noise. The empirical and Blackman–Tukey methods were computed using the 'nonpar' function of the UNIT toolbox [26].

7.5.1 FSF Estimation

This section illustrates the use and behaviour of FSF using the data set d as input and data set y as output to identify the FSF parameters corresponding to the transfer function $G(s)$ (7.29). The results are displayed (Figures 7.9–7.11) in two forms, the modulus of the frequency response ($|G(i\omega)|$) and the corresponding impulse response $g(t)$; the figures are organised so that the frequency response is to the left and the time response to the right.

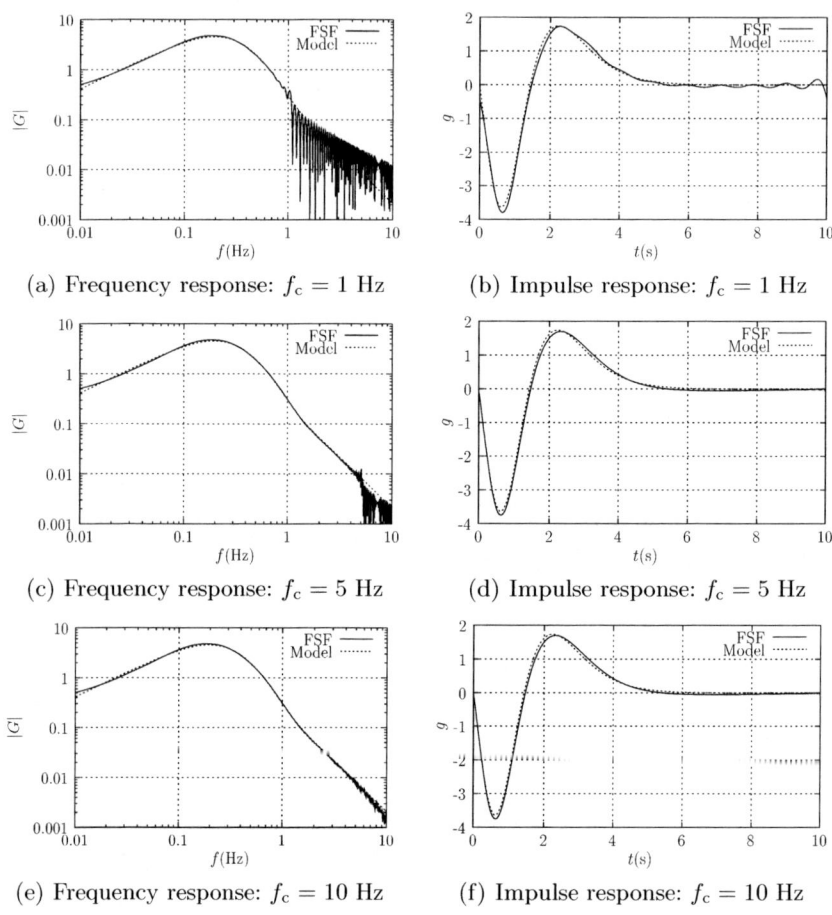

Fig. 7.9. FSF properties: effect of cut-off frequency f_c ($T_m = 10$ s)

As discussed at the end of Section 7.2, the FSF is parameterised by the cut-off frequency f_c and the settling time T_m.

- f_c is essentially a *frequency-domain* parameter and determines the largest *frequency* of interest. It also has time domain implications. The effect of f_c is shown in Figure 7.9 for three values of f_c. Figures 7.9(a), 7.9(c) and 7.9(e) illustrate the fact that f_c determines the upper bound of the frequency for which the frequency response is matched by the FSFs.
- T_m is essentially a time-domain parameter and determines the largest *time* of interest. It also has frequency domain implications insofar as it fixes the *frequency-domain* sampling interval $\Omega = \frac{2\pi}{T_m}$. Figures 7.10(b), 7.10(d) and 7.10(f) illustrate the fact that T_m determines the upper bound of the time for which the time response is matched by the FSFs. These figures also

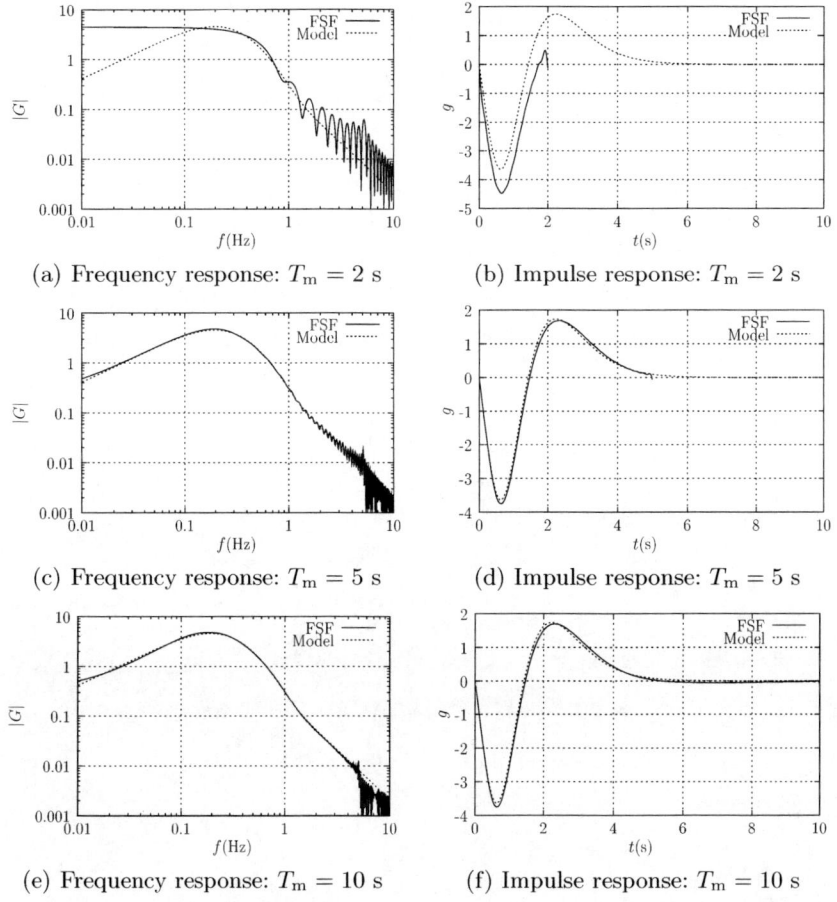

Fig. 7.10. FSF properties: effect of settling time T_m ($f_c = 5$ Hz)

show that if T_m is less than the actual settling time of the system, the estimated response is not accurate.

As with any identification technique, the FSF method is affected by measurement noise. The effect of measurement noise is illustrated by artificially adding noise to the data (Figures 7.7(b)–7.7(d)) to give Figure 7.11. As would be expected, the accuracy of both the time and frequency responses declines with increased measurement noise.

7.5.2 PMB Estimation

The resulting estimated parameters are shown in Table 7.1. As discussed in Section 7.4 the impulse and frequency responses estimated by the FSF can be

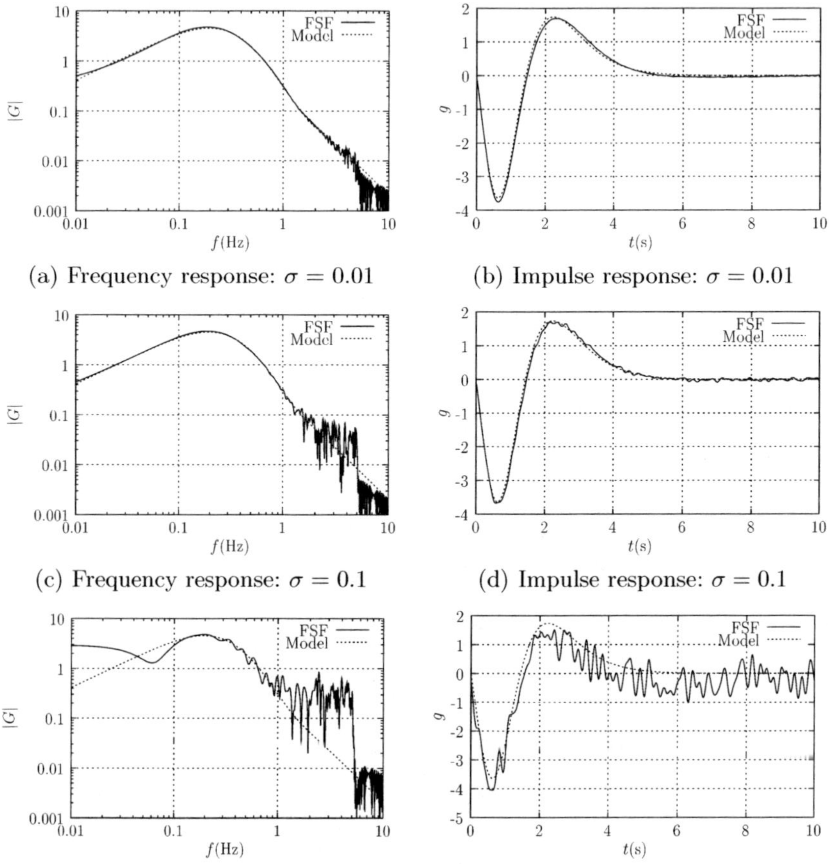

(a) Frequency response: $\sigma = 0.01$ (b) Impulse response: $\sigma = 0.01$

(c) Frequency response: $\sigma = 0.1$ (d) Impulse response: $\sigma = 0.1$

(e) Frequency response: $\sigma = 1$ (f) Impulse response: $\sigma = 1$

Fig. 7.11. FSF properties: effect of noise level σ ($f_c = 5$Hz, $T_m = 10$ s)

Table 7.1. Estimated physical parameters (true values $\boldsymbol{\alpha} = 0.85$, $J_p = 15$)

Domain	σ	$\hat{\boldsymbol{\alpha}}$	$\hat{J}_p$
time	0.01	0.84	14.83
freq	0.01	0.84	15.05
time	0.10	0.84	14.78
freq	0.10	0.84	15.08
time	1.00	0.86	15.83
freq	1.00	0.87	16.71

7 Estimation Using Data Compression 209

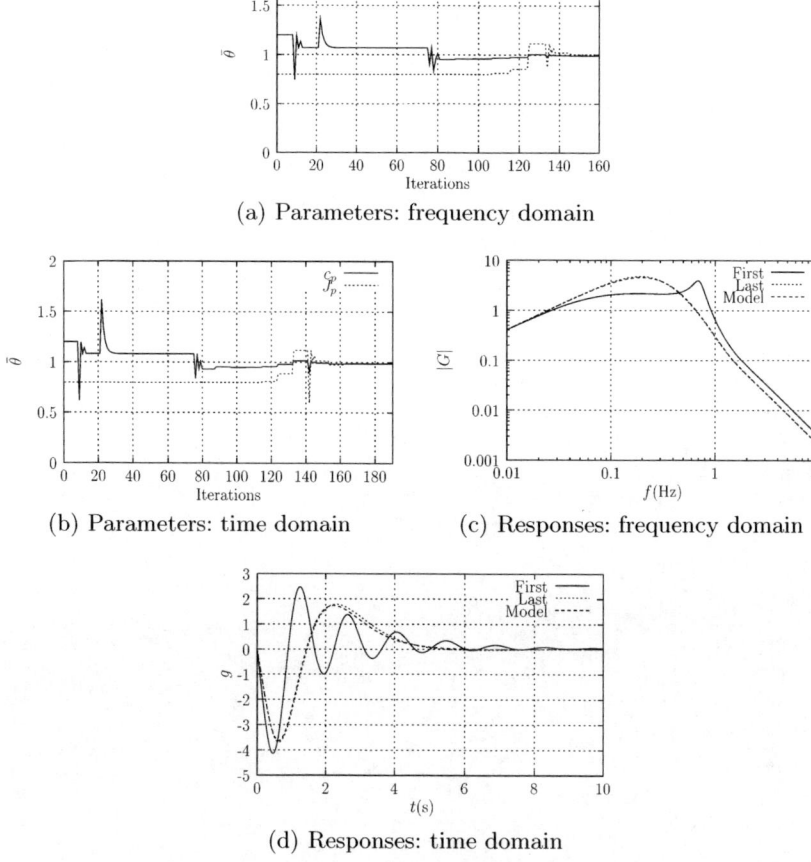

Fig. 7.12. PMB estimation (low noise): $\sigma = 0.01$, $T_\mathrm{m} = 10$ s and $f_\mathrm{c} = 5$ Hz

transformed into a set of *physical* parameters Θ using a non-linear optimisation approach such as that of Broyden–Fletcher–Goldfarb–Shanno (BFGS) [9] (here, the Octave [8] implementation `bfgsmin` is used). This is illustrated in this chapter by estimating two (α and J_p) of the three physical parameters (α, J_p k_g) of the experimental system of Figure 7.6 from each of the 6 FSF responses of Figure 7.11.

The observations from the above results are summarised as below.

- Figure 7.12 is based on the low-noise FSF responses of Figures 7.11(a) and (b). The left-hand figures correspond to frequency-domain optimisation (7.27) and the right-hand figures to time-domain optimisation (7.26). The top row shows how the parameters evolve during the BFGS optimisation

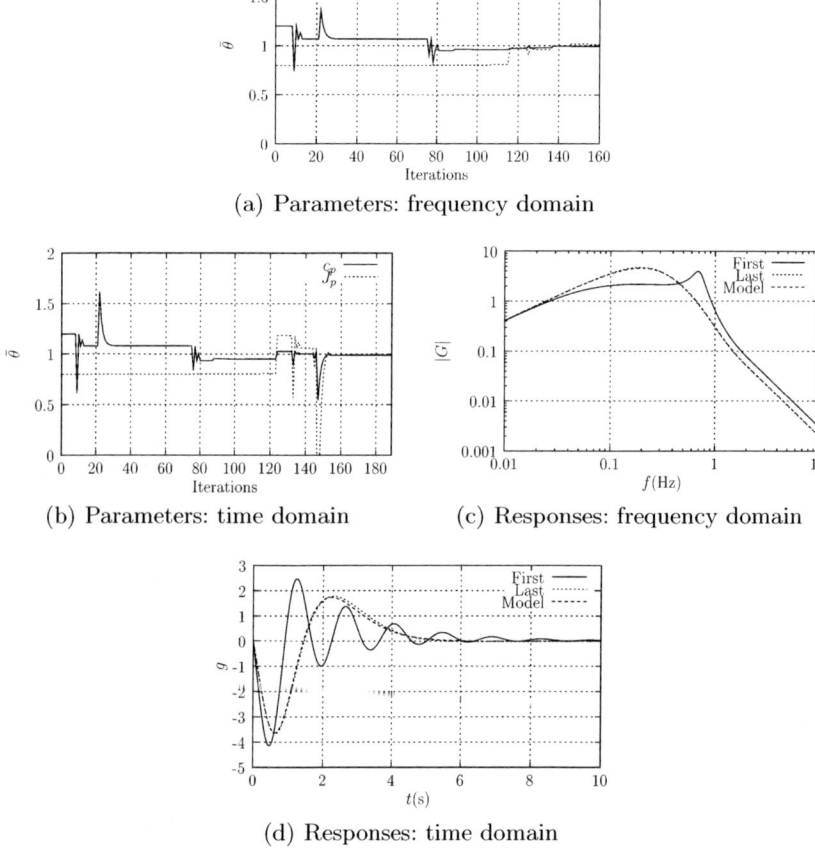

Fig. 7.13. PMB estimation (medium noise): $\sigma = 0.1$, $T_m = 10$ s and $f_c = 5$ Hz

process; the bottom row shows responses corresponding to the first and last iterations together with the correct response for comparison.
- Figure 7.13 is similar to Figure 7.12 except that it is based on the medium noise responses of Figures 7.11(c) and (d).
- Figure 7.14 is similar to 7.12 except that it is based on the high noise responses of Figures 7.11(e) and (f).

7.6 Conclusions

This chapter presented an approach to continuous-time system identification using data compression, where in the first step of the method the raw experimental data were compressed into a set of frequency-response coefficients of

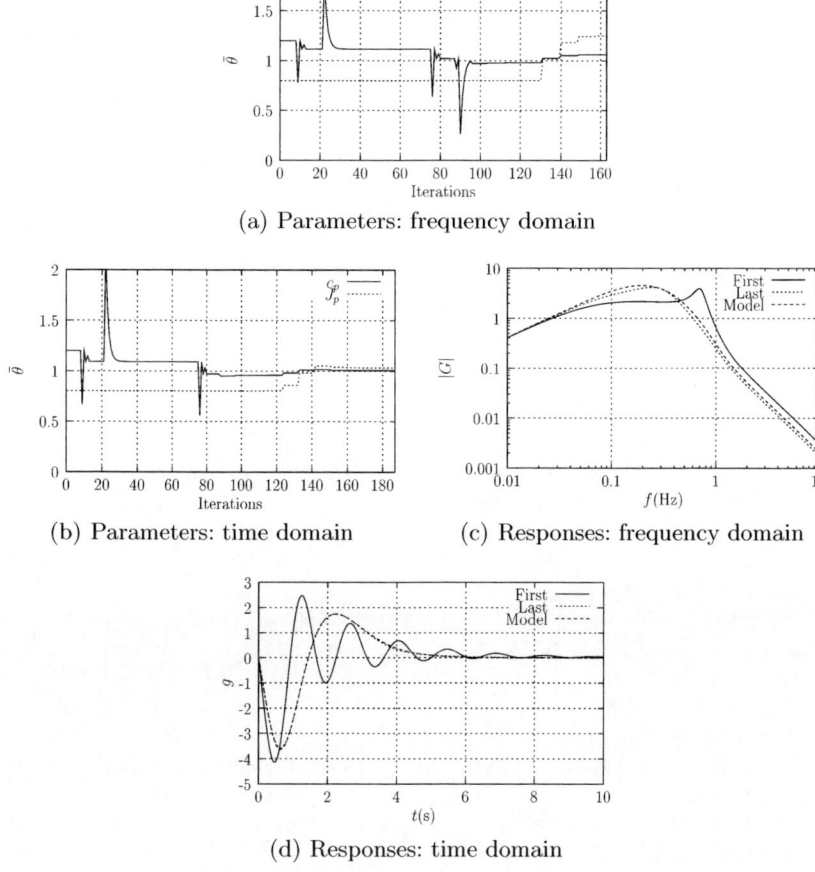

Fig. 7.14. PMB estimation (high noise): $\sigma = 1.0$, $T_{\mathrm{m}} = 10$ s and $f_{\mathrm{c}} = 5$ Hz

the FSF model, while the second step of the method used a non-linear optimisation scheme to find the parameters of a partially known physical model. Furthermore, in the data compression procedure, the frequency-response coefficients of the FSF model were estimated with respect to constraints in which *a priori* knowledge is incorporated to improve the estimation results. A Monte Carlo simulation study was used to demonstrate the improvement of the estimation in a noise environment. As an illustration, a partially known system consisting of an unknown unstable system with a known stabilising controller in the feedback loop was identified using the methods of this chapter.

References

1. K.J. Åström, P. Hagander, and J. Sternby. Zeros of sampled systems. *Automatica*, 20:31–38, 1984.
2. K.J. Åström and T. Hagglund. Automatic tuning of simple regulators with specifications on phase and amplitude margins. *Automatica*, 20:645–651, 1984.
3. C.H. An, C.G. Atkeson, and J.M. Hollerbach. *Model-based Control of Robot Manipulators*. The MIT Press, Cambridge, MA, USA, 1988.
4. R.R. Bitmead and B.D.O. Anderson. Adaptive frequency sampling filters. *IEEE Transactions on Circuits and Systems*, 28:524–533, 1981.
5. C. Canudas de Wit. *Adaptive Control for Partially Known Systems*. Elsevier, Amsterdam, 1988.
6. D.W. Clarke. Generalized least squares estimation of parameters of a dynamic model. *First IFAC Symposium on Identification and Automatic Control Systems*, Prague, 1967.
7. S. Dasgupta, B.D.O. Anderson, and R.J. Kaye. Output error identification methods for partially known systems. *International Journal of Control*, 43(1):177–191, 1986.
8. J.W. Eaton. *GNU Octave Manual*. Network Theory Limited, 2002.
9. R. Fletcher. *Practical Methods of Optimization*. 2nd edition, Wiley, Chichester, 1987.
10. P.J. Gawthrop. Physical model-based control: A bond graph approach. *Journal of the Franklin Institute*, 332B(3):285–305, 1995.
11. P.J. Gawthrop, J. Ježek, R.W. Jones, and I. Sroka. Grey-box model identification. *Control-Theory and Advanced Technology*, 9(1):139–157, 1993.
12. P.J. Gawthrop, R.W. Jones, and S.A. MacKenzie. Identification of partially-known systems. *Automatica*, 28(4):831–836, 1992.
13. P.J. Gawthrop and L.P.S. Smith. *Metamodelling: Bond Graphs and Dynamic Systems*. Prentice Hall, Hemel Hempstead, Herts, England., 1996.
14. P.J. Gawthrop. Sensitivity bond graphs. *Journal of the Franklin Institute*, 337(7):907–922, November 2000.
15. P.J. Gawthrop and E. Ronco. Estimation and control of mechatronic systems using sensitivity bond graphs. *Control Engineering Practice*, 8(11):1237–1248, November 2000.
16. P.J. Gawthrop and G.P. Bevan. Bond-graph modeling: A tutorial introduction for control engineers. *IEEE Control Systems Magazine*, 27(2):24–45, April 2007.
17. P.J. Gawthrop and L. Wang. Transfer function and frequency response estimation using resonant filters. *Proceedings of the Institution of Mechanical Engineers Pt. I: Journal of Systems and Control Engineering*, 216(I6):441–453, 2002.
18. P.J. Gawthrop and L. Wang. Data compression for estimation of the physical parameters of stable and unstable linear systems. *Automatica*, 41(8):1313–1321, August 2005.
19. P.J. Gawthrop and L. Wang. Estimation of bounded physical parameters. *14th IFAC Symposium on System Identification*, Newcastle, Australia, March 2006.
20. D.C. Karnopp, D.L. Margolis, and R.C. Rosenberg. *System Dynamics: A Unified Approach*. John Wiley, New-York, USA, 1990.
21. C.T. Kelley. *Iterative Methods for Optimization*. Frontiers in Applied Mathematics. SIAM, Philadelphia, USA, 1999.

22. I.D. Loram and M. Lakie. Direct measurement of human ankle stiffness during quiet standing: the intrinsic mechanical stiffness is insufficient for stability. *Journal of Physiology*, 545(3):1041–1053, 2002.
23. I.D. Loram and M. Lakie. Human balancing of an inverted pendulum: position control by small, ballistic-like, throw and catch movements. *Journal of Physiology*, 540(3):1111–1124, 2002.
24. D.G. Luenberger. *Linear and Nonlinear Programming*. 2nd edition, Addison-Wesley Publishing Company, Reading, USA, 1984.
25. N. Munro (ed). *Symbolic Methods in Control Systems Analysis and Design*. Number 56 in Control Engineering Series. IEE, Stevenage, UK, 1999.
26. B. Ninness and A. Wills. An identification toolbox for profiling novel techniques. *14th IFAC Symposium on System Identification*, Newcastle, Australia, March 2006.
27. T. Söderström. Convergence properties of the generalized least squares method. *Automatica*, 10:617–626, 1974.
28. L. Wang and W.R. Cluett. Frequency sampling filters: an improved model structure for step-response identification. *Automatica*, 33(5):939–944, 1997.
29. L. Wang, M. Desarmo, and W.R. Cluett. Real-time estimation of process frequency response and step response from relay feedback experiments. *Automatica*, 35:1427–1436, 1999.
30. L. Wang and W.R. Cluett. *From Plant Data to Process Control*. Taylor and Francis, London and New York, 2000.
31. L. Wang, P.J. Gawthrop, C. Chessari, T. Podsiadly, and A. Giles. Indirect approach to continuous-time system identification of food extruder. *Journal of Process Control*, 14(6):603–615, September 2004.
32. L. Wang and P.J. Gawthrop. On the estimation of continuous-time transfer functions. *International Journal of Control*, 74(9):889–904, June 2001.

8

Frequency-domain Approach to Continuous-time System Identification: Some Practical Aspects

Rik Pintelon, Johan Schoukens, and Yves Rolain

Vrije Universiteit Brussel, Belgium

8.1 Introduction

Since the end of the 1950s – beginning of the 1960s – the control society developed for its control designs a technique to build discrete-time models of continuous-time processes. Due to its overwhelming success a classical time-domain school emerged, and its authority in the field of system identification was soon widely recognised. The continuous-time identification methods developed in the early days of system identification [30, 67] got into a tight corner, and were 'forgotten' for several decades. Nowadays many people select discrete-time models and classical time-domain identification methods to solve their particular modelling problems. If the input is zero-order-hold, then discrete-time models are the natural choice, however, in all other cases continuous-time models might be preferred. Also, if the final goal is physical interpretation, then continuous-time modelling is the prime choice.

Since physical interpretation is mostly the main motivation for continuous-time modelling, special attention is paid in this chapter to some – often implicitly made – basic assumptions: the inter-sample behaviour of the excitation (zero-order-hold or band-limited), the measurement setup (zero-order-hold or band-limited, calibration of the systematic errors, open loop versus closed loop ...), the noise model (discrete-time or continuous-time, parametric or non-parametric), the stochastic framework (generalised output error or errors-in-variables), the sampling scheme (uniform or non-uniform), and the linearity (influence of non-linear distortions). Within this framework the advantages and drawbacks of the existing continuous-time identification methods are discussed. Several practical aspects are illustrated on two real measurement examples. The chapter concludes with some guidelines for the user.

8.2 The Inter-sample Behaviour and the Measurement Setup

This section reveals the inter-relations between the signal input inter-sample behaviour (zero-order-hold or band-limited), the model that relates the observed input samples to the observed output samples (discrete time or continuous time), the stochastic framework (generalised output error or errors-in-variables), and the measurement setup (infinite bandwidth or band-limited). We first handle the case where both the input and output of the continuous-time process are available (= plant modelling), and next study the case where only the output is observed (= noise modelling).

8.2.1 Plant Modelling

In both the zero-order-hold (see Figure 8.1) and band-limited (see Figure 8.2) measurement setup, the continuous-time plant is typically excited via an actuator, which is driven by an arbitrary waveform generator or a digital controller. The output of an arbitrary waveform generator or a digital controller is mostly a piecewise-constant signal $u_{\text{ZOH}}(t)$

$$u_{\text{ZOH}}(t) = \sum_{n=-\infty}^{+\infty} u_d(n)\text{zoh}(t - nT_s) \text{ with } \text{zoh}(t) = \begin{cases} 1 & 0 \leq t < T_s \\ 0 & \text{elsewhere} \end{cases} \quad (8.1)$$

T_s the sampling period, and $u_d(n)$ the discrete-time signal stored in the memory. The dynamics of the input–output acquisition channels are represented by G_u and G_y, respectively, and they include the signal conditioning (buffers, and amplifiers/attenuators) and possibly the anti-alias filters. Each continuous-time process produces some noise: the plant generates process noise $n_p(t)$, and the input–output acquisition channels are subject to measurement errors $m_u(t)$ and $m_y(t)$, respectively. The sequel of this section studies in detail the differences between and the implications of the zero-order-hold (ZOH) and band-limited (BL) setup.

The Zero-order-hold Setup

In the ZOH setup (see Figure 8.1) a model is built from the *known* discrete-time sequence $u_d(n)$ (sampled piecewise-constant input of the actuator) to the *measured* output samples of the plant

$$y(nT_s) = y_0(nT_s) + n_y(nT_s) \quad (8.2)$$

with $y_0(nT_s)$ the noiseless output samples and $n_y(nT_s)$ the output noise. The latter depends on the process noise $n_p(nT_s)$ and measurement errors $m_y(nT_s)$. This identification problem fits within a generalized output error stochastic framework [32, 77]. The model obtained includes the dynamics of the actuator

8 Practical Aspects of Frequency-domain CT Modelling 217

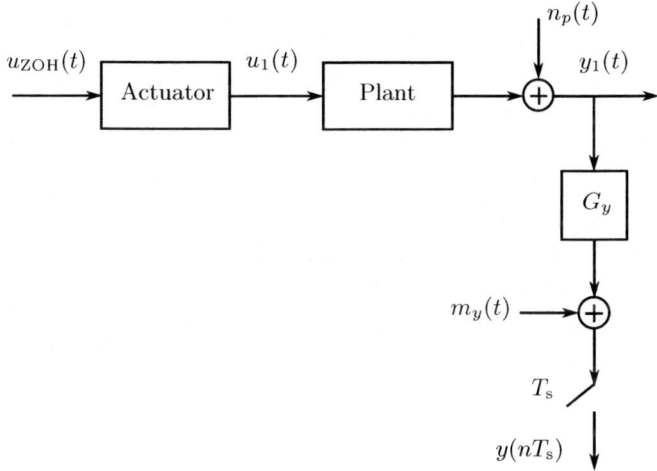

Fig. 8.1. Zero-order-hold measurement setup

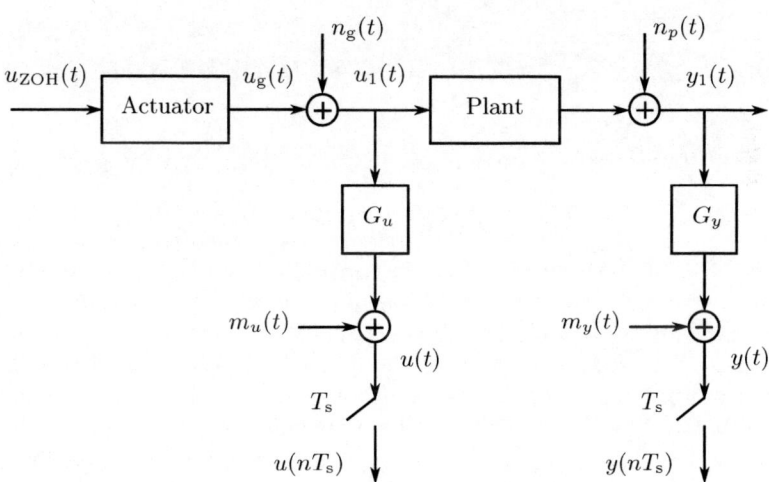

Fig. 8.2. Band-limited measurement setup

$G_{\text{act}}(i\omega)$, the plant $G(i\omega)$, and the acquisition channel $G_y(i\omega)$. Since the input of the actuator is piecewise-constant, the noiseless output samples $y_0(nT_s)$ are *exactly* related to $u_d(n) = u_{\text{ZOH}}(nT_s)$ by the following discrete-time transfer function

$$G_{\text{ZOH}}(z^{-1}) = \frac{Z\{y_0(nT_s)\}}{Z\{u_{\text{ZOH}}(nT_s)\}} = (1 - z^{-1}) Z\left\{ L^{-1}\left\{ \frac{G_{\text{act}}(s)G(s)G_y(s)}{s} \right\} \right\} \quad (8.3)$$

with $Z\{\ \}$ the Z-transform and $L^{-1}\{\ \}$ the inverse Laplace transform [32,43]. In control applications the goal is to predict the output $y_1(nT_s)$ (see Figure 8.1) given the past input and output samples $u_d(k)$ and $y_1(kT_s)$, $k = n - 1, n - 2, \ldots$; while in physical modelling the goal is to study the plant dynamics. In both cases the dynamics of the acquisition channel $G_y(i\omega)$ introduce a systematic error that should be eliminated. This is possible if the transfer function $G_c(s) = G_{\text{act}}(s)G(s)G_y(s)$ is lowpass with sufficiently small amplitude at half the sampling frequency

$$|G_c(i\omega_s/2)| \ll \max_\omega |G_c(i\omega)| \quad (8.4)$$

Indeed, under assumption (8.4) the frequency-response function $G_{\text{ZOH}}(e^{-i\omega T_s})$ (8.3) becomes

$$G_{\text{ZOH}}(e^{-i\omega T_s}) = \sum_{k=-\infty}^{+\infty} G_{\text{act}}(i\Omega_k)G(i\Omega_k)G_y(i\Omega_k)\text{ZOH}(\Omega_k/\omega_s)|_{\Omega_k=\omega-k\omega_s} \quad (8.5)$$

where $\text{ZOH}(x) = T_s e^{-j\pi x} \sin(\pi x)/\pi x$ [53], can be approximated as

$$G_{\text{ZOH}}(e^{-i\omega T_s}) \approx G_{\text{act}}(i\omega)G(i\omega)G_y(i\omega)\text{ZOH}(\omega/\omega_s) \text{ for } |\omega| < \omega_s/2 \quad (8.6)$$

From (8.6) it follows that dividing $G_{\text{ZOH}}(e^{-i\omega T_s})$ by $G_y(i\omega)$ eliminates the systematic errors introduced by the acquisition channel. Measuring $G_y(i\omega)$ requires a calibrated power meter, and a phase-calibrated broadband excitation signal. Obtaining the latter is the difficult step (bottleneck) of the absolute calibration procedure. Note that in physical modelling the same lines can be followed to eliminate the actuator dynamics $G_{\text{act}}(i\omega)$ in (8.6). If (8.4) is *not* satisfied, then correction for the acquisition and actuator dynamics is, in general, *impossible*.

One could think that the influence of the actuator characteristic is eliminated by constructing a model from the sampled input of the plant $u_1(nT_s)$ to the sampled output $y_1(nT_s)$. This is not true. Indeed, the noiseless output samples $y_{10}(nT_s)$ are *exactly* related to the input samples $u_1(nT_s)$ by the following discrete-time transfer function

$$G_1(z^{-1}) = \frac{Z\{y_{10}(nT_s)\}}{Z\{u_1(nT_s)\}} = \frac{\frac{Z\{y_{10}(nT_s)\}}{Z\{u_{\text{ZOH}}(nT_s)\}}}{\frac{Z\{u_1(nT_s)\}}{Z\{u_{\text{ZOH}}(nT_s)\}}} = \frac{Z\{L^{-1}\{G_{\text{act}}(s)G(s)/s\}\}}{Z\{L^{-1}\{G_{\text{act}}(s)/s\}\}} \quad (8.7)$$

(apply formula (8.3) twice), which clearly depends on $G_{\text{act}}(s)$. It also shows that $G_1(z^{-1})$ depends on the *inter-sample behaviour* of the input $u_1(t)$ described by the following impulse response [32, 53]

$$L^{-1}\left\{\left(1 - e^{-sT_s}\right) G_{\text{act}}(s)/s\right\} \tag{8.8}$$

For an ideal ZOH setup ($G_{\text{act}} = 1$ and $G_y = 1$) the continuous-time plant dynamics can be recovered from the identified discrete-time model via the inverse of the step-invariant transformation (8.3), if the sampling frequency is larger than twice the bandwidth of the plant, and if the plant contains no delay [27]. While the transformation of the zeros is quite complicated [3] that of the poles is given by the impulse invariant transformation $s = \log z/T_s$.

Band-limited Setup

In the BL setup (see Figure 8.2) a model is built from the *measured* input samples $u(nT_s)$ to the *measured* output samples $y(nT_s)$

$$\begin{aligned} y(nT_s) &= y_0(nT_s) + n_y(nT_s) \\ u(nT_s) &= u_0(nT_s) + n_u(nT_s) \end{aligned} \tag{8.9}$$

with $u_0(nT_s)$, $y_0(nT_s)$ the noiseless input–output samples and $n_u(nT_s)$, $n_y(nT_s)$ the input–output noise sources, which depend on the process noise $n_p(nT_s)$, the input–output measurement errors $m_u(nT_s)$, $m_y(nT_s)$, and possibly the generator noise $n_g(nT_s)$ (see Section 8.4 for a detailed discussion). This identification problem fits within an errors-in-variables stochastic framework [75]. The model obtained includes the dynamics of the plant $G(i\omega)$ and the input–output acquisition channels $G_u(i\omega)$ and $G_y(i\omega)$, and is independent of the actuator characteristics. The latter may even be non-linear. The noiseless input–output spectra $u_0(t)$, $y_0(t)$ are *exactly* related by the following continuous-time transfer function

$$G_c(i\omega) = \frac{F\{y_0(t)\}}{F\{u_0(t)\}} = \frac{G_y(i\omega)}{G_u(i\omega)} G(i\omega) \tag{8.10}$$

where $F\{\ \}$ stands for the Fourier transform.

Since the goal of the BL setup is to study the plant physics, the systematic error introduced by the dynamics of the input–output acquisition channels should be eliminated. This is possible via a relative calibration of the data-acquisition channels: a signal covering the frequency band of interest ($\omega < \omega_s/2$) is applied simultaneously to both acquisition channels ($G = 1$ in (8.10)) and the ratio of the measured spectra is exactly equal to $G_y(i\omega)/G_u(i\omega)$. Contrary to the absolute calibration, the relative calibration does not require a calibrated power meter nor a phase-calibrated broadband excitation.

To avoid aliasing during the sampling process, the anti-alias filters $G_u(i\omega)$ and $G_y(i\omega)$ in the data-acquisition channels should ideally be zero for $|\omega| > \omega_s/2$. In practice, attenuations of a hundred dB and more are realisable, possibly at the price of increased passband ripple and phase distortion. The latter can, however, easily be corrected for via a relative calibration.

8.2.2 Noise Modelling

Identification of the system characteristics from output observations only (see Figure 8.3), is referred to as time-series analysis in econometrics, blind identification in signal processing, noise modelling in system identification, and operational modal analysis in mechanical engineering. Although it is typically assumed that the excitation $e_c(t)$ is a random process, sometimes a mixture of random and periodic components is allowed [44]. Here, we will limit ourselves to random processes. Figure 8.3 shows a typical measurement setup with $H(i\omega)$ the process (noise) dynamics, $G_v(i\omega)$ the characteristics of the acquisition unit and $m_v(t)$ the measurement errors. In the rest of this section three cases of continuous-time random processes $e_c(t)$ are handled: (i) Wiener processes, (ii) piecewise-constant processes, and (iii) band-limited white noise processes.

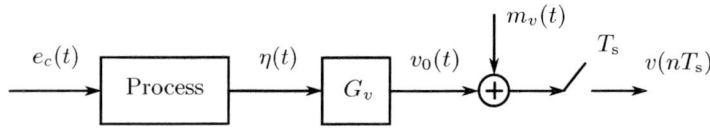

Fig. 8.3. Measurement of a continuous-time random process

It is important to realise that only the noise power spectrum can be identified from output measurements only (see Section 8.5.3). As a consequence the phase of $G_v(i\omega)$ does not affect the identified noise model, and only an absolute amplitude calibration of the acquisition channel is needed. Another consequence is that the sum of the power spectra of $v_0(nT_s)$ and $m_v(nT_s)$ is modelled. To remove the bias introduced by the measurement errors, the power spectrum of $m_v(nT_s)$ is measured in the absence of excitation ($\eta(t) = 0$ in Figure 8.3), and subsequently subtracted from the initial measurements (see [58] for the details).

Zero-order-hold Setup

A Wiener process (also called Brownian motion) is a stochastic process with continuous-time white Gaussian noise increments. The variance of such a process increases linearly in time. Assuming that the noise-generating mechanism $e_c(t)$ is a Wiener process and that the signal $\eta(t)$ in Figure 8.3 is sampled without anti-alias protection ($G_v = 1$), it has been shown in [2] and [23] that an rth-order continuous-time stochastic process can be described exactly at the sampling instances by an rth-order discrete-time process. The poles of

8 Practical Aspects of Frequency-domain CT Modelling

a discrete-time process are related to those of the original continuous-time process by the impulse invariant transformation $z = \exp(sT_s)$.

If the driving noise source $e_c(t)$ is a piecewise-constant stochastic process (8.1), with $e_c(nT_s)$ discrete-time white noise, then all the results of Section 8.2.1, subsection zero-order-hold setup, are valid. The continuous-time noise dynamics should then be recovered from the identified discrete-time model via the inverse of the step-invariant transformation (8.3).

One can wonder whether Wiener processes, which have asymptotically (time $\to \infty$) infinite variance, or piecewise-constant noise, which is a non-stationary continuous-time process (see [58]), are realistic descriptions (approximations) of the true noise-generating mechanism. This question is especially relevant if the ultimate goal of the identification experiment is physical interpretation. However, if the application is prediction or control, then, no matter what the true inter-sample behaviour of the noise-generating mechanism is, it suffices to have a good approximation of the observed noise power spectrum.

Band-limited Setup

The concept of continuous-time (CT) band-limited (BL) white noise has been introduced in [2]. By definition $e_c(t)$ is CT-BL white noise if its power spectral density $\Phi_{e_c}(\omega)$ satisfies

$$\Phi_{e_c}(\omega) = \begin{cases} \Phi_{e_c}(0) & |\omega| \leq \omega_B \\ 0 & \text{elsewhere} \end{cases} \tag{8.11}$$

Assuming that the acquisition channel is an ideal BL setup

$$|G_v(i\omega)| = \begin{cases} 1 & |\omega| \leq \omega_s/2 \\ 0 & \text{elsewhere} \end{cases} \tag{8.12}$$

and that $\omega_B \geq \omega_s/2$, it can be seen from Figure 8.3 that

$$\Phi_{v_0}(\omega) = |H(i\omega)|^2 |G_v(i\omega)|^2 \Phi_{e_c}(\omega) \equiv |H(i\omega)|^2 \Phi_e(\omega) \tag{8.13}$$

with $H(i\omega)$ the noise dynamics, and where $\Phi_e(\omega)$ is the power spectral density of CT-BL white noise $e(t)$ with bandwidth $\omega_s/2$. Since $e(t)$ has no power above $f_s/2$, it follows from (8.13) that a continuous-time noise model is the natural choice in a BL measurement setup. Since condition (8.11) can be relaxed to

$$\Phi_{e_c}(\omega) = \begin{cases} \Phi_{e_c}(0) & |\omega| \leq \omega_B \\ O(\omega^{-(1+\delta)}) & \text{elsewhere} \end{cases} \text{with } \delta > 0 \tag{8.14}$$

(see [58]), and since the non-idealities of the anti-alias filter can be easily compensated for (see the discussion in the introduction of this section), the concept of CT-BL white noise within a BL measurement setup is well suited for physical modelling.

In some applications such as, for example, sunspots data in astrophysics [46], econometric data [5], and order tracking in rotating machinery [19], it is impossible to lowpass filter the signals before sampling. Hence, it makes sense to consider continuous-time noise modelling without anti-alias protection. However, one should realise that at low sampling rates, the noise power spectral density of the observed samples may then strongly depend on the true intersample behaviour of the driving noise source.

8.2.3 Summary

The requirements for physical modelling are summarised in Table 8.1.

Table 8.1. Requirements for physical modelling

	Band-limited setup	Zero-order-hold setup
Measurement device	• plant modelling: relative amplitude/phase calibration • noise modelling: absolute amplitude calibration • anti-alias filters required • analogue bandwidth $\geq f_s/2$ • two-channel measurement	• plant modelling: absolute amplitude/phase calibration • noise modelling: absolute amplitude calibration • no anti-alias filters allowed • analogue bandwidth many times f_s • one-channel measurement
Actuator	• no calibration needed • may be non-linear	• calibration needed • should be linear
Stochastic framework	• noisy input, noisy output (= errors-in-variables)	• input known, noisy output (= generalised output error)
Model	• continuous-time	• discrete-time

Exact reconstruction of the continuous-time (CT) plant characteristics using a ZOH setup is possible only if $G_y(i\omega) = 1$ (infinite bandwidth acquisition channel), $G_{\text{act}}(i\omega) = 1$ (the plant input is exactly ZOH), $f_s/2$ is larger than the bandwidth of the plant, and the plant contains no delay. The continuous-time transfer function $G(s)$ is then reconstructed from the discrete-time (DT) model $G_{\text{ZOH}}(z^{-1})$ via the inverse of the step-invariant transformation (8.3). If the two conditions $G_y = 1$ and $G_{\text{act}} = 1$ are not fulfilled, then *approximate* reconstruction of $G(s)$ is possible if $f_s/2$ is many times (typically ten) the plant bandwidth (condition (8.4) is then satisfied). In addition, an absolute calibration of the actuator and the acquisition channel is needed.

Exact reconstruction of the CT plant dynamics using a BL setup is possible only if $G_y(i\omega)/G_u(i\omega) = 1$ for $|\omega| \leq \omega_s/2$, and $G_y(i\omega) = G_u(i\omega) = 0$ for $|\omega| > \omega_s/2$. If these two conditions are not fulfilled, then *approximate* reconstruction is possible via a relative calibration of the data-acquisition channels in the band $|\omega| \leq \omega_s/2$, and a sufficiently high attenuation of the anti-alias filters for $|\omega| > \omega_s/2$.

The only drawback of the BL setup is that it leads to an errors-in-variables problem that, for arbitrary excitations, is much more difficult to solve than the generalised output error problem of the ZOH setup. Fortunately, this is *not* the case for periodic excitations (see Section 8.5.2). The drawbacks of the ZOH setup are that, compared with the BL case, the ideal setup is much harder to realise, and the reconstruction of the CT dynamics is much more involved.

8.3 Parametric Models

In Section 8.2 it has been shown that, depending on the measurement setup (ZOH or BL), either a discrete-time model (difference equation), or a continuous-time model (differential equation) should be used to describe the relationship between the sampled input and output signals. In this section a time- and frequency-domain description will be given. With some abuse of notation the argument t will denote the continuous-time variable for CT systems and the discrete-time variable $\ldots, t-1, t, t+1, \ldots$ for DT systems. The difference will be clear from the context.

8.3.1 Plant Models

Discrete-time

A finite-dimensional, linear time-invariant discrete-time system, is described by a linear difference equation with constant coefficients, that can be written as

$$y(t) = G(q^{-1})u(t) \tag{8.15}$$

with q^{-1} the backward shift operator ($q^{-1}x(t) = x(t-1)$), and $G(z^{-1})$ a rational form in z^{-1}

$$G(z^{-1}) = \frac{B(z^{-1})}{A(z^{-1})} = \frac{\sum_{r=0}^{n_b} b_r z^{-r}}{\sum_{r=0}^{n_a} a_r z^{-r}} \tag{8.16}$$

with z the Z-transform variable. Note that the time-domain representation (8.15) contains implicitly the influence of the initial conditions.

Assume now that N samples of the input–output signal $u(t)$, $y(t)$, $t = 0, 1, \ldots, N-1$ are available. The relationship between the discrete Fourier transform (DFT) spectra $U(k)$, $Y(k)$ of these samples

$$X(k) = \frac{1}{\sqrt{N}} \sum_{t=0}^{N-1} x(t) e^{-2\pi i k t / N} \tag{8.17}$$

with $x = u, y$ and $X = U, Y$, is *exactly* given by

$$Y(k) = G(z_k^{-1})U(k) + T_G(z_k^{-1}) \text{ with } z_k = e^{2\pi i k/N} \quad (8.18)$$

where $G(z^{-1})$ is defined in (8.16), and where the transient term T_G

$$T_G(z^{-1}) = \frac{I_G(z^{-1})}{A(z^{-1})} = \frac{\sum_{r=0}^{n_{i_g}} i_{g_r} z^{-r}}{\sum_{r=0}^{n_a} a_r z^{-r}} \text{ with } n_{i_g} = \max(n_a, n_b) - 1 \quad (8.19)$$

contains the influence of the initial and final conditions of the experiment [53, 62]. The transient term $T_G(z_k^{-1})$ decreases to zero as an $O(N^{-\frac{1}{2}})$ w.r.t. the main term $G(z_k^{-1})U(k)$. It is *exactly zero* for experiments where the initial state equals the final state, for example, periodic input–output signals, or time-limited input–output signals that are completely captured within the measurement window.

From (8.19) it follows that the transient term T_G in (8.18), which is responsible for the leakage error in frequency-response function measurements

$$\hat{G}_N(e^{-i\omega_k}) = \begin{cases} \frac{Y(k)}{U(k)} = G(z_k^{-1}) + \frac{T_G(z_k^{-1})}{U(k)} \\ \frac{\hat{\Phi}_{yu}^N(\omega_k)}{\hat{\Phi}_u^N(\omega_k)} = G(z_k^{-1}) + \frac{E\{T_G(z_k^{-1})U(k)\}}{E\{|U(k)|^2\}} \end{cases} \quad (8.20)$$

has a lot of structure. It is a rational form in z^{-1} with the same poles as the plant transfer function G. Using the smooth behaviour of $T_G(e^{-i\omega})$, non-parametric frequency-response function estimators can be constructed that suppress the leakage error more effectively than the classical time-domain windows [71].

Continuous-time

A finite dimensional, linear time-invariant continuous-time system, is described by a linear differential equation with constant coefficients, that can be written as

$$y(t) = G(p)u(t) \quad (8.21)$$

with p the derivative operator $(px(t) = dx(t)/dt)$, and $G(s)$ a rational form in s

$$G(s) = \frac{B(s)}{A(s)} = \frac{\sum_{r=0}^{n_b} b_r s^r}{\sum_{r=0}^{n_a} a_r s^r} \quad (8.22)$$

with s the Laplace transform variable. Note that the time domain representation (8.21) contains implicitly the influence of the initial conditions. Assuming that N samples of the input–output signal $u(nT_s)$, $y(nT_s)$, $n = 0, 1, \ldots, N - 1$ are available, the relationship between the discrete Fourier transform (DFT) spectra $U(k)$, $Y(k)$ (8.17) of these samples is *exactly* given by

$$Y(k) = G(s_k)U(k) + T_G(s_k) + \Delta(s_k) \text{ with } s_k = 2\pi i k/N \quad (8.23)$$

where $G(s)$ is defined in (8.22), $\Delta(s_k)$ is the residual alias error, and where the transient term T_G

$$T_G(s) = \frac{I_G(s)}{A(s)} = \frac{\sum_{r=0}^{n_{i_g}} i_{g_r} s^r}{\sum_{r=0}^{n_a} a_r s^r} \text{ with } n_{i_g} = \max(n_a, n_b) - 1 \tag{8.24}$$

contains the influence of the initial and final conditions of the experiment [52, 53]. Both the transient term $T_G(s_k)$ and the residual alias error $\Delta(s_k)$ decrease to zero as an $O(N^{-\frac{1}{2}})$ w.r.t. the main term $G(s_k)U(k)$. For periodic input–output signals $T_G(s_k)$ and $\Delta(s_k)$ are *exactly zero*. Practice has shown that the alias error $\Delta(s_k)$ can be made arbitrarily small by increasing the order of $I_G(s)$: $n_{i_g} \geq \max(n_a, n_b) - 1$ [52]. Hence, (8.23) can be written as

$$Y(k) = G(s_k)U(k) + T_G(s_k) \text{ with } s_k = 2\pi i k/N \tag{8.25}$$

where, similarly to n_a and n_b, $n_{i_g} \geq \max(n_a, n_b) - 1$ should be determined via model order selection (see Section 8.5).

Similarly to the DT case, the transient term $T_G(s_k)$ in (8.25) is responsible for the leakage error in frequency response function measurements $\hat{G}_N(i\omega_k)$ (replace z_k^{-1} by s_k in (8.20)). Since $T_G(i\omega)$ is also a smooth function of the frequency, the same non-parametric frequency-response function estimators as in [71] can be used to suppress the leakage errors in $\hat{G}_N(i\omega_k)$.

8.3.2 Noise Models

The time- and frequency-domain descriptions of the noise models are similar to that of the plant models; the only difference being the stochastic properties of the driving noise source.

Discrete-time

Equations (8.15) and (8.18) become

$$\begin{aligned} v(t) &= H(q^{-1})e(t) \\ V(k) &= H(z_k^{-1})E(k) + T_H(z_k^{-1}) \text{ with } z_k = e^{2\pi i k/N} \end{aligned} \tag{8.26}$$

where $H(z^{-1})$ and $T_H(z^{-1})$ are rational forms in z^{-1}

$$H(z^{-1}) = \frac{C(z^{-1})}{D(z^{-1})} = \frac{\sum_{r=0}^{n_c} c_r z^{-r}}{\sum_{r=0}^{n_d} d_r z^{-r}}, \text{ and } T_H(z^{-1}) = \frac{I_H(z^{-1})}{D(z^{-1})} = \frac{\sum_{r=0}^{n_{i_h}} i_{h_r} z^{-r}}{\sum_{r=0}^{n_d} d_r z^{-r}} \tag{8.27}$$

with $n_{i_h} = \max(n_c, n_d) - 1$. The discrete Fourier transform (DFT) $E(k)$ of the driving noise source $e(t)$ has the following properties [6, 53]. If $e(t)$ is discrete-time white noise, then $E(k)$ is uncorrelated, and circular complex ($E\{E^2(k)\} = 0$). If in addition $e(t)$ is normally distributed, then $E(k)$ is independent, circular complex normally distributed. For non-Gaussian i.i.d. noise $e(t)$, $E(k)$ is asymptotically ($N \to \infty$) independent, circular complex normally distributed.

Continuous-time

Equations (8.21) and (8.25) become

$$v(t) = H(p)e(t)$$
$$V(k) = H(s_k)E(k) + T_H(s_k) \text{ with } s_k = 2\pi ik/N \quad (8.28)$$

where $H(s)$ and $T_H(s)$ are rational forms in s

$$H(s) = \frac{C(s)}{D(s)} = \frac{\sum_{r=0}^{n_c} c_r s^r}{\sum_{r=0}^{n_d} d_r s^r}, \text{ and } T_H(s) = \frac{I_H(s)}{D(s)} = \frac{\sum_{r=0}^{n_{i_h}} i_{h_r} s^r}{\sum_{r=0}^{n_d} d_r s^r} \quad (8.29)$$

with $n_{i_h} \geq \max(n_c, n_d) - 1$. The driving noise source $e(t)$ is a continuous-time, band-limited white noise process, with bandwidth equal to half the sampling frequency (see Section 8.2.2). Since the autocorrelation of $e(t)$

$$R_e(\tau) = F^{-1}\{\Phi_e(\omega)\} = \frac{1}{2\pi} \int_{-\omega_s/2}^{+\omega_s/2} \sigma^2 e^{i\omega\tau} d\omega = \sigma^2 \sin(\pi f_s \tau)/(\pi f_s \tau) \quad (8.30)$$

with $\sigma^2 = \text{var}(e(t))$, is zero at the sampling instances $\tau = nT_s$, $n \neq 0$, the discrete-time sequence $e(nT_s)$ is uncorrelated. Hence, $E(k)$ is uncorrelated and circular complex. If $e(t)$ is normally distributed, then $e(nT_s)$ is a white Gaussian sequence, and $E(k)$ is independent, circular complex normally distributed. Note that $E(k)$ has the same properties as in the discrete-time case.

8.3.3 Summary

In a generalised output error framework the plant and noise models are combined as

$$y(t) = G(w)u(t) + H(w)e(t) \text{ with } w = q^{-1} \text{ or } p$$
$$Y(k) = G(\Omega_k)U(k) + H(\Omega_k)E(k) + T_G(\Omega_k) + T_H(\Omega_k) \text{ with } \Omega = z^{-1} \text{ or } s \quad (8.31)$$

where $u(t)$ is the *known* input, $y(t)$ the *measured* output, and $e(t)$ the *unobserved* driving noise source; and with $Y(k)$, $U(k)$, $E(k)$ the DFT spectra of N samples of $u(t)$, $y(t)$, $e(t)$ respectively. The plant $G = B/A$ and noise $H = C/D$ transfer functions, and the plant $T_G = I_G/A$ and noise $T_H = I_H/D$ transient terms, are rational functions in Ω. According to the parametrisation of the plant and the noise model one distinguishes different model structures such as OE, ARMAX, BJ, ... (see Table 8.2). For the ARMAX model T_G and T_H are indistinguishable. Therefore, T_H is set to zero and the order of T_G is chosen as $n_{i_g} \geq \max(n_a, n_b, n_c) - 1$.

One could also think to combine a continuous-time plant model with a discrete-time noise model [24, 61, 81]. This is called hybrid modelling. The disadvantage of this approach is that it leads to biased estimates for identification in feedback [53].

Table 8.2. Model structures in a generalised output error framework (the OE, ARMA, ARMAX, and BJ model structures have Ω as argument).

	OE	ARMA	ARMAX	BJ	hybrid BJ
G	B/A	0	B/A	B/A	$B(s)/A(s)$
T_G	I_G/A	0	I_G/A	I_G/A	$I_G(s)/A(s)$
H	1	C/D	C/A	C/D	$C(z^{-1})/D(z^{-1})$
T_H	0	I_H/D	0	I_H/D	$I_H(z^{-1})/D(z^{-1})$

Diffusion phenomena such as heat or mass transfer are often described by non-even irrational transfer functions in $\sqrt{s}$. Such transfer functions can be approximated very well by a rational function in $\sqrt{s}$ [60]. Also, some noise processes are better approximated by rational forms in $\sqrt{s}$ rather than in s. Think, for example, of $1/f$ noise whose power spectral density is proportional to $1/f$. The corresponding noise filter H must be proportional to $1/\sqrt{s}$. To conclude, the frequency domain description (8.31) can be extended to $\Omega = \sqrt{s}$ for describing diffusion phenomena. The corresponding time-domain equation with $w = p^{\frac{1}{2}}$ uses fractional derivatives of order $n+1/2$, $n = 0, 1, \ldots$ (see [45]).

8.4 The Stochastic Framework

Figure 8.4 shows a general stochastic framework for the identification of a plant model. The plant is captured in a feedback loop with controller M, and all kinds of disturbances are present: generator noise N_g, process noise N_P, controller noise N_C, and input–output measurement noise M_U and M_Y. The measured input–output DFT spectra U, Y equal the noiseless values U_0, Y_0 plus some noise N_U, N_Y

$$U(k) = U_0(k) + N_U(k)$$
$$Y(k) = Y_0(k) + N_Y(k)$$
(8.32)

According to the type of reference signal R – periodic or arbitrary – a particular noise source acts as a disturbance (part of N_U, N_Y) or contributes to the excitation (part of U_0, Y_0). This is discussed in the rest of this section. Without any loss of generality the transient terms are neglected in frequency domain equations.

8.4.1 Periodic Excitations

If the reference signal $r(t)$ is periodic, then any deviation from the periodic behaviour is considered as noise. Hence, the true values U_0, Y_0 and the disturbances N_U, N_Y in (8.32) are given by (for notational simplicity the arguments are omitted)

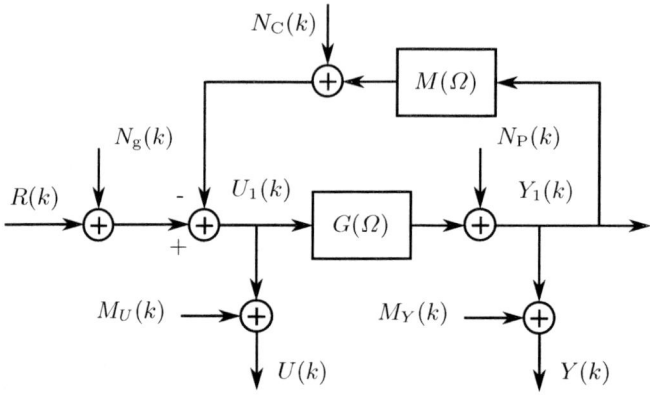

Fig. 8.4. Identification of a plant G captured in a feedback loop with controller M

$$\begin{cases} U_0 = \frac{1}{1+GM}R \\ Y_0 = \frac{G}{1+GM}R \end{cases} \text{and} \begin{cases} N_U = M_U + \frac{1}{1+GM}(N_g - N_C) - \frac{M}{1+GM}N_P \\ N_Y = M_Y + \frac{G}{1+GM}(N_g - N_C) + \frac{1}{1+GM}N_P \end{cases} \quad (8.33)$$

By construction, the disturbances N_U and N_Y are independent of the true values U_0 and Y_0. Due to the generator noise N_g, the controller noise N_C, and the process noise N_P, the input–output errors N_U and N_Y are correlated. Since these noise sources contribute to the true input U_1 of the plant (see Figure 8.4)

$$U_1 = \frac{1}{1+GM}(R + N_g - N_C - MN_P) \quad (8.34)$$

a part of the information is lost, which is a *drawback* of the periodic framework (see also Section 8.5.2). However, N_g and N_C do not increase the variability of the plant estimates (see Section 8.5.2). Another possible source for correlation between the input–output errors is a common disturbance picked up by both acquisition channels of the measurement device. In that case the measurement errors M_U and M_Y are no longer independent.

8.4.2 Arbitrary Excitations

If the reference signal $r(t)$ is arbitrary, then U_1 is the noiseless input of the plant (see Figure 8.4), and only M_U, M_Y, and N_P act as disturbances. Hence, the true values U_0, Y_0 and the disturbances N_U, N_Y in (8.32) are given by

$$\begin{cases} U_0 = \frac{1}{1+GM}(R + N_g - N_C - MN_P) \\ Y_0 = \frac{G}{1+GM}(R + N_g - N_C - MN_P) \end{cases} \text{and} \begin{cases} N_U = M_U \\ N_Y = M_Y + \frac{1}{1+GM}N_P \end{cases} \quad (8.35)$$

Due to the feedback loop ($M \neq 0$) and the process noise N_P, the output disturbance N_Y is correlated with the true values U_0 and Y_0. This is the *major difficulty* of identification in feedback with arbitrary excitations. Note that the process noise N_P does not introduce a correlation between the input–output disturbances N_U and N_Y. The latter are correlated only if the measurement errors M_U and M_Y are correlated.

8.5 Identification Methods

This section gives a brief overview of the time-domain methods and discusses the frequency-domain maximum likelihood solutions in more detail. The distinction between time- and frequency-domain methods may – at first glance – be artificial because most of the things that can be done in one domain are also possible in the other domain. This is certainly true for (non-linear) least squares based algorithms (apply Parseval's theorem). However, the calculations may be much simpler in one domain than in the other. For example, the covariance matrix of filtered white noise is asymptotically ($N \to \infty$) Toeplitz in the time domain, and asymptotically ($N \to \infty$) diagonal in the frequency domain [70]. Hence, performing calculations with a non-parametric representation of the noise model is much easier in the frequency domain. Other examples are non-uniform sampling [9] and recursive identification [34] which can much more be handled easily in the time domain.

A class of methods that act totally differently in both domains are least absolute-value algorithms, which intend to suppress outliers in the data. Indeed, an outlier in the time-domain (extreme value) is smeared out in the frequency domain; while an outlier in the frequency domain (large peak due to a periodic disturbance), is smeared out in the time domain (sinewave). Hence, a least squares time-domain algorithm robustified for outliers by adding a least absolute-value part (see [32]) is not equivalent to the corresponding robustified least squares frequency-domain algorithm.

An important issue in identification is the filtering of the measured input–output signals. Often, one is only interested in the plant characteristics on a part of the unit circle (or imaginary axis), or one would like to remove the effect of trends (low-frequency range) or disturbances (mains, high-frequency noise, . . .), or one would like to reduce the complexity of the identification problem by focusing separately on different frequency bands. The classical time-domain approach consists in prefiltering the sampled input–output signals $u(t)$ and $y(t)$ with $F(z^{-1})$. The filtered signals $u_f(t)$ and $y_f(t)$ still satisfy the same input–output relationship, if also all the noise terms are filtered by $F(z^{-1})$. For example, in a generalised output error framework (8.31), we get

$$y_f(t) = G(q^{-1})u_f(t) + F(q^{-1})H(q^{-1})e(t) \qquad (8.36)$$

To preserve the consistency and efficiency of the identified plant model, the noise models should be flexible enough to follow the noise power spectrum, $\lambda \left| F(e^{-i\omega})H(e^{-i\omega}) \right|^2$ with $\lambda = \text{var}(e(t))$, accurately. As such, it will try to cancel the effect of the prefilter. Hence, through the prefilter and the noise model selection a compromise must be made between the suppression of the undesired frequency bands, and the loss in efficiency and consistency of the plant estimates [32]. These conflicting demands, which are inherent to *all* time-domain methods, are avoided in the frequency domain where filtering consists in removing the undesired DFT frequencies in (8.31). This *exact* filter operation does not increase the complexity of the noise model. On the contrary, the noise (and plant) model should be identified in the frequency band(s) of interest only. Note that exactly the same conclusions hold in an errors-in-variables framework. In the rest of this chapter it will be assumed that the frequency domain data $U(k)$, $Y(k)$ is available at DFT frequencies $f_k = kf_s/N$, $k \in \mathbb{K}$ where $\mathbb{K}$ is a subset of the DFT line numbers

$$\mathbb{K} \subseteq \{0, 1, 2, \ldots, N/2\} \tag{8.37}$$

and with N the number of time-domain points used to calculate the DFT spectra.

Another issue is the identification of unstable plants. For example, the best (in the mean-square sense) linear approximation of a non-linear system may be unstable. Another example is the identification of an unstable system enclosed in a stabilising feedback loop. In the frequency domain, irrespective of the position of the poles (stable or unstable) of the transfer function $G(\Omega)$, the response of $G(\Omega)$ to an input with DFT spectrum $U(k)$ is calculated as

$$G(\Omega_k)U(k) + T_G(\Omega_k) \tag{8.38}$$

(see, for example, (8.31)). Hence, stable and unstable models are handled in exactly the same way. This is not the case in the time domain. For example, for $\Omega = z^{-1}$ the time-domain response corresponding to (8.38) is obtained by splitting $G(z^{-1})$ and $T_G(z^{-1})$ in a stable causal part (poles inside the unit circle) and a stable anti-causal part (poles outside the unit circle), and next adding the two convolution products with $u(t)$.

A final issue is the scaling of the coefficients in continuous-time models. It is well known that continuous-time modelling is numerically ill-conditioned if no special precautions are taken – even for modest orders of the transfer function [48]. An easy way to improve the numerical conditioning significantly consists in scaling the angular frequencies. Although the scaling that minimises the condition number depends on the system, the model, the excitation signal, and the estimator used, the median of the angular frequencies is a good compromise [48]. For example, the $(r+1)$th term in the denominator of (8.22) becomes

$$a_r s^r = (a_r \omega_{\text{med}}^r) \left(\frac{s}{\omega_{\text{med}}}\right)^r = a_{\text{norm}r} s_{\text{norm}}^r \text{ with } \omega_{\text{med}} = \text{median}\{\omega_k | k \in \mathbb{K}\} \tag{8.39}$$

and where $\mathbb{K}$ is defined in (8.37). If the scaling (8.39) is not sufficient to obtain reliable estimates, then the powers of s/ω_{med} in (8.22) and (8.23) are replaced by (vector) orthogonal polynomials (see [64] and [10]). Note that the latter can also be necessary for discrete-time models, especially when the frequency band of interest covers only a small fraction of the unit circle.

8.5.1 Asymptotic Properties of the Frequency-domain Gaussian Maximum Likelihood Estimators

We list here the basic asymptotic ($F = \text{ord}(\mathbb{K}) \to \infty$, with $\mathbb{K}$ defined in (8.37)) properties of the frequency-domain Gaussian maximum likelihood (ML) estimators discussed in Sections 8.5.2 to 8.5.4. The reader is referred to [53,55,56] for the detailed proofs. Beside estimates of the model parameters, the Gaussian ML estimators also provide an estimate of the covariance matrix.

The ML estimators are all constructed under the assumption that the input–output noise $N_U(k)$ and $N_Y(k)$ on the measured input–output DFT spectra $U(k)$ and $Y(k)$ (see (8.33)), has zero mean, and is independent (over k), circular complex normally distributed. Under some standard assumptions it can be shown that the Gaussian ML estimators have the following asymptotic properties: strongly consistent, asymptotically normally distributed, and asymptotically efficient, except for the errors-in-variables problem with non-parametric noise models where the covariance matrix is very close but not equal to the Cramér–Rao lower bound. The Gaussian ML estimators are also *robust* w.r.t. the basic assumptions made to construct them: the strong consistency and asymptotic normality remain valid for non-Gaussian distributed input–output errors $N_U(k)$ and $N_Y(k)$, and correlation over the frequency is even allowed as long as it tends sufficiently fast to zero (mixing condition). The asymptotic properties of the estimates have also been studied when the true model does not belong to the considered model set (*e.g.*, unmodelled dynamics, and non-linear distortions).

8.5.2 Periodic Excitations

General

Although any periodic signal can be handled, typically a sum of harmonically related sinewaves is applied to the plant

$$u(t) = \frac{1}{\sqrt{N}} \sum_{k=1}^{N/2-1} A_k \sin(2\pi f_k t + \phi_k) \quad (8.40)$$

with $f_k = k f_s / N$, and A_k the user-defined amplitudes. According to the intended goal of the experiment, the identification of the true underlying linear system (if it exists) or the best (in the mean-square sense) linear approximation of the non-linear system, the phases ϕ_k are chosen to minimise the peak value [21] or are chosen randomly such that $E\left\{e^{i\phi_k}\right\} = 0$ [54, 68]. Note that the amplitudes of the sinewaves in (8.40) are scaled by $\sqrt{N}$ such that the rms value of $u(t)$ remains the same when increasing N. The number of excited frequencies F in (8.40) equals the number of non-zero amplitudes and increases with N: $F = O(N)$.

The *big advantage* of periodic excitations w.r.t. arbitrary excitations is that they allow us to distinguish between noise and signal by comparing consecutive periods of the steady-state response. For example, this separation property allows us to verify whether output energy at a non-excited frequency is due to noise or to non-linear behaviour of the plant (see Section 8.6.1). Another advantage of this separation property is that errors-in-variables identification (in feedback) with correlated input–output disturbances is as easy as identification within an open-loop, generalised output error framework (see Section 8.5.2). Assuming that M periods of the steady-state response are available the sample mean and sample (co-)variances of the M input–output DFT spectra can be calculated as

$$\hat{X}(k) = \frac{1}{M} \sum_{m=1}^{M} X^{[m]}(k)$$
$$\hat{\sigma}^2_{\hat{X}\hat{Z}}(k) = \frac{1}{M(M-1)} \sum_{m=1}^{M} \left(X^{[m]}(k) - \hat{X}(k)\right) \overline{\left(Z^{[m]}(k) - \hat{Z}(k)\right)} \quad (8.41)$$

with $X, Z = U$ and/or Y, $X^{[m]}(k)$ the DFT spectrum of the mth period of $x(t)$, and $\overline{X}$ the complex conjugate of X. Because an estimate of the variance of the sample mean is needed, an extra factor M appears in the expression of $\hat{\sigma}^2_{\hat{X}\hat{Z}}(k)$. Using (8.41), the quality of the input–output measurements can be verified via the input–output signal-to-noise ratios

$$\text{SNR}_U(k) = \left|\hat{U}(k)\right|/\hat{\sigma}_{\hat{U}}(k) \text{ and } \text{SNR}_Y(k) = \left|\hat{Y}(k)\right|/\hat{\sigma}_{\hat{Y}}(k) \quad (8.42)$$

where $\hat{\sigma}_{\hat{X}} = \hat{\sigma}_{\hat{X}\hat{X}}$ with $X = U, Y$, and a non-parametric estimate of the frequency-response function and its variance is obtained as

$$\hat{G}(\Omega_k) = \frac{\hat{Y}(k)}{\hat{U}(k)}$$
$$\hat{\sigma}^2_{\hat{G}}(k) = \left|\hat{G}(\Omega_k)\right|^2 \left(\frac{\hat{\sigma}^2_{\hat{Y}}(k)}{|\hat{Y}(k)|^2} + \frac{\hat{\sigma}^2_{\hat{U}}(k)}{|\hat{U}(k)|^2} - 2\text{Re}\left(\frac{\hat{\sigma}^2_{\hat{Y}\hat{U}}(k)}{\hat{Y}(k)\overline{\hat{U}(k)}}\right)\right) \quad (8.43)$$

with $\Omega_k = i\omega_k$ for CT and $\Omega_k = e^{-i\omega_k T_s}$ for DT [53]. For input signal-to-noise ratios $\text{SNR}_U(k)$ of 40 dB and more, $\hat{G}(i\omega_k)$ is approximately circular complex normally distributed, and a $p\%$ confidence bound is given by a circle with centre $\hat{G}$ and radius $\sqrt{-\log(1-p)}\hat{\sigma}_{\hat{G}}$. If $\text{SNR}_U(k)$ is smaller than 40 dB, then the precise calculation of the radius is somewhat more involved (see [49]). The FRF and its uncertainty (8.43) reveals the complexity of the parametric modelling, and is used to validate the identified model.

The *two drawbacks* of periodic excitations w.r.t. arbitrary excitations are the loss in frequency resolution of a factor M, and the fact that the non-periodic parts of the true excitation are discarded in (8.41).

Overview

Since the early days of system identification [30,67] the advantages of periodic excitation signals have almost exclusively been exploited in frequency-domain algorithms. Only very recently have some time-domain algorithms been developed that use explicitly the periodic nature of the excitation [15, 22, 40]. A possible reason for this is that, in the time domain, non-parametric noise models are more difficult to handle than the parametric ones.

The Frequency-domain Gaussian Maximum likelihood Solution

Since periodic excitations can be used for both ZOH and BL measurement setups, both discrete-time and continuous-time modelling are considered here. The Gaussian maximum likelihood estimates of the numerator and denominator coefficients θ of the plant model (8.22), are found by minimising

$$V_{\mathrm{ML}}(\theta, Z) = \sum_{k \in \mathbb{K}} \frac{|Y(k) - G(\Omega_k, \theta)U(k)|^2}{\sigma_Y^2(k) + \sigma_U^2(k)|G(\Omega_k, \theta)|^2 - 2\mathrm{Re}(\sigma_{YU}^2(k)\overline{G(\Omega_k, \theta)})} \quad (8.44)$$

w.r.t. θ, where $\mathbb{K}$ is defined in (8.37), and where $\sigma_{XZ}^2(k)$, with $X, Z = U$ and/or Y, are the true noise (co-)variances [53]. For the ZOH setup we have $\Omega = z^{-1}$, $\sigma_U = 0$, and $\sigma_{YU} = 0$; while for the BL setup $\Omega = s$. Some starting-value algorithms are available for the non-linear minimisation problem (8.44): (weighted) linear least squares [30, 67], total least squares [78], subspace [41, 80], ... (see [47] and [53] for an overview).

Besides the model parameters, the orders n_a and n_b of the numerator and denominator polynomials in (8.22) must also be estimated. This is done via comparison of the identified model with the non-parametric FRF estimate (8.43), a whiteness test of the residuals, and an analysis of the ML cost function via the AIC and MDL criteria [32, 53]. An automatic procedure for selecting the model orders n_a and n_b can be found in [65].

Although the generator, controller, and process noise sources are suppressed by the averaging procedure (8.41), they are not considered as noise in the ML cost function (8.44). Indeed, in frequency bands where the generator noise N_g and/or the controller noise N_C are dominant (see Figure 8.4), the denominator of the cost function (8.44) can be approximated by

$$\left| \frac{G_0(\Omega_k) - G(\Omega_k, \theta)}{1 + G_0(\Omega_k)M_0(\Omega_k)} \right|^2 (\sigma_\mathrm{g}^2(k) + \sigma_\mathrm{C}^2(k)) \quad (8.45)$$

with G_0 and M_0 the true plant and controller transfer functions. In the neighbourhood of $\theta = \theta_0$, (8.45) becomes very small and, hence, the generator and controller noise contribute to the knowledge of the plant model via a large weighting in the cost function. In frequency bands where the process noise N_P is dominant the denominator of (8.44) can be approximated by

$$\sigma_{\mathrm{P}}^2(k) \left| 1 - \frac{G_0(\Omega_k) - G(\Omega_k, \theta)}{1 + G_0(\Omega_k) M_0(\Omega_k)} \right|^2 \tag{8.46}$$

In the neighbourhood of $\theta = \theta_0$, (8.46) equals $\sigma_{\mathrm{P}}^2(k)$, which is exactly the variance of noisy plant output $Y_1(k)$ given the value of the true input $U_1(k)$ of the plant (see Figure 8.4). It shows that the process noise part of the excitation disappears in the denominator of (8.44) and, hence, contributes to the knowledge of the plant model.

In practice the true noise (co-)variances are unknown and are replaced by the sample (co-)variances (8.41). The minimiser of the corresponding cost function is called the sample maximum likelihood estimator (SML). For $M \geq 7$ it has exactly the same asymptotic properties as the ML estimator (see [53, 73] for the details), except that its covariance matrix is $(M-2)/(M-3)$ times larger. The latter is due to the extra variability introduced by the sample (co-)variances (8.41).

Extensions to multivariable systems are available in [20] and [37].

8.5.3 Arbitrary Excitations: Generalised Output Error

General

This section covers the case where the identification starts from an exactly known input and noisy output observations. We explicitly assume that (see Figure 8.4 and (8.35)) $M_U = 0$, $N_C = 0$, and $N_g = 0$ in open loop ($M = 0$), while $M_U = 0$, $N_g = 0$, $N_C = 0$, and $M_Y = 0$ in closed loop ($M \neq 0$). The classical time-domain methods implicitly assume that the acquisition bandwidth is infinitely large. Hence, they are suitable for physical modelling only if the input of the plant is either piecewise-constant or band limited (see Section 8.2).

Overview

If the plant input $u(t)$ is *zero-order-hold* (piecewise-constant), then the continuous-time equations (8.21) are transformed without systematic errors to a difference equation (see Section 8.2). This difference equation can be parametrised in the original continuous-time model parameters, and the identification can be done using standard time-domain prediction error methods (see [32] and the references therein). These methods minimise

$$V_{\mathrm{PE}}(\theta, Z) = \sum_{t=0}^{N-1} \left(H^{-1}(q^{-1}, \theta) \left(y(t) - G(q^{-1}, \theta) \right) \right)^2 \tag{8.47}$$

w.r.t. θ, and a parametric noise model is identified simultaneously with the plant model [32, 77].

If the plant input $u(t)$ is *band limited*, then differential equation (8.21) is transformed to a discrete-time equation by explicit or implicit approximation of the time derivatives using digital filters. According to the digital filters used these methods are known as the Poisson moment functional approach, the integrated sampling approach, the instantaneous sampling approach, the Laguerre approach ... (see [12,74] and the references therein). These modelling approaches work reasonably well (small systematic errors) for frequencies below $f_s/4$, but in the neighbourhood of the Nyquist frequency ($f > f_s/4$) they either introduce large errors or require (very) high-order digital filters [79]. Weighted (non-)linear least squares [74], maximum likelihood [79], instrumental variable [74,81] and subspace identification [4,25] methods have been developed. All these methods, except the ARMAX-modelling in [24] and the Box–Jenkins modelling in [81] (see also Chapter 4 in this book), either do not identify a noise model or assume that it is known.

The direct methods for identifying CT noise models [19, 29, 35] implicitly assume that the signals are not lowpass filtered before sampling (= ZOH measurement setup). Hence, at low sampling rates the identified model may strongly depend on the true inter-sample behaviour of the driving noise source (see Section 8.2.2). Therefore, alias-correcting methods have been proposed in [19].

A non-parametric estimate of the frequency-response function and its variance is obtained as

$$\hat{G}(\Omega_k) = \frac{\hat{\Phi}_{yu}^N(\omega_k)}{\hat{\Phi}_u^N(\omega_k)}$$
$$\hat{\sigma}_{\hat{G}}^2(k) = \left|\hat{G}(\Omega_k)\right|^2 \left(\frac{\hat{\Phi}_u^N(\omega_k)\hat{\Phi}_y^N(\omega_k)}{|\hat{\Phi}_{yu}^N(\omega_k)|^2} - 1\right) \quad (8.48)$$

where the leakage error is classically suppressed via time-domain windows. Better results are obtained via the so-called Taylor window (see [69, 71]).

The Frequency-domain Gaussian Maximum likelihood Solution

Since the ZOH and BL measurement setups can be used for identifying continuous-time systems (see Section 8.2), both DT and CT modelling are considered here. The Gaussian maximum likelihood estimates of the plant and noise model parameters (8.31), are found by minimising

$$V_{\mathrm{ML}}(\theta, Z) = \sum_{k \in \mathbb{K}} |\varepsilon(\Omega_k, \theta) g_F(\theta)|^2 \quad (8.49)$$

w.r.t. the model parameters θ, with $\mathbb{K}$ defined in (8.37), $\varepsilon(\Omega_k, \theta)$ the prediction error

$$\varepsilon(\Omega_k, \theta) = H^{-1}(\Omega_k, \theta)\left(Y(k) - G(\Omega_k, \theta)U(k) - T_G(\Omega_k, \theta) - T_H(\Omega_k, \theta)\right) \quad (8.50)$$

and $g_F(\theta)$ a scalar function depending on the plant model G, the noise model H, and the true controller transfer function M_0

$$g_F(\theta) = \exp\left(\frac{1}{F}\sum_{k\in\mathbb{K}}\log\frac{H(\Omega_k,\theta)}{1+G(\Omega_k,\theta)M_0(\Omega_k)}\right) \quad (8.51)$$

where $F = \text{ord}(\mathbb{K})$ (see [31, 38, 39, 42, 50, 55]). For the ZOH setup we have $\Omega = z^{-1}$, while for the BL setup $\Omega = s$.

A *first surprising consequence* of (8.49) is that the knowledge of the controller contributes to the knowledge of the plant and noise models ($M \neq M_0$ in (8.51) leads to biased estimates), which is not the case for the time-domain prediction error method (8.47) (see [32]). This was mentioned for the first time in [38]. The apparent contradiction can be explained by the fact that cutting out a part of the unit circle corresponds to non-causal filtering in the time domain (*e.g.*, convolution with a sinc-function). The latter invalidates the classical construction of the likelihood function based on time-domain data captured in feedback [11]. For discrete-time models, a monic parametrisation of $H(z^{-1}, \theta)$, $\mathbb{K}$ covering the whole unit circle, and $G(z^{-1}, \theta)$ and/or $M_0(z^{-1})$ having at least one sample delay, $g_F(\theta)$, (8.51) converges to one for $N \to \infty$ and, hence, the cost function (8.49) reduces to the classical time domain prediction error method (8.47) (for the proof see [55]).

The following practical advice results. Besides the true input $u(t)$ and noisy output $y(t)$ of the plant, it is strongly recommended to store also the true reference signal $r(t)$ in a feedback experiment. Indeed, assuming that $N_C = 0$, $N_g = 0$, $M_U = 0$, and $M_Y = 0$ in Figure 8.4, the true controller transfer function can easily be reconstructed from these three signals. It allows the plant and process noise in the relevant frequency band(s) to be modelled via minimisation of (8.49). If the controller is unknown, then consistent estimation is only possible at the price of modelling simultaneously the plant, the process noise, the controller, and the reference signal [55].

A *second surprising consequence* of (8.49) is that, in contrast to the time-domain prediction error method (8.47), consistent estimation of the plant model parameters in open loop ($M_0 = 0$ in (8.51)) *always* requires the correct noise model structure.

Extension of these results to multi-variable systems can be found in [42] and [57].

8.5.4 Arbitrary Excitations: Errors-in-variables

In this section it is explicitly assumed that the system operates in open loop ($M = 0$ and $N_C = 0$ in Figure 8.4), that the input–output errors N_U and N_Y in (8.35) are independent, and that the true input $u_0(t)$ can be written as filtered white noise

$$u_0(t) = L(q^{-1})e_L(t) \text{ or } u_0(t) = L(p)e_L(t) \quad (8.52)$$

with $L(z^{-1})$ and $L(s)$, respectively, the discrete-time (DT) and continuous-time (CT) signal model. Even under these simplified conditions, the errors-in-variables (EIV) problem with arbitrary excitation is the *most difficult* linear system identification problem to be solved. An exhaustive overview of the different time- and frequency- domain methods can be found in [75].

In Section 8.2.1 it has been shown that the discrete-time model built from the input samples $u_1(nT_s)$ to the output samples $y_1(nT_s)$ (see Figure 8.2) depends on the actuator characteristics (see (8.7)). As such, discrete-time modelling in an EIV framework is, in general, not suited for physical interpretation. Note that only a few time-domain algorithms exist for direct continuous-time EIV modelling: in [36, 76] continuous-time models are identified in the presence of white input–output errors.

Identifiability is an important issue in EIV modelling [75]. For example, in the CT case, the plant $G(s)$, signal $L(s)$, input noise $H_U(s)$, and output noise $H_Y(s)$ models can be uniquely identified if the following conditions are satisfied: (i) the numerator and denominator polynomials of each transfer function G, L, H_U, and H_Y separately have no common poles and zeros; (ii) monic parameterisation of L, H_U, H_Y, and the denominator of G; (iii) G has no quadrant symmetric poles nor zeros; (iv) no pole or zero of $G(s)$ is respectively a zero or pole of $L(s)L(-s)$; and (v) one of the following conditions is fulfilled [1]

$$\lim_{s \to s_0} \frac{H_U(s)H_U(-s)}{L(s)L(-s)} = 0 \quad \text{for} \quad \begin{cases} s_0 = \infty \\ \text{or a pole of } L(s)L(-s) \end{cases} \quad (8.53)$$

$$\lim_{s \to s_0} \frac{H_Y(s)H_Y(-s)}{G(s)L(s)G(-s)L(-s)} = 0 \quad \text{for} \quad \begin{cases} s_0 = \infty \\ \text{or a pole of } G(s)L(s)G(-s)L(-s) \end{cases} \quad (8.54)$$

A frequency-domain Gaussian maximum likelihood solution has been developed in [56] that can handle coloured input–output errors and DT as well as CT models. Similarly to all time-domain methods the following *open problems* still need to be solved: the sensitivity to model errors, the generation of sufficiently good starting values for the ML solution in the general case of coloured input–output errors, and the validation of the identified models (*e.g.*, the non-parametric FRF estimate (8.48) is biased for EIV problems).

8.6 Real Measurement Examples

8.6.1 Operational Amplifier

To avoid saturation during the open-loop gain measurement, the output of the operational amplifier (opamp) is fed back to its the negative input via a buffer in series with a resistor (see Figure 8.5). The buffer in the feedback loop prevents loading of the opamp, and the choice of the resistor values $R_1 = 300 \; \Omega$

and $R_2 = 12$ kΩ is a compromise between the risk of driving the opamp in saturation and a sufficiently large input signal-to-noise ratio [51]. The experiments are performed with an odd random phase multi-sine excitation $u_g(t)$ (see (8.40) with $A_{2k} = 0$) with logarithmic frequency distribution between 9.5 Hz and 95 kHz. Of each group of three consecutive odd harmonics, one odd harmonic is randomly eliminated. The generator and acquisition units are synchronised, and their sampling frequencies are derived from the same mother clock. The acquisition inputs and the generator output are lowpass filtered. Five consecutive periods of the steady-state response are measured, and the results are shown in Figure 8.6.

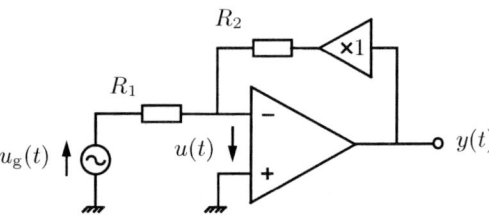

Fig. 8.5. Basic scheme for measuring the open-loop gain of an operational amplifier

From the top row of Figure 8.6 it can be seen that the input and output DFT spectra contain energy at the non-excited harmonics (the 'o' and grey '*' are well above the noise standard deviation). Due to the feedback loop (see Figure 8.5), the distortion at the input can be due to the non-linear behaviour of the generator and/or the operational amplifier. Hence, to interpret correctly the level of the non-linear distortion at the output $Y(k)$, it must be compensated for the linear contribution of the energy at the non-excited frequencies in $U(k)$ (see [63] for the details). After correction (see Figure 8.6, bottom), it can be seen that the opamp indeed behaves non-linearly, and that the non-linear distortions are (i) dominantly odd (the 'o' are well above the grey '*'), (ii) very large below 1 kHz, and (iii) decrease with increasing frequency. Using the input and corrected output DFT spectra one can calculate the open-loop gain, its noise standard deviation, and the total standard deviation due to the input–output noise and the non-linear distortions (see [63]). The results are shown in Figure 8.7. Clearly, the 'noisy' behaviour of the open-loop gain is almost completely due to the non-linear distortions. Note that all this information is obtained from an initial non-parametric preprocessing step. This is the *power* of the periodic excitation signals (8.40).

Since the non-linear distortions are dominant in the measurements, the grey '+' level in Figure 8.7 is used as variance $\hat{\sigma}_G^2(k)$ in the ML cost function (8.44),

8 Practical Aspects of Frequency-domain CT Modelling 239

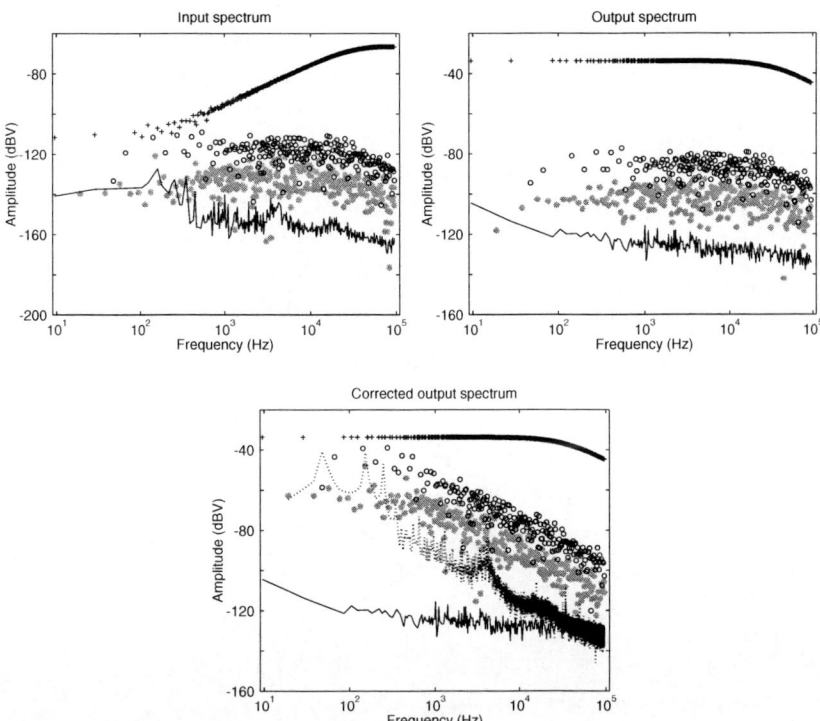

Fig. 8.6. Measured input and output DFT spectra ($u(t) = 12.3$ mVrms). Excited harmonics '+'; non-excited even harmonics grey '*'; non-excited odd harmonics 'o'; noise standard deviation excited harmonics '−'; and noise standard deviation non-excited harmonics '−' (coincides with '−' for the top figures).

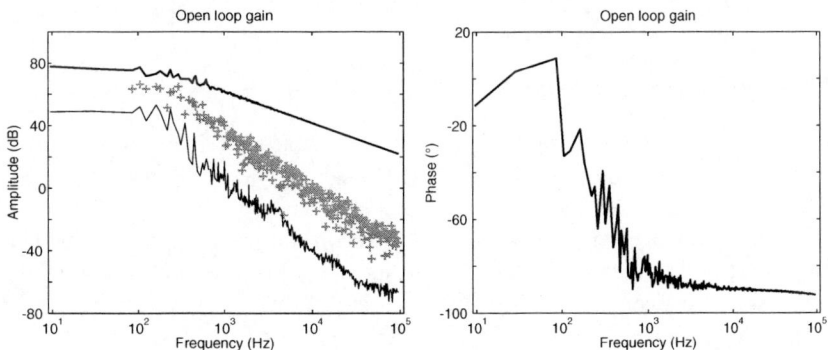

Fig. 8.7. Measured open-loop gain. Left: bold solid line: amplitude, grey '+': total standard deviation (non-linear distortion + noise), black line: noise standard deviation. Right: phase.

where

$$Y(k) = \hat{G}(k), U(k) = 1, \sigma_Y^2(k) = \hat{\sigma}_{\hat{G}}^2(k), \sigma_U^2(k) = 0, \text{ and } \sigma_{YU}^2(k) = 0 \quad (8.55)$$

with $\hat{G}(k)$ defined in (8.43). It turns out that a continuous-time model (8.22) of order $n_a = 6$ and $n_b = 4$ explains all dynamics in the measurements (see Figure 8.8). This model is to be interpreted as the best linear approximation (in the mean-square sense) of the non-linear system, for the class of Gaussian excitation signals with power spectrum defined by the excited lines in Figure 8.6, top left [14, 68]. Note that a full EIV approach in the presence of non-linear distortions requires experiments with different realisations of the random phase multi-sine (8.40), see [59].

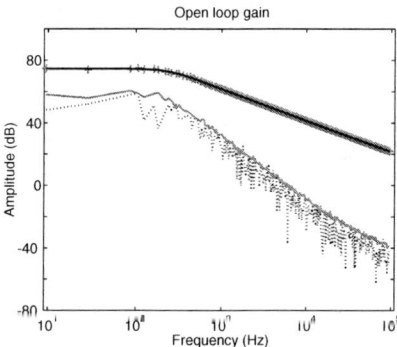

Fig. 8.8. Comparison measured (grey '+') and modelled (bold solid line) open-loop gain. Bold gray line: 95% uncertainty bound of the measurement, '–' magnitude of the complex difference between model and measurement.

8.6.2 Flight-flutter Analysis

To excite an airplane during flight, a perturbation signal is injected in the control loop of the flap mechanism at the tip side of the right wing. The angle perturbation of the flap is used as a measure of the applied force, and the acceleration is measured at two different positions on the left wing. Beside the applied perturbation, the airplane is also excited during flight by the natural turbulence. The resulting turbulent forces acting on the airplane cannot be measured and are assumed to be white in the frequency band of interest. Since the input–output signals are lowpass filtered before sampling, a continuous-time plant and noise model is the natural choice. The signals are measured during about 128 s at the sampling rate $f_s = 256$ Hz, giving $N = 32,768$ data points per channel. Figure 8.9 shows the 2×1 force to acceleration frequency-response function (FRF) in the band [4.00 Hz, 7.99 Hz] (DTF lines $k = 513, 514, \ldots, 1023 \Longrightarrow F = 511$). From Figure 8.9 it can be seen that a continuous-time CARMAX model (see (8.31) with $\Omega = s$ and Table 8.2) of order $n_a = 8$, $n_b = 8$, $n_{i_g} = 9$, and $n_c = 8$, selected by the MDL criterion [32], explains the measurements very well. The non-parametric FRF and noise power spectra shown in these figures (grey bullets) are calculated as

$$\hat{G}(s_k) = \frac{Y(k) - T_G(s_k, \hat{\theta}) - T_H(s_k, \hat{\theta})}{U(k)}$$
$$\hat{V}(k)\hat{V}^H(k) \text{ with } \hat{V}(k) = Y(k) - G(s_k, \hat{\theta})U(k) - T_G(s_k, \hat{\theta}) - T_H(s_k, \hat{\theta})$$
(8.56)

where $\hat{q}$ is the ML estimate obtained by minimising the multi-variable version of cost function (8.49).

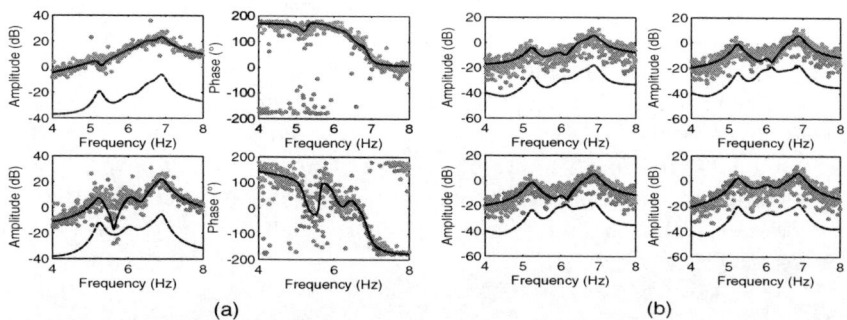

Fig. 8.9. Flight-flutter data. (a): plant frequency-response function, (b): noise power spectrum. Grey bullet: non-parametric estimate, bold solid line: estimated model, bold dashed line: standard deviation estimated model.

8.7 Guidelines for Continuous-time Modelling

The methods are listed in order of increasing difficulty.

8.7.1 Prime Choice: Uniform Sampling, Band-limited Measurement Setup, Periodic Excitation

Using periodic excitation signals a non-parametric noise model is obtained in a preprocessing step prior to the identification of the parametric plant model (see Section 8.5.2). Errors-in-variables identification, and identification in closed loop is as easy as identification in a generalised output error stochastic framework. Other features are (i) simplified model validation via improved non-parametric transfer function estimate, and (ii) detection, qualification and quantification of non-linear distortions (see Section 8.6.1).

To fully exploit the power of periodic excitation signals, the generator and acquisition units should be synchronised; otherwise spectral leakage errors will be introduced. If synchronisation is impossible, then an additional preprocessing is needed to suppress the leakage errors [72]. It requires that more than two periods are measured.

Software tools: FDIDENT for single-input, single-output, open- and closed-loop errors-in-variables problems [26, 28]; and FREQID for multi-variable generalised output error problems [13].

8.7.2 Second Choice: Uniform Sampling, Band-limited Measurement Setup, Arbitrary Excitation

Open Loop – Generalised Output Error

A parametric noise model is identified simultaneously with a parametric noise model. The prime choice is the combination of a continuous-time (CT) plant model with a continuous-time noise model (see Section 8.5.3). The second choice is hybrid modelling (a CT plant model combined with a DT noise model), because it is inconsistent in feedback, and only allows for a Box–Jenkins model structure (see Table 8.2). Contrary to the frequency-domain approach, the time-domain algorithms require the design of digital filters.

Software tool: CONTSID for hybrid modelling of multiple-input, single-output systems [16–18].

Closed Loop – Generalised Output Error

In addition to the open-loop case, the controller or the reference signal should be known; otherwise the controller and signal models should also be identified (see Section 8.5.3).

Open Loop – Errors-in-variables

This is the most difficult linear system identification problem and it should be avoided whenever possible: simultaneous identification of the plant, signal, input noise, and output noise models.

8.7.3 Third Choice: Uniform Sampling, Zero-order-hold Measurement Setup

In addition to the band-limited case (see Section 8.7.2), the discrete-time model should be transformed to the continuous-time domain. This is fairly easy for the poles (impulse invariant transformation) but can be rather tricky for the zeros [3, 27]. The drawback of this approach is that it relies heavily on the perfect realisation of the ZOH characteristic of the excitation, which may be poor in practice.

Software tools: SID [33] and CONTSID [16] for discrete-time modelling parameterised in the original continuous-time parameters; and ARMARA for time-series analysis [7, 8].

8.7.4 Last Resort: Non-uniform Sampling

If the input–output signals are oversampled by a factor ten, then the non-uniform grid can be transformed to a uniform grid by cubic interpolation [66], and then the methods of the previous sections can be applied. In the case of a smaller oversampling factor, dedicated time-domain methods should be used. Software tools: ARMASA for time-series analysis [7]; and CONTSID for generalised output error problems [16].

8.7.5 To be Avoided

Mixing the measurement setup (ZOH or BL) and the model (DT or CT), such as a BL setup with a DT model or a ZOH setup with a CT model, introduces bias errors. These errors create additional mathematical poles and zeros that complicates or even jeopardises the physical interpretation of the identified model. The identified models can still be used for simulation/prediction purposes as long as the true inter-sample behaviour of the simulation/prediction setup is identical to that of the identification setup.

8.8 Conclusions

Accurate physical modelling requires a careful design and calibration of the experimental setup and, if one has full control of the input, a careful choice of the excitation signal. The band-limited framework is the least sensitive to deviations of the true input and the real measurement setup w.r.t. the assumptions implicitly made by the identification algorithms. It leads in a natural way to continuous-time plant and noise modelling, and allows use of the full bandwidth from DC to Nyquist without introducing systematic errors. The latter is especially important in applications where the measurement bandwidth is costly (e.g., in the GHz range). If the input can freely be chosen then one should use periodic excitation signals.

Indeed, these signals allow us to estimate the disturbing noise power spectra in a preprocessing step and, as a consequence, the errors-in-variables problem (noisy input and output observations) – which is the natural framework for continuous-time modelling – is as easy to solve as the generalised output error problem (known input and noisy output observations).

References

1. J.C. Agüero and G.C. Goodwin. Identifiability of errors in variables dynamic systems. *14th IFAC Symposium on System Identification (SYSID'2006)*, pages 196–201, Newcastle (Australia), March 2006.

2. K.J. Åström. *Introduction to Stochastic Control Theory*. Academic Press, New York, 1970.
3. K.J. Åström, P. Hagander, and J. Sternby. Zeros of sampled systems. *Automatica*, 20(1):31–38, 1984.
4. T. Bastogne, H. Garnier, and P. Sibille. A PMF-based subspace method for continuous-time model identification. Application to a multivariable winding process. *International Journal of Control*, 74(2):118–132, 2001.
5. A.R. Bergström. *Continuous Time Econometric Modeling*. Oxford University Press, Oxford, UK, 1990.
6. D.R. Brillinger. *Time Series: Data Analysis and Theory*. McGraw-Hill, New York, 1981.
7. P.M.T. Broersen. ARMASA: automatic spectral analysis of time series. http://www.mathworks.com/matlabcentral/fileexchange/loadFile.do?objectId=1330&objectType=file, 2006.
8. P.M.T. Broersen. *Automatic Autocorrelation and Spectral Analysis*. Springer, Berlin, Germany, 2006.
9. P.M.T. Broersen and R. Bos. Time-series analysis if data are randomly missing. *IEEE Transactions on Instrumentation and Measurement*, IM-55(1):79–84, 2006.
10. A. Bultheel, M. Van Barel, Y. Rolain, and R. Pintelon. Numerically robust transfer function modeling from noisy frequency response data. *IEEE Transactions on Automatic Control*, AC-50(11):1835–1839, 2005.
11. P.E. Caines. *Linear Stochastic Systems*. Wiley, New York, 1988.
12. C.T. Chou, M. Verhaegen, and R. Johansson. Continuous-time identification of SISO systems using Laguerre functions. *IEEE Transactions on Signal Processing*, 47(?):349–362, 1999.
13. R.A. de Callafon. FREQID: a graphical user interface for frequency domain system identification. http://mechatronics.ucsd.edu/freqid/, 2006.
14. M. Enqvist and L. Ljung. Linear approximations of nonlinear FIR systems for separable input processes. *Automatica*, 41(3):459–473, 2004.
15. U. Forssell, F. Gustafsson, and T. McKelvey. Time-domain identification of dynamic errors-in-variables systems using periodic excitation signals. *14th World IFAC Congress*, Beijing, China, 1999.
16. H. Garnier and M. Gilson. CONTSID: continuous-time system identification toolbox for MATLAB®. http://www.cran.uhp-nancy.fr/contsid/, 2006.
17. H. Garnier, M. Gilson, and O. Cervellin. Latest developments for the MATLAB® CONTSID toolbox. *14th IFAC Symposium on System Identification*, pages 714–719, Newcastle, Australia, March 2006.
18. H. Garnier, M. Mensler, and A. Richard. Continuous-time model identification from sampled data. Implementation issues and performance evaluation. *International Journal of Control*, 76(13):1337–1357, 2003.
19. J. Gillberg and L. Ljung. Frequency-domain identification of continuous-time ARMA models from sampled data. *16th IFAC World Congress*, Prague, Czech Republic, July 2005.
20. P. Guillaume, R. Pintelon, and J. Schoukens. Parametric identification of two-port models in the frequency domain. *IEEE Transactions on Instrumentation and Measurement*, IM-41(2):233–239, 1992.
21. P. Guillaume, J. Schoukens, R. Pintelon, and I. Kollár. Crest factor minimization using non-linear Chebycheff approximation methods. *IEEE Transactions on Instrumentation and Measurement*, IM-40(6):982–989, 1991.

22. F. Gustafsson and J. Schoukens. Utilizing periodic excitation in prediction error based system identification. *37th IEEE Conference on Decision and Control*, volume 4, pages 3296–3931, Tampa, Florida, USA, December 1998.
23. A.H. Jazwinski. *Stochastic Processes and Filtering Theory*. Academic Press, New York, 1970.
24. R. Johansson. Identification of continuous-time models. *IEEE Transactions on Signal Processing*, 42(4):887–896, 1994.
25. R. Johansson, M. Verhaegen, and C.T. Chou. Stochastic theory of continuous-time state-space identification. *IEEE Transactions on Signal Processing*, 47(1):41–50, 1999.
26. I. Kollár. FDIDENT: Frequency domain system identification toolbox for MATLAB®. http://elecwww.vub.ac.be/fdident/, 2006.
27. I. Kollár, G. Franklin, and R. Pintelon. On the equivalence of z-domain and s-domain models in system identification. *IEEE Instrumentation and Measurement Technology Conference*, pages 14–19, Brussels, Belgium, June 1996.
28. I. Kollár, J. Schoukens, and R. Pintelon. Frequency domain system identification toolbox for MATLAB®: characterizing nonlinear errors of linear models. *14th Symposium on System Identification*, pages 726–731, Newcastle, Australia, March 2006.
29. E.K. Larsson and T. Söderström. Identification of continuous-time AR processes from unevenly sampled data. *Automatica*, 38(4):709–718, 2002.
30. E.C. Levy. Complex curve fitting. *IEEE Transactions on Automatic Control*, AC-4:37–43, 1959.
31. L. Ljung. Some results on identifying linear systems using frequency domain data. *32nd IEEE Conference on Decision and Control*, volume 4, pages 3534–3538, San Antonio, Texas, USA, December 1993.
32. L. Ljung. *System Identification. Theory for the User*. Prentice Hall, Upper Saddle River, NJ, USA, 2nd edition, 1999.
33. L. Ljung. SID: System identification toolbox for MATLAB®. http://www.mathworks.com/access/helpdesk/help/toolbox/ident/ident.shtml, 2006.
34. L. Ljung and T. Söderström. *Theory and Practice of Recursive Identification*. The MIT Press, Cambridge, MA, USA, 1983.
35. K. Mahata and M. Fu. Modeling continuous-time processes via input-to-state filters. *Automatica*, 42(7):1073–1084, 2006.
36. K. Mahata and H. Garnier. Identification of continuous-time errors-in-variables models. *Automatica*, 42(9):1477–1490, September 2006.
37. K. Mahata, R. Pintelon, and J. Schoukens. On parameter estimation using non-parametric noise models. *IEEE Transactions on Automatic Control*, 51(10):1602–1612, October 2006.
38. T. McKelvey. Frequency domain identification. *12th IFAC Symposium on System Identification*, Santa Barbara, USA, June 2000.
39. T. McKelvey. Frequency domain identification methods. *Circuits, Systems & Signal Processing*, 21(1):39–55, 2002.
40. T. McKelvey and H. Akçay. Subspace based system identification with periodic excitation signals. *System & Control Letters*, 26:349–361, 1995.
41. T. McKelvey, H. Akçay, and L. Ljung. Subspace-based multivariable system identification from frequency response data. *IEEE Transactions on Automatic Control*, 41(7):960–979, 1996.

42. T. McKelvey and L. Ljung. Frequency domain maximum likelihood identification. *11th IFAC Symposium on System Identification*, volume 4, pages 1741–1746, Kitakyushu, Japan, July 1997.
43. R.H. Middleton and G.C. Goodwin. *Digital Control and Estimation - A Unified Approach*. Prentice Hall, London, UK, 1990.
44. P. Mohanty and D.J. Rixen. Operational modal analysis in the presence of harmonic excitation. *Journal of Sound and Vibration*, 270:93–109, 2004.
45. K.B. Oldham and J. Spanier. *The Fractional Calculus*. Academic Press, New York, USA, 1974.
46. M.S. Phadke and S.M. Wu. Modelling of continuous stochastic processses from discrete observations with application to sunspots data. *Journal of the American Statistical Association*, 69(346):325–329, 1974.
47. R. Pintelon, P. Guillaume, Y. Rolain, J. Schoukens, and H. Van Hamme. Parametric identification of transfer functions in the frequency domain - a survey. *IEEE Transactions on Automatic Control*, AC-39(11):2245–2260, 1994.
48. R. Pintelon and I. Kollár. On the frequency scaling in continuous-time modeling. *IEEE Transactions on Instrumentation and Measurement*, IM-53(5):318–321, 2005.
49. R. Pintelon, Y. Rolain, and W. Van Moer. Probability density function for frequency response function measurements using periodic signals. *IEEE Transactions on Instrumentation and Measurement*, IM-52(1):61–68, 2003.
50. R. Pintelon, Y. Rolain, and J. Schoukens. Box-Jenkins identification revisited - Part II: applications. *Automatica*, 42(1):77–84, 2006.
51. R. Pintelon, Y. Rolain, G. Vandersteen, and J. Schoukens. Experimental characterization of operational amplifiers: a system identification approach - Part II: calibration and measurements. *IEEE Transactions on Instrumentation and Measurement*, IM-53(3):863–876, 2004.
52. R. Pintelon and J. Schoukens. Identification of continuous-time systems using arbitrary signals. *Automatica*, 33(5):991–994, 1997.
53. R. Pintelon and J. Schoukens. *System Identification: a Frequency Domain Approach*. IEEE Press, Piscataway, USA, 2001.
54. R. Pintelon and J. Schoukens. Measurement and modeling of linear systems in the presence of non-linear distortions. *Mechanical Systems and Signal Processing*, 16(5):785–801, 2002.
55. R. Pintelon and J. Schoukens. Box-Jenkins identification revisited - Part I: theory. *Automatica*, 42(1):63–75, January 2006.
56. R. Pintelon and J. Schoukens. Frequency domain maximum likelihood estimation of linear dynamic errors-in-variables. *Automatica*, 43(4):621–630, April 2007.
57. R. Pintelon, J. Schoukens, and P. Guillaume. Box-Jenkins identification revisited - Part III: multivariable systems. *Automatica*, 43(5):868–875, May 2007.
58. R. Pintelon, J. Schoukens, and P. Guillaume. Continuous-time noise modelling from sampled data. *IEEE Transactions on Instrumentation and Measurement*, IM-55(6):2253–2258, December 2006.
59. R. Pintelon, J. Schoukens, W. Van Moer, and Y. Rolain. Identification of linear systems in the presence of nonlinear distortions. *IEEE Transactions on Instrumentation and Measurement*, IM-50(4):855–863, 2001.
60. R. Pintelon, J. Schoukens, L. Pauwels, and E. Van Gheem. Diffusion systems: stability, modeling and identification. *IEEE Transactions on Instrumentation and Measurement*, IM-54(5):2061–2067, 2005.

61. R. Pintelon, J. Schoukens, and Y. Rolain. Box-Jenkins continuous-time modeling. *Automatica*, 36(7):983–991, 2000.
62. R. Pintelon, J. Schoukens, and G. Vandersteen. Frequency domain system identification using arbitrary signals. *IEEE Transactions on Automatic Control*, AC-42(12):1717–1720, 1997.
63. R. Pintelon, G. Vandersteen, L. de Locht, Y. Rolain, and J. Schoukens. Experimental characterization of operational amplifiers: a system identification approach - Part I: theory and simulations. *IEEE Transactions on Instrumentation and Measurement*, IM-53(3):854–862, 2004.
64. Y. Rolain, R. Pintelon, K.Q. Xu, and H. Vold. Best conditioned parametric identification of transfer function models in the frequency domain. *IEEE Transactions on Automatic Control*, AC-40(11):1954–1960, 1995.
65. Y. Rolain, J. Schoukens, and R. Pintelon. Order estimation for linear time-invariant systems using frequency domain identification methods. *IEEE Transactions on Automatic Control*, AC-42(10):1408–1417, 1997.
66. Y. Rolain, J. Schoukens, and G. Vendersteen. Signal reconstruction for non-equidistant finite length sample sets: a KIS approach. *IEEE Transactions on Instrumentation and Measurement*, IM-47(5):1046–1049, 1998.
67. J. Sanathanan and C.K. Koerner. Transfer function synthesis as a ratio of two complex polynomials. *IEEE Transactions on Automatic Control*, AC-8:56–58, 1963.
68. J. Schoukens, T. Dobrowiecki, and R. Pintelon. Parametric and non-parametric identification of linear systems in the presence of nonlinear distortions. A frequency domain approach. *IEEE Transactions on Automatic Control*, AC-43(3):176–190, 1998.
69. J. Schoukens and R. Pintelon. Estimation of non-parametric noise models. *IEEE Instrumentation and Measurement Technology Conference*, pages 102–106, Sorrento, Italy, April 2006.
70. J. Schoukens, R. Pintelon, and Y. Rolain. Study of conditional ML estimators in time and frequency domain system identification. *Automatica*, 35(1):91–100, 1999.
71. J. Schoukens, Y. Rolain, and R. Pintelon. Leakage reduction in frequency response function measurements. *IEEE Transactions on Instrumentation and Measurement*, IM-55(6):2286–2291, December 2006.
72. J. Schoukens, Y. Rolain, G. Simon, and R. Pintelon. Fully automated spectral analysis of periodic signals. *IEEE Transactions on Instrumentation and Measurement*, IM-52(4):1021–1024, 2003.
73. J. Schoukens, G. Vandersteen, R. Pintelon, and P. Guillaume. Frequency domain system identification using non-parametric noise models estimated from a small number of data sets. *Automatica*, 33(6):1073–1086, 1997.
74. N.K. Sinha and G.P. Rao (eds). *Identification of Continuous-time Systems. Methodology and Computer Implementation*. Kluwer Academic Publishers, Dordrecht, Holland, 1991.
75. T. Söderström. Errors-in-variable methods in system identification. *14th IFAC Symposium on System Identification*, pages 1–19, Newcastle, Australia, March 2006.
76. T. Söderström, E.K. Larsson, K. Mahata, and M. Mossberg. Using continuous-time modeling for errors-in-variables identification. *14th IFAC Symposium on System Identification*, pages 428–433, Newcastle, Australia, March 2006.

77. T. Söderström and P. Stoica. *System Identification*. Series in Systems and Control Engineering. Prentice Hall, Englewood Cliffs, USA, 1989.
78. J. Swevers, B. de Moor, and H. Van Brussel. Stepped sine system identification, errors-in-variables and the quotient singular value decomposition. *Mechanical Systems and Signal Processing*, 6(2):121–134, 1992.
79. P. Van Hamme, R. Pintelon, and J. Schoukens. *Discrete-time modeling and identification of continuous-time systems: a general framework*, in N.K. Sinha and G.P. Rao (eds), Identification of Continuous-time Systems. Methodology and Computer Implementation, pages 17–77, Kluwer Academic Publishers, Dordrecht, Holland, 1991.
80. P. Van Overschee and B. de Moor. Continuous-time frequency domain subspace system identification. *Signal Processing*, 52(2):179–194, 1996.
81. P.C. Young, H. Garnier, and M. Gilson. An optimal instrumental variable approach for identifying hybrid continuous-time Box-Jenkins model. *14th IFAC Symposium on System Identification*, pages 225–230, Newcastle, Australia, March 2006.

9

The CONTSID Toolbox: A Software Support for Data-based Continuous-time Modelling

Hugues Garnier[1], Marion Gilson[1], Thierry Bastogne[1] and Michel Mensler[2]

[1] Nancy-Université, CNRS, France
[2] Direction de la Recherche, Etudes Avancées, Matériaux - Renault, France

9.1 Introduction

This chapter describes the continuous-time system identification (CONTSID) toolbox for MATLAB®, which supports continuous-time (CT) transfer function and state-space model identification directly from regularly or irregularly time-domain sampled data, without requiring the determination of a discrete-time (DT) model. The motivation for developing the CONTSID toolbox was first to fill in a gap, since no software support was available to serve the cause of direct time-domain identification of continuous-time linear models but also to provide the potential user with a platform for testing and evaluating these data-based modelling techniques. The CONTSID toolbox was first released in 1999 [15]. It has gone through several updates, some of which have been reported at recent symposia [11,12,16]. The key features of the CONTSID toolbox can be summarised as follows:

- it supports most of the time-domain methods developed over the last thirty years [17] for identifying linear dynamic continuous-time parametric models from measured input/output sampled data;
- it provides transfer function and state-space model identification methods for single-input single-output (SISO) and multiple-input multiple-output (MIMO) systems, including both traditional and more recent approaches;
- it can handle irregularly sampled data in a straightforward way;
- it may be seen as an add-on to the system identification (SID) toolbox for MATLAB® [26]. To facilitate its use, it has been given a similar setup to the SID toolbox;
- it provides a flexible graphical user interface (GUI) that lets the user analyse the experimental data, identify and evaluate models in an easy way.

The chapter is organised in the following way. Section 9.2 outlines the main steps of the procedure for direct continuous-time model identification. An overview of the identification tools available in the toolbox is given in Section 9.3. An introductory example to the command mode along with a brief description of the GUI are then presented in Section 9.4. In Section 9.5, the advantages of CT model identification approaches are discussed and illustrated. A few successful application results from real-life process data are described in Section 9.6. Finally, Section 9.7 presents conclusions of the chapter and highlights future developments for the toolbox.

9.2 General Procedure for Continuous-time Model Identification

The procedure to directly determine a continuous-time model of a dynamical system directly from observed time-domain input/output data is similar to the general approach used for traditional DT model identification and involves three basic ingredients:

- the time-domain sampled input/output data;
- a set of candidate models (the model structure);
- a criterion to select a particular model in the set, based on the information in the data (the parametric model estimation method).

The identification procedure consists then in repeatedly selecting a model structure, computing the best model in the chosen structure, and evaluating the identified model. More precisely, the iterative procedure involves the following steps:

1. Design an experiment and collect time-domain input/output data from the process to be identified.
2. Examine the data. Remove trends and outliers, and select useful portions of the original data.
3. Select and define a model structure (a set of candidate system descriptions) within which a model is to be estimated.
4. Estimate the parameters in the chosen model structure according to the input/output data and a given criterion of fit.
5. Examine the finally estimated model properties.

If the model is good enough, then stop; otherwise go back to Step 3 and try another model set. Possibly also try other estimation methods (Step 4) or work further on the input/output data (Steps 1 and 2).

As described in the following section, the CONTSID toolbox includes tools for applying the general data-based modelling procedure summarised above.

9.3 Overview of the CONTSID Toolbox

9.3.1 Parametric Model Estimation

The CONTSID toolbox offers a variety of parametric model estimation methods for the most common input/output and state-space model structures.

CT ARX Models

The CT autoregressive with external input (ARX) model structure considered here, takes the form

$$A(p)y(t_k) = B(p)u(t_k - \tau) + e(t_k) \qquad (9.1)$$

with

$$B(p) = b_1 p^{n_b-1} + b_2 p^{n_b-2} + \cdots + b_{n_b}, \qquad (9.2)$$

$$A(p) = p^{n_a} + a_1 p^{n_a-1} + \cdots + a_{n_a}, \quad n_a \geq n_b - 1 \qquad (9.3)$$

where p denotes the differential operator; $u(t_k)$ and $y(t_k)$ represent the deterministic input and noisy output signals at time instant t_k, respectively; $e(t_k)$ is a zero-mean DT white Gaussian sequence[3] with variance σ_e^2; τ is a pure time delay in time units. The latter will be assumed in the following to be an integer number related to the sampling time T_s, i.e., $\tau = t_{n_k} = n_k T_s$. Note however that this is not essential in this CT environment, 'fractional' time delays can be introduced if required (e.g., see Chapter 11 and [27, 49]).

The CT ARX model structure, when the time delay is supposed to be an integer multiple of the sampling period, is denoted in the SISO case by the triad [n$_a$ n$_b$ n$_k$].

Here, the integers n$_a$ and n$_b$ are the number of parameters to be estimated in each polynomial while n$_k$ is the number of delays from input to output. Note that (9.1) also applies in a straightforward manner to the multiple-input case, with n_u input channels

$$A(p)y(t_k) = B_1(p)u_1(t_{k-n_{k_1}}) + \ldots + B_{n_u}(p)u_{n_u}(t_{k-n_{k_{n_u}}}) + e(t_k) \qquad (9.4)$$

The CT MISO ARX model structure is denoted by [n$_a$ n$_{b_1}$ ... n$_{b_{nu}}$ n$_{k_1}$ ... n$_{k_{nu}}$].

Equation (9.1) can also be explicitly written as

$$y^{(n_a)}(t_k) + a_1 y^{(n_a-1)}(t_k) + \cdots + a_{n_a} y(t_k) =$$
$$b_1 u^{(n_b-1)}(t_{k-n_k}) + \cdots + b_{n_b} u(t_{k-n_k}) + e(t_k) \qquad (9.5)$$

where $x^{(i)}(t_k)$ denotes the ith time derivative of the continuous-time signal $x(t)$ at time instant $t_k = kT_s$.

It may be noted that in contrast to the difference-equation model, the differential-equation model (9.5) is not a linear combination of samples of only the measurable process input and output signals. It also contains time-derivative terms that are not available as measurement data in most practical cases. The general scheme for CT ARX model estimation then requires two stages [17]:

---

[3] The disturbance term is modelled here as a zero-mean discrete-time Gaussian noise sequence. This avoids mathematical difficulties associated with continuous-time stochastic process modelling (see, e.g., [1] and Chapter 2).

- the primary stage that consists in using a preprocessing method to generate some measures of the process signals and their time derivatives. This stage also includes finding an approximating or discretizing technique so that the preprocessing operation can be performed in a purely digital way from sampled input/output data;
- the secondary stage in which the CT parameters are estimated within the framework of a LS or IV-based linear regression methods. Most of the well-known LS or IV-based methods developed for DT parameter estimation can be extended to the CT case with slight modifications.

Therefore, the main difference from conventional DT ARX model identification lies in the primary stage. There is a range of choice for the preprocessing required in the primary stage. Each method is characterised by specific advantages such as mathematical convenience, simplicity in numerical implementation and computation, physical insight, accuracy and others. However, all perform some prefiltering on the process signals. Process signal prefiltering is indeed a very useful and important way to improve the statistical efficiency in system identification and yields lower variance of the parameter estimates. Preprocessing methods developed over the last thirty years are traditionally grouped into three main classes of methods that are summarised below. The main references that have been used as the basis for their implementation along with their acronym used in the toolbox, are also given (see [17]):

- for the linear filters: the state-variable filter (SVF) [52] and the generalised Poisson moment functionals (GPMF) [18, 42];
- for the modulating functions: the Fourier [34] and Hartley modulating functions (HMF) [43];
- for the integral methods:
  - *among the numerical integration methods*: the block-pulse functions (BPF) [9], the trapezoidal-pulse functions (TPF) [9] and the Simpson's rule of integration (SIMPS) [7];
  - *among the orthogonal functions*: the Fourier trigonometric functions (FOURIE) [33], the Walsh functions (WALSH) [8], for the orthogonal polynomials: Hermite (HERMIT) [33], Laguerre (LAGUER) [33], Legendre (LEGEND) [33], first and second kind of Chebychev polynomials (CHEBY1 and CHEBY2) [32];
  - *among the others methods*: the linear integral filter (LIF) [39] and the re-initialised partial moments (RPM) [35].

Several parameter estimation algorithms associated with all implemented preprocessing techniques are available for identifying CT ARX models of the form of (9.1) or (9.5). First, conventional least squares (LS)-based methods have been implemented. In order to overcome the bias problem associated with simple LS-based estimation in the presence of noisy output data, a two-step instrumental variable (IV) estimator where the instruments are built up from an auxiliary model, has also been coupled with all available preprocessing

Table 9.1. Available methods for CT ARX model identification

Preprocessing methods			LS SISO	LS MISO	IV SISO	IV MISO	BCLS SISO	WLS SISO
linear filters		SVF	✓	✓	✓	✓		
		GPMF	✓	✓	✓	✓	✓	
modulating functions		FMF	✓	✓	✓	✓	✓	✓
		HMF	✓	✓	✓	✓		✓
integral methods	specific	RPM	✓	✓	✓	✓		
		LIF	✓	✓	✓	✓		
	numerical integration	BPF	✓		✓			
		TPF	✓		✓			
		SIMPS	✓		✓			
	orthogonal functions	FOURIE	✓		✓			
		WALSH	✓		✓			
	orthogonal polynomials	CHEBY1	✓		✓			
		CHEBY2	✓		✓			
		HERMIT	✓		✓			
		LAGUER	✓		✓			
		LEGEND	✓		✓			

techniques [17]. A few specific bias-reduction algorithms are also included, like a weighted least squares (WLS) associated with the modulating functions or a bias-compensating least squares (BCLS) technique coupled with the GPMF approach [19]. For details on the different parametric estimation methods for the CT ARX model identification, the reader is referred to [17]. Table 9.1 lists the methods available in the CONTSID toolbox for SISO and MISO CT ARX model identification. The performances of the sixteen prefiltering methods mentioned above have been thoroughly analysed by Monte Carlo simulation. Simulation [17,30] and real-life [28] studies have shown that integral methods (this does not include the RPM and LIF techniques) can have quite poor performances in the presence of medium to high measurement output noise. In particular, they require the estimation of additional initial condition parameters and have a high sensitivity to their user parameters. However, six of the methods exhibit very good overall performance: these are based on linear filters (GPMF and SVF), on modulating functions (FMF and HMF), and on the two particular types of integral methods (LIF and RPM). The final choice for a particular approach will probably depend on the taste or experience of the user since their global performances are very close. It is, therefore, not necessary to be able to tell which of the approaches is 'best'. Experience says that each may have its advantages. It is, however, good practice to have them all in one's toolbox. Furthermore, these methods can be used to get an initial high-quality estimate for iterative parametric estimation methods, presented in the next paragraphs.

CT Hybrid OE Models

The so-called CT hybrid output-error (OE) model structure is given by

$$y(t_k) = \frac{B(p)}{F(p)} u(t_{k-n_k}) + e(t_k) \qquad (9.6)$$

with

$$F(p) = p^{n_f} + f_1 p^{n_f - 1} + \cdots + f_{n_f}, \quad n_f \geq n_b - 1 \qquad (9.7)$$

where the noise $e(t_k)$ is assumed to be a zero-mean DT white Gaussian sequence so that no explicit noise modelling is necessary, except in relation to the estimation of the variance of the DT white noise process.

The CT hybrid OE model structure is denoted in the SISO case by the triad [n_b n_f n_k].

Here, the integers n_b and n_f are the number of parameters to be estimated in each polynomial ; n_k is the number of delays from input to output. For a multiple-input systems, (9.6) becomes

$$y(t_k) = \frac{B_1(p)}{F_1(p)} u_1(t_{k-n_{k_1}}) + \ldots + \frac{B_{n_u}(p)}{F_{n_u}(p)} u_{n_u}(t_{k-n_{k_u}}) + e(t_k) \qquad (9.8)$$

The CT hybrid MISO OE model structure is denoted by [n_{b_1} ... $n_{b_{nu}}$ n_{f_1} ... $n_{f_{nu}}$ n_{k_1} ... $n_{k_{nu}}$].

Two methods for identifying MISO OE structure-based models with different denominators are available in the toolbox.

The first is based on the iterative simplified refined instrumental variable method for continuous-time model identification (SRIVC: see Chapter 4). This approach involves a method of adaptive prefiltering based on an optimal statistical solution to the problem in this white noise case. This SRIVC method has been recently extended to handle MISO systems described by multiple CT transfer functions with different denominators [13] of the form of (9.8). It is important to mention that for day-to-day usage, the SRIVC algorithm provides a quick and reliable approach to CT model identification and has been used for many years as the algorithm of choice for this in the CAPTAIN toolbox[4] and, more recently, in the CONTSID toolbox. The application results of the SRIVC method to different real-life processes are further presented in this chapter.

The second method abbreviated by COE (continuous-time output error) implements the Levenberg–Marquardt or Gauss–Newton algorithm via sensitivity functions [38]. In contrast to LS- and IV-based methods, these algorithms rely on a numerical search procedure with a risk to get stuck in local minima and also require a larger amount of computation.

---

[4] See http://www.es.lancs.ac.uk/cres/captain/.

Table 9.2. Available methods for CT hybrid OE and BJ model identification

Methods	OE SISO	OE MISO	BJ SISO
COE	✓	✓	
SRIVC	✓	✓	
RIVC			✓

CT Hybrid BJ Models

The so-called CT hybrid Box–Jenkins (BJ) model structure in the SISO case is given by

$$y(t_k) = \frac{B(p)}{F(p)}u(t_{k-n_k}) + \frac{C(q^{-1})}{D(q^{-1})}e(t_k) \tag{9.9}$$

with

$$C(q^{-1}) = 1 + c_1 q^{-1} + \cdots + c_q q^{-q} \tag{9.10}$$
$$D(q^{-1}) = 1 + d_1 q^{-1} + \cdots + d_p q^{-p} \tag{9.11}$$

where $e(t_k)$ is a zero-mean DT white Gaussian sequence. Here, the model of the basic dynamic system is in continuous time, while the associated additive noise model is a discrete-time, autoregressive moving-average (ARMA) process (see Chapter 4). This CT hybrid BJ model structure is denoted by the following model order structure [n_b n_c n_d n_f n_k].

One of the main advantage of the CT hybrid BJ model is the asymptotic independence of the process and noise estimates. An approach based on the refined optimal IV, denoted by RIVC, has been derived to estimate the parameters of such models (see Chapter 4).

Table 9.2 lists the methods available in the CONTSID toolbox for CT hybrid OE and BJ model identification.

CT State-space Models

Continuous-time state-space models considered in the CONTSID toolbox take the form

$$\begin{cases} \dot{x}(t_k) = Ax(t_k) + Bu(t_k) \\ y(t_k) = Cx(t_k) + Du(t_k) + \xi(t_k) \end{cases} \tag{9.12}$$

where $u(t_k) \in \mathbb{R}^{n_u}$ is the input vector and $y(t_k) \in \mathbb{R}^{n_y}$ the output vector and $x(t_k) \in \mathbb{R}^n$ is the state vector at time t_k, $\xi(t_k) \in \mathbb{R}^{n_y}$ is the possibly coloured output noise vector.

Table 9.3. Available methods for CT state-space model identification

	Canonical model			Fully parameterised model
	LS	IV	BCLS	N4SID
GPMF	✓	✓	✓	✓
FMF				✓
HMF				✓
RPM				✓
LIF				✓

Two types of approaches for CT state-space model identification are available in the CONTSID toolbox. A first family of techniques relies on the *a priori* knowledge of structural indices, and considers the estimation of CT canonical state-space models. From the knowledge of the observability indices, the canonical state-space model can, in a straightforward way, be first transformed into an equivalent input–output polynomial description that is linear-in-its-parameters and therefore more suitable for the parameter estimation problem. A preprocessing method may then be used to convert the differential equation into a set of linear algebraic equations in a similar way to that for CT ARX type of models. The unknown model parameters can finally be estimated by LS-, IV- or BCLS-based algorithms [18, 20]. This scheme has been implemented for the GPMF approach only.

A second class of state-space model identification schemes is based on the subspace-estimation techniques. Most efficient data-based modelling methods, discussed so far, rely on iterative, non-linear optimisation or IV-type methods to fit parameters in a preselected model structure, so as to best fit the observed data. Subspace methods are an alternative class of identification methods that are 'one-shot' rather than iterative, and rely on linear algebra.

Moreover, these subspace methods are attractive since canonical forms are not required, while fully parameterised state-space models are estimated directly from sampled I/O data. Most commonly known subspace methods were developed for DT model identification [44]. The association of the more efficient preprocessing methods with subspace methods of the 4SID family [44] has been implemented in the toolbox [4, 29] (see also Chapter 10 in this book).

Table 9.3 summarises the methods available in the CONTSID toolbox for CT state-space model identification.

The application results of the GPMF-based subspace algorithm to a multiple-input multiple-output winding process are presented in Section 9.6.3.

9.3.2 Model Order Selection and Validation

The toolbox also includes tools for selecting the model orders as well as for evaluating the estimated model properties.

Model Order Selection

Model order selection is one of the difficult tasks in the system identification procedure. A natural way to find the most appropriate model orders is to compare the results obtained from model structures with different orders and delays. A model order selection algorithm associated to the SRIVC model estimation method allows the user to automatically search over a whole range of different model orders. Two statistical measures are then used for the analysis. The first is the simulation coefficient of determination R_T^2, defined as follows

$$R_T^2 = 1 - \frac{\sigma_{\hat{e}}^2}{\sigma_y^2} \qquad (9.13)$$

where $\sigma_{\hat{e}}^2$ is the variance of the estimated noise $\hat{e}(t_k)$ and σ_y^2 is the variance of the measured output $y(t_k)$. This should be differentiated from the standard coefficient of determination R^2, where the $\sigma_{\hat{e}}^2$ in (9.13) is replaced by the variance of the final noise model residuals $\hat{\sigma}^2$. R_T^2 is clearly a normalised measure of how much of the output variance is explained by the deterministic system part of the estimated model. However, it is well known that this measure, on its own, is not sufficient to avoid overparametrisation and identify a parsimonious model, so that other order identification statistics are required. In this regard, because the SRIVC method exploits optimal instrumental variable methodology, it is able to utilise the special properties of the instrumental product matrix (IPM) [45,53]; in particular, the YIC statistic [47] is defined as follows

$$\text{YIC} = \log_e \frac{\hat{\sigma}^2}{\sigma_y^2} + \log_e \{\text{NEVN}\}; \quad \text{NEVN} = \frac{1}{n_\theta} \sum_{i=1}^{n_\theta} \frac{\hat{p}_{ii}}{\hat{\theta}_i^2} \qquad (9.14)$$

Here, n_θ is the number of estimated parameters; $\hat{p}_{ii}$ is the ith diagonal element of the block-diagonal SRIVC covariance matrix and so is an estimate of the variance of the estimated uncertainty on the ith parameter estimate. $\hat{\theta}_i^2$ is the square of the ith SRIVC parameter estimate, so that the ratio $\hat{p}_{ii}/\hat{\theta}_i^2$ is a normalised measure of the uncertainty on the ith parameter estimate.

From the definition of R_T^2, we see that the first term in the YIC is simply a relative measure of how well the model explains the data: the smaller the model residuals, the more negative the term becomes. The normalised error variance norm (NEVN) term, on the other hand, provides a measure of the conditioning of the IPM, which needs to be inverted when the IV normal equations are solved (see *e.g.*, [46]): if the model is overparameterised, then it can be shown that the IPM will tend to singularity and, because of its ill-conditioning, the elements of its inverse will increase in value, often by several orders of magnitude. When this happens, the second term in the YIC tends to dominate the criterion function, indicating overparametrisation.

Although heuristic, the YIC has proven very useful in practical identification terms. It should not, however, be used as a sole arbiter of model order: rather the combination of R_T^2 and YIC provides an indication of the best parsimonious models that can be evaluated by other standard statistical measures (*e.g.*, the autocovariance of the model residuals, the cross-covariance of the residuals with the input signal $u(t_k)$, *etc.*). Also, within a 'data-based mechanistic' (DBM) model setting (see, *e.g.*, [48]), the physical interpretation of the model can often provide valuable information on the model adequacy: for instance, a model with complex eigenvalues caused by overparametrisation may prove incompatible with the non-oscillatory nature of the physical system under study.

The CONTSID toolbox includes a `srivcstruc` routine that allows the user to automatically search over a range of different orders by using the SRIVC algorithm and computes the two loss functions YIC and R_T^2. The in-line help specifies the required input parameters for the `srivcstruc` function

```
data=iddata(y,u,Ts);
V=srivcstruc(data,[],modstruc);
```

The routine collects in a matrix `modstruc` all the CT hybrid OE model to be investigated so that each row of `modstruc` is of the type $[n_{b_1} \ldots n_{b_{nu}} \; n_{f_1} \ldots n_{f_{nu}} \; n_{k_1} \ldots n_{k_{nu}}]$, where n_{b_j} and n_{f_j} are the number of parameters for the numerator and denominator, respectively, and n_{k_j} represents the number of samples for the delay. Then, a continuous-time model is fitted to the iddata set `data` for each of the structures in `modstruc`. For each of these estimated models, the two loss functions YIC and R_T^2 are computed from this estimation data set. The best model structures sorted according to the chosen criterion ('YIC' or 'RT2') are displayed with

```
selcstruc(V,criterion,nu);
```

where `nu` indicates the number of inputs. The application results of this model order selection procedure are illustrated further in this chapter.

Experiment Design, Model Validation and Simulation

In addition to the parameter estimation and model order determination routines, the toolbox provides several functions in order to generate excitation signals, simulate and examine CT models (see Table 9.4).

A few functions are available to generate excitation signals: `prbs` allows the design of a pseudo-random binary signal of maximum length, while `sineresp` returns the exact steady-state response of a continuous-time model for a sum of sine signals.

Simulated data can then be generated by using the function `simc` that allows the simulation of a CT model under an `idss` or `idpoly` format from a given `iddata` input object from regularly or irregularly sampled data.

Two functions are available for model validation purposes: `comparec` displays the measured output with the identified model output, while `residc` plots the auto-covariance function of the model residuals and the cross-covariance function of the residuals with the input signal.

Note that most of the functions included in the SID toolbox for the computation and presentation of frequency functions and zeros/poles (`bode`, `zpplot`) can be used with the identified CONTSID models.

The main demonstration program called `idcdemo` provides several examples illustrating the use and the relevance of the CONTSID toolbox approaches. These demos also illustrate what might be typical sessions with the CONTSID toolbox.

Table 9.4. CONTSID toolbox programs for experiment design, model simulation, order selection and validation

Program	Description
idcdemo	is the main routine for the CONTSID toolbox demonstration programs
prbs	generates a pseudo-random binary sequence of maximum length
sineresp	generates the exact steady-state response of a CT model for a sum of sine signals
sineresp1	generates the exact steady-state response of a CT first-order model for a sine given arbitrary initial conditions
sineresp2	generates the exact steady-state response of a CT second-order model for a sine given arbitrary initial conditions
simc	simulates a system under its CT `idpoly` or `idss` form
comparec	compares measured and model outputs
residc	computes and plots the residuals of a CT model. Plots the autocovariance function of the model residuals, the cross-covariance function of the residuals with the input signal
srivcstruc	computes the fit between simulated and measured outputs for a set of model structure of CT hybrid OE type estimated using the `srivc` method
selcstruc	helps to choose a model structure from the information obtained as the output from `srivcstruc`

Recursive Estimation

In many situations, there is a need to estimate the model at the same time as the data is collected during the measurement. The model is then 'updated' at each time instant some new data become available. The updating is performed by a recursive algorithm. Recursive versions RLSSVF, RIVSVF and RSRIVC of the LS, IV-based SVF methods and optimal IV technique for CT hybrid OE models are available in the CONTSID toolbox.

Identification from Non-uniformly Sampled Data

The problem of system identification from non-uniformly sampled data is of importance as this case occurs in several applications (see, *e.g.*, Chapter 11). The case of irregularly sampled data is not easily handled by discrete-time model identification techniques, but as illustrated further in this chapter (see Sections 9.4.1 and 9.6.2), mild irregularity can be easily handled by some of the CONTSID toolbox methods. This is because the differential-equation model is valid whatever the time instants considered and, in particular, it does not assume regularly sampled data, as required in the case of the standard difference-equation model.

Table 9.5 lists the functions available for data-based modelling from irregularly sampled data.

Table 9.5. Available functions for CT model identification from irregularly sampled data

Program	Description
LSSVF	LS-based state-variable filter method for CT ARX models
IVSVF	IV-based state-variable filter method for CT ARX models
COE	non-linear optimisation method for CT hybrid OE models
SRIVC	optimal instrumental variable method for CT hybrid OE models
SIDGPMF	subspace-based generalised Poisson moment functionals method for CT state space models

9.4 Software Description

The CONTSID toolbox is compatible with MATLAB® versions 6.x and 7.x. Two external commercial toolboxes are required: the Control toolbox and the SID toolbox. The current version can be considered as an add-on to the SID toolbox and makes use of the `iddata`, `idpoly` and `idss` objects used in the SID toolbox. It is freely available for academic researchers and can be downloaded from
http://www.cran.uhp-nancy.fr/contsid/

All available parametric model estimation functions share the same command structure

    m = function(data,modstruc)
    m = function(data,modstruc,specific_parameters)

The input argument `data` is an `iddata` object that contains the output- and input-data sequences along with the sampling time and inter-sample behaviour for the input, while `modstruc` specifies the particular structure of the model to be estimated. The specific parameters depend on the preprocessing method used. The resulting estimated model is contained in m, which is a model object that stores the various usual information. The function name is defined by the abbreviation for the estimation method and the abbreviation for the associated preprocessing technique, as for example, IVSVF for the instrumental variable-based state-variable filter approach or SIDGPMF for subspace-based state-space model identification GPMF approach.

Note that help on any CONTSID toolbox function may be obtained from the command window by invoking classically help name_function.

9.4.1 Introductory Example to the Command Mode

A part of the first demonstration program is presented in this section. This demo is designed to get the new user started quickly with the CONTSID toolbox: it is straightforward to run the demo by typing idcdemo1 in the MATLAB® command window) and follow along. This example considers a second-order SISO CT system without delay. The complete equation for the data-generating system has the following form

$$y(t_k) = \frac{3}{p^2 + 4p + 3} u(t_k) + e(t_k) \qquad (9.15)$$

where $e(t_k)$ is a zero-mean DT white Gaussian noise sequence. Let us first create an idpoly model structure object describing the model. The polynomials are entered in descending powers of the differential operator

```
m0=idpoly(1,[3],1,1,[1 4 3],'Ts',0);
```

'Ts' and 0 indicate here that the system is time continuous.
We take a PRBS of maximum length with 1016 points as input u. The sampling period is chosen to be 0.05 s

```
u = prbs(7,8);
Ts = 0.05;
```

We then create an iddata object for the input signal with no output, the input u and sampling interval Ts. The input inter-sample behaviour is specified by setting the property 'Intersample' to 'zoh' since the input is piecewise-constant here

```
datau = iddata([],u,Ts,'InterSample','zoh');
```

The noise-free output is simulated with the simc CONTSID routine and stored in ydet. We then create an iddata object with output ydet, input u and sampling interval Ts

```
ydet = simc(m0,datau);
datadet = iddata(ydet,u,Ts,'InterSample','zoh');
```

We then identify a CT ARX model for this system from the deterministic iddata object datadet with the conventional least squares-based state-variable filter (lssvf) method. The extra pieces of information required are

- the number of denominator and numerator parameters and number of samples for the delay of the model [n_a n_b n_k] =[2 1 0];
- the cut-off frequency (in rad/s) of the SVF filter, set to 2 here.

The lssvf routine can now be used as follows

```
mlssvf = lssvf(datadet,[2 1 0],2)
```

which leads to[5]

```
CT IDPOLY model: A(s)y(t) = B(s)u(t) + e(t)
A(s) = s² + 3.999 s + 2.999
B(s) = 2.999
Estimated using LSSVF
Loss function 6.03708e-15 and FPE 6.07284e-15
```

It will be noted that, not surprisingly, the estimated model coefficients are very close to the true parameters. This is, of course, because the measurements are not noise corrupted. Note that even in the noise-free case, the true parameters are not estimated exactly here. This is due to small simulation errors introduced in the numerical implementation of the continuous-time state-variable filtering for the output signal.

Let us now consider the case when a white Gaussian noise is added to the output samples. The variance of $e(t_k)$ is adjusted to obtain a signal-to-noise ratio (SNR) of 10 dB. The SNR is defined as

$$\text{SNR} = 10 \log \frac{P_{y_{det}}}{P_e} \qquad (9.16)$$

where P_e represents the average power of the zero-mean additive noise on the system output (*e.g.*, the variance) while $P_{y_{det}}$ denotes the average power of the noise-free output fluctuations.

```
snr=10;
y = simc(m0,datau,snr);
data = iddata(y,u,Ts);
```

The input/output data are displayed in Figure 9.1. The use of this noisy output in the basic lssvf routine will inevitably lead to biased estimates. A bias-reduction algorithm based on the two-step instrumental variable technique where the instruments are built up from an auxiliary model (ivsvf) can be used instead

---

[5] Note that in the Matlab® System Identification (SID) toolbox, the variable 's' instead of 'p' is used to denote the differential operator. The CONTSID toolbox makes use of the SID object models ; therefore the CONTSID estimated models are displayed with the 's' variable.

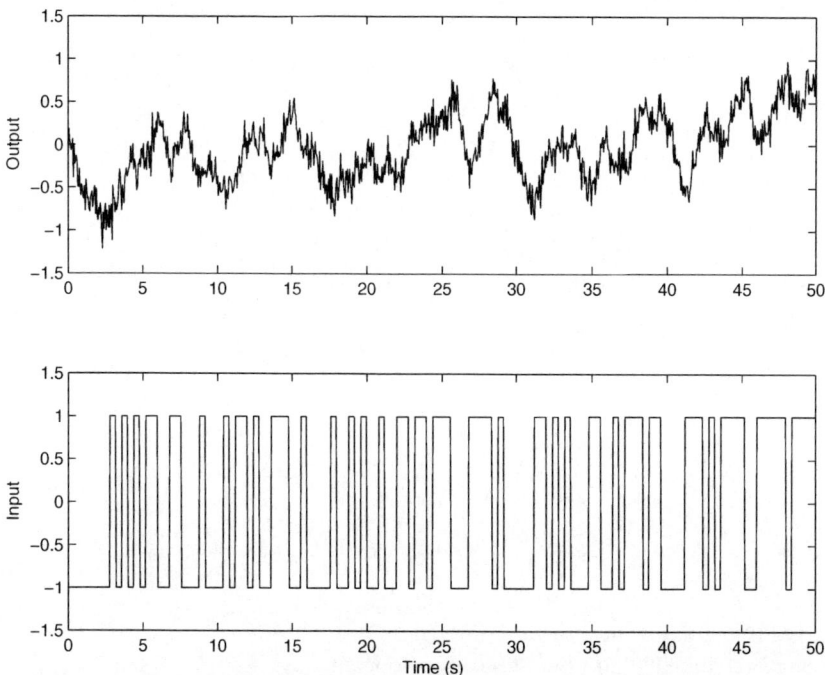

Fig. 9.1. Input–output data (noisy case) – SNR=10 dB

```
mivsvf=ivsvf(data,[2 1 0],2)
```

which leads to

```
CT IDPOLY model: A(s)y(t) = B(s)u(t) + e(t)
A(s) = s² + 3.988 s + 3.076
B(s) = 3.008
Estimated using IVSVF
Loss function 0.217742 and FPE 0.219032
```

It will be noted now that the parameters are close to the true ones. However, this basic IV-based SVF method suffers from two drawbacks, even if it is asymptotically unbiased:

- it requires the *a priori* knowledge of a user parameter: the cut-off frequency of the SVF filter here;
- it is suboptimal, in the sense that the variance of the estimates is not minimal (it depends of the SVF filter mainly).

It is better, therefore, to use the optimal (in this white output measurement noise context) iterative IV method (`srivc`) that overcomes the two latter drawbacks. The searched model now takes the form of a CT hybrid OE model

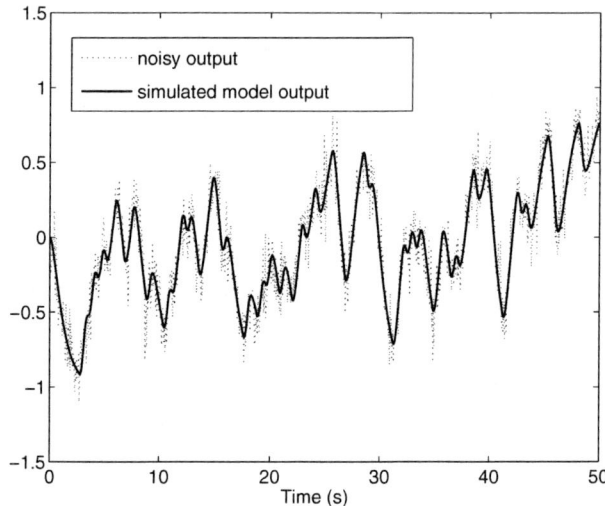

Fig. 9.2. Noisy and simulated SRIVC model outputs

(9.6). The model structure becomes $[n_b\, n_f\, n_k] = [1\ 2\ 0]$.
The srivc routine can now be used as follows

```
msrivc = srivc(data,[1 2 0]);
```

The estimated parameters together with the estimated standard deviations can be displayed

```
present(msrivc);
```

which leads to

```
CT IDPOLY model: y(t) = [B(s)/F(s)]u(t) + e(t)
B(s) = 3.002 (+-0.1113)
F(s) = s² + 3.992 (+-0.1619) s + 3.067 (+-0.1061)
Estimated using SRIVC
Loss function 0.0135763 and FPE 0.0136567
```

Let us now compare the model output for the input signal with the measured output. This can be done easily by using the comparec CONTSID routine

```
comparec(data,msrivc,1:1000);
```

which plots the measured and the simulated model outputs. As can be seen in Figure 9.2, they coincide very well. We can also check the residuals of this model, and plot the autocovariance of the residuals and the cross-covariance between the input and the residuals by using the CONTSID residc routine

```
residc(data,msrivc);
```

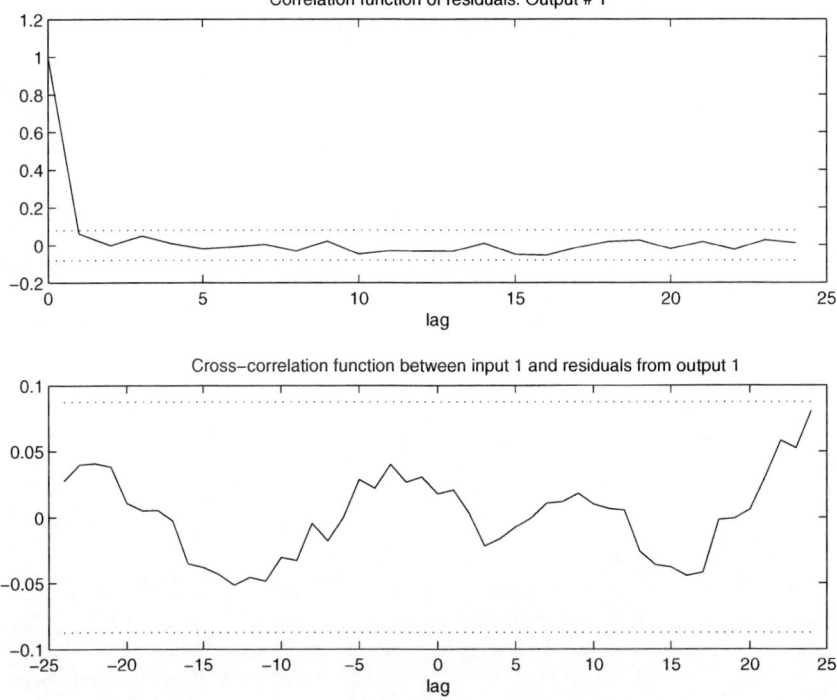

Fig. 9.3. Correlation test for the SRIVC model

From Figure 9.3, it may be seen that the residuals are white and totally uncorrelated with the input signal. We can thus be satisfied with the model. Let us finally compare the Bode plots of the estimated model and the true system

```
bode(msrivc,'sd',3,'fill',m0,'y--')
```

The confidence regions corresponding to three standard deviations are also displayed. From Figure 9.4, it may be observed that the Bode plots coincide very well with narrow confidence regions.

As previously mentioned, non-uniformly sampled data can be handled easily by some CONTSID toolbox methods [21]. This is now illustrated here. The data-generating system has the following form

$$y(t_k) = \frac{5}{p^2 + 2.8p + 4} u(t_k) + e(t_k) \qquad (9.17)$$

The input signal is chosen as the sum of three sines,

$$u(t) = \sin(0.714t) + \sin(1.428t) + \sin(2.142t)$$

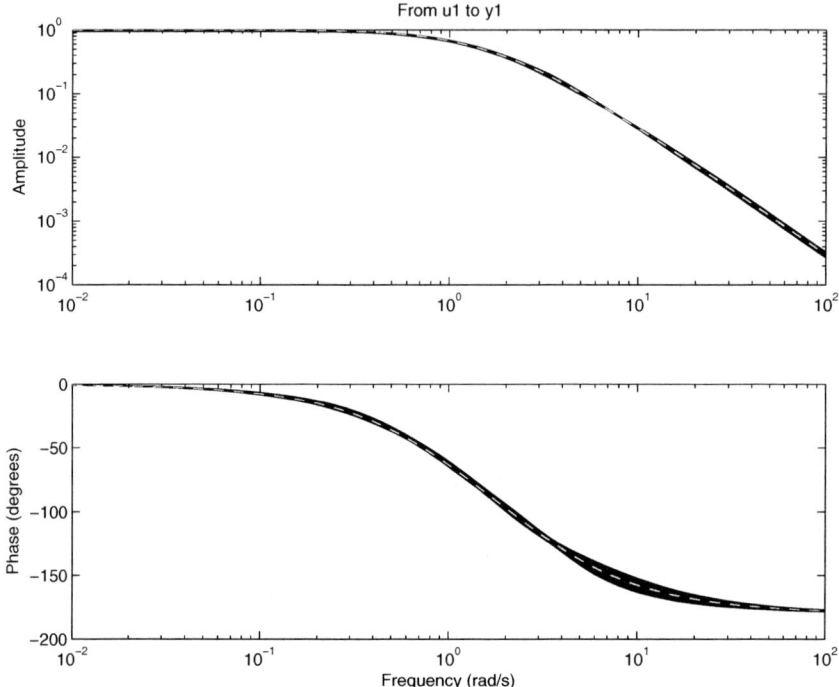

Fig. 9.4. Bode plot of the SRIVC model together with a 3 standard deviation uncertainty region

A non-uniform sampling setup similar to the one used in [23] is chosen. The distance between two sampling instants is denoted by $h_k = t_{k+1} - t_k$. We assume that $\underline{h} \leq h_k \leq \overline{h}$, where $\underline{h} > 0$ and $\overline{h}$ are the finite lower and upper bounds, respectively. A uniform probability density function $U(\underline{h}, \overline{h})$ is used to describe the variations of the sampling interval, *i.e.*, $h_k \sim U(0.01s, 0.1s)$. 3000 data points are used for the identification. Analytic expressions are used to compute the noise-free output in order to avoid errors due to numerical simulations. A zero-mean white noise is then added to the system output in order to get a signal-to-noise ratio of 10 dB. The simulated output is stored in y. Figure 9.5 displays a short section of 3 s of the sampled records and reveals the non-uniform sampling intervals.

We first create a `iddata` object with output y, input u, and the available time instant stored in the vector t. The input inter-sample behaviour is specified by setting the property 'Intersample' to 'foh' since the input is not piecewise-constant here

```
data = iddata(y,u,[],'SamplingInstants',t,'InterSample','foh');
```

The optimal IV algorithm `srivc` can now be used for the appropriate model structure

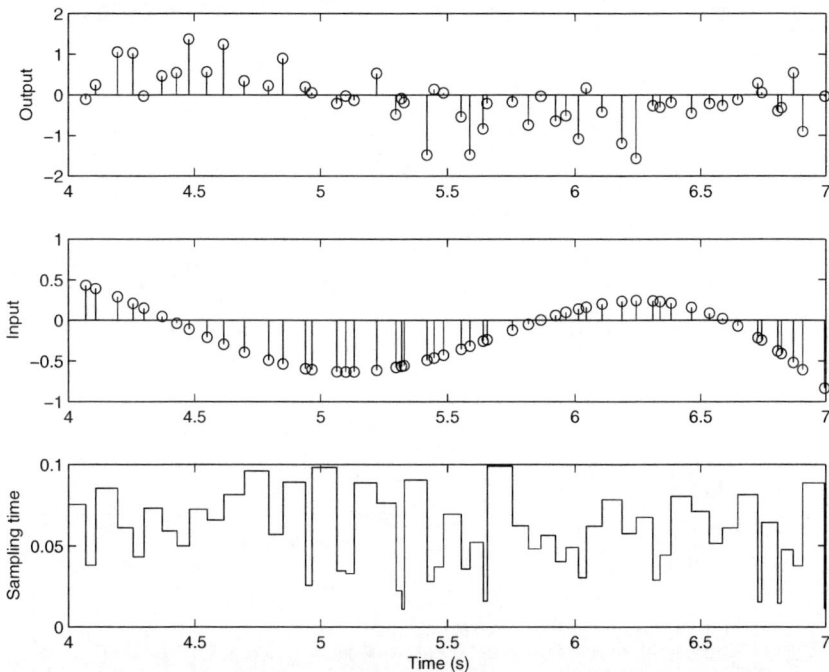

Fig. 9.5. A section of the noisy output, input signal and sampling interval value in the case of irregularly sampled data – SNR=10 dB

```
msrivc=srivc(data,[1 2 0]);
```

The estimated parameters and standard errors can then be displayed with

```
present(msrivc);
```

which leads to

```
Continuous-time IDPOLY model:
y(t) = [B(s)/F(s)]u(t)
B(s) = 4.958 (+-0.0909)
F(s) = s² + 2.796 (+-0.05637) s + 3.994 (+-0.05056)
Loss function 0.1748 and FPE 0.1752
Estimated using SRIVC
```

The estimated parameters are very close to the true ones.
As previously, traditional model validation tests can be performed to further examine the quality of the estimated model.

9.4.2 The Graphical User Interface

The graphical user interface (GUI) for the CONTSID toolbox provides a main window, as shown in Figure 9.6, which is divided into three basic parts:

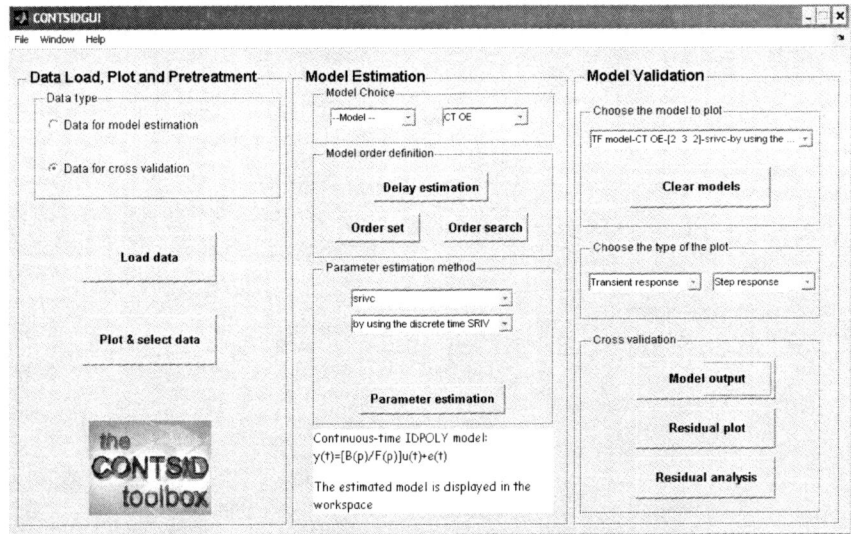

Fig. 9.6. The main window of the CONTSID GUI

- a data panel on the left part where data-sets can be imported, plotted, pretreated and selected;
- a model estimation panel in the middle where CT ARX and CT hybrid OE transfer function model structures and associated parameter estimation methods can be tested;
- a model validation panel in the right part where basic properties of the identified model can be examined.

The CONTSID GUI can be started by typing contsidgui in the MATLAB® command window.

The Data Panel

As shown in Figure 9.6, the GUI lets the user define two data-sets: one for identifying the model and one, if sufficient data are available, for cross-validation purposes.

Importing Measured Data

By clicking on the *Load data* button, time-domain sampled data from a *.mat* file can be imported for systems with single- or multiple- input and single-output channels. From this window, the input and output variables can be specified along with the type of sampling scheme (regular or irregular), the sampling time (T_s) and the assumption on the input inter-sample behaviour (piecewise-constant (zoh) or continuous).

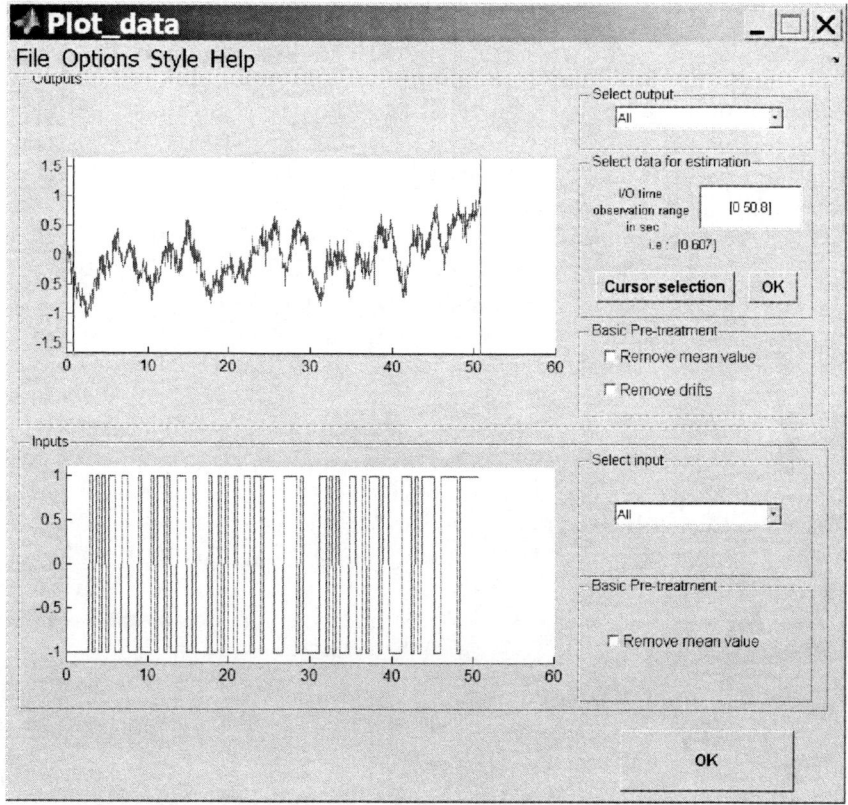

Fig. 9.7. Data plot and pretreatment GUI window

Preprocessing and Selecting Observed Data

After the data have been imported, basic operations for data analysis and preprocessing can be applied. An example of the window obtained after a click on the button *Plot & select data* is displayed in Figure 9.7. This window also allows the preprocessing of data including offset, drift removal and the display of the results after the operation.

It is often the case that the whole data record is not directly suitable for identification. This is mainly for two reasons:

- the data-sets include erroneous values that it is essential to eliminate;
- if only one data-set is available, it is advisable to divide the data-set into two parts, the first for model estimation purposes and the second reserved for cross-validation purposes.

The *Cursor selection* button allows the insertion of two vertical axes on the output plot that can be used to define the selected portion of measured data.

Model Estimation Panel

While the CONTSID toolbox supports transfer function and state-space model identification methods, the GUI lets the user estimate CT ARX and hybrid OE models only. The user is thus invited to choose the type of model structure in the unrolling menu at the top right of the *model estimation* panel, as shown in Figure 9.6.

After selecting the model structure, the user has to specify the polynomial orders and the time delay of the model to be estimated.

A first option is to deduce an estimate of the number of samples for the time delay from an estimation of the impulse response by correlation analysis.

Then, if the TF model orders are not known *a priori*, the *Order search* button allows the user to automatically search over a whole range of different model orders. The user can choose several available criteria to sort and display the estimation results in the MATLAB® workspace. From these results, the user can select the best model orders and then set the order of the final model to be estimated by clicking on the *Order set* button from the main window.

Once the number of samples for the time delay and the number of coefficients for the polynomial model have been set, the model parameters can then be estimated by using one of the available parametric estimation methods chosen from an unrolling menu

- in the case of a CT hybrid OE model structure, the user can choose to use the continuous-time output error (COE) method or the simplified refined instrumental variable (SRIVC) method;
- in the case of a CT ARX model structure, the user can select one of the six preprocessing-based methods that have proven successful. These preprocessing methods are coupled with conventional least squares or basic auxiliary model-based instrumental variable methods. These all require a user parameter to be specified by the user [17] which should be chosen in order to emphasise the frequency band of interest.

Once the parameter estimation method is chosen, the identified model is displayed in the command window after a click on the *Parameter estimation* button.

Model Validation Panel

Once a model is estimated, it appears in the drop-down menu located at the top part of the *Model Validation* panel (see Figure 9.6). Several basic model properties can then be evaluated from an unrolling menu by using first the data that were used for model identification

- *model-output comparison*: plots and compares the simulated model output with the measured output. This indicates how well the system dynamics are captured;

- *residual plot*: displays the residuals;
- *transient response*: displays the model response to an impulse or step excitation signal;
- *frequency response*: displays the Nyquist or Bode plots to show damping levels and resonance frequencies;
- *zeros and poles*: plots the zeros and poles of the identified models and tests for zero-pole cancelation indicating overparameterised modelling;
- *correlation test*: displays the autocovariance function of the residuals and the cross-covariance function between the input signal and the residuals.

If a cross-validation data-set is available, then traditional cross-validation tests consist of comparisons between the measured and simulated model outputs and analysis of the residuals.

9.5 Advantages and Relevance of the CONTSID Toolbox Methods

There are two fundamentally different time-domain approaches to the problem of obtaining a CT model of a naturally CT system from its sampled input/output data:

- the indirect approach that involves two steps. First, a DT model for the original CT system is obtained by applying the DT model estimation methods and the DT model is then transformed into CT form;
- the direct approach where a CT model is obtained straightaway using CT model identification methods discussed in this chapter.

The indirect approach has the advantage that it uses well-established DT model identification methods [24, 41]. Examples of such methods, which are known to give consistent and statistically efficient estimates under very general conditions, are gradient-optimisation procedures, such as the maximum likelihood and prediction error methods in the SID toolbox; and iterative, relaxation procedures, such as the refined instrumental variable (RIV) method in the CAPTAIN toolbox.

On the surface, the choice between the two approaches may seem trivial. However, some recent studies have shown some serious shortcomings of the indirect route through DT models. Indeed, an extensive analysis aimed at comparing direct and indirect approaches has been discussed recently. The simulation model used in this analysis provides a very good test for CT and DT model estimation methods: it was first suggested by Rao and Garnier [36] (see also [17,37,38]); further investigations by Ljung [25] confirmed the results. This example illustrates some of the well-known difficulties that may appear in DT modelling under less standard conditions such as rapidly sampled data or relatively wide-band systems:

- relatively high sensitivity to the initialisation. DT model identification often requires computationally costly minimisation algorithms without even guaranteeing convergence (to the global optimum). In fact, in many cases, the initialisation procedure for the identification scheme is a key factor to obtain satisfactory estimation results compared to direct methods (see, e.g., [50] for a recent reference);
- numerical issues in the case of fast sampling because the eigenvalues lie close to the unit circle in the complex domain, so that the model parameters are more poorly defined in statistical terms;
- a priori knowledge of the relative degree is not easy to accommodate;
- non-inherent data prefiltering in the gradient-based methods (adaptive prefiltering is an inherent part of the DT RIV method in CAPTAIN).

Further, the question of parameter transformation between a DT description and a CT representation is non-trivial. First, the zeros of the DT model are not as easily translatable to CT equivalents as the poles [2]; second, due to the discrete nature of the measurements they do not contain all the information about the CT signals. To describe the signals between the sampling instants some additional assumptions have to be made, for example, assuming that the excitation signal is constant between the sampling intervals (zero-order hold assumption). Violation of these assumptions may lead to severe estimation errors [40].

The advantages of direct CT identification approaches over the alternative DT identification methods can be summarised as follows:

- they directly provide differential-equation models whose parameters can be interpreted immediately in physically meaningful terms. As a result, they are of direct use to scientists and engineers who most often derive models in differential-equation terms based on natural laws and who are much less familiar with 'black-box' discrete-time models;
- they can estimate a fractional time-delay system;
- the estimated model is defined by a unique set of parameter values that are not dependent on the sampling interval T_s;
- there is no need for conversion from discrete to continuous time, as required in the indirect route based on initial DT model estimation;
- the direct continuous-time methods can easily handle the case of mild irregularity in the sampled data;
- the a priori knowledge of the relative degree is easy to accommodate and therefore allows the identification of more parsimonious models than in discrete time;
- they also offer advantages when applied to systems with widely separated modes;
- they include inherent data filtering;
- they are well suited in the case of very rapid sampling. This is particularly interesting since modern data-acquisition equipment can provide nearly continuous-time measurements and, therefore, make it possible to acquire

data at very high sampling frequencies. Note that, as mentioned in [25], the use of prefiltering and decimation strategies may lead to better results in the case of DT modelling, but these may not be so obvious for practitioners. The CONTSID toolbox methods are free of these difficulties.

All these advantages will facilitate for the user the application of the general data-based modelling procedure. In the following, these advantages are illustrated with the help of a simulated benchmark system.

The SYSID'2006 Benchmark System

Here, the performance of the CONTSID toolbox techniques is illustrated by applying them to a benchmark example that was prepared for the 14th IFAC Symposium on System Identification (SYSID'06) in Newcastle, Australia[6].
The intent of the benchmark was to set up a format in which rigorous comparisons between competing techniques for the identification of CT models from sampled data, including time- and frequency-domain approaches, could be undertaken. The goal was also to collect and analyse quantitative results in order to understand similarities and differences among the approaches and to highlight the strengths and weaknesses of each approach.
Two benchmark data sets were generated. Both were simulated continuous-time systems based closely on mechatronic systems. Data corresponding to these two benchmarks were sent to participants to apply their preferred technique.
Unfortunately, the associated Benchmark Session at SYSID was cancelled because referees felt that insufficient submitted papers were acceptable (only one of the papers submitted to the proposed benchmark session got even close to the correct model, demonstrating the difficulty of the benchmark exercise). This section presents the CONTSID toolbox results obtained on the benchmark data set 1, in which the additive measurement noise is a simple white additive noise. The second benchmark data set is more difficult since the white measurement noise is replaced by non-stationary noise (similar results were obtained using the CAPTAIN toolbox routines [50], where a modified example with coloured additive noise is considered).
The SYSID Benchmark data set 1 is obtained from

$$\begin{cases} x(t) = G_o(p)u(t), \quad \text{subject to zero initial conditions} \\ y(t_k) = x(t_k) + e(t_k) \end{cases} \quad (9.18)$$

where the measurement noise $e(t_k)$ is a zero-mean DT white Gaussian sequence.
The system is a linear, fourth-order, non-minimum phase system with complex poles where the Laplace transfer function is given by

---

[6] The data can be downloaded from
http://sysid2006benchmark.cran.uhp-nancy.fr/.

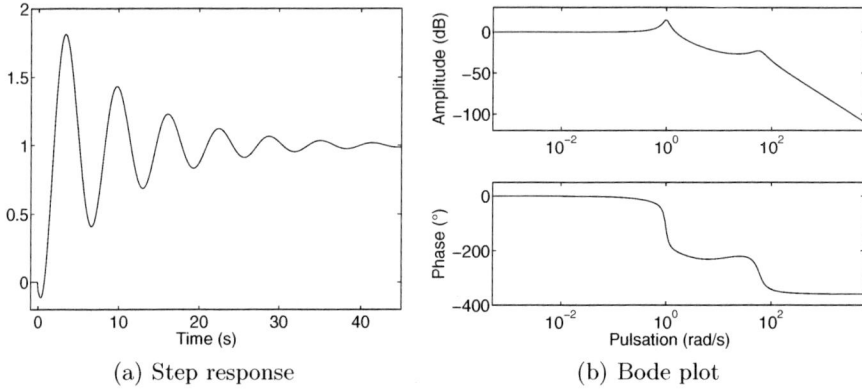

(a) Step response (b) Bode plot

Fig. 9.8. Step response and Bode plot for the SYSID'2006 benchmark system

$$G_o(s) = \frac{K(-T_1 s + 1)(T_2 s + 1)e^{-\tau s}}{\left(\frac{s^2}{\omega_{n,1}^2} + \frac{2\zeta_1 s}{\omega_{n,1}} + 1\right)\left(\frac{s^2}{\omega_{n,2}^2} + \frac{2\zeta_2 s}{\omega_{n,2}} + 1\right)} \quad (9.19)$$

with $\tau = 0.035$ s, $K = 1$, $T_1 = 1/2$ s, , $T_2 = 1/15$ s, $\omega_{n,1} = 1$ rad/s, $\zeta_1 = 0.1$, $\omega_{n,2} = 60$ rad/s, $\zeta_2 = 0.25$.

The system has one *fast* oscillatory mode with relative damping 0.25 and one *slow* oscillatory mode with relative damping 0.1 spread over one decade and a half. The system has a small time delay of 35 ms and is non-minimum phase, with a zero in the right half-plane.

The step response and the Bode plot of the system are shown in Figure 9.8. The settling time of the system is about 40 s.

The variance of the additive noise is adjusted to obtain a signal-to-noise ratio (see (9.16)) SNR= 8 dB.

The sampling period has been chosen as $T_s = 5$ ms (sampling frequency $\omega_s = 1256$ rad/s) that corresponds to about 20 times the bandwidth and is therefore higher than the usual rule of thumb given for discrete-time model identification[7].

The input signal is a pseudo-random binary sequence of maximum length (± 1.0) and the complete data set consists of 6138 input/output samples. A section of the input/output data is plotted in Figure 9.9.

The model order selection procedure (see Section 9.3.2) was applied for different model structures. The best models sorted according to YIC are presented in Table 9.6. From this table, the correct model order structure [n_b n_f n_k]= [3 4 7] is quite clear cut. It has indeed the fourth most negative YIC=-11.33 with the highest associated $R_T^2 = 0.962$. The subsequent single-run SRIVC model estimation results are shown in Table 9.7, where it can

---

[7] The usual rule of the thumb for discrete-time model identification is to choose the sampling frequency about ten times the bandwidth.

Table 9.6. Best SRIVC model orders for the SYSID'2006 benchmark system. N_{iter} denotes the number of iterations for the SRIVC algorithm to converge

n_b	n_f	n_k	YIC	R_T^2	N_{iter}
2	2	7	−12.65	0.937	2
2	2	8	−12.61	0.936	2
2	4	7	−12.24	0.936	4
3	4	7	−11.33	0.962**	4
3	4	8	−11.18	0.961	4
4	5	7	−10.89	0.962	4
2	4	8	−10.67	0.912	4
4	4	8	−9.95	0.962	4

be seen that the algorithm provides good parameter estimates with relatively small standard error (SE) estimates. Figure 9.10 shows that the step and frequency responses of the single run SRIVC estimated model are hardly distinguishable from those of the true model. It is interesting to mention here that indirect estimation using the discrete-time ARX, IV4 and PEM algorithms, followed by conversion to continuous time using the MATLAB® d2cm function, failed at this fast sampling rate because the algorithms did not converge on acceptable discrete-time models.

9.6 Successful Application Examples

In this section, identification results for three real-life processes selected from robotic, biological and electro-mechanical fields are summarised. They illustrate the use of the CONTSID toolbox methods to identify both continuous-time transfer function and state-space models directly from regularly and irregularly sampled data.

Additional successful experiences of the CONTSID toolbox methods with other real-life data processes [10, 13, 14, 28, 31] have already been reported including an industrial binary distillation column [10] but also in the case of biomedical systems [14] and environmental process data [51].

9.6.1 Complex Flexible Robot Arm

Process Description and Modelling Purpose

The robot arm is installed on an electrical motor. The modelling aim is here to design a control law based on a model between the measured reaction torque of the structure on the ground to the acceleration of the flexible arm. The robot arm is described in more detail in [22].

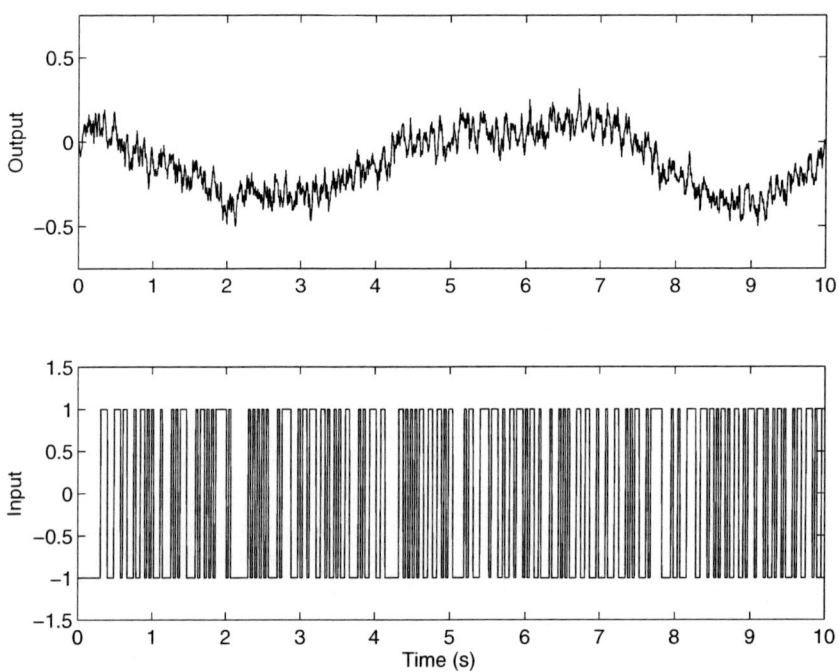

Fig. 9.9. A section of the SYSID'2006 benchmark data set 1

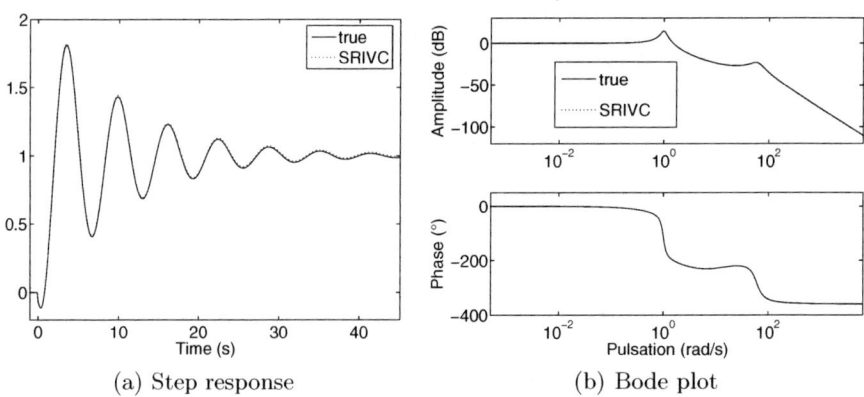

(a) Step response (b) Bode plot

Fig. 9.10. Comparison of true and SRIVC model step and frequency responses (SYSID'2006 benchmark)

Table 9.7. SYSID'2006 benchmark system estimation results

Method	Value	$\hat{b}_0$	$\hat{b}_1$	$\hat{b}_2$	$\hat{a}_1$	$\hat{a}_2$	$\hat{a}_3$	$\hat{a}_4$
	True	−120	−1560	3600	30.2	3607	750	3600
SRIVC	$\hat{\theta}$	−114.6	−1493	3543	29.32	3524	732.9	3509
	SE	2.7	21.2	38.7	0.92	35.8	9.0	36.2

Experiment Design

The excitation signal is a multi-sine. The sampling period is set to 2 ms. Measurements are made with anti-aliasing filters. $K = 10$ periods each of length $M = 4096$ are exactly measured and a record of $N = KM = 40,960$ data points is collected. The data set over the 3rd period is displayed in Figure 9.11.

Model Order Determination

The empirical transfer function estimate (ETFE) obtained from the 3rd period data set is displayed in Figure 9.12. From this figure, one can have a good indication about the model orders of the system. Indeed, one can see from the ETFE that the system has at least 3 resonant modes and 4 zeros in the frequency band $\omega \in [0; 350]$ rad/s.

Different model structures in the range $[n_b \; n_f \; n_k] = [4\;4\;0]$ to $[7\;6\;0]$ have been computed for the 3rd period data set. The other data set periods were kept for model validation purposes[8].

The 7 best models sorted according to YIC are given in Table 9.8. From this table, the first model with $[n_b \; n_f \; n_k] = [6\;6\;0]$ seems to be quite clear cut (it has the most negative YIC=−9.19, with the highest associated $R_T^2 = 0.977$).

Identification Results

The process identification is performed with the SRIVC algorithm on the third-period data set. The identification result is given as the [6 6 0] Laplace transfer function model

$$\hat{G}(s) = \frac{20.87(s - 618.5)(s^2 - 1.698s + 710.6)(s^2 + 8.435s + 2.012e4)}{(s^2 + 1.033s + 2094)(s^2 + 0.9808s + 9905)(s^2 + 2.693s + 7.042e4)} \tag{9.20}$$

This estimated model is characterised by three, lightly damped dynamic modes, as defined in Table 9.9.

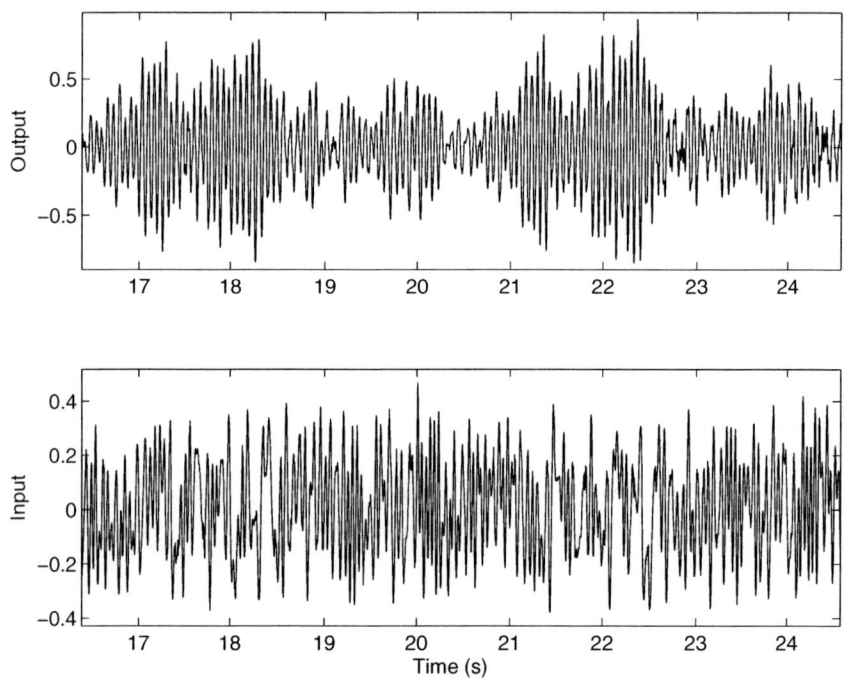

Fig. 9.11. The data set over the 3rd-period robot arm data set

Model Validation

Figure 9.13(a) compares the simulated SRIVC model output with the measured output series, over a short section of 0.4 s in the 8th-period data set. It can be noticed that the simulated output matches the measured data quite well, with $R_T^2 > 0.95$. There is also a very good agreement between the ETFE and the frequency response of the estimated SRIVC model, as shown in Figure 9.13(b).

9.6.2 Uptake Kinetics of a Photosensitising Agent into Cancer Cells

Process Description and Modelling Purpose

Figure 9.14(a) depicts the basic material used in *in vitro* experiments for studying the uptake kinetics of a photosensitising drug into living cancer cells. Cells are seeded in culture wells and are exposed at time $t_0 = 0$ to a photosensitising drug D. The purpose of this study is the estimation of the uptake yield

---

[8] Similar identification results have been obtained from all of the other 9 period data sets.

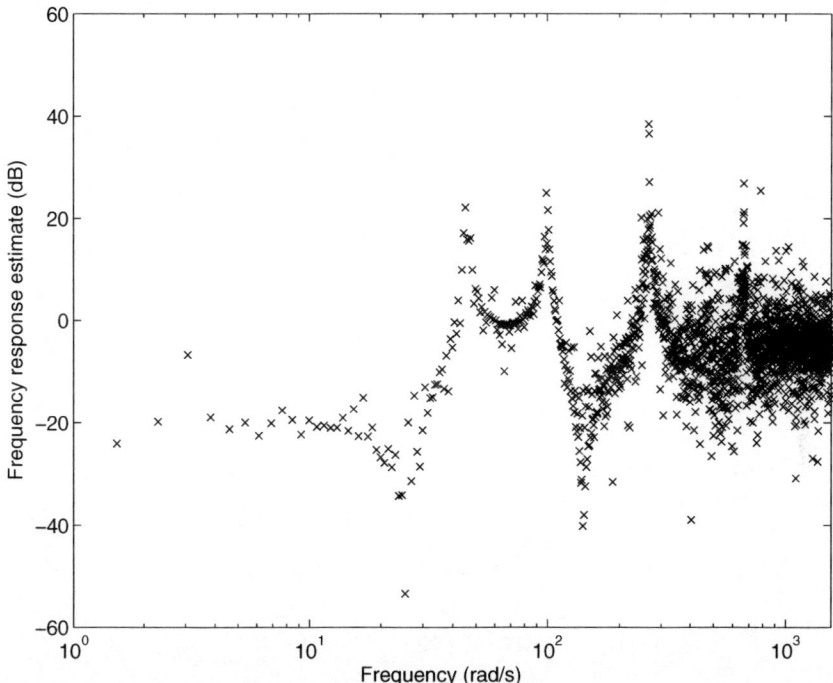

Fig. 9.12. Empirical transfer function estimate for the robot arm

(ρ) and the initial uptake rate (v_0). These biological parameters allow the biologists to discriminate the uptake characteristics of different photosensitisers and thus to choose the suitable photosensitising agent for the treatment of a given cancer cell line [3].

Experiment Design

The input variable $u(t)$ of this biological process is a step signal that corresponds to the amount of drug injected into the well from time t_0. The magnitude of the step is given by $u_0 = 5 \times 10^{-3} \mu\text{mol} \cdot \text{L}^{-1}$. $x(t)$ and $y(t)$ denote the extracellular quantity of photosensitising drug and the amount of drug absorbed by the cells, respectively. The process output is $y_m(t)$, the measurement of $y(t)$, given by a spectrofluorimeter at times $\{t_k\} = \{1, 2, 4, 6, 8, 14, 18, 24h\}$. Therefore, we are confronted with a model identification problem from non-uniformly sampled data. In this study, it is assumed that $y_m(t_k) = y(t_k) + e(t_k)$ where $e(t_k) \sim \mathcal{N}(0, \sigma_e^2)$, is the measurement noise. Two experiments have been carried out with two different protein concentrations $[Se] = 0\%$ et $[Se] = 9\%$ in the culture medium.

Table 9.8. Best SRIVC model orders for the robot arm data set

n_b	n_f	n_k	YIC	R_T^2
6	6	0	−9.19	0.977**
4	4	0	−8.56	0.940
7	6	0	−8.03	0.977
5	6	0	−7.41	0.976
5	4	0	−7.01	0.940
6	5	0	−5.56	0.966
4	5	0	−4.86	0.959
4	6	0	−3.49	0.950

Table 9.9. Eigenvalues and dynamic modes for the robot arm SRIVC model

Real	Imag.	Damping	Nat. Freq. (rad/s)
−0.52	+ 45.76i	0.0113	45.76
−0.52	− 45.76i	0.0113	45.76
−0.49	+ 99.52i	0.0049	99.52
−0.49	− 99.52i	0.0049	99.52
−1.35	+ 265.37i	0.0051	265.37
−1.35	− 265.37i	0.0051	265.37

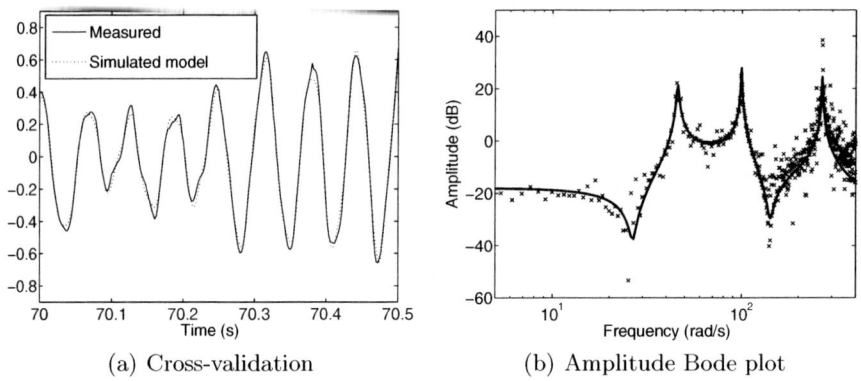

(a) Cross-validation   (b) Amplitude Bode plot

Fig. 9.13. Cross-validation results on a short section of the 8th-period data set and comparison of ETFE ('×') and SRIVC model (solid line) frequency responses for the flexible robot arm

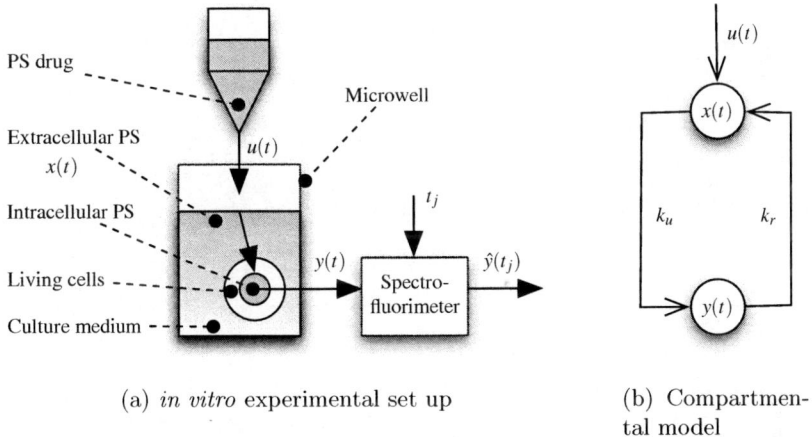

(a) *in vitro* experimental set up   (b) Compartmental model

Fig. 9.14. Biological process

Model Structure Selection

The *in vitro* uptake of the photosensitising agent into cancer cells can be described by a linear model with two compartments, as shown in Figure 9.14(b). The two compartments are associated with the extracellular and intracellular volumes, respectively. Parameters k_u and k_r are the uptake and release rates respectively. Differential equations of this compartmental model are defined as follows

$$\frac{dx(t)}{dt} = k_r y(t) - k_u x(t) + \frac{du(t)}{dt} \tag{9.21}$$

$$\frac{dy(t)}{dt} = k_u x(t) - k_r y(t) \tag{9.22}$$

with $x(0) = y(0) = 0$. Substitution of $x(t)$ from (9.21) into (9.22), yields

$$\frac{1}{k_u + k_r} \frac{dy(t)}{dt} + y(t) = \frac{k_u}{k_u + k_r} u(t) \tag{9.23}$$

Accordingly, the equivalent first-order model used in the parameter estimation step is

$$G_{[Se]}(p) = \frac{b_0}{p + f_0} \tag{9.24}$$

with $b_0 = k_u$ and $f_0 = k_u + k_r$. The uptake yield rate and the initial uptake rate of the photosensitiser uptake process are given by $\rho = b_0/f_0$ and $v_0 = b_0$.

Identification Results

The process identification is performed with the srivc algorithm on two *in vitro* data sets. The estimated transfer function models for the two protein concentrations take the form

$$\hat{G}_{0\%}(p) = \frac{162.7(\pm 21.45)}{p + 0.333(\pm 0.056)}; \quad \hat{G}_{9\%}(p) = \frac{41.1(\pm 7.18)}{p + 0.535(\pm 0.12)} \quad (9.25)$$

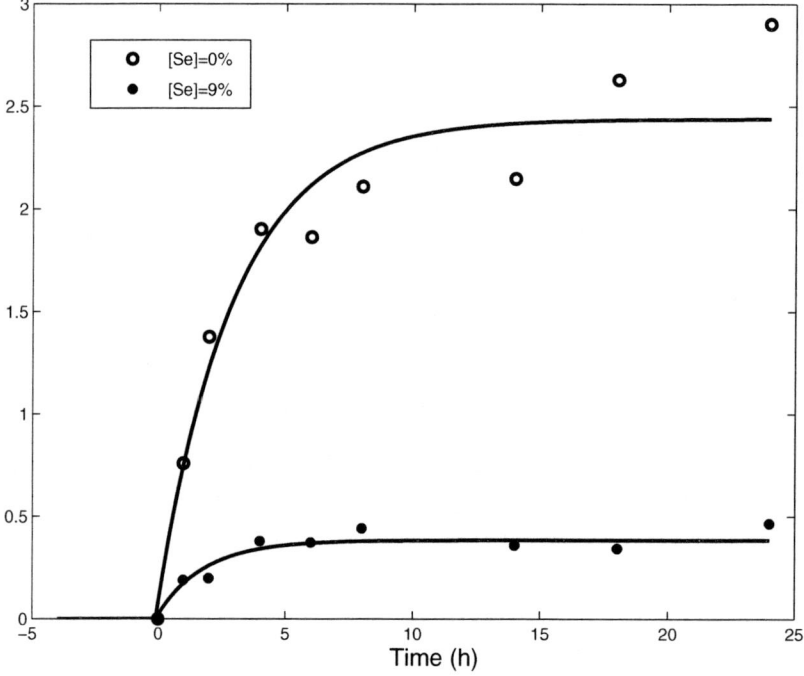

Fig. 9.15. Measured ('•' and 'o') and simulated uptake responses (solid lines)

Model Validation

Figure 9.15 shows the two estimation data sets and the simulated uptake responses obtained from $\hat{G}_{0\%}$ and $\hat{G}_{9\%}$. In experimental biology, identical experiments often produce different outcomes. This variability of the measurements is mainly due to the high sensitivity of living cells to external disturbances. Accordingly, cross-validation tests are usually not applicable in such a biological context. Here, however, the identified model has been validated by biologists [6].

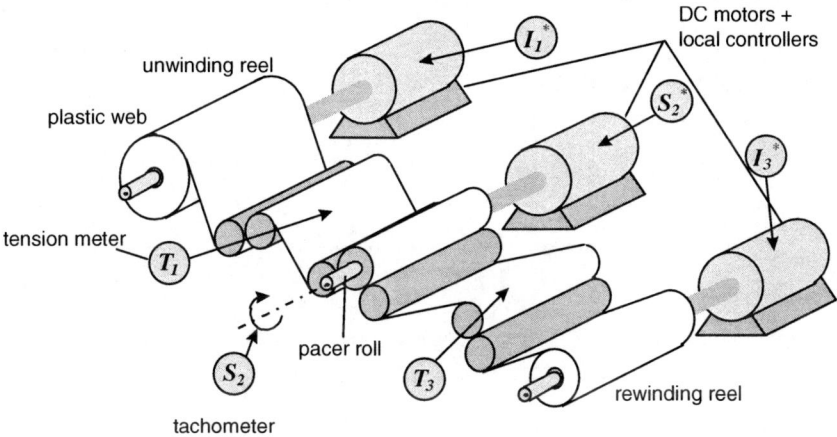

Fig. 9.16. Winding process

9.6.3 Multi-variable Winding Process

Pilot description

The present section turns to a multiple-input, multiple-output system based on a winding pilot plant [5]. Winding systems are, in general, continuous, non-linear processes. They are encountered in a wide variety of industrial plants such as rolling mills in the steel industry, plants involving web conveyance including coating, papermaking and polymer film extrusion processes. The main role of a winding process is to control the web conveyance in order to avoid the effects of friction and sliding, as well as the problems of material distortion that can also damage the quality of the final product.

As illustrated in Figure 9.16, the main part of this MIMO pilot plant is a winding process composed of a plastic web and three reels. Each reel is coupled with a direct-current motor via gear reduction. The angular speed of each reel (S_1, S_2, S_3) is measured by a tachometer, while the tensions between the reels (T_1, T_3) are measured by tension meters. At a second level, each motor is driven by a local controller. Two PI control loops adjust the motor currents (I_1) and (I_3) and a double PI control loop drives the angular speed (S_2). The setpoints of the local controllers (I_1^*, S_2^*, I_3^*) constitute the manipulated inputs of the winding system $u(t) = \begin{bmatrix} I_1^*(t) & S_2^*(t) & I_3^*(t) \end{bmatrix}^T$. Essentially, driving a winding process comes down to controlling the web linear velocity and the web tensions (T_1) and (T_3) around a given operating point. Consequently, the output variables of the winding system are $y(t) = \begin{bmatrix} T_1(t) & T_3(t) & S_2(t) \end{bmatrix}^T$. The process is described in more detail in [4]. The relevant MATLAB® files in the CONTSID toolbox are *idcdemo7.m* and *winding.mat*.

Experiment Design

Discrete-time internal binary sequences were used as excitation signals. The sampling period is set to 10 ms. The mean and linear trend of the signals were removed and the resulting input/output signals are shown in Figures 9.18 and 9.17.

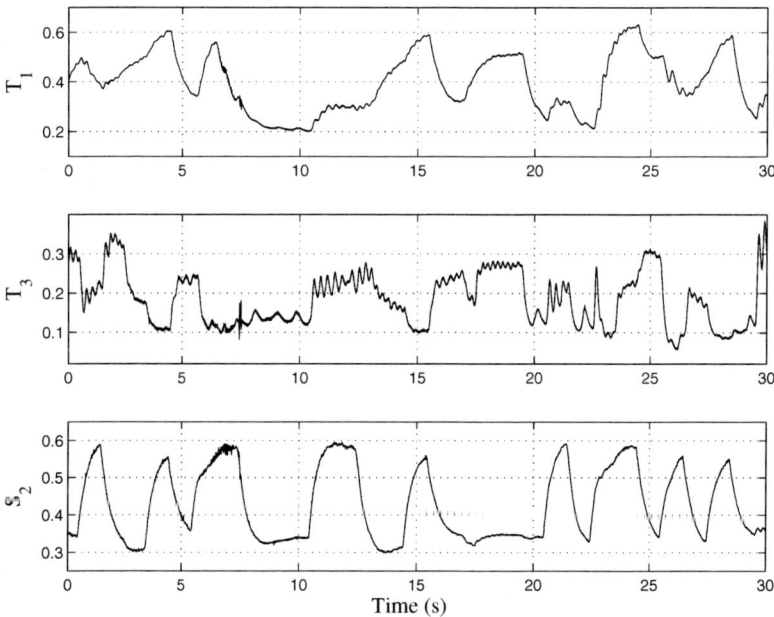

Fig. 9.17. Output signals for the winding process

Model Structure Selection

The system order $n = 3$ has been chosen by analysing the evolution of the mean square error between the process and model outputs with respect to n. No significant decrease has been observed for $n > 3$. Note, however, that the algorithm makes it possible to estimate the system order along with the model parameters if it is not known *a priori*.

Identification Results

The process identification is performed with the 4SID-based GPMF algorithm `sidgpmf`. The identification result is given as a CT state-space model that takes the form

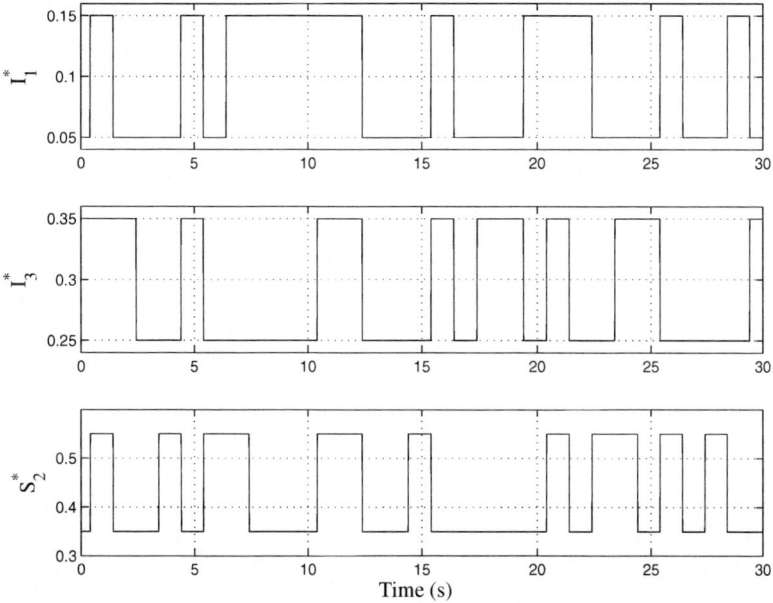

Fig. 9.18. Input signals for the winding process

$$\begin{cases} \dot{x}(t_k) = Ax(t_k) + Bu(t_k) \\ y(t_k) = Cx(t_k) + Du(t_k) + \xi(t_k) \end{cases} \quad (9.26)$$

with

$$\left(\begin{array}{c|c} A & B \\ \hline C & D \end{array}\right) = \left(\begin{array}{ccc|ccc} -1.6414 & -0.9874 & -0.4773 & 4.4994 & -3.1047 & -4.0889 \\ -0.1261 & -2.7725 & -1.3205 & 2.0652 & -3.3796 & -9.0513 \\ 0.4517 & 2.1746 & -4.2674 & 11.7889 & 9.6855 & -15.4186 \\ \hline -1.1073 & 0.4345 & -0.0536 & & & \\ 0.1442 & -0.1717 & -0.2537 & & 0 & \\ -0.2047 & -0.4425 & 0.1120 & & & \end{array}\right) \quad (9.27)$$

Model Validation

Cross-validation results are plotted in Figure 9.19 where it may be observed that there is a very good agreement with quite high values for the coefficient of determination.

9.7 Conclusions

This chapter has outlined the main features of the MATLAB® CONTSID toolbox and illustrated its potential in practical applications. The toolbox,

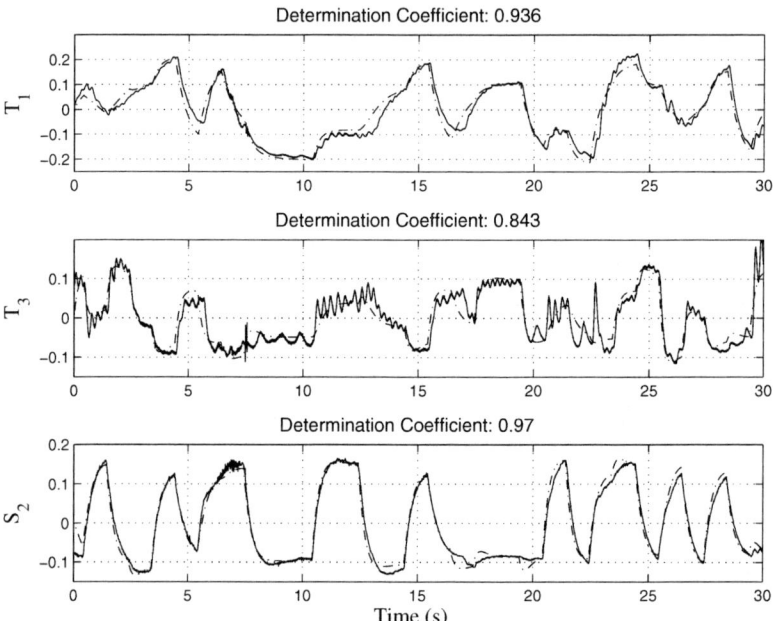

Fig. 9.19. Cross-validation results for the winding process. Measured (solid line) and model (dashdot line) outputs.

which provides access to most of the time-domain continuous-time model identification techniques that allow for the identification of continuous-time models from discrete-time data, is in continual development. Planned new releases will incorporate routines to solve errors-in-variables and closed-loop identification problems, as well as non-linear continuous-time model identification techniques.

Acknowledgements

The authors wish to thank Eric Huselstein and Hamza Zbali for the developments of the CONTSID toolbox and the CONTSID graphical user interface, respectively.
They are also grateful to Professor Istvan Kollar from the Budapest University of Technology and Economics, Hungary who kindly provided us with the robot-arm data used in Section 9.6.1.
The authors are very grateful to Professor Peter Young for his help in reading and commenting on this chapter during its preparation. Of course, the authors remain responsible for any errors or omissions in this final version.

References

1. K.J. Åström. *Introduction to Stochastic Control Theory*. Academic Press, New York, 1970.
2. K.J. Åström, P. Hagander, and J. Sternby. Zeros of sampled systems. *Automatica*, 20(1):31–38, 1984.
3. M. Barberi-Heyob, P.-O. Védrine, J.-L. Merlin, R. Millon, J. Abecassis, M.-F. Poupon, and F. Guillemin. Wild-type p53 gene transfer into mutated p53 HT29 cells improves sensitivity to photodynamic therapy via induction of apoptosis. *International Journal of Oncology*, 24:951–958, 2004.
4. T. Bastogne, H. Garnier, and P. Sibille. A PMF-based subspace method for continuous-time model identification. Application to a multivariable winding process. *International Journal of Control*, 74(2):118–132, 2001.
5. T. Bastogne, H. Noura, P. Sibille, and A. Richard. Multivariable identification of a winding process by subspace methods for a tension control. *Control Engineering Practice*, 6(9):1077–1088, 1998.
6. T. Bastogne, L. Tirand, M. Barberi-Heyob, and A. Richard. System identification of photosensitiser uptake kinetics in photodynamic therapy. *6th IFAC Symposium on Modelling and Control in Biomedical System*, Reims, France, September 2006.
7. Y.C. Chao, C.L. Chen, and H.P. Huang. Recursive parameter estimation of transfer function matrix models via Simpson's integrating rules. *International Journal of Systems Science*, 18(5):901–911, 1987.
8. C.F. Chen and C.H. Hsiao. Time-domain synthesis via Walsh functions. *IEE Proceedings*, 122(5):565–570, 1975.
9. H. Dai and N.K. Sinha. *Use of numerical integration methods*, in N.K. Sinha and G.P. Rao (eds), Identification of Continuous-Time Systems. Methodology and Computer Implementation, pages 79–121, Kluwers Academic Publishers: Dordrecht, Holland, 1991.
10. H. Garnier. Continuous-time model identification of real-life processes with the CONTSID toolbox. *15th IFAC World Congress*, Barcelona, Spain, July 2002.
11. H. Garnier, M. Gilson, and O. Cervellin. Latest developments for the MATLAB® CONTSID toolbox. *14th IFAC Symposium on System Identification*, Newcastle, Australia, pages 714–719, March 2006.
12. H. Garnier, M. Gilson, and E. Huselstein. Developments for the MATLAB® CONTSID toolbox. *13th IFAC Symposium on System Identification*, Rotterdam, The Netherlands, pages 1007–1012, August 2003.
13. H. Garnier, M. Gilson, P.C. Young, and E. Huselstein. An optimal IV technique for identifying continuous-time transfer function model of multiple input systems. *Control Engineering Practice*, 46(15):471–486, April 2007.
14. L. Cuvillon, E. Laroche, H. Garnier, J. Gangloff, and M. de Mathelin. Continuous-time model identification of robot flexibilities for fast visual servoing. *14th IFAC Symposium on System Identification*, Newcastle, Australia, pages 1264–1269, March 2006.
15. H. Garnier and M. Mensler. CONTSID: a continuous-time system identification toolbox for Matlab. *5th European Control Conference*, Karlsruhe, Germany, September 1999.
16. H. Garnier and M. Mensler. The CONTSID toolbox: a MATLAB® toolbox for CONtinuous-Time System IDentification. *12th IFAC Symposium on System Identification*, Santa Barbara, USA, June 2000.

17. H. Garnier, M. Mensler, and A. Richard. Continuous-time model identification from sampled data. Implementation issues and performance evaluation. *International Journal of Control*, 76(13):1337–1357, 2003.
18. H. Garnier, P. Sibille, and T. Bastogne. A bias-free least squares parameter estimator for continuous-time state-space models. *36th IEEE Conference on Decision and Control*, San Diego, USA, Vol. 2, pages 1860–1865, December 1997.
19. H. Garnier, P. Sibille, H.L. NGuyen, and T. Spott. A bias-compensating least-squares method for continuous-time system identification via Poisson moment functionals. *10th IFAC Symposium on System Identification*, Copenhagen, Denmark, pages 3675–3680, July 1994.
20. H. Garnier, P. Sibille, and A. Richard. Continuous-time canonical state-space model identification via Poisson moment functionals. *34th IEEE Conference on Decision and Control*, New Orleans, USA, Vol. 2, pages 3004–3009, December 1995.
21. E. Huselstein and H. Garnier. An approach to continuous-time model identification from non-uniformly sampled data. *41st IEEE Conference on Decision and Control*, Las Vegas, USA, December 2002.
22. I. Kollar. *Frequency Domain System Identification Toolbox Users's Guide*. The Mathworks, Inc., Mass., 1994.
23. E.K. Larsson and T. Söderström. Identification of continuous-time AR processes from unevenly sampled data. *Automatica*, 38(4):709–718, 2002.
24. L. Ljung. *System Identification. Theory for the User*. Prentice Hall, Upper Saddle River, NJ, USA, 2nd edition, 1999.
25. L. Ljung. Initialisation aspects for subspace and output-error identification methods. *European Control Conference*, Cambridge, UK, September 2003.
26. L. Ljung. SID: System identification toolbox for MATLAB®. http://www.mathworks.com/access/helpdesk/help/toolbox/ident/ident.shtml, 2006.
27. K. Mahata and H. Garnier. Identification of continuous-time Box-Jenkins models with arbitrary time delay. *Submitted to the 46th Conference on Decision and Control*, New Orleans, USA, December 2007.
28. M. Mensler, H. Garnier, and E. Huselstein. Experimental comparison of continuous-time model identification methods on a thermal process. *12th IFAC Symposium on System Identification*, Santa Barbara, USA, June 2000.
29. M. Mensler, K. Wada. Subspace method for continuous-time system identification. *32nd ISCIE International Symposium on Stochastic Systems Theory and Its Applications*, Tottori, Japan, November 2000.
30. M. Mensler, H. Garnier, A. Richard, and P. Sibille. Comparison of sixteen continuous-time system identification methods with the CONTSID toolbox. *5th European Control Conference*, Karlsruhe, Germany, September 1999.
31. M. Mensler, S. Joe, and T. Kawabe. Identification of a toroidal continuously variable transmission using continuous-time system identification methods. *Control Engineering Practice*, 14(1):45–58, January 2006.
32. B.M. Mohan and K.B. Datta. Analysis of time-delay systems via shifted Chebyshev polynomials of the first and second kinds. *International Journal of Systems Science*, 19(9):1843–1851, 1988.
33. P.N. Paraskevopoulos. System analysis and synthesis via orthogonal polynomial series and Fourier series. *Mathematics and Computers in Simulation*, 27:453–469, 1985.

34. A.E. Pearson and Y. Shen. Weighted least squares / MFT algorithms for linear differential system identification. *32nd IEEE Conference on Decision and Control*, San Antonio, USA, pages 2032–2037, 1993.
35. M. Djamai, E. Tohme, R. Ouvrard, and S. Bachir. Continuous-time model identification using reinitialized partial moments. Application to power amplifier modeling. *14th IFAC Symposium on System Identification*, Newcastle, Australia, March 2006.
36. G.P. Rao and H. Garnier. Numerical illustrations of the relevance of direct continuous-time model identification. *15th IFAC World Congress*, Barcelona, Spain, July 2002.
37. G.P. Rao and H. Garnier. Identification of continuous-time models: direct or indirect? *Invited semi-plenary paper for the XV International Conference on Systems Science*, Wroclaw, Poland, September 2004.
38. G.P. Rao and H. Unbehauen, Identification of continuous-time systems, *IEE Proceedings - Control Theory and Applications*, 153(2):185-220, March 2006.
39. S. Sagara and Z.Y. Zhao. Numerical integration approach to on-line identification of continuous-time systems. *Automatica*, 26(1):63–74, 1990.
40. J. Schoukens, R. Pintelon, and H. Van Hamme. Identification of linear dynamic systems using piecewise constant excitations: use, misuse and alternatives. *Automatica*, 30(7):1953–1169, 1994.
41. T. Söderström and P. Stoica. *System Identification*. Series in Systems and Control Engineering. Prentice Hall, Englewood Cliffs, 1989.
42. H. Unbehauen and G.P. Rao. *Identification of Continuous Systems*. Systems and control series. North-Holland, Amsterdam, 1987.
43. H. Unbehauen and G.P. Rao. Identification of continuous-time systems: a tutorial. *11th IFAC Symposium on System Identification*, Kitakyushu, Japan, Vol. 3, pages 1023–1049, July 1997.
44. P. Van Overschee and B. De Moor. *Subspace Identification for Linear Systems - Theory, Implementation, Applications*. Kluwer Academic Publishers, Boston, USA, 1996.
45. P.E. Wellstead. An instrumental product moment test for model order estimation. *Automatica*, 14:89–91, 1978.
46. P.C. Young. *Recursive Estimation and Time-series Analysis*. Springer-Verlag, Berlin, 1984.
47. P.C. Young. Recursive estimation, forecasting and adaptive control. in C.T. Leondes (ed), Control and Dynamic Systems: Advances in Theory and Applications, pages 119–166, Vol 31, Academic Press, 1989.
48. P.C. Young. Data-based mechanistic modeling of engineering systems. *Journal of Vibration and Control*, 4:5–28, 1998.
49. P.C. Young. The data-based mechanistic approach to the modelling, forecasting and control of environmental systems. *Annual Reviews in Control*, 30:169–182, 2006.
50. P.C. Young. The Refined Instrumental Variable (RIV) method: unified estimation of discrete and continuous-time transfer function models. *Journal Européen des Systèmes Automatisés*, in press, 2008.
51. P.C. Young and H. Garnier. Identification and estimation of continuous-time data-based mechanistic (DBM) models for environmental systems. *Environmental Modelling and Software*, 21(8):1055–1072, August 2006.

52. P.C. Young and A.J. Jakeman. Refined instrumental variable methods of time-series analysis: Part III, extensions. *International Journal of Control*, 31:741–764, 1980.
53. P.C. Young, A.J. Jakeman, and R. McMurtries. An instrumental variable method for model order identification. *Automatica*, 16:281–296, 1980.

10

Subspace-based Continuous-time Identification

Rolf Johansson

Lund University, Sweden

10.1 Introduction

The last few years have witnessed a strong interest in system identification using realisation-based algorithms. The use of Markov parameters as suggested by Ho and Kalman [18] Akaike [1], and Kung [28], of a system can be effectively applied to the problem of state-space identification; see Verhaegen *et al.* [43, 44], van Overschee and de Moor [41], Juang and Pappa [26], Moonen *et al.* [36], Bayard [3, 4, 33, 34]. Suitable background for the discrete-time theory supporting stochastic subspace model identification is to be found in [1, 14, 41]. As for model structures and realisation theory, see the important contributions [12, 31]. As these subspace-model identification algorithms deal with the case of fitting a discrete-time model, it remains as an open problem how to extend these methods for continuous-time (CT) systems. A great deal of modelling in natural sciences and technology is made by means of continuous-time models and such models require suitable methods of system identification [19]. To this end, a theoretical framework of continuous-time identification and statistical model validation is needed. In particular, as experimental data are usually provided as time series, it is relevant to provide continuous-time theory and algorithms that permit application to discrete-time data.

This chapter[1] treats the problem of continuous-time system identification based on discrete-time data and provides a framework with algorithms presented in preliminary forms in [16, 21–23]. The approach adopted is that of subspace-model identification [25, 41, 43, 44], though elements of continuous-time identification are similar to those previously presented for the prediction-error identification [19, 20]. Some relevant numerical aspects are treated in references [5, 45, 46].

---

[1] This chapter is partly based on the paper [23] co-authored with M. Verhaegen and C. T. Chou.

10.2 Problem Formulation

Consider a continuous-time time-invariant system $\Sigma_n(A, B, C, D)$ with the state-space equations

$$\dot{x}(t) = Ax(t) + Bu(t) + v(t)$$
$$y(t) = Cx(t) + Du(t) + e(t) \quad (10.1)$$

with input $u \in \mathbb{R}^m$, output $y \in \mathbb{R}^p$, state vector $x \in \mathbb{R}^n$ and zero-mean disturbance stochastic processes $v \in \mathbb{R}^n$, $e \in \mathbb{R}^p$ acting on the state dynamics and the output, respectively. The continuous-time system identification problem is to find estimates of system matrices A, B, C, D from finite sequences $\{u_k\}_{k=0}^N$ and $\{y_k\}_{k=0}^N$ of input–output data.

10.2.1 Discrete-time Measurements

Assume periodic sampling to be made with period h at a time sequence $\{t_k\}_{k=0}^N$, with $t_k = t_0 + kh$ with the corresponding discrete-time input–output data $\{y_k\}_{k=0}^N$ and $\{u_k\}_{k=0}^N$ sampled from the continuous-time dynamic system of (10.1). Alternatively, data may be assumed generated by the time-invariant discrete-time state-space system

$$x_{k+1} = A_z x_k + B_z u_k + v_k; \quad A_z = e^{Ah}, \quad B_z = \int_0^h e^{As} B \, ds \quad (10.2)$$
$$y_k = Cx_k + Du_k + e_k \quad (10.3)$$

with equivalent input–output behaviour to that of (10.1) at the sampling-time sequence. The underlying discretised state sequence $\{x_k\}_{k=0}^N$ and discrete-time stochastic processes $\{v_k\}_{k=0}^N$, $\{e_k\}_{k=0}^N$ correspond to disturbance processes v and e that can be represented by the components

$$v_k = \int_{t_{k-1}}^{t_k} e^{A(t_k - s)} v(s) ds, \quad k = 1, 2, ..., N \quad (10.4)$$
$$e_k = e(t_k) \quad (10.5)$$

with the covariance

$$E\left\{ \begin{bmatrix} v_i \\ e_i \end{bmatrix} \begin{bmatrix} v_j \\ e_j \end{bmatrix}^T \right\} = Q\delta_{ij} = \begin{bmatrix} Q_{11} & Q_{12} \\ Q_{12}^T & Q_{22} \end{bmatrix} \delta_{ij}, \quad Q \geq 0, \quad q = \mathrm{rank}(Q) \quad (10.6)$$

Consider a discrete-time time-invariant system $\Sigma_n(A, B, C, D)$ with the state-space equations with input $u_k \in \mathbb{R}^m$, output $y_k \in \mathbb{R}^p$, state vector $x_k \in \mathbb{R}^n$ and noise sequences $v_k \in \mathbb{R}^n$, $e_k \in \mathbb{R}^p$ acting on the state dynamics and the output, respectively.

Remark: As computation and statistical tests deal with discrete-time data, we assume the original sampled stochastic disturbance sequences to be uncorrelated with a uniform spectrum up to the Nyquist frequency. The underlying continuous-time stochastic processes will have an autocorrelation function according to Figure 10.1, thereby avoiding the mathematical problems associated with the stochastic processes of Brownian motion.

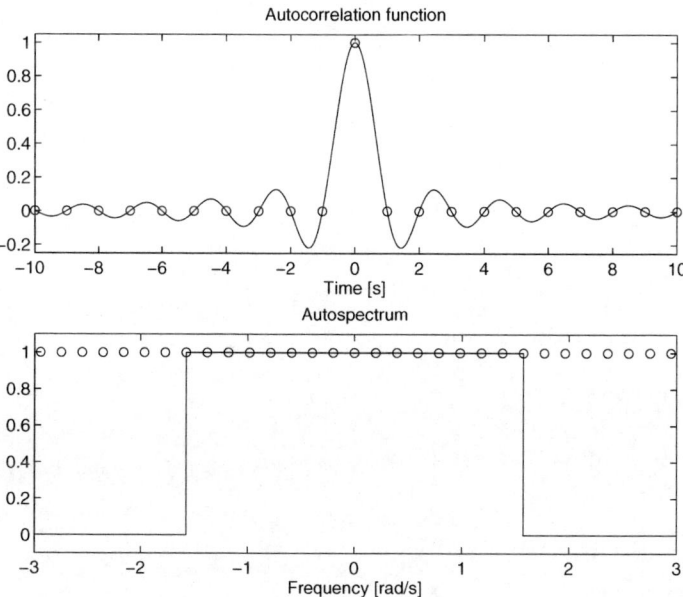

Fig. 10.1. Autocorrelation functions (*upper diagram*) and autospectra (*diagram below*) of a continuous-time (*solid line* stochastic variable $w(t)$ and a discrete-time ('*o*') sample sequence $\{w_k\}$. The CT process is bandwidth limited to the Nyquist frequency $\omega_N = \pi/2$ [rad/s] of a sampling process with sampling frequency 1 Hz. Properties of the sampled sequence $\{w_k\}$ confirm that the sampled sequence is an uncorrelated stochastic process with a uniform autospectrum.

10.2.2 Continuous-time State-space Linear System

From the set of first-order linear differential equations of (10.1) one finds the Laplace transform

$$sX = AX + BU + V + sx_0; \qquad x_0 = x(t_0) \qquad (10.7)$$
$$Y = CX + DU + E \qquad (10.8)$$

Introduction of the complex-variable transform

$$\lambda(s) = \frac{1}{1+s\tau} \qquad (10.9)$$

corresponding to a stable, causal operator permits an algebraic transformation of the model

$$X = (I + \tau A)[\lambda X] + \tau B[\lambda U] + \tau[\lambda V] + (1-\lambda)x_0 \qquad (10.10)$$
$$Y = CX + DU + E \qquad (10.11)$$

Reformulation, while ignoring the initial conditions to linear system equations gives

$$\begin{bmatrix}\xi\\y\end{bmatrix} = \begin{bmatrix}I+\tau A & \tau B\\ C & D\end{bmatrix}\begin{bmatrix}x\\u\end{bmatrix} + \begin{bmatrix}\tau v\\e\end{bmatrix}, \qquad x(t) = [\lambda\xi](t) \qquad (10.12)$$

$$= \begin{bmatrix}A_\lambda & B_\lambda\\ C & D\end{bmatrix}\begin{bmatrix}x\\u\end{bmatrix} + \begin{bmatrix}\tau v\\e\end{bmatrix}, \qquad \begin{cases}A_\lambda = I + \tau A\\ B_\lambda = \tau B\end{cases} \qquad (10.13)$$

the mapping between (A, B) and (A_λ, B_λ) being bijective. Provided that a standard positive semi-definiteness condition of Q is fulfilled so that the Riccati equation has a solution, it is possible to replace the linear model of (10.13) by the innovations model

$$\begin{bmatrix}\xi\\y\end{bmatrix} = \begin{bmatrix}A_\lambda & B_\lambda\\ C & D\end{bmatrix}\begin{bmatrix}x\\u\end{bmatrix} + \begin{bmatrix}K_\lambda\\I\end{bmatrix}w, \qquad K_\lambda = \tau K \qquad (10.14)$$

By recursion, it is found that

$$y = Cx + Du + w \qquad (10.15)$$
$$= CA_\lambda[\lambda x] + CB_\lambda[\lambda u] + Du + CK_\lambda[\lambda w] + w \qquad (10.16)$$
$$\vdots$$
$$= CA_\lambda^k[\lambda^k x] + \sum_{j=1}^{k} CA^{k-j}B_\lambda[\lambda^{k-j}u] + Du$$
$$+ \sum_{j=1}^{k} CA^{k-j}K_\lambda[\lambda^{k-j}w] + w \qquad (10.17)$$

For the purpose of subspace model identification, it is straightforward to formulate extended linear models for the original models and its innovations form

$$\mathcal{Y} = \Gamma_x \mathcal{X} + \Gamma_u \mathcal{U} + \Gamma_v \mathcal{V} + \mathcal{E} \qquad (10.18)$$
$$\mathcal{Y} = \Gamma_x \mathcal{X} + \Gamma_u \mathcal{U} + \Gamma_w \mathcal{W} \qquad (10.19)$$

with input–output and state variables

$$\mathcal{Y} = \begin{bmatrix} [\lambda^{i-1}y] \\ [\lambda^{i-2}y] \\ \vdots \\ [\lambda^1 y] \\ y(t) \end{bmatrix}, \quad \mathcal{U} = \begin{bmatrix} [\lambda^{i-1}u] \\ [\lambda^{i-2}u] \\ \vdots \\ [\lambda^1 u] \\ u(t) \end{bmatrix}, \quad \mathcal{X} = [\lambda^{i-1}x] \qquad (10.20)$$

and stochastic processes of disturbance

$$\mathcal{V} = \begin{bmatrix} [\lambda^{i-1}v] \\ [\lambda^{i-2}v] \\ \vdots \\ [\lambda^1 v] \\ v(t) \end{bmatrix}, \quad \mathcal{E} = \begin{bmatrix} [\lambda^{i-1}e] \\ [\lambda^{i-2}e] \\ \vdots \\ [\lambda^1 e] \\ e(t) \end{bmatrix}, \quad \mathcal{W} = \begin{bmatrix} [\lambda^{i-1}w] \\ [\lambda^{i-2}w] \\ \vdots \\ [\lambda^1 w] \\ w(t) \end{bmatrix} \qquad (10.21)$$

and parameter matrices of state variables and input–output behaviour

$$\Gamma_x = \begin{bmatrix} C \\ CA_\lambda \\ \vdots \\ CA_\lambda^{i-1} \end{bmatrix} \in \mathbb{R}^{ip \times n} \qquad (10.22)$$

$$\Gamma_u = \begin{bmatrix} D & 0 & \cdots & 0 \\ CB_\lambda & D & \ddots & \vdots \\ \vdots & \vdots & \ddots & 0 \\ CA_\lambda^{i-2}B_\lambda & CA_\lambda^{i-3}B_\lambda & \cdots & D \end{bmatrix} \in \mathbb{R}^{ip \times im} \qquad (10.23)$$

and for stochastic input–output behaviour

$$\Gamma_v = \begin{bmatrix} 0 & 0 & \cdots & 0 & 0 \\ \tau C & 0 & & 0 & 0 \\ \tau CA_\lambda & \tau C & \ddots & \vdots & \vdots \\ \vdots & \vdots & \ddots & 0 & 0 \\ \tau CA_\lambda^{i-2} & \tau CA_\lambda^{i-3} & \cdots & \tau C & 0 \end{bmatrix} \in \mathbb{R}^{ip \times im} \qquad (10.24)$$

and

$$\Gamma_w = \begin{bmatrix} I & 0 & \cdots & 0 & 0 \\ CK_\lambda & I & \ddots & \vdots & \vdots \\ \vdots & CK_\lambda & \ddots & 0 & 0 \\ CA_\lambda^{i-3}K_\lambda & \vdots & \ddots & I & 0 \\ CA_\lambda^{i-2}K_\lambda & CA_\lambda^{i-3}K_\lambda & \cdots & CK_\lambda & I \end{bmatrix} \qquad (10.25)$$

It is clear that Γ_x of (10.22) represents the extended observability matrix as known from linear system theory and subspace model identification [41,43,44].

10.3 System Identification Algorithms

The theory provided permits formulation of a variety of algorithms with the same algebraic properties as the original discrete-time version, though with application to continuous-time modeling and identification. Below is presented one realisation-based algorithm (Alg. 1) and two subspace-based algorithms (Alg. 2–3) with application to time-domain data and frequency-domain data, respectively. Theoretical justification for each one of these algorithms follows separate from the algorithms.

Algorithm 1 (System Realisation *ad modum* Ho–Kalman) [3, 18, 21, 26]

1. Use least squares identification to find a multi-variable transfer function

$$G(\lambda(s)) = D_L^{-1}(\lambda) N_L(\lambda) = \sum_{k=0}^{\infty} G_k \lambda^k \qquad (10.26)$$

where $D_L(\lambda)$, $N_L(\lambda)$ are polynomial matrices obtained by means of some identification method such as linear regression with

$$\varepsilon(t, \theta) = D_L(\lambda) y(t) - N_L(\lambda) u(t) \qquad (10.27)$$
$$G(\lambda) = D_L^{-1}(\lambda) N_L(\lambda) \qquad (10.28)$$
$$D_L(\lambda) = I + D_1 \lambda + \cdots + D_n \lambda^n \qquad (10.29)$$
$$N_L(\lambda) = N_0 + N_1 \lambda + \cdots N_n \lambda^n \qquad (10.30)$$

2. Solve for the transformed Markov parameters to give

$$G_k = N_k - \sum_{j=1}^{k} D_j G_{k-j}, \qquad k = 0, \ldots, n \qquad (10.31)$$

$$G_k = -\sum_{j=1}^{n} D_j G_{k-j}, \qquad k = n+1, \ldots, N \qquad (10.32)$$

3. For suitable numbers q, r, s such that $r + s \leq N$ arrange the Markov parameters in the Hankel matrix

$$G_{r,s}^{(q)} = \begin{bmatrix} G_{q+1} & G_{q+2} & \cdots & G_{q+s} \\ G_{q+2} & G_{q+3} & \cdots & G_{q+s+1} \\ \vdots & \vdots & \ddots & \vdots \\ G_{q+r} & G_{q+r+1} & \cdots & G_{q+r+s-1} \end{bmatrix} \qquad (10.33)$$

4. Determine rank n and resultant system matrices

$$G_{r,s}^{(0)} = U\Sigma V^T \quad \text{(singular-value decomposition)} \tag{10.34}$$

$$E_y^T = [I_{p \times p} \ \ 0_{p \times (r-1)p}] \tag{10.35}$$

$$E_u^T = [I_{m \times m} \ \ 0_{m \times (s-1)m}] \tag{10.36}$$

$$\Sigma_n = \text{diag}\{\sigma_1, \sigma_2, \ldots, \sigma_n\} \tag{10.37}$$

$$U_n = \text{matrix of first } n \text{ columns of } U \tag{10.38}$$

$$V_n = \text{matrix of first } n \text{ columns of } V \tag{10.39}$$

$$A_n = \Sigma_n^{-1/2} U_n^T G_{r,s}^{(1)} V_n \Sigma_n^{-1/2}, \quad \widehat{A} = \frac{1}{\tau}(A_n - I) \tag{10.40}$$

$$B_n = \Sigma_n^{1/2} V_n^T E_u, \quad \widehat{B} = \frac{1}{\tau} B_n \tag{10.41}$$

$$C_n = E_y^T U_n \Sigma_n^{1/2}, \quad \widehat{C} = C_n \tag{10.42}$$

$$D_n = G_0, \quad \widehat{D} = D_n \tag{10.43}$$

which yields the nth-order state-space realisation

$$\dot{x}(t) = \widehat{A}x(t) + \widehat{B}u(t)$$
$$y(t) = \widehat{C}x(t) + \widehat{D}u(t) \tag{10.44}$$

Algorithm 2 (Subspace Model Identification (MOESP)) *[43, 44]*

1. Arrange data matrices $\mathcal{U}_N$, $\mathcal{Y}_N$ by using the following notation for sampled filtered data

$$[\lambda^j u]_k = [\lambda^j u](t_k), \quad [\lambda^j y]_k = [\lambda^j y](t_k), \quad \text{etc.} \tag{10.45}$$

where

$$\mathcal{Y}_N = \begin{bmatrix} [\lambda^{i-1} y]_1 & [\lambda^{i-1} y]_2 & \cdots & [\lambda^{i-1} y]_N \\ [\lambda^{i-2} y]_1 & [\lambda^{i-2} y]_2 & \cdots & [\lambda^{i-2} y]_N \\ \vdots & \vdots & & \vdots \\ [\lambda y]_1 & [\lambda y]_2 & \cdots & [\lambda y]_N \\ y_1 & y_2 & \cdots & y_N \end{bmatrix} \in \mathbb{R}^{ip \times N} \tag{10.46}$$

and a similar construction for $\mathcal{U}_N$.

2. Make a QR-factorisation such that

$$\begin{bmatrix} \mathcal{U}_N \\ \mathcal{Y}_N \end{bmatrix} = \begin{bmatrix} R_{11} & 0 \\ R_{21} & R_{22} \end{bmatrix} \begin{bmatrix} Q_1 \\ Q_2 \end{bmatrix} \tag{10.47}$$

3. Make a SVD of the matrix $R_{22} \in \mathbb{R}^{ip \times ip}$ approximating the column space of Γ_x

$$R_{22} = [U_n \ U_0] \begin{bmatrix} S_n & 0 \\ 0 & S_0 \end{bmatrix} [V_n \ V_0]^T \tag{10.48}$$

4. Determine estimates $\widehat{A}$, $\widehat{C}$ of system matrices A, C from equations

$$U_n^{(1)} = \text{rows 1 through } (s-1)p \text{ of } U_n \quad (10.49)$$
$$U_n^{(2)} = \text{rows } p+1 \text{ through } sp \text{ of } U_n \quad (10.50)$$
$$U_n^{(1)} \widehat{A}_\lambda = U_n^{(2)}, \quad \widehat{A} = \frac{1}{\tau}(\widehat{A}_\lambda - I) \quad (10.51)$$
$$\widehat{C} = \text{rows 1 through } p \text{ of } U_n \quad (10.52)$$

5. Determine estimate $\widehat{B}$, $\widehat{D}$ of system matrices B, D from relationship

$$\widehat{\Gamma}_u = R_{21} R_{11}^{-1} \quad (10.53)$$

An algorithmic modification to accommodate frequency-domain data can be made by replacing Step 1 of Algorithm 2 by the following:

1' Arrange data matrices $\mathcal{U}_N$, $\mathcal{Y}_N$ using the filtered fequency-domain data

$$[\Lambda^j U]_k = [\Lambda^j U(s)]_{s=i\omega_k}, \quad [\Lambda^j Y]_k = [\Lambda^j Y(s)]_{s=i\omega_k}, \ldots \quad (10.54)$$

evaluated for

$$\omega_k = k\frac{2\pi}{N}\omega_s \quad (10.55)$$

and arrange a matrix equation of frequency-sampled data as

$$\mathcal{Y}_N = \begin{bmatrix} [\Lambda^{i-1}Y]_1 & [\Lambda^{i-1}Y]_2 & \cdots & [\Lambda^{i-1}Y]_N \\ [\Lambda^{i-2}Y]_1 & [\Lambda^{i-2}Y]_2 & \cdots & [\Lambda^{i-2}Y]_N \\ \vdots & \vdots & & \vdots \\ [\Lambda Y]_1 & [\Lambda Y]_2 & \cdots & [\Lambda Y]_N \\ Y_1 & Y_2 & \cdots & Y_N \end{bmatrix} \in \mathbb{R}^{ip \times N} \quad (10.56)$$

and with similar construction for $\mathcal{U}_N$ and proceed as from Step 2 of Algorithm 2.

Algorithm 3 (Subspace Correlation Method) *Along with the data matrices $\mathcal{U}_N$, $\mathcal{Y}_N$ of Algorithm 2, introduce the correlation variable*

$$\mathcal{Z}_N = \frac{1}{N} \begin{bmatrix} [\lambda^{j-1}u]_1 & [\lambda^{j-1}u]_2 & \cdots & [\lambda^{j-1}u]_N \\ [\lambda^{j-2}u]_1 & [\lambda^{j-2}u]_2 & \cdots & [\lambda^{j-2}u]_N \\ \vdots & \vdots & & \vdots \\ [\lambda u]_1 & [\lambda u]_2 & \cdots & [\lambda u]_N \\ u_1 & u_2 & \cdots & u_N \end{bmatrix} \in \mathbb{R}^{jm \times N} \quad (10.57)$$

for $j > m+p+n$ chosen sufficiently large. Proceed as from Step 2 of Algorithm 2 with application of QR-factorisation to the matrix

$$\begin{bmatrix} \mathcal{U}_N \mathcal{Z}_N^T \\ \mathcal{Y}_N \mathcal{Z}_N^T \end{bmatrix} \in \mathbb{R}^{(m+p)i \times jm} \quad (10.58)$$

10.3.1 Theoretical Remarks on the Algorithms

In this section, we provide some theoretical justification for the algorithms suggested.

Remarks on Algorithm 1—Continuous-time State-space Realisation

After operator reformulation, and a least squares transfer function estimate, the algorithm follows the Ho–Kalman algorithm step by step.

1. The first step aims towards system identification. The (high-order) least squares identification serves to find a non-minimal input–output model with good prediction-error accuracy as the first priority;
2. Step 2 serves to provide transformed Markov parameter where the

$$G_k = CA_\lambda^{k-1} B_\lambda, \quad k \geq 1 \qquad (10.59)$$

The recursion to obtain $\{G_k\}$ may be replaced by a linear equation;
3. Organisation of the Markov parameter in the Hankel matrices $G_{r,s}^{(q)}$ of block row dimension r and block column dimension s, respectively, permits

$$G_{r,s}^{(q)} = \mathcal{O}_r A_\lambda^q \cdot \mathcal{C}_s \qquad (10.60)$$

where

$$\mathcal{O}_r = \begin{bmatrix} C \\ CA_\lambda \\ \vdots \\ CA_\lambda^{r-1} \end{bmatrix}, \quad \mathcal{C}_s = \begin{bmatrix} B_\lambda & A_\lambda B_\lambda & \ldots & A_\lambda^{s-1} B_\lambda \end{bmatrix} \qquad (10.61)$$

Thus, for $A_\lambda \in \mathbb{R}^{n \times n}$ the rank of $\mathcal{O}_r$, A_λ^q and $\mathcal{C}_s$ cannot exceed n, which justifies the determination of model order from a rank test of $G_{r,s}^{(q)}$;
4. The last algorithmic step involves a singular-value decomposition that accomplishes the factorisation into the extended observability matrix and extended controllability matrix, which permits rank evaluation of $G_{r,s}^{(q)}$ and, hence, estimation of system order n. From the full-rank matrix factors U_n, Σ_n, V_n, estimates of A_λ, B_λ, C and D, are found. The final transformation to parameter matrices in the s–domain provides the state-space realisation.

Remarks on Algorithm 2—Continuous-time Subspace Model Identification

This algorithm is similar to the MOESP algorithm of discrete-time subspace model identification

1. The arrangement of input–output data matrices $\mathcal{Y}_N, \mathcal{U}_N$ of sampled data serves to express data in the form of (10.19) so that

$$\mathcal{Y}_N = \Gamma_x \mathcal{X}_N + \Gamma_u \mathcal{U}_N + \Gamma_w \mathcal{W}_N \qquad (10.62)$$

where $\mathcal{W}_N$ is the disturbance sample matrix (not available to measurement) and

$$\mathcal{X}_N = \left[[\lambda^{i-1} x]_1 \; [\lambda^{i-1} x]_2 \; \cdots \; [\lambda^{i-1} x]_N \right] \qquad (10.63)$$

2. The QR-factorisation serves to retrieve the matrix product $\Gamma_x \mathcal{X}_N$ that is found as the column space of R_{22} in the case of disturbance-free data;
3. The singular-value factorisation of the matrix R_{22} serves to find the left factor U_n of rank n corresponding to Γ_x (up to a similarity transformation). The rank condition is evaluated by means of the non-zero singular values of Σ_n;
4. As the estimate $\widehat{\Gamma}_x = U_n$ contains products of the C–matrix and powers of A_λ, it is straightforward to find an estimate of C from the p first rows. Next, an estimate $\widehat{A}$ is found. Subsequent transformation of A_λ to the s-domain is required;
5. Given $\widehat{A}, \widehat{C}$, then $\widehat{B}, \widehat{D}$ can be found to fit the input–output relationship provided by Γ_u.

Algorithm 2 and its frequency-domain modification are very closely related as their data matrices with different interpretation obey the relationship

$$\mathcal{Y}_N = \Gamma_x \mathcal{X}_N + \Gamma_u \mathcal{U}_N + \Gamma_w \mathcal{W}_N \qquad (10.64)$$

By definition, the discrete-time Fourier transform is formulated as the linear transformation

$$\begin{bmatrix} Y_1^T \\ Y_2^T \\ \vdots \\ Y_N^T \end{bmatrix} = \mathcal{T} \begin{bmatrix} y_1^T \\ y_2^T \\ \vdots \\ y_N^T \end{bmatrix}, \quad \mathcal{T} = \begin{bmatrix} 1 & e^{i\omega_0 h} & \cdots & e^{i\omega_0 (N-1)h} \\ 1 & e^{i\omega_1 h} & \cdots & e^{i\omega_1 (N-1)h} \\ \vdots & \vdots & & \vdots \\ 1 & e^{i\omega_{N-1} h} & \cdots & e^{i\omega_{N-1} (N-1)h} \end{bmatrix} \qquad (10.65)$$

For the standard FFT set of frequency points $\omega_k = k \cdot (2\pi/Nh)$, $k = 0, 1, 2, \ldots, N-1$, we have $\mathcal{T}^* \mathcal{T} = N \cdot I_N$ so that $\mathcal{Y}_N, \mathcal{U}_N, \ldots$ of Algorithm 2 and its frequency-domain version only differ by a right invertible factor $\mathcal{T}^T$ as found from

$$\begin{bmatrix} [\Lambda^{i-1} Y]_1 & \cdots & [\Lambda^{i-1} Y]_N \\ [\Lambda^{i-2} Y]_1 & \cdots & [\Lambda^{i-2} Y]_N \\ \vdots & & \vdots \\ [\Lambda Y]_1 & \cdots & [\Lambda Y]_N \\ Y_1 & \cdots & Y_N \end{bmatrix} = \begin{bmatrix} [\lambda^{i-1} y]_1 & \cdots & [\lambda^{i-1} y]_N \\ [\lambda^{i-2} y]_1 & \cdots & [\lambda^{i-2} y]_N \\ \vdots & & \vdots \\ [\lambda y]_1 & \cdots & [\lambda y]_N \\ y_1 & \cdots & y_N \end{bmatrix} \mathcal{T}^T \qquad (10.66)$$

The right factor $\mathcal{T}^T$ does not affect the observability subspace that is always extracted from a left matrix factor and that is the quantity of primary interest in subspace model identification.

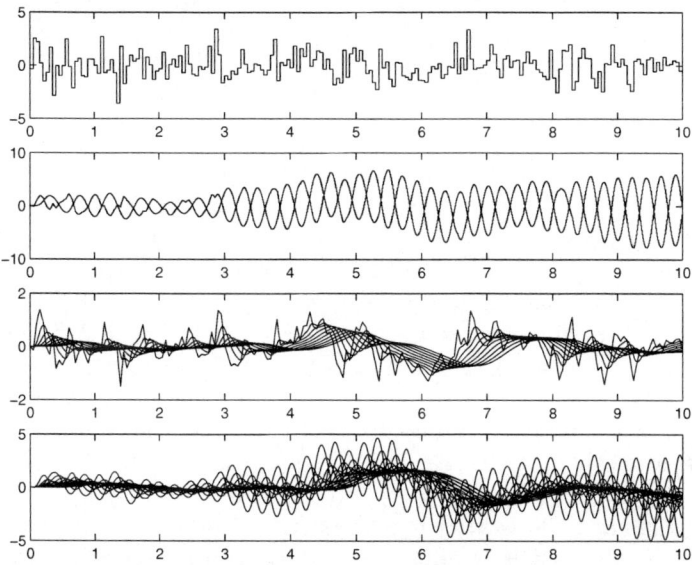

Fig. 10.2. Input-output data (*upper two graphs*) and filter data used for identification with sampling period $h = 0.01$, filter order $i = 5$ and operator time constant $\tau = 0.05$

Remarks on Algorithm 3—Subspace Correlation Method

The subspace correlation method is similar to Algorithm 2 but differs in the linear dependencies

$$\mathcal{Y}_N \mathcal{Z}_N^T = \Gamma_x \mathcal{X}_N \mathcal{Z}_N^T + \Gamma_u \mathcal{U}_N \mathcal{Z}_N^T + \Gamma_w \mathcal{W}_N \mathcal{Z}_N^T, \quad \mathcal{Y}_N \mathcal{Z}_N^T \in \mathbb{R}^{pi \times mj} \quad (10.67)$$

The left matrix factor extracted in estimation of observability subspace is not affected by the right multiplication of $\mathcal{Z}_N^T$. However, the algorithm output is not identical to that of Algorithm 2 due to the change of relative magnitude of the disturbance term as a result of the right multiplication. Another property is the reduction of the matrix column dimension of the data matrix applied QR-factorisation.

When input and disturbance are uncorrelated, this algorithm serves to reduce disturbance-related bias in parameter estimates. Statistical properties are analysed in greater detail below.

10.3.2 Numerical Example

The algorithms were applied to $N = 1000$ samples of input–output data generated by simulation of the linear system

$$\frac{dx}{dt} = \begin{bmatrix} 0 & 0 & 100.0 \\ 0 & -0.10 & -100.0 \\ -1.00 & 1.00 & 0 \end{bmatrix} x(t) + \begin{bmatrix} 10.0 \\ 0 \\ 0 \end{bmatrix} u(t) \quad (10.68)$$

$$y(t) = \begin{bmatrix} 1 & 0 & 0 \\ 0 & 1 & 0 \end{bmatrix} x(t) + v(t) \quad (10.69)$$

with input of variance $\sigma_u^2 = 1$ and a zero-mean stochastic disturbance v of variance σ_v^2; see input–output data in Figure 10.2. A third-order model was identified with very good accuracy for purely deterministic data ($\sigma_v^2 = 0$) and with good accuracy for $\sigma_v^2 = 0.01$; see transfer-function properties (Figure 10.3) and prediction performance (Figure 10.4). The influence of the choices of algorithmic parameters (number of block rows i or r and operator time constant τ) on relative prediction error ($\|\varepsilon\|_2/\|y\|_2$) and parameter error as measured by the gap metric are found in Figure 10.5. The identification was considered to be failing for a relative prediction error norm of value larger than one. Figure 10.5 has been drawn accordingly without representing a relative error larger than one, thus showing the effective range of the choice of τ and i. This figure also serves to illustrate sensitivity to stochastic disturbance and sensitivity to the choice of the free algorithm parameters (operator time constant τ and number of block rows i or r). The level surfaces indicate that τ may be chosen in a suitable range over, perhaps, two orders of magnitude for Alg. 2–3 and one order of magnitude for Alg. 1; see Figure 10.5 that includes contours of level surfaces, the central part corresponding to 1% error with degradation for inappropriate values of τ and i.

Another application of the realisation algorithm (Algorithm 1) to experimental impulse-response data obtained as ultrasonic echo data for object identification detection in robotic environments has proved successful; see [21].

10.4 Statistical Model Validation

Statistical model validation accompanies parameter estimation to provide confidence in a model obtained. An important aspect of statistical model validation is evaluation of the mismatch between input–output properties of a model and data. Statistical hypothesis tests applied to the autocorrelation of residuals as well as cross correlation between residuals and input are instrumental in such model validation, partially relying upon the algorithmic property of $\mathcal{U}_N^\perp$ that

$$\mathcal{Y}_N \mathcal{U}_N^\perp = (\Gamma_x \mathcal{X}_N + \Gamma_u \mathcal{U}_N + \Gamma_w \mathcal{W}_N)\mathcal{U}_N^\perp$$
$$= \Gamma_x \mathcal{X}_N \mathcal{U}_N^\perp + \Gamma_w \mathcal{W}_N \mathcal{U}_N^\perp \quad (10.70)$$
$$\mathsf{E}\{\mathcal{Y}_N \mathcal{U}_N^\perp\} = \Gamma_x \mathcal{X}_N \mathcal{U}_N^\perp + \Gamma_w \mathsf{E}\{\mathcal{W}_N \mathcal{U}_N^\perp\} \quad (10.71)$$

where $\mathcal{U}_N \mathcal{U}_N^\perp = 0$ by construction (*i.e.*, by the projection property of the QR-factorisation of (10.47)) whereas statistical properties of $\mathsf{E}\{\mathcal{W}_N \mathcal{U}_N^\perp\}$ are more

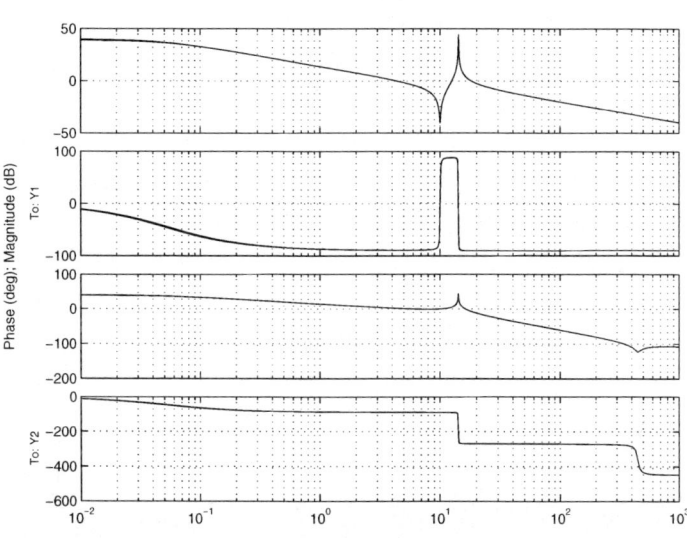

Fig. 10.3. Transfer function (*solid*) and estimate (*dashed*) using a third-order model with sampling period $h = 0.01$, filter order $i = 5$ and operator time constant $\tau = 0.05$ for $N = 1000$ samples of data with $\sigma_v^2 = 0.01$

difficult to evaluate also under assumptions of uncorrelated disturbances and control inputs. In the case of uncorrelated disturbance and input, multiplication of the right factor $\mathcal{Z}_N^T$ before the QR-factorisation in Algorithm 3 serves to reduce the disturbance-related bias of parameter estimates as

$$\begin{aligned}
(\mathcal{Y}_N \mathcal{Z}_N^T)(\mathcal{U}_N \mathcal{Z}_N^T)^\perp &= (\Gamma_x \mathcal{X}_N \mathcal{Z}_N^T + \Gamma_u \mathcal{U}_N \mathcal{Z}_N^T + \Gamma_w \mathcal{W}_N \mathcal{Z}_N^T)(\mathcal{U}_N \mathcal{Z}_N^T)^\perp \\
&= \Gamma_x (\mathcal{X}_N \mathcal{Z}_N^T)(\mathcal{U}_N \mathcal{Z}_N^T)^\perp + \Gamma_w (\mathcal{W}_N \mathcal{Z}_N^T)(\mathcal{U}_N \mathcal{Z}_N^T)^\perp \\
\mathsf{E}\{(\mathcal{Y}_N \mathcal{Z}_N^T)(\mathcal{U}_N \mathcal{Z}_N^T)^\perp\} &= \Gamma_x (\mathcal{X}_N \mathcal{Z}_N^T)(\mathcal{U}_N \mathcal{Z}_N^T)^\perp \\
&\quad + \Gamma_w \mathsf{E}\{(\mathcal{W}_N \mathcal{Z}_N^T)(\mathcal{U}_N \mathcal{Z}_N^T)^\perp\}
\end{aligned} \quad (10.72)$$

By the correlation properties of input and disturbance, the last term tends to be small, similar to the spectrum analysis and the instrumental variable method of identification. The consistency properties of this algorithm will be analysed in detail in future work.

Model Misfit Evaluation

Identification according to Algorithms 1–3 gives the model

$$\begin{bmatrix} \widehat{\xi} \\ y \end{bmatrix} = \begin{bmatrix} \widehat{A}_\lambda & \widehat{B}_\lambda \\ \widehat{C} & \widehat{D} \end{bmatrix} \begin{bmatrix} \widehat{x} \\ u \end{bmatrix}, \qquad \widehat{x}(t) = [\lambda \widehat{\xi}](t) \qquad (10.73)$$

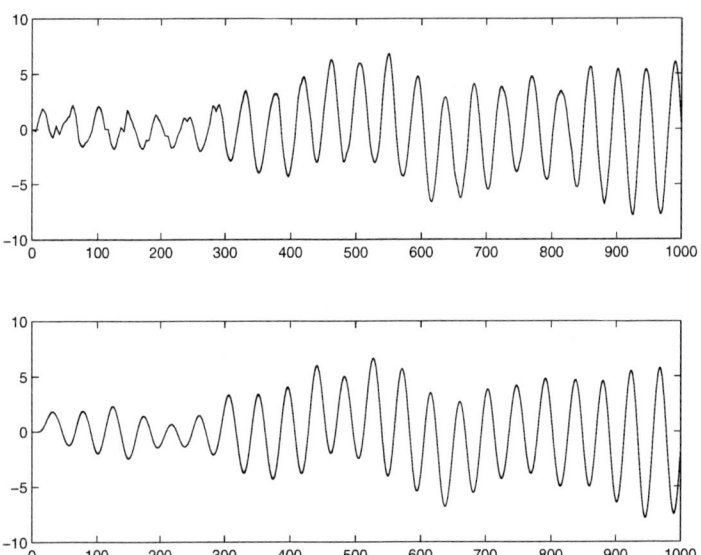

Fig. 10.4. Output data (*solid*) and estimate (*dashed*) using Alg. 2 and a third-order model with sampling period $h = 0.01$, filter order $i = 5$ and operator time constant $\tau = 0.05$ for $N = 1000$ samples of data with $\sigma_v^2 = 0.01$

A reconstruction $\widehat{r}$ of the state x for some matrix K such that $\widehat{A} - K\widehat{C}$ is stable (*i.e.*, Re $\lambda < 0$) can be done as

$$\begin{bmatrix} \dot{\widehat{x}} \\ \widehat{y} \end{bmatrix} = \begin{bmatrix} \widehat{A} - K\widehat{C} & \widehat{B} - K\widehat{D} \\ \widehat{C} & \widehat{D} \end{bmatrix} \begin{bmatrix} \widehat{x} \\ u \end{bmatrix} + \begin{bmatrix} K \\ 0 \end{bmatrix} y \qquad (10.74)$$

Model-error dynamics of $\widetilde{x} = x - \widehat{x}$ and $\varepsilon = y - \widehat{y}$

$$\begin{bmatrix} \dot{\widetilde{x}} \\ \varepsilon \end{bmatrix} = \begin{bmatrix} A - KC & B - KD \\ C & D \end{bmatrix} \begin{bmatrix} \widetilde{x} \\ 0 \end{bmatrix}$$

$$+ \begin{bmatrix} \widetilde{A} - K\widetilde{C} & \widetilde{B} - K\widetilde{C} \\ \widetilde{C} & \widetilde{D} \end{bmatrix} \begin{bmatrix} \widehat{x} \\ u \end{bmatrix} + \begin{bmatrix} I & -K \\ 0 & I \end{bmatrix} \begin{bmatrix} v \\ e \end{bmatrix} \qquad (10.75)$$

The stochastic realisation problem can be approached by Kalman filter theory and covariance-matrix factorisation ("spectral factorisation") [2, 8], and provided that a continuous-time Riccati equation can be solved to find an optimal K, one finds that the model mismatch can be expressed by either of the spectral factors

$$\varepsilon(s) = C(sI - A)^{-1}V(s) + E(s) \qquad (10.76)$$
$$\varepsilon(s) = (C(sI - A)^{-1}K + I)W(s) = H(s)W(s) \qquad (10.77)$$

where $\varepsilon(s)$, $V(s)$, $E(s)$, $W(s)$ are the Laplace transforms of the residuals, disturbance and innovations processes, respectively. The discrete-time counterpart is

$$\varepsilon(s) = C(zI - A_z)^{-1}V(z) + E(z) \tag{10.78}$$
$$\varepsilon(z) = (C(zI - A_z)^{-1}K_z + I)W(z) = H(z)W(z) \tag{10.79}$$

To solve for identification residuals it is suitable to use the transfer operator inverses

$$H^{-1}(s) = -C(sI - (A - KC))^{-1}K + I \tag{10.80}$$
$$= -C(I - (A_\lambda - K_\lambda C)\lambda)^{-1}K_\lambda \lambda + I \tag{10.81}$$
$$H^{-1}(z) = -C(zI - (A_z - K_z C))^{-1}K_z + I \tag{10.82}$$

For nominal system parameter matrices A, B, C, D and a solution K and $v = Ke = Kw$ from the Riccati equation of the Kalman filter, one would have

$$\begin{bmatrix} \dot{\tilde{x}} \\ \varepsilon \end{bmatrix} = \begin{bmatrix} A - KC & B - KD \\ -C & -D \end{bmatrix} \begin{bmatrix} \tilde{x} \\ 0 \end{bmatrix} + \begin{bmatrix} 0 \\ I \end{bmatrix} w \tag{10.83}$$

so that the output ε reproduce w of Σ except for a transient arising from the initial condition of $\tilde{x}(t_0)$. However, as no covariance data are *a priori* known and as the system identification including its validation procedure is assumed to utilise discrete-time data, it is generally necessary to resort to the residual realisation algorithm

$$\begin{bmatrix} \widehat{x}_{k+1} \\ \varepsilon_k \end{bmatrix} = \begin{bmatrix} \widehat{A}_z - K_z \widehat{C} & \widehat{B}_z - K_z \widehat{D} \\ \widehat{C} & \widehat{D} \end{bmatrix} \begin{bmatrix} \widehat{x}_k \\ u_k \end{bmatrix} + \begin{bmatrix} K_z \\ I \end{bmatrix} y \tag{10.84}$$

Reformulation of the Riccati equation, see [13], is

$$\begin{bmatrix} I_n & K \\ 0 & I_p \end{bmatrix} \begin{bmatrix} S & 0 \\ 0 & R \end{bmatrix} \begin{bmatrix} I_n & K \\ 0 & I_p \end{bmatrix}^T = \begin{bmatrix} A & \beta \\ C & \delta \end{bmatrix} \begin{bmatrix} S & 0 \\ 0 & I_q \end{bmatrix} \begin{bmatrix} A_z & \beta \\ C & \delta \end{bmatrix}^T \tag{10.85}$$

where the full-rank matrices β, δ arise from the factorisation

$$Q = \begin{bmatrix} \beta \\ \delta \end{bmatrix} \begin{bmatrix} \beta \\ \delta \end{bmatrix}^T, \quad \beta \in \mathbb{R}^{n \times q}, \quad \delta \in \mathbb{R}^{p \times q} \tag{10.86}$$

and where (10.85) represents factorisation of the covariance matrix of the variables

$$\begin{bmatrix} I_n & K_z \\ 0 & I_p \end{bmatrix} \begin{bmatrix} \tilde{x}_k \\ w_k \end{bmatrix}, \quad \mathsf{E}\{w_k w_k^T\} = R \in \mathbb{R}^{p \times p}, \tag{10.87}$$

$$\begin{bmatrix} A_z & \beta \\ C & \delta \end{bmatrix} \begin{bmatrix} \tilde{x}_k \\ \omega_k \end{bmatrix}, \quad \mathsf{E}\{\omega_k \omega_k^T\} = I_q \in \mathbb{R}^{q \times q} \tag{10.88}$$

Then, use of the full-rank matrices β, δ of (10.85) suggests that the stochastic state-space model be provided as

$$x_{k+1} = A_z x_k + B_z u_k + \beta w_k$$
$$y_k = C x_k + D u_k + \delta w_k$$
$$z_k = \delta^\dagger y_k = C_1 x_k + D_1 u_k + w_k, \qquad z_k, w_k \in \mathbb{R}^q \qquad (10.89)$$

with a matrix $\delta^\dagger$ chosen as the pseudo-inverse of δ and with

$$\delta^\dagger \delta = I_q, \quad C_1 = \delta^\dagger C, \quad D_1 = \delta^\dagger D \qquad (10.90)$$

An innovations-like model pseudoinverse is provided as

$$\begin{bmatrix} \widehat{x}_{k+1} \\ \epsilon_k \end{bmatrix} = \begin{bmatrix} A_z - \beta \delta^\dagger C & B_z - \beta \delta^\dagger D \\ \delta^\dagger C & \delta^\dagger D \end{bmatrix} \begin{bmatrix} \widehat{x}_k \\ u_k \end{bmatrix} + \begin{bmatrix} \beta \\ I_q \end{bmatrix} \delta^\dagger y_k \qquad (10.91)$$

where A_z, B_z are discrete-time versions of A and B, respectively, and with $\beta \delta^\dagger$ for rank-deficient covariance matrices Q replacing the K_z of the standard Kalman filter. Then, the output $\{\epsilon_k\}$ reproduces the rank-deficient innovations sequence.

10.5 Discussion

This chapter has treated the problem of continuous-time system identification based on discrete-time data and provides a framework with algorithms presented in preliminary forms in [16, 21] thereby extending subspace model identification to continuous-time models. We have provided both subspace-based algorithms and realisation-based algorithms with application both in the time domain and in the frequency domain. Whereas the first time-domain algorithms were presented in [16, 23], subspace-based frequency-domain algorithms were previously presented [33, 35]. Several issues remain open and we cannot claim to have any complete treatment. The accuracy of estimates, effects of stochastic disturbance, performance comparison and robustness of algorithms—i.e., algorithmic effects and behaviour when data cannot be generated by a model in the model class—need further attention; see [6, 27, 39–41] for discussion on these issues for the discrete-time case. As for implementation and application issues, see [5, 15, 17]; [10, 11, 32]; [37] and [30].

A relevant question is, of course, how general is the choice λ and if it can, for instance, be replaced by some other bijective mapping

$$\mu = \frac{bs + a}{s + a}, \quad b \in \mathbb{R}, \quad a \in \mathbb{R}^+, \quad s = \frac{a\mu - a}{b - \mu} \qquad (10.92)$$

with the Laplace-transformed linear model

10 Continuous-time Subspace Identification

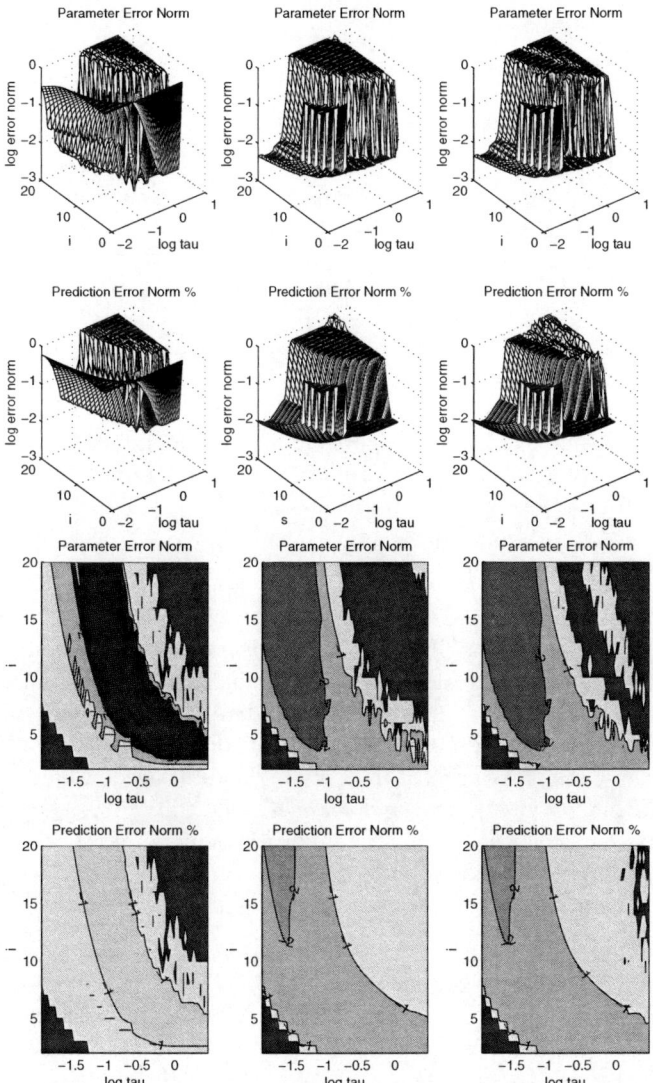

Fig. 10.5. Relative prediction error norm $\|\varepsilon\|_2/\|y\|_2$ and parameter error norm as measured vs. choices of the number of block columns block and operator time constant for Alg. 1 (*left*), Alg. 2 (*middle*), and Alg. 3 (*right*). Level surfaces (*diagram below*) and magnitude plot (*upper diagram*) using a third-order model with sampling period $h = 0.01$ for $N = 1000$ samples of data with $\sigma_v^2 = 0.01$ illustrate algorithm robustness and degradation properties for inappropriate τ and i.

$$\begin{bmatrix} sX \\ Y \end{bmatrix} = \begin{bmatrix} A & B \\ C & D \end{bmatrix} \begin{bmatrix} X \\ U \end{bmatrix} \qquad (10.93)$$

and by the operator transformation

$$\begin{bmatrix} \mu X \\ Y \end{bmatrix} = \begin{bmatrix} (aI + A)^{-1}(aI + Ab) & (aI + A)^{-1}Bb & -(aI + A)^{-1}B \\ C & D & 0 \end{bmatrix} \begin{bmatrix} X \\ U \\ \mu U \end{bmatrix}$$

Obviously, such an operator transformation entails a non-linear parameter transformation with an inverse

$$\widehat{A} = a(I - \widehat{A}_\lambda)(\widehat{A}_\lambda - bI)^{-1} \qquad (10.94)$$

which, of course, may be error-prone or otherwise sensitive due to singularities or poor numerical properties of the matrix inverse. By comparison, a model transformation using λ is linear, simple and does not exhibit such parameter-matrix singularities, a circumstance that motivates the attention given the favourable properties of this transformation. Actually, further studies to cover other linear fractional transformations are in progress [16] including advice on the choice of the additional parameters involved.

We have considered the problem of finding appropriate stochastic realisation to accompany estimated input–output models in the case of multi-input, multi-output subspace model identification. The case considered includes the problem of rank-deficient residual covariance matrices, a case that is encountered in applications with mixed stochastic-deterministic input–output properties as well as for cases where outputs are linearly dependent [41]. The inverse of the output covariance matrix is generally needed both for formulation of an innovations model and for a Kalman filter [25, 38, 42]. Our approach has been the formulation of an innovations model for the rank-deficient model output that generalises previously used methods of stochastic realisation [7, 9, 24, 29, 31].

The modified pseudo-inverse of (10.91) provides means to evaluate a residual sequence from the mismatch between an identified continuous-time model and discrete-time data in such a way that standard statistical validation tests can be applied [19]. Such statistical tests include: autocorrelation test of residual sequence $\{\varepsilon_k\}$; cross-correlation test of input $\{u_k\}$ and residual sequence $\{\varepsilon_k\}$; test of normal distribution (zero crossings, distribution, skewness, kurtosis, etc.).

10.6 Conclusions

This chapter has treated the problem of continuous-time system identification based on discrete-time data and provides a framework with algorithms presented in preliminary forms in [16,21]. The methodology involves a continuous-time operator translation [19, 20], permitting an algebraic reformulation and the use of subspace and realisation algorithms. We have provided subspace-based algorithms as well as realisation-based algorithms with application both to time-domain and to frequency-domain data. Thus, the algorithms and the theory presented here provide extensions both of the continuous-time identification and of subspace model identification.

A favourable property is the following. Whereas the model obtained is a continuous-time model, statistical tests can proceed in a manner that is standard for discrete-time models [19]. Conversely, as validation data are generally available as discrete-time data, it is desirable to provide means for validation of continuous-time models to available data.

Acknowledgement

This research was supported by the Swedish Research Council under the grants 2005-4182 and 2006-5243.

References

1. H. Akaike. Markovian representation of stochastic processes by canonical variables. *SIAM Journal of Control*, 13:162–173, 1975.
2. B.D.O. Anderson and P.J. Moylan. Spectral factorization of a finite-dimensional nonstationary matrix covariance. *IEEE Transactions on Automatic Control*, AC-19(6):680–692, 1974.
3. D.S. Bayard. An algorithm for state-space frequency domain identification without windowing distortions. *IEEE Transactions on Automatic Control*, 39(9):1880–1885, 1994.
4. S. Bigi, T. Söderström, and B. Carlsson. An IV-scheme for estimating continuous-time stochastic models from discrete-time data. *10th IFAC Symposium on System Identification (SYSID'94)*, volume 3, pages 645–650, Copenhagen, Denmark, 1994.
5. C.T. Chou, M. Verhaegen, and R. Johansson. Continuous-time identification of SISO systems using Laguerre functions. *IEEE Transactions on Signal Processing*, 47:349–362, 1999.
6. M. Deistler, K. Peternell, and W. Scherrer. Consistency and relative efficiency of subspace methods. *Automatica*, 31(12):1865–1875, 1995.
7. U.B. Desai and D. Pal. A realisation approach to stochastic model reduction and balanced stochastic realisations. *IEEE Conference on Decision and Control (CDC'1982)*, pages 1105–1112, Orlando, FL, USA, 1982.
8. B.W. Dickinson, T. Kailath, and M. Morf. Canonical matrix fraction and state-space description for deterministic and stochastic linear systems. *IEEE Transactions on Automatic Control*, AC-19:656–667, 1974.
9. P. Faurre. Stochastic realisation algorithms. In R.K. Mehra and D. Lainiotis (eds), *System identification: Advances and Case Studies*. Academic Press, New York, USA, 1976.

10. H. Garnier, M. Gilson, and E. Huselstein. Developments for the Matlab CONTSID Toolbox. *13th IFAC Symposium on System Identification (SYSID'2003)*, Rotterdam, The Netherlands, 2003.
11. H. Garnier, M. Mensler, and A. Richard. Continuous-time model identification from sampled data: Implementation issues and performance evaluation. *International Journal of Control*, 76(13):1337–1357, 2003.
12. R. Guidorzi. Canonical structures in the identification of multivariable systems. *Automatica*, 11:361–374, 1975.
13. P. Hagander and A. Hansson. How to solve singular discrete-time Riccati equations. *13th IFAC World Congress*, pages 313–318, San Francisco, CA, USA, July 1996.
14. E.J. Hannan and M. Deistler. *The Statistical Theory of Linear Systems*. Wiley, New York, 1988.
15. B.R.J. Haverkamp, M. Verhaegen, C.T. Chou, and R. Johansson. Tuning of the continuous-time Kalman filter from sampled data. *IEEE American Control Conference (ACC'99)*, pages 3895–3899, San Diego, CA, USA, June 1999.
16. B.R.J. Haverkamp, C.T. Chou, M. Verhaegen, and R. Johansson. Identification of continuous-time MIMO state space models from sampled data, in the presence of process and measurement noise. *36th IEEE Conference on Decision and Control (CDC'96)*, pages 1539–1544, Kobe, Japan, December 1996.
17. L.R.J. Haverkamp. *State Space Identification—Theory and Practice*. PhD thesis, T. U. Delft, Delft, The Netherlands, 2001.
18. B.L. Ho and R.E. Kalman. Effective construction of linear state-variable models from input/output functions. *Regelungstechnik*, 14:545–548, 1966.
19. R. Johansson. *System Modeling and Identification*. Prentice Hall, Englewood Cliffs, NJ, USA, 1993.
20. R. Johansson. Identification of continuous-time models. *IEEE Transactions on Signal Processing*, 42(4):887–897, 1994.
21. R. Johansson and G. Lindstedt. An algorithm for continuous-time state space identification. *34th IEEE Conference on Decision and Control (CDC'95)*, pages 721–722, New Orleans, LA, USA, 1995.
22. R. Johansson, M. Verhaegen, and C.T. Chou. Stochastic theory of continuous-time state-space identification. *37th IEEE Conference on Decision and Control (CDC'97)*, pages 1866–1871, San Diego, CA, USA, 1997.
23. R. Johansson, M. Verhaegen, and C.T. Chou. Stochastic theory of continuous-time state-space identification. *IEEE Transactions on Signal Processing*, 47:41–51, January 1999.
24. R. Johansson, M. Verhaegen, C.T. Chou, and A. Robertsson. Residual models and stochastic realisation in state-space identification. *International Journal of Control*, 74(10):988–995, 2001.
25. J.N. Juang. *Applied System Identification*. Prentice Hall, Englewood Cliffs, NJ, USA, 1994.
26. J.N. Juang and R.S. Pappa. An eigensystem realisation algorithm for modal parameter identification and model reduction. *Journal of Guidance, Control and Dynamics*, 8:620–627, 1985.
27. T. Katayama. *Subspace Methods for System Identification*. Springer-Verlag, London, UK, 2005.
28. S.Y. Kung. A new identification and model reduction algorithm via singular value decomposition. *12th Asilomar Conference on Circuits, Systems and Computers*, pages 705–714, Pacific Grove, CA, USA, 1978.

29. W. Larimore. Canonical variate analysis in identification, filtering and adaptive control. *29th IEEE Conference Decision and Control (CDC'90)*, pages 596–604, Hawaii, USA, 1990.
30. W. Li, H. Raghavan, and S. Shah. Subspace identification of continuous-time models for process fault detection and isolation. *Journal of Process Control*, 13(5):407–421, 2003.
31. A. Lindquist and G. Picci. Realisation theory for multivariate stationary Gaussian processes. *SIAM Journal of Control and Optimisation*, 23(6):809–857, 1985.
32. K. Mahata and H. Garnier. Identification of continuous-time errors-in-variables models. *Automatica*, 42(9):1470–1490, 2006.
33. T. McKelvey and H. Akçay. An efficient frequency domain state-space identification algorithm. *33rd IEEE Conference on Decision and Control (CDC'1994)*, pages 3359–3364, 1994.
34. T. McKelvey and H. Akçay. An efficient frequency domain state-space identification algorithm: Robustness and stochastic analysis. *33rd IEEE Conference on Decision and Control (CDC'1994)*, pages 3348–3353, 1994.
35. T. McKelvey, H. Akçay, and L. Ljung. Subspace-based multivariable system identification from frequence response data. *IEEE Transactions on Automatic Control*, 41:960–979, 1996.
36. M. Moonen, B. de Moor, L. Vandenberghe, and J. Vandewalle. On- and off-line identification of linear state-space models. *International Journal of Control*, 49:219–232, 1989.
37. A. Ohsumi, K. Kameyama, and K.-I. Yamaguchi. Subspace identification for continuous-time stochastic systems via distribution-based approach. *Automatica*, 38(1):63–79, 2002.
38. P. Van Overschee and B. De Moor. N4SID: subspace algorithm for the identification of combined deterministic-stochastic systems. *Automatica*, 30:75–93, 1994.
39. K. Peternell, W. Scherrer, and M. Deistler. Statistical analysis of novel subspace identification methods. *Signal Processing*, 52(2):161–177, 1996.
40. T. Ribarits, M. Deistler, and B. Hanzon. On new parametrization methods for the estimation of linear state-space models. *International Journal of Adaptive Control and Signal Processing*, 18(9-10):717–743, 2004.
41. P. van Overschee and B. de Moor. *Subspace Identification for Linear Systems— Theory, Implementation, Applications*. Kluwer Academic Publishers, Boston-London-Dordrecht, 1996.
42. M. Verhaegen. Identification of the deterministic part of MIMO state space models given in innovation form from input–output data. *Special Issue on Statistical Signal Processing and Control, Automatica*, 30(1):61–74, 1994.
43. M. Verhaegen and P. Dewilde. Subspace model identification—Analysis of the elementary output-error state-space model identification algorithm. *International Journal of Control*, 56:1211–1241, 1992.
44. M. Verhaegen and P. Dewilde. Subspace model identification—The output-error state-space model identification class of algorithms. *International Journal of Control*, 56:1187–1210, 1992.
45. L. Wang. Continuous time model predictive control design using orthonormal functions. *International Journal of Control*, 74:1588–1600, 2001.
46. L. Wang. Discrete model predictive controller design using Laguerre function. *Journal Process of Control*, 14:131–142, 2004.

11

Process Parameter and Delay Estimation from Non-uniformly Sampled Data

Salim Ahmed, Biao Huang, and Sirish L. Shah

University of Alberta, Canada

11.1 Introduction

Time-delay estimation is an important part of system identification. In process industries, it is even more important to consider the delay because of its common occurrence and significant impact on limiting the achievable performance of control systems. However, both in continuous-time (CT) and discrete-time (DT) model identification, the development of time delay-estimation methods lags behind the advancement of the estimation techniques for other model parameters. For example, linear filter methods are commonly used for continuous-time identification and significant developments have taken place in this field over the last few decades, see, *e.g.*, [5,7,31,35,38,40]. In the linear filter approach, the most commonly used algorithm to estimate the time delay is based on a comprehensive search routine as used in [30,31,40] where process parameters are estimated for a set of time delays within a certain range and a predefined cost function is calculated for every set of estimated parameters corresponding to each delay term. Finally, the delay that gives the optimum value of the cost function is chosen. This procedure is computationally expensive especially for rapidly sampled data. Another popular approach is approximation of the delay by a polynomial or by a rational transfer function such as the Padé approximation as in [1] or by the use of the Laguerre expansion. Such an approach requires estimation of more parameters and an unacceptable approximation error may occur for systems having large delays [36]. Most of the methods to directly estimate the delay along with other model parameters are based on the step test [16,22,25,36], the so-called piecewise step test [23] or the pulse test [13].

An important issue in time-delay estimation is that in many methods the delay is expressed in terms of the number of sampling intervals. The problem arises when the sampling interval is not constant. For such a case, the time delay becomes time varying and most of the methods fail to estimate such a parameter.

Uniform sampling has been a standard assumption in system identification. Assumption of single-rate sampling of different variables has also been standard. These assumptions imply that the set of sensors under consideration are delivering measurements synchronously and at the same constant rate. However this ideal situation often remains unpractised in real plants. In process industries, the strategy of sampling may be different for different variables. For example, from the time and cost considerations, concentrations are less frequently measured than flow rates, temperatures and pressures. Also, some variables may be measured at constant sampling intervals, while others may be measured at different intervals. The latter is often the case when manual sampling or laboratory analysis is required. These practices result in non-uniformly sampled data matrices. Multi-rate data is one form of non-uniform data where different variables are sampled at different intervals with the less frequently sampled variables having sampling intervals as integer multiples of that of the most frequently sampled one. Another form of non-uniformity is data with missing elements where measurements of all variables are available at some time instants, but at others, measurements of only a few variables are available. In chemical processes, data can be missing for two basic reasons: failure in the measurement devices and errors in data management. The most common failures are sensor breakdown, measurement outside the range of the sensor, malfunction of data-acquisition system, energy blackout, interruption of transmission lines, *etc.* The common errors in data-management are wrong format in logged data, crashes in data-management software, data-storage errors, incorrect or missing time stamps particularly in manually sampled data and so on [14,15]. In robust analysis of data, observed values that lie far from the normal trend are considered as outliers and often discarded or treated as missing. Also, highly compressed data or unequal length batch data, which may not immediately appear as non-uniform data, can be treated as non-uniform. An extreme form of irregular data may be asynchronised data for which different variables are sampled at different time instants.

The problem of non-uniform data has been considered in discrete-time identification literature using the expectation maximisation (EM) algorithm, *e.g.*, in [17] for ARX models and in [11, 28, 29] for state-space models. Use of the lifting technique for identification from multi-rate data has been reported in a number of articles, *e.g.*, in [21,33]. In continuous-time identification literature, methods have been proposed for unevenly sampled data in [12,19] where the problem of non-uniform sampling is handled by adopting numerical algorithms suitable for the data type. Methods for frequency-domain identification from non-uniformly sampled data have been presented for continuous-time autoregressive (CAR) models in [8], for autoregressive moving average (CARMA) models in [9] and for output error (COE) models in [10]. A frequency-domain identification technique for continuous-time models from data with missing elements in both of the input and the output signals has been presented in [27]. In fact, continuous-time identification methods can handle the uneven sampling problem by their nature, provided that appropriate numerical

11 Parameter and Delay Estimation from Non-uniformly Sampled Data

techniques are used. However, the inherent assumption of the numerical methods on the inter-sample behaviour of the variables, typically assumed to be stepwise-constant, may introduce errors in the parameter estimates.

Generally, input variables of an identification exercise are the manipulated variables of the process and are available regularly and at a faster rate. On the other hand, some quality variables, such as product compositions that may require laboratory analysis, may be sampled at slower rates or may be missing at some time instants. More often, the quality variables are the outputs for the models to be identified.

In this chapter we first introduce a linear filter method for simultaneous estimation of parameters and delay. An algorithm for system identification from non-uniformly sampled data is presented next followed by evaluation of the proposed methods on simulated and experimental data.

11.2 Estimation of Parameters and Delay

The mathematical procedure of the linear filter method is detailed in this section. The necessary mathematical formulation is presented first with an example of a second-order plus time delay (SOPTD) model. The nth-order generalisation of the method is subsequently outlined.

11.2.1 Second-order Modelling

Consider a SOPTD[1] model of a single-input single-output (SISO) system described by the following differential equation

$$a_2 y^{(2)}(t) + a_1 y^{(1)}(t) + a_0 y(t) = b_0 u(t - \tau) + e(t) \tag{11.1}$$

$y^{(i)}$ and $u^{(i)}$ are ith-order time derivatives of y and u. $e(t)$ is the error term. Without loss of generality it can be assumed that $a_0 = 1$. So, the objective of this exercise is to derive an estimate of the parameter vector, $[a_2 \; a_1 \; b_0 \; \tau]^T$, from a given set of measurements of input, $u(t)$ and output, $y(t)$. For the sake of simplicity in presentation, we assume zero initial conditions, *i.e.*, the input and the output are initially at steady states. Parameter estimation in the presence of non-zero initial conditions is discussed in the next section. With this assumption applying Laplace transform on both sides of (11.1) we get

$$a_2 s^2 Y(s) + a_1 s Y(s) + Y(s) = b_0 e^{-\tau s} U(s) + E(s) \tag{11.2}$$

$Y(s)$, $U(s)$ and $E(s)$ are the Laplace transforms of $y(t)$, $u(t)$ and $e(t)$, respectively. Now, consider a causal filter described in the Laplace domain as $F(s)$.

---

[1] A similar procedure can be followed for a first-order plus time delay (FOPTD) model.

If we apply the filtering operation on both sides of (11.2) we end up with the formulation

$$a_2 s^2 F(s)Y(s) + a_1 s F(s)Y(s) + F(s)Y(s) = b_0 e^{-\tau s} F(s)U(s) + F(s)E(s) \tag{11.3}$$

Parameter estimation using this filtering approach has been used in the literature for over four decades; however, to estimate only the parameters but not the delay. The delay estimation is mathematically different from the estimation of other parameters because of the fact that the parameters appear explicitly in the model equation (11.1) while the delay appears implicitly. To be solvable using the idea of regression analysis, the time delay should appear as an explicit element in the parameter vector. To get the delay in the parameter vector it is necessary that it appears explicitly in the estimation equation.

To have an explicit appearance of the delay term in the estimation equation and get it as an element in the parameter vector, we introduce a linear filter method with a novel structure of the filter having a first-order integral dynamics along with a lag term. The order of the lag term is the same as that of the process denominator. This structure of the filter was first proposed by the authors in [4]. The filter transfer function $F(s)$ may have different forms [2]. In the ensuing discussion we adopt the following

$$F(s) = \frac{1}{sA(s)} \tag{11.4}$$

where, $A(s)$ is the denominator of the process transfer function. So for the process under consideration $A(s) = a_2 s^2 + a_1 s + 1$. This filter has been used in [3] for identification from step response. The reason to include an integrator in the filter is to generate an integration term of delayed input in the estimation equation. This integrated delayed input signal, which represents a certain area under the input curve, can be expressed by subtracting two subareas from the integrated input signal as shown in (11.15) and in Figure 11.1. By doing so, the delay, τ, becomes an explicit parameter in the estimation equation.

For the filter (11.4), the parameter estimation equation becomes

$$a_2 s \frac{Y(s)}{A(s)} + a_1 \frac{Y(s)}{A(s)} + \frac{Y(s)}{sA(s)} = b_0 e^{-\tau s} \frac{U(s)}{sA(s)} + \frac{E(s)}{sA(s)} \tag{11.5}$$

The filtered input can be expressed as

$$\frac{U(s)}{sA(s)} = \left[\frac{C(s)}{A(s)} + \frac{1}{s}\right] U(s) \tag{11.6}$$

where, $C(s) = -(a_2 s + a_1)$. Now using the notations

$$\underline{Y}(s) = \frac{Y(s)}{A(s)} \qquad Y_I(s) = \frac{Y(s)}{s} \tag{11.7}$$

11 Parameter and Delay Estimation from Non-uniformly Sampled Data

and adopting similar notations for $U(s)$, (11.5) can be written as

$$a_2 s \underline{Y}(s) + a_1 \underline{Y}(s) + \underline{Y}_I(s) = b_0 \left[C(s) \underline{U}(s) + U_I(s) \right] e^{-\tau s} + \underline{E}_I(s) \quad (11.8)$$

This equation can be rearranged in standard least squares form

$$\underline{Y}_I(s) = -a_2 s \underline{Y}(s) - a_1 \underline{Y}(s) + b_0 \underline{U}_C(s) e^{-\tau s} + b_0 U_I(s) e^{-\tau s} + \eta(s) \quad (11.9)$$

where, $U_C(s) = C(s)U(s)$ and $\eta(s) = \underline{E}_I(s)$. Taking the inverse Laplace transform, (11.9) can be written for any sampling instant $t = t_k$

$$\underline{y}_I(t_k) = -a_2 \underline{y}^{(1)}(t_k) - a_1 \underline{y}(t_k) + b_0 \underline{u}_C(t_k - \tau) + b_0 u_I(t_k - \tau) + \eta(t_k) \quad (11.10)$$

with

$$\underline{y}_I(t_k) = \mathcal{L}^{-1}\left[\frac{Y(s)}{s} \right] \quad (11.11)$$

$$\underline{u}_C(t_k - \tau) = \mathcal{L}^{-1}\left[C(s)\underline{U}(s) e^{-\tau s} \right] \quad (11.12)$$

$$u_I(t_k - \tau) = \mathcal{L}^{-1}\left[\frac{1}{s} U(s) e^{-\tau s} \right] \quad (11.13)$$

where, $\mathcal{L}^{-1}$ represents the inverse Laplace transform. The integrals of the input and the delayed input for any time $t = t_k$ are given by

$$u_I(t_k) = \int_0^{t_k} u(t) dt \quad (11.14)$$

$$u_I(t_k - \tau) = u_I(t_k) - \int_{t_k - \tau}^{t_k} [u(t) - u(t_k)] dt - u(t_k) \tau \quad (11.15)$$

Equation (11.15) can be presented graphically as in Figure 11.1. From the figure, it is seen that the integrated delayed input, $u_I(t_k - \tau)$, represents the area under the input signal up to time $(t_k - \tau)$, while the integrated input signal, $u_I(t_k)$, represents the area under the input curve up to time t_k. Also, the 2nd and 3rd terms on the right hand-side of (11.15) represent areas as shown by the legends in Figure 11.1. From the figure, it is seen that by subtracting these two areas from $u_I(t_k)$, we get $u_I(t_k - \tau)$.

Applying (11.15) in (11.10) and rearranging it to give a standard least squares form we get

$$\underline{y}_I(t_k) = -a_2 \underline{y}^{(1)}(t_k) - a_1 \underline{y}(t_k)$$
$$+ b_0 \left[\underline{u}_C(t_k - \tau) + u_I(t_k) - \int_{t_k - \tau}^{t_k} [u(t) - u(t_k)] dt \right]$$
$$- b_0 \tau u(t_k) + \eta(t_k) \quad (11.16)$$

Denoting

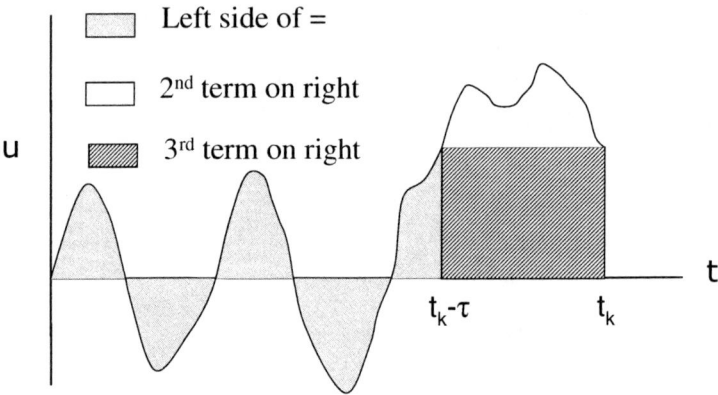

Fig. 11.1. Graphical representation of (11.15)

$$u_+(t_k - \tau) = \underline{u}_C(t_k - \tau) + u_I(t_k) - \int_{t_k-\tau}^{t_k} [u(t) - u(t_k)]dt \quad (11.17)$$

(11.16) can be written as

$$\underline{y}_I(t_k) = \begin{bmatrix} -\underline{y}^{(1)}(t_k) & \underline{y}(t_k) & u_+(t_k - \tau) & -u(t_k) \end{bmatrix} \begin{bmatrix} a_2 \\ a_1 \\ b_0 \\ b_0\tau \end{bmatrix} + \eta(t_k) \quad (11.18)$$

Or equivalently

$$\gamma(t_k) = \boldsymbol{\varphi}^T(t_k)\boldsymbol{\theta} + \eta(t_k) \quad (11.19)$$

where,

$$\gamma(t_k) = \underline{y}_I(t_k), \quad \boldsymbol{\varphi}(t_k) = \begin{bmatrix} -\underline{y}^{(1)}(t_k) \\ -\underline{y}(t_k) \\ u_+(t_k - \tau) \\ -u(t_k) \end{bmatrix}, \quad \boldsymbol{\theta} = \begin{bmatrix} a_2 \\ a_1 \\ b_0 \\ b_0\tau \end{bmatrix} \quad (11.20)$$

11.2.2 Higher-order Modelling

To describe the method for a model with numerator order n and denominator order m with $n > m > 0$, let us consider a linear single-input, single-output (SISO) system with time delay described by

$$\mathbf{a}_n \mathbf{y}^{(n)}(t) = \mathbf{b}_m \mathbf{u}^{(m)}(t - \tau) + e(t) \quad (11.21)$$

where,

11 Parameter and Delay Estimation from Non-uniformly Sampled Data

$$\mathbf{a}_n = [a_n \; a_{n-1} \cdots a_0] \in \mathbb{R}^{1 \times (n+1)}$$
$$\mathbf{b}_m = [b_m \; b_{m-1} \cdots b_0] \in \mathbb{R}^{1 \times (m+1)}$$
$$\mathbf{y}^{(n)}(t) = \left[y^{(n)}(t) \; y^{(n-1)}(t) \cdots y^{(0)}(t)\right]^T \in \mathbb{R}^{(n+1) \times 1}$$
$$\mathbf{u}^{(m)}(t - \tau) = \left[u^{(m)}(t - \tau) \cdots u^{(0)}(t - \tau)\right]^T \in \mathbb{R}^{(m+1) \times 1}$$

Applying Laplace transformation on both sides of (11.21), considering that both input and output are initially at rest, we can write

$$\mathbf{a}_n \mathbf{s}^n Y(s) = \mathbf{b}_m \mathbf{s}^m U(s) e^{-\tau s} + E(s) \tag{11.22}$$

where,

$$\mathbf{s}^n = \left[s^n \; s^{n-1} \cdots s^0\right]^T \in \mathbb{R}^{(n+1) \times 1} \tag{11.23}$$

If we apply the filtering operation on both sides of (11.22) we get

$$\mathbf{a}_n \mathbf{s}^n F(s) Y(s) = \mathbf{b}_m \mathbf{s}^m F(s) U(s) e^{-\tau s} + F(s) E(s) \tag{11.24}$$

Using the notations in (11.7), we can express (11.24) in terms of the filtered input and output as

$$\mathbf{a}_n \mathbf{s}_+^{n-1} \underline{Y}(s) = \mathbf{b}_m \mathbf{s}_+^{m-1} \underline{U}(s) e^{-\tau s} + \eta(s) \tag{11.25}$$

where the subscript $(\bullet_+)$ means that the $\mathbf{s}^{n-1}$ vector has been augmented by $\frac{1}{s}$, i.e.,

$$\mathbf{s}_+^{n-1} = \left[s^{n-1} \; s^{n-2} \cdots s^0 \; \frac{1}{s}\right] \tag{11.26}$$

Next we use the expression in (11.6). For nth-order models $C(s) = -(a_n s^{n-1} + a_{n-1} s^{n-2} + \cdots + a_1)$. Restructuring (11.25) by applying (11.6) gives a standard least squares form

$$\underline{Y}_I(s) = -\bar{\mathbf{a}}_n \mathbf{s}^{n-1} \underline{Y}(s) + \bar{\mathbf{b}}_m \mathbf{s}^{m-1} \underline{U}(s) e^{-\tau s}$$
$$+ b_0 \left[C(s) \underline{U}(s) + U_I(s)\right] e^{-\tau s} + \eta(s) \tag{11.27}$$

where,
$\bar{\mathbf{a}}_n$: $\mathbf{a}_n$ with its last column removed, $\bar{\mathbf{a}}_n \in \mathbb{R}^{1 \times n}$
$\bar{\mathbf{b}}_m$: $\mathbf{b}_m$ with its last column removed, $\bar{\mathbf{b}}_m \in \mathbb{R}^{1 \times m}$

Taking the inverse Laplace transform, we get the equivalent time-domain expression for any sampling instant $t = t_k$

$$\underline{y}_I(t_k) = -\bar{\mathbf{a}}_n \underline{\mathbf{y}}^{(n-1)}(t_k) + \bar{\mathbf{b}}_m \underline{\mathbf{u}}^{(m-1)}(t_k - \tau)$$
$$+ b_0 \left[\underline{u}_C(t_k - \tau) + u_I(t_k - \tau)\right] + \eta(t_k) \tag{11.28}$$

Applying (11.15) in (11.28) and re-arranging the equation we get

$$\underline{y}_I(t_k) = -\bar{\mathbf{a}}_n \mathbf{y}^{(n-1)}(t_k) + \mathbf{b}_m \underline{\mathbf{u}}_+^{(m-1)}(t_k - \tau) - b_0 u(t_k)\tau + \eta(t_k) \quad (11.29)$$

where,

$$\underline{\mathbf{u}}_+^{(m-1)}(t_k - \tau) = \begin{bmatrix} u^{(m-1)}(t_k - \tau) \\ \vdots \\ u(t_k - \tau) \\ u_+(t_k - \tau) \end{bmatrix} \quad (11.30)$$

$u_+(t_k - \tau)$ has been defined by (11.17). Equivalently, we can write

$$\gamma(t_k) = \varphi^T(t_k)\boldsymbol{\theta} + \eta(t_k) \quad (11.31)$$

where,

$$\gamma(t_k) = \underline{y}_I(t_k) \quad (11.32)$$

$$\varphi(t_k) = \begin{bmatrix} -\mathbf{y}^{(n-1)}(t_k) \\ \underline{\mathbf{u}}_+^{(m-1)}(t_k - \tau) \\ -u(t_k) \end{bmatrix} \quad (11.33)$$

$$\boldsymbol{\theta} = [\bar{\mathbf{a}}_n \ \mathbf{b}^m \ b_0 \tau]^T \quad (11.34)$$

11.2.3 Treatment of Initial Conditions

So far we have assumed zero initial conditions of both the input and output variables for the sake of simplicity in the presentation. However, the estimation equation can be formulated considering non-zero initial conditions of the output as shown in [4]. Initial conditions of both input and output have been considered in [2, 3]. Here, we outline the procedure for parameter estimation in the presence of non-zero initial conditions of the output.

When the output is not initially at a steady-state application of Laplace transformation on both sides of (11.21) gives

$$\mathbf{a}_n s^n Y(s) = \mathbf{b}_m s^m U(s)e^{-\tau s} + \mathbf{c}_{n-1} s^{n-1} + E(s) \quad (11.35)$$

where the elements of $\mathbf{c}_{n-1}$ capture the initial conditions of the output and are defined by

$$\mathbf{c}_{n-1} = [c_{n-1} \ c_{n-2} \cdots c_0] \in \mathbb{R}^{1 \times n} \quad (11.36)$$

$$c_{n-i} = \mathbf{h}_i \mathbf{y}^{n-1}(0), \quad i = 1 \cdots n \quad (11.37)$$

$$\mathbf{h}_i = [\mathbf{0}^{1 \times (n-i)} \ a_n \cdots a_{n-(i-1)}] \in \mathbb{R}^{1 \times n} \quad (11.38)$$

$$\mathbf{y}^{(n-1)}(0) = \left[y^{(n-1)}(0) \ y^{(n-2)}(0) \cdots y(0)\right]^T \in \mathbb{R}^{n \times 1} \quad (11.39)$$

If we apply the filtering operation on both sides of (11.35) we get

$$\mathbf{a}_n s^n F(s)Y(s) = \mathbf{b}_m s^m F(s)U(s)e^{-\tau s} + \mathbf{c}_{n-1} s^{n-1} F(s) + F(s)E(s) \quad (11.40)$$

It is straightforward to follow that the inclusion of the initial conditions does not change the mathematical derivations that follows (11.24) except that n additional terms appear in the estimation equation. Now, following the above derivation, for non-zero initial conditions we can write (11.29) as

$$\underline{y}_I(t_k) = -\bar{\mathbf{a}}_n \mathbf{y}^{(n-1)}(t_k) + \mathbf{b}_m \underline{\mathbf{u}}_+^{(m-1)}(t_k - \tau) - b_0 u(t_k)\tau$$
$$+ \mathbf{c}_{n-1} \mathbf{f}^{(n-1)}(t_k) + \eta(t_k) \qquad (11.41)$$

where,

$$\mathbf{f}^{(n-1)}(t_k) = \left[f^{(n-1)}(t_k) \ f^{(n-2)}(t_k) \cdots f(t_k) \right]^T \in \mathbb{R}^{n \times 1} \qquad (11.42)$$

$$f^{(i)}(t) = \mathcal{L}^{-1}[s^i F(s)] \qquad (11.43)$$

So for this case we can write

$$\gamma(t_k) = \boldsymbol{\varphi}^T(t_k)\boldsymbol{\theta} + \eta(t_k) \qquad (11.44)$$

with

$$\gamma(t_k) = \underline{y}_I(t_k) \qquad (11.45)$$

$$\boldsymbol{\varphi}(t_k) = \begin{bmatrix} -\mathbf{y}^{(n-1)}(t_k) \\ \underline{\mathbf{u}}_+^{(m-1)}(t_k - \tau) \\ -u(t_k) \\ \mathbf{f}^{(n-1)}(t_k) \end{bmatrix} \qquad (11.46)$$

$$\boldsymbol{\theta} = [\bar{\mathbf{a}}_n \ \mathbf{b}_m \ b_0\tau \ \mathbf{c}_{n-1}]^T \qquad (11.47)$$

11.2.4 Parameter Estimation

To estimate the parameter (11.19), (11.31) or (11.44) can be written for $t_k = t_{d+1}, t_{d+2} \cdots t_N$, where, $t_d \leq \tau < t_{d+1}$ and combined as

$$\boldsymbol{\Gamma} = \begin{bmatrix} \gamma(t_{d+1}) \\ \gamma(t_{d+2}) \\ \vdots \\ \gamma(t_N) \end{bmatrix}, \quad \boldsymbol{\Phi} = \begin{bmatrix} \boldsymbol{\varphi}^T(t_{d+1}) \\ \boldsymbol{\varphi}^T(t_{d+2}) \\ \vdots \\ \boldsymbol{\varphi}^T(t_N) \end{bmatrix} \qquad (11.48)$$

to give

$$\boldsymbol{\Gamma} = \boldsymbol{\Phi}\boldsymbol{\theta} + \boldsymbol{\eta} \qquad (11.49)$$

Solution of (11.49) gives the parameter vector $\boldsymbol{\theta}$. From $\boldsymbol{\theta}$ one can directly get $\bar{\mathbf{a}}_n$ and $\mathbf{b}_m$. τ is obtained as $\tau = \boldsymbol{\theta}(n+m+2)/\boldsymbol{\theta}(n+m+1)$. In the case where initial conditions are considered, to retrieve $\mathbf{y}^{(n-1)}(0)$ from $\mathbf{c}_{n-1}$, (11.37) can be written for $i = 1 \cdots n$ to give

$$(\mathbf{c}_{n-1})^T = \mathbf{H}\mathbf{y}^{(n-1)}(0) \tag{11.50}$$

where, $\mathbf{H} = [(\mathbf{h}_1)^T \ (\mathbf{h}_2)^T \cdots (\mathbf{h}_n)^T]^T \in \mathbb{R}^{n \times n}$. Finally

$$\mathbf{y}^{(n-1)}(0) = (\mathbf{H})^{-1}(\mathbf{c}_{n-1})^T \tag{11.51}$$

However, in estimating $\boldsymbol{\theta}$ by solving (11.49), there are two problems. First, we need to know $A(s)$ and τ, both of which are unknowns. This obvious problem can be solved by applying an iterative procedure that adaptively adjusts initial estimates of $A(s)$ and τ until they converge. Secondly, the least squares (LS) estimate of $\boldsymbol{\theta}$, obtained by minimising the sum of the squared errors and given by

$$\hat{\boldsymbol{\theta}}_{\text{LS}} = \left[\boldsymbol{\Phi}^T\boldsymbol{\Phi}\right]^{-1}\boldsymbol{\Phi}^T\boldsymbol{\Gamma} \tag{11.52}$$

is biased in the presence of general forms of measurement noise such as the coloured noise. Even if the measurement noise is assumed to be white with zero mean, the filtering operation makes it coloured. So, the LS solution is not unbiased even for a white measurement noise. To remove this bias we need another step. A popular bias-elimination procedure is to use the instrumental variable (IV) method. A bootstrap estimation of IV type where the instrumental variable is built from an auxiliary model [39] is considered here. The instrument vector in the case of steady initial conditions is given by

$$\boldsymbol{\zeta}(t_k) = \begin{bmatrix} \hat{\mathbf{y}}^{(n-1)}(t_k) \\ \mathbf{u}^{(m-1)}(t_k - \tau) \\ -u(t_k) \end{bmatrix} \tag{11.53}$$

where,

$$\hat{Y}(s) = \frac{B_{\text{LS}}(s)}{A_{\text{LS}}(s)} e^{-\tau_{\text{LS}} s} U(s) \tag{11.54}$$

$A(s) = \mathbf{a}_n s^n$, $B(s) = \mathbf{b}_m s^m$ and $A_{\text{LS}}, B_{\text{LS}}(s)$ being their least squares estimates. $\hat{y}(t) = \mathcal{L}^{-1}\left[\hat{Y}(s)\right]$. The IV-based bias-eliminated parameters are given by

$$\hat{\boldsymbol{\theta}}_{IV} = \left[\boldsymbol{\Psi}^T\boldsymbol{\Phi}\right]^{-1}\boldsymbol{\Psi}^T\boldsymbol{\Gamma} \tag{11.55}$$

where,

$$\boldsymbol{\Psi} = \begin{bmatrix} \boldsymbol{\zeta}^T(t_{d+1}) \\ \boldsymbol{\zeta}^T(t_{d+2}) \\ \vdots \\ \boldsymbol{\zeta}^T(t_N) \end{bmatrix} \tag{11.56}$$

The iterative identification algorithm for simultaneous estimation of the delay and other parameters from a uniform data set is summarised in Algorithm 1. Extensive simulation studies show that the iterative estimation converges monotonically except for non-minimum phase (NMP) processes. However, for NMP processes it always exhibits monotonic divergence. For such processes, we suggest to use the following procedure as presented in [4].

11 Parameter and Delay Estimation from Non-uniformly Sampled Data

Algorithm 1: Linear filter algorithm for simultaneous estimation of the delay and model parameters from uniformly sampled data.

Step 1-Initialisation: Choose the initial estimates $\hat{A}_0(s)$ and $\hat{\tau}_0$.
Step 2-LS step: Construct $\mathbf{\Gamma}$ and $\mathbf{\Phi}$ by replacing $A(s)$ and τ by $\hat{A}_0(s)$ and $\hat{\tau}_0$ and get the LS solution of $\boldsymbol{\theta}$

$$\hat{\boldsymbol{\theta}}_{\text{LS}} = (\mathbf{\Phi}^T \mathbf{\Phi})^{-1} \mathbf{\Phi}^T \mathbf{\Gamma} \tag{11.57}$$

Set $i = 1$. $\hat{\boldsymbol{\theta}}_1 = \hat{\boldsymbol{\theta}}_{\text{LS}}$. Get $\hat{A}_1(s)$, $\hat{B}_1(s)$ and $\hat{\tau}_1$ from $\hat{\boldsymbol{\theta}}_1$.
Step 3-IV step: $i = i+1$. Construct $\mathbf{\Gamma}$, $\mathbf{\Phi}$ and $\mathbf{\Psi}$ by replacing $A(s)$, $B(s)$ and τ by $\hat{A}_{i-1}(s)$, $\hat{B}_{i-1}(s)$ and $\hat{\tau}_{i-1}$ and get the IV solution of $\boldsymbol{\theta}$

$$\hat{\boldsymbol{\theta}}_i = (\mathbf{\Psi}^T \mathbf{\Phi})^{-1} \mathbf{\Psi}^T \mathbf{\Gamma} \tag{11.58}$$

Obtain $\hat{A}_i(s)$, $\hat{B}_i(s)$ and $\hat{\tau}_i$ from $\hat{\boldsymbol{\theta}}_i$ and repeat step 3 until $\hat{A}_i$ and $\hat{\tau}_i$ converge.
Step 4-Termination: When $\hat{A}_i$ and $\hat{\tau}_i$ converge, the corresponding $\hat{\boldsymbol{\theta}}_i$ is the final estimate of parameters.

11.2.5 Non-minimum Phase Processes

The iteration procedure described in the previous section is a fixed-point iteration scheme expressed as $\hat{\tau} = g(\tau) = \hat{\boldsymbol{\theta}}(n+m+2)/\hat{\boldsymbol{\theta}}(n+m+1)$, where, $\hat{\boldsymbol{\theta}}$ is given by (11.52) and (11.55) with $\mathbf{\Phi} = \mathbf{\Phi}(\tau)$ and $\mathbf{\Psi} = \mathbf{\Psi}(\tau)$. For minimum phase processes, $g(\tau)$ maps τ within the region of monotonic convergence while for NMP processes, it maps τ in the region of monotonic divergence. To make the diverging scheme converging, the following result is used.
If a fixed-point iteration scheme $\hat{x} = g(x)$ diverges monotonically, another scheme $\hat{x} = x + \frac{1}{r}[x - g(x)]$ with $r > 0$, will converge monotonically if $g(x)$ is bounded by the region $g(x) = x$ and $g(x) = (r+1)x + c$ where c is a constant satisfying that $g(x)$ passes through the fixed point.
Simulation results for a large number of process models show that for NMP processes $g(\tau)$ maps τ within the region $g(\tau) = \tau$ and $g(\tau) = (r+1)\tau + c$ with $r = 1$. Hence, expressing the estimation equation as $\hat{\tau} = \tau + [\tau - g(\tau)]$ will lead to convergence. So, if the diverging scheme gives

$$\hat{\tau}_{i+1}^d = g(\hat{\tau}_i) \tag{11.59}$$

To make the scheme converging, one needs to choose

$$\hat{\tau}_{i+1}^c = \hat{\tau}_i + [\hat{\tau}_i - g(\hat{\tau}_i)] = \hat{\tau}_i + [\hat{\tau}_i - \hat{\tau}_{i+1}^d] \tag{11.60}$$

We define $\Delta\hat{\tau}_i = \hat{\tau}_i - \hat{\tau}_{i+1}^d$ and for successive iteration for a value of $\hat{\tau}_i$, $\hat{\tau}_{i+1}$ is computed as

$$\hat{\tau}_{i+1} = \hat{\tau}_i + \Delta\hat{\tau}_i \tag{11.61}$$

The iteration steps otherwise remain the same.

11.2.6 Choice of $\hat{A}_0(s)$ and $\hat{\tau}_0$

The initiation of the iteration procedure involves choice of $\hat{A}_0(s)$ and $\hat{\tau}_0$. In theory, there is no constraint on the choice of $\hat{A}_0(s)$ except that the filter should not be unstable. Moreover, as the filter is updated in every step, the final estimate of the parameters is found not to be very sensitive to the initial choice. The proposed filter is the same as the filter used in the simplified refined instrumental variable method for continuous-time (SRIVC) models [40] except for the integral dynamics. Choice of $\hat{A}_0(s)$ for the SRIVC method with reference to the CAPTAIN toolbox has been discussed in [40]. Following the guidelines in [40], we suggest to choose $\hat{A}_0(s) = \frac{1}{(s+\lambda)^n}$ where λ is an estimate of the process cut-off frequency. Similarly, for $\hat{\tau}_0$ a choice based on process information would save computation. In cases where process information is unavailable we suggest choosing a small positive value for $\hat{\tau}_0$.

11.3 Identification from Non-uniformly Sampled Data

The linear filter method described in the previous section can be applied to non-uniformly sampled data provided that appropriate numerical techniques are used. However, the inherent assumption of the numerical methods on the inter-sample behaviour of the signals, typically step wise-constant, may introduce significant error in the estimates. This is true especially if the sampling intervals vary over a wide range or data are missing for long periods. This section details an iterative algorithm for more accurate estimation of the parameters and delay from such data sets.

By non-uniform data we refer to data sets where inputs and outputs are sampled at irregular intervals. However, we assume that the time instants at which the variables are sampled are the same, *i.e.*, the available data is synchronous. We also consider that measurements may be missing at some time instants. Regarding missing elements we assume that the inputs are available at all sampling instants while the outputs are available at some instants but missing at others. This is a more general form of synchronised data. Multi-rate data can be considered as a special form of this non-uniform structure.

11.3.1 The Iterative Prediction Algorithm

The idea of iterative prediction is used for identification from non-uniform and missing data. In the initialisation step of the iterative procedure, a so-called input-only model is used. Examples of this are the finite impulse response (FIR) model and basis-function models. A distinguishing feature of such models is that the current output is expressed in terms of only current and previous inputs. So, the parameter estimation equation can be formulated only at those time instants when the output is available. The estimated input-only model is then used to predict the missing values to get a complete data set. Next, this complete data set is used to estimate the parameters of the continuous-time transfer function model using the procedure described in the previous section. In the next step the estimated transfer function model is used to predict the missing outputs. This procedure is carried on iteratively by replacing the prediction from a previous model by that from the current one until some convergence criteria are met. The iteration algorithm is presented graphically in Figure 11.2 and the necessary steps of the iteration procedure are detailed below.

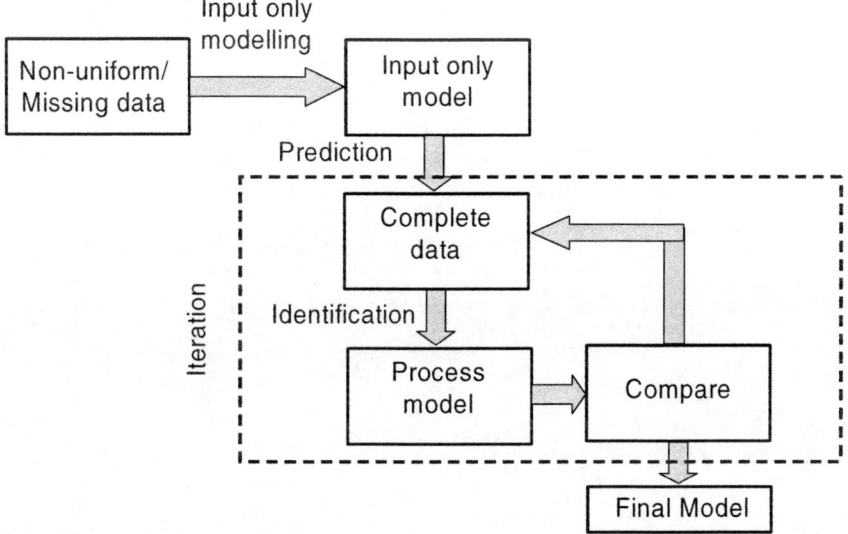

Fig. 11.2. Graphical representation of the iterative prediction algorithm for identification from non-uniformly sampled data

11.3.2 Input-only Modelling Using Basis-function Model

For the purpose of initial prediction, we consider a model that expresses the output in terms of only the past inputs. A number of such input-only approaches, both in discrete time and continuous time, are available in the literature. For reasons of parsimony, different basis-function methods can be used to serve this purpose. In this work, we use one of the orthogonal basis-function models, namely a Laguerre polynomial model in continuous time for the initial prediction when the system response is non-oscillatory. For oscillating systems a two-parameter Kautz model is used. The following section details both of the model types.

Basis-function Models

The use of Laguerre functions in identification goes back to [37]. In the Laplace domain, the Laguerre functions are given by [20]

$$L_j(s,\kappa) = \frac{\sqrt{2\kappa}}{(s+\kappa)} \left[\frac{(s-\kappa)}{(s+\kappa)}\right]^{j-1}, \quad j=1,2\cdots \quad \kappa > 0 \quad (11.62)$$

where κ is the parameter of the Laguerre model to be specified by the user. Kautz [18] introduced a set of orthonormal basis functions known as the Kautz functions. We will use here a special form of the Kautz functions presented in [32] and given by

$$K_{2j-1}(s,\beta,\nu) = \frac{\sqrt{2\beta}s}{s^2+\beta s+\nu} \left[\frac{s^2-\beta s+\nu}{s^2+\beta s+\nu}\right]^{j-1} \quad (11.63)$$

$$K_{2j}(s,\beta,\nu) = \frac{\sqrt{2\beta\nu}}{s^2+\beta s+\nu} \left[\frac{s^2-\beta s+\nu}{s^2+\beta s+\nu}\right]^{j-1} \quad (11.64)$$

$$\beta > 0, \quad \nu > 0, \quad j=1,2,\cdots$$

Let $z_i(t)$ be the output of the ith orthogonal function, with $u(t)$ as its input, i.e., for the Laguerre model

$$Z_i(s) = L_i(s)U(s) \quad (11.65)$$

and for the Kautz model

$$Z_i(s) = K_i(s)U(s) \quad (11.66)$$

where, $Z_i(s)$ and $U(s)$ represent the Laplace transform of $z_i(t)$ and $u(t)$, respectively. The output of a stable plant with input $u(t)$ can be approximated by a truncated pth-order Laguerre or Kautz model

$$y(t) = \sum_{i=1}^{p} z_i(t)\alpha_i \quad (11.67)$$

where, $\boldsymbol{\alpha} = [\alpha_1, \alpha_2 \cdots \alpha_p]^T$, is the parameter vector for the basis-function model. Theories and proofs of the convergence of the Laguerre model can be found in [24, 26, 34] and details of the properties of the two-parameter Kautz model can be found in [32].

We denote the time instants when the output is available by $t_{\text{obs},k}$ with $k = 1, 2 \cdots M$, where, M is the number of available output data. Also, the time instants when the output is missing are denoted by $t_{\text{mis},k}$ with $k = 1, 2 \cdots N - M$, with N being the length of the input vector. So, we have

$$\mathbf{Y}_{\text{obs}} = \begin{bmatrix} y(t_{\text{obs},1}) \\ y(t_{\text{obs},2}) \\ \vdots \\ y(t_{\text{obs},M}) \end{bmatrix} \quad \mathbf{Y}_{\text{mis}} = \begin{bmatrix} y(t_{\text{mis},1}) \\ y(t_{\text{mis},2}) \\ \vdots \\ y(t_{\text{mis},N-M}) \end{bmatrix} \quad (11.68)$$

11 Parameter and Delay Estimation from Non-uniformly Sampled Data

In the initial prediction stage, to obtain the basis-function model, the estimation equation (11.67) is formulated only at the time instants when the output is available. This gives

$$y(t_{\text{obs},k}) = \sum_{i=1}^{p} z_i(t_{\text{obs},k})\alpha_i \qquad (11.69)$$

Next (11.69) can be formulated for $t_{\text{obs},k}$ with $k = 1, 2 \cdots M$ and combined to give an equation in least squares form

$$\mathbf{Y}_{\text{obs}} = \mathbf{Z}_{\text{obs}}\alpha \qquad (11.70)$$

where,

$$\mathbf{Z}_{\text{obs}} = \begin{bmatrix} z_0(t_{\text{obs},1}) & z_1(t_{\text{obs},1}) & \cdots & z_p(t_{\text{obs},1}) \\ z_0(t_{\text{obs},2}) & z_1(t_{\text{obs},2}) & \cdots & z_p(t_{\text{obs},2}) \\ \cdots & \cdots & \cdots & \cdots \\ z_0(t_{\text{obs},M}) & z_1(t_{\text{obs},M}) & \cdots & z_p(t_{\text{obs},M}) \end{bmatrix} \qquad (11.71)$$

Finally, the parameter vector can be obtained as

$$\hat{\alpha} = (\mathbf{Z}_{\text{obs}}^T \mathbf{Z}_{\text{obs}})^{-1} \mathbf{Z}_{\text{obs}}^T \mathbf{Y}_{\text{obs}} \qquad (11.72)$$

The missing elements of the output can be predicted using

$$\hat{y}(t_{\text{mis},k}) = \sum_{i=0}^{p} z_i(t_{\text{mis},k})\hat{\alpha}_i \qquad (11.73)$$

The estimated value of the missing elements can then be inserted into the output vector to get a complete data set

$$\mathbf{Y}_{\text{complete}} = \{\mathbf{Y}_{\text{obs}} \ \hat{\mathbf{Y}}_{\text{mis}}\} \qquad (11.74)$$

This complete data is then used for the identification of the transfer function model of the process. The iterative prediction algorithm for identification from non-uniformly sampled and missing data is summarised in Algorithm 2.

11.3.3 Choice of Basis-function Parameters

The initial prediction step of the iterative algorithm involves choice of the parameters of the basis-function model, namely κ and p for the Laguerre model and β, ν and p for the Kautz model. The order of the polynomial, p, is chosen as a few orders higher than the order of the transfer function model. Generally, we choose κ on the basis of the knowledge of the process cut-off frequency. A value slightly higher than the cut-off frequency is chosen. The use of two parameters in the Kautz model facilitates its application for an oscillating process. However, more process information is needed to choose the model parameters. The parameters can be obtained from the knowledge of the natural period of oscillation of the process and its damping coefficient. However, lack of process knowledge can be compensated by estimating a higher-order model.

Algorithm 2: Iterative prediction algorithm for parameter estimation from non-uniform and missing data.

Step 1-Initial prediction: Using only the observed output, estimate the parameters of the input only model using (11.72). Predict the missing element of the output using (11.73). Use these predicted values, $\hat{Y}^0_{\text{mis}}$, to replace $\hat{Y}_{\text{mis}}$ in (11.74). $i = 0$.

Step 2-Iterative prediction: $i = i + 1$. Estimate the continuous-time model parameters using the complete data set $Y_{complete} = \{Y_{\text{obs}} \ \hat{Y}^{i-1}_{\text{mis}}\}$ and applying Algorithm 1. Use the estimated model, $\hat{\boldsymbol{\theta}}_i$, to get the ith prediction of the missing values, $\hat{Y}^i_{\text{mis}}$. Replace $\hat{Y}^{i-1}_{\text{mis}}$ by $\hat{Y}^i_{\text{mis}}$.

Step 3-Comparison: Compare MSE^i_{obs} with MSE^{i-1}_{obs}. If there is significant improvement, go back to step 2 and repeat the iteration.

Step 4-Termination: When MSE^i_{obs} converges, the corresponding $\hat{\boldsymbol{\theta}}_i$ is considered as the final estimates.

11.3.4 Criterion of Convergence

The proposed iterative procedure is based on the idea of iterative prediction. Consequently, a natural option for convergence criterion is the prediction error. As the output has missing elements, we can define the mean squared error at the ith stage of the iteration based on the observed output and their predicted values

$$\text{MSE}^i_{\text{obs}} = \frac{1}{M} \sum_{k=1}^{M} [y(t_{\text{obs,k}}) - \hat{y}_i(t_{\text{obs,k}})]^2 \tag{11.75}$$

where, $\hat{y}_i$ is the prediction of the model obtained in the ith stage of iteration. Convergence of this MSE criterion is equivalent to the convergence of the model prediction and the model parameters.

11.4 Simulation Results

To demonstrate the applicability of the proposed methods, different first- and second-order processes are considered. For the simulation study, the inputs are either random binary signals (RBS) or multi-sine signals generated using the *idinput* command in MATLAB® with levels $[-1 \ 1]$. Simulink® is used to generate the data and numerically simulate the filtered input and output. The process and filters are represented by continuous-time transfer function blocks. The sampled noise-free outputs are corrupted by discrete-time white noise sequences. The noise-to-signal ratio (NSR) is defined as the ratio of the variance of the noise to that of the noise-free signal[2].

---

[2] $\text{NSR} = \frac{\text{variance of noise}}{\text{variance of signal}}$

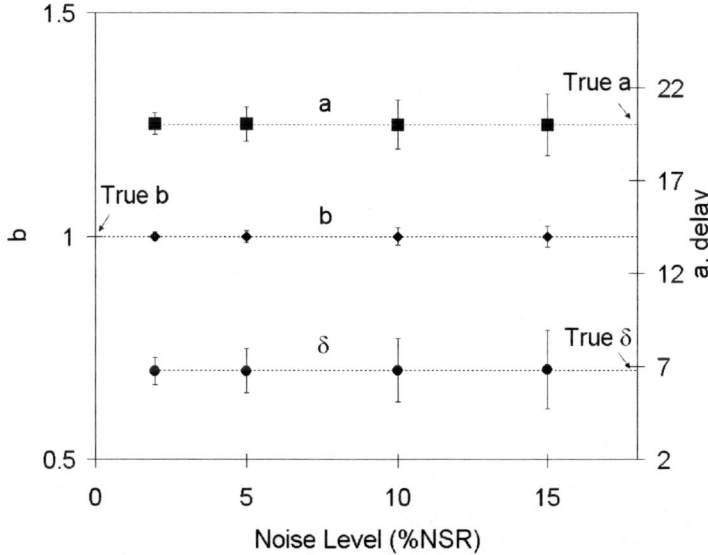

Fig. 11.3. Effect of noise on parameter estimates

11.4.1 Estimation from Uniformly Sampled Data

In this section results on the performance of the linear filter method for uniformly sampled data are presented.

Example 11.1. A first-order process having the following transfer function

$$G(s) = \frac{1}{20s+1}e^{-6.8s} \qquad (11.76)$$

is used to demonstrate the performance of the method in the presence of noise. The sampling interval is 2 s. The input is a multi-sine signal with frequency band [0 0.05]. The band contains the lower and upper limits of the passband, expressed in fractions of the Nyquist frequency.

Figure 11.3 shows the quality of the parameter estimates with different levels of noise. These results are from a Monte Carlo simulation (MCS) based on 100 random realisations. The mean values of 100 estimates are plotted and bounded by the estimated ± values of the sample standard deviation. From the figure it is seen that for noise as high as NSR= 15%, reasonably good estimates are obtained.

Example 11.2. In this example a number of second-order processes are considered. The NSR for all cases are 10%. Table 11.1 shows the true and estimated models of the second-order processes ranging from slow to fast ones

Table 11.1. True and estimated models for different second-order processes

True model	Estimated model
$\dfrac{1.25e^{-0.234s}}{0.25s^2 + 0.7s + 1}$	$\dfrac{1.25(\pm 0.02)e^{-0.234(\pm 0.039)s}}{0.25(\pm 0.025)s^2 + 0.7(\pm 0.019)s + 1}$
$\dfrac{2e^{-4.1s}}{100s^2 + 25s + 1}$	$\dfrac{2(\pm 0.04)e^{-4.13(\pm 0.742)s}}{99.4(\pm 19.7)s^2 + 25(\pm 0.67)s + 1}$
$\dfrac{(4s+1)e^{-0.615s}}{9s^2 + 2.4s + 1}$	$\dfrac{(4(\pm 0.5)s + 1(\pm 0.04))e^{-0.62(\pm 0.07)s}}{9(\pm 0.66)s^2 + 2.4(\pm 0.18)s + 1}$

and from underdamped to overdamped. A model with a zero in the numerator of the transfer function is also considered. A non-minimum phase process is considered in the next example where the special procedure described in Section 11.2.5 has been applied. The parameters shown here are the mean of the 100 parameter estimates. The numbers in the parentheses are the estimated standard deviation of the 100 estimates.

11.4.2 Estimation from Non-uniformly Sampled Data

We consider a second-order process having the following transfer function to demonstrate the performance of the iterative prediction algorithm

$$G(s) = \frac{-4s + 1}{9s^2 + 2.4s + 1} e^{-0.615s} \qquad (11.77)$$

A complete data set of 2000 samples with non-uniform sampling intervals over the range $30-60$ milliseconds (ms) is generated using a random binary signal (RBS) as the input. The NSR is 10%. Table 11.2 summarises the parameter estimation results from a Monte Carlo simulation based on 100 random realisations when the entire data set is used for identification.

To test the performance of the algorithm proposed for non-uniform and missing data, we generate three sets of data that differ in terms of percentage of missing elements. For case (i) every 3rd sample is taken out to generate a data set with 33% missing elements; for (ii) every 2nd sample is removed to generate a data set with 50% missing elements and for case (iii) every 2nd and 3rd samples are removed to generate a data set with 67% missing elements. The model estimated using the iterative algorithm is compared with the model estimated using only the available data, $i.e.$, data at the time instants when both input and output are available. To compare different models with a single index we define a total-error criterion that is a combined measure of bias and variance. We denote it by E_{total}

$$E_{\text{total}} = \frac{1}{N_\theta} \sum_{i=1}^{N_\theta} \frac{(\hat{\boldsymbol{\theta}}(i) - \boldsymbol{\theta}(i))^2 + \text{var}(\hat{\boldsymbol{\theta}}(i))}{\boldsymbol{\theta}(i)^2} \qquad (11.78)$$

Table 11.2. Estimation results using the entire data set

	$\hat{a}_2(9.00)$	$\hat{a}_1(2.40)$	$\hat{b}_1(-4.00)$	$\hat{b}_0(1.00)$	$\hat{\tau}(0.615)$
Mean	9.01	2.44	-4.04	1.03	0.63
Variance	0.040	0.049	0.063	0.005	0.020

where, $\boldsymbol{\theta}(i)$ represents the true values of the ith parameter and $\hat{\boldsymbol{\theta}}(i)$ is its estimated value. N_θ is the number of parameters. Figure 11.4 shows the total error for the results from 100 MCS runs. The estimated values of the parameters are the means of 100 estimates. The error corresponding to 0% missing data refers to the model estimated using the entire data set and can be taken as the benchmark. When 33% of the data are missing, the model estimated using only the available data has error comparable with the benchmark value and the iterative algorithm has little room to improve. This suggests that the available data are enough to give a good model. Consequently, the error level of the model estimated using the iterative algorithm remains almost the same. However, when more data are missing the error corresponding to the model estimated using the available data is much higher than the benchmark value and the iterative algorithm reduces the error to levels comparable with the benchmark.

11.5 Experimental Evaluation

11.5.1 Identification of a Dryer

To show the performance of the linear filter method in real processes, identification results of a laboratory process are presented in this section. This exercise is carried out using the data set from a dryer (*dryer.mat*) available in the CONTSID toolbox. Details of the process and experiment are obtained from [6] and described below. The SISO laboratory setup is a bench-scale hot air-flow device. Air is pulled by a fan into a 30-cm tube through a valve and heated by a mesh of resistor wires at the inlet. The output is the air temperature at the outlet of the tube measured as voltage delivered by a thermocouple. The input is the voltage over the heating device. The input signal is a pseudo-random binary signal (PRBS) with maximum length. The sampling period is 100 ms. There are two data sets, one for identification and the other for validation, each containing 1905 measurements collected under the same conditions. A first-order model with time delay is estimated for this process. Figure 11.5 shows the validation data set and the output from the estimated model. It can be seen that the simulated output matches the measured one quite well. Here, no *a priori* knowledge of the time delay is used and an initial guess of 0.1 s converged to the final estimate of 0.53 s.

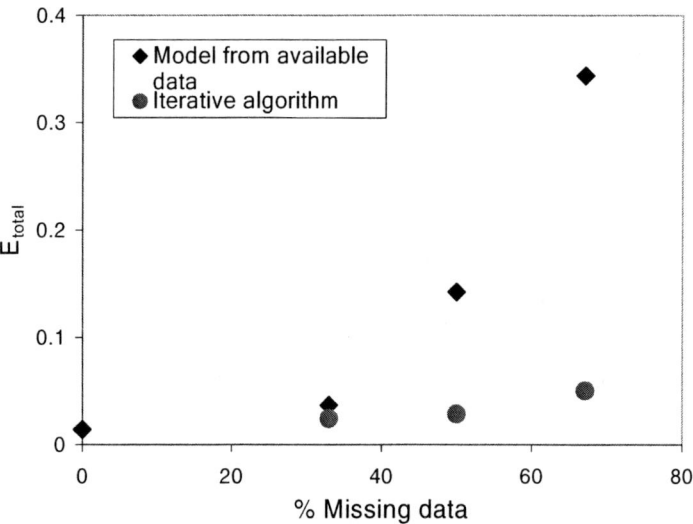

Fig. 11.4. Improvement of model quality using the iterative algorithm for the simulation example

11.5.2 Identification of a Mixing Process

The performance of the iterative prediction algorithm is evaluated using an experimental data set from a mixing process. The setup consists of a continuous stirred tank used as a mixing chamber. Two input streams are fed to the tank from two feed tanks. A salt solution and pure water run from the feed tanks and are mixed together in the mixing chamber. The volume and temperature of the solution in the mixing tank are maintained at constant values. Also, the total inlet flow is kept constant. The input to the process is the flow rate of the salt solution as a fraction of total inlet flow. The output is the concentration of salt in the mixing tank. We assume a uniform concentration of salt throughout the solution in the tank. The concentration is measured in terms of the electrical conductivity of the solution. A photograph of the setup is shown in Figure 11.6.

The input signal is a random binary signal. The sampling period is 20 s. A total of 955 data points are used for this study. To study the effect of % data missing and evaluate the performance of the iterative prediction algorithm, missing data are chosen on a random basis and the algorithm is applied. The study is carried out for 30%, 50% and 70% missing data sets. To generate a particular data set, say with 30% of its elements missing, 30% of the available output data are taken out on a random basis. The identification algorithm is then applied with the remaining 70% data points. The same procedure is applied 100 times with a different data set chosen each time containing

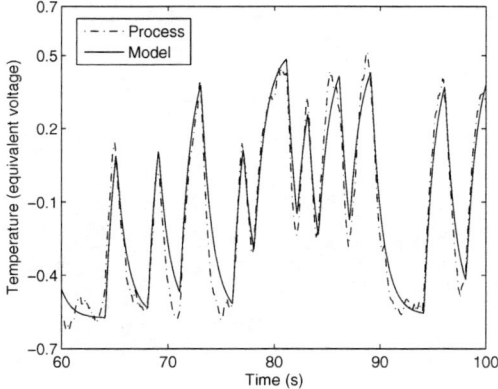

Fig. 11.5. Validation data for dryer

only 70% of the total data. Finally, we get 100 estimates of the parameters. The total error is then calculated from the estimated mean and variance of the 100 estimates. To calculate the bias error, the model estimated using the entire data set is taken as the nominal or true value. Figure 11.7 shows the performance of the proposed iterative algorithm for the mixing process data. While the error levels for models estimated only from the available data points are high, the iterative algorithm gives final estimates of the parameters with a much lower level of error.

11.6 Conclusions

Identification from non-uniformly sampled data has been considered in discrete-time identification but mainly for multi-rate data. In continuous-time identification, it is assumed that the methods are capable of dealing with data non-uniformity provided that appropriate numerical techniques are used. However, the inherent assumption of the numerical methods on the inter-sample behaviour of the signals that results in certain arbitrary interpolation, may introduce errors in the estimation of the parameters. In this work we have presented a simple algorithm to deal with non-uniformly sampled output data. It has been demonstrated using simulated and experimental data that the quality of the model estimated using the proposed model based prediction algorithm is much better than the quality of the model estimated using only the available output data. Also we introduce a novel linear filter method that simultaneously estimates the parameters and the delay of continuous-time models following an iterative procedure.

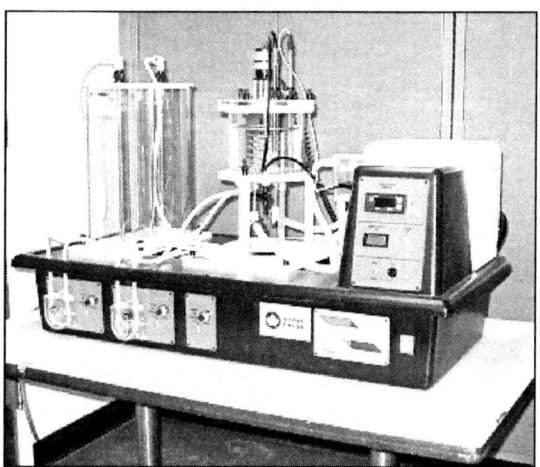

Fig. 11.6. Photograph of the mixing process

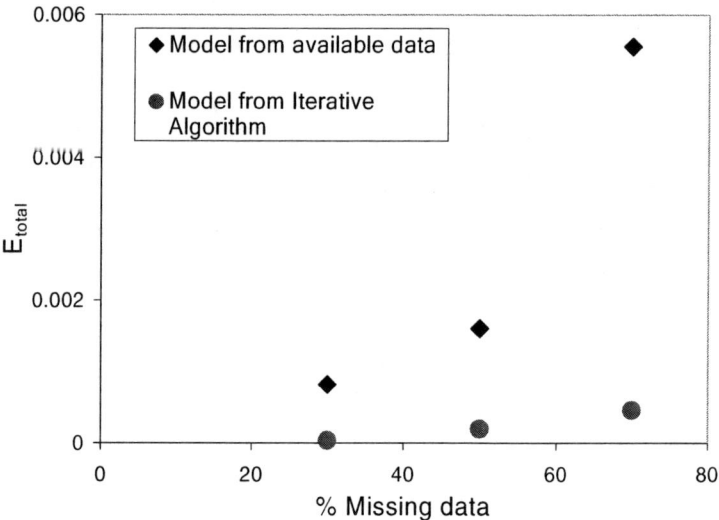

Fig. 11.7. Improvement of model quality using the iterative algorithm for the mixing process

References

1. M. Agarwal and C. Canudas. On-line estimation of time-delay and continuous-time process parameters. *International Journal of Control*, 46(1):295–311, 1987.
2. S. Ahmed. *Parameter and delay estimation of continuous-time models from uniformly and non-uniformly sampled data.* PhD thesis, University of Alberta, Edmonton, Canada, July 2006.
3. S. Ahmed, B. Huang, and S.L. Shah. Novel identification method from step response. *Control Engineering Practice*, 15(5):545–556, May 2007.
4. S. Ahmed, B. Huang, and S.L. Shah. Parameter and delay estimation of continuous-time models using a linear filter. *Journal of Process Control*, 16(4):323–331, 2006.
5. F.W. Fairman. Parameter estimation for a class of multivariate nonlinear processes. *International Journal of Control*, 1(3):291–296, 1971.
6. H. Garnier. Continuous-time model identification of real-life processes with the CONTSID toolbox. *15th IFAC World Congress*, Barcelona, Spain, July 2002.
7. H. Garnier, M. Mensler, and A. Richard. Continuous-time model identification from sampled data: Implementation and performance evaluation. *International Journal of Control*, 76(13):1337–1357, 2003.
8. J. Gillberg and F. Gustafsson. Frequency-domain continuous-time AR modeling using non-uniformly sampled measurements. *IEEE Conference on Acoustics, Speech and Signal Processing*, Philadelphia, USA, volume 4, pages 105 – 108, March 2005.
9. J. Gillberg and L. Ljung. Frequency domain identification of continuous-time ARMA models: Interpolation and non-uniform sampling. Technical Report - LiTH-ISY-R-2625, Department of Electrical Engineering, Linköping University, Linköping, Sweden, Sept. 2004.
10. J. Gillberg and L. Ljung. Frequency-domain identification of continuous-time output error models from nonuniformly sampled data. *14th IFAC Symposium on System Identification*, Newcastle, Australia, March 2006.
11. R.B. Gopaluni. *Iterative identification and predictive control.* PhD thesis, University of Alberta, Edmonton, Canada, 2003.
12. E. Huselstein and H. Garnier. An approach to continuous-time model identification from non-uniformly sampled data. *41st IEEE Conference on Decision and Control*, Las Vegas, USA, pages 622–623, December 2002.
13. S. Hwang and S. Lai. Use of two-stage least-squares algorithms for identification of continuous systems with time delay based on pulse response. *Automatica*, 40:1591–1568, 2004.
14. S.A. Imtiaz, S.L. Shah, and S. Narasimhan. Missing data treatment using IPCA and data reconciliation. *7th IFAC Symposium on Dynamics and Control of Process Systems*, Boston, USA, July 2004.
15. S.A. Imtiaz. *The treatment of missing data in process monitoring and identification.* PhD thesis, University of Alberta, Edmonton, Canada, April 2007.
16. A. Ingimundarson and T. Hagglund. Robust tuning procedure of dead-time compensating controllers. *Control Engineering Practice*, 9:1195–1208, 2001.
17. A.J. Isaksson. Identification of ARX-models subject to missing data. *IEEE Transactions on Automatic Control*, 38(5):813–819, 1993.
18. W.H. Kautz. Transient synthesis in time domain. *IRE Transactions on Circuit Theory*, CT-1(3):29–39, 1954.

19. E.K. Larsson and T. Söderström. Identification of continuous-time AR processes from unevenly sampled data. *Automatica*, 38(4):709–718, 2002.
20. Y.W. Lee. Synthesis of electrical networks by means of Fourier transform of Laguerre functions. *Journal of Mathematical Physics*, 11:83–113, 1932.
21. D. Li, S.L. Shah, and T. Chen. Identification of fast-rate models from multirate data. *International Journal of Control*, 74(7):680–689, 2001.
22. S.Y. Li, W.J. Cai, H. Mei, and Q. Xiong. Robust decentralized parameter identification for two-input two-output processes from closed-loop step responses. *Control Engineering Practice*, 13:519–531, 2005.
23. M. Liu, Q.G. Wang, B. Huang, and C. Hang. Improved identification of continuous-time delay processes from piecewise step test. *Journal of Process Control*, 17(1):51–57, January 2007.
24. P.M. Makila. Approximation of stable systems by Laguerre filters. *Automatica*, 26(2):333–345, 1990.
25. H. Mei, S. Li, W.J. Cai, and Q. Xiong. Decentralized closed-loop parameter identification for multivariable processes from step responses. *Mathematics and Computer in Simulation*, 68(2):171–192, 2005.
26. J.R. Parington. Approximation of delay systems by Fourier-Laguerre series. *Automatica*, 27(3):569–572, 1991.
27. R. Pintelon and J. Schoukens. Identification of continuous-time systems with missing data. *IEEE Transactions on Instrumentation and Measurement*, 48(3):736–740, 1999.
28. H. Raghavan. *Quantitative approaches for fault detection and diagnosis in process industries*. PhD thesis, Dept. Chemical & Materials Engineering, University of Alberta, Edmonton, Canada, 2004.
29. H. Raghavan, A. Tangirala, R.B. Gopaluni, and S.L. Shah. Identification of chemical processes with irregular output sampling. *Control Engineering Practice*, 14(5):467–480, 2005.
30. G.P. Rao and L. Sivakumar. Identification of deterministic time-lag systems. *IEEE Transactions on Automatic Control*, 21:527–529, 1976.
31. D.C. Saha and G.P. Rao. *Identification of Continuous Dynamical Systems - The Poisson Moment Functional Approach*. Springer-Verlag, 1st edition, Berlin, Germany, 1983.
32. B. Wahlberg and P.M. Makila. On approximation of stable linear dynamic systems by Laguerre and Kautz functions. *Automatica*, 32(5):693–708, 1996.
33. J. Wang, T. Chen, and B. Huang. Multi-rate sampled data systems: computing fast models. *Journal of Process Control*, 14(1):79–88, 2004.
34. L. Wang and W.R. Cluett. Building transfer function models from noisy step response data using Laguerre network. *Chemical Engineering Science*, 50(1):149–161, 1995.
35. L. Wang and P. Gawthrop. On the estimation of continuous-time transfer functions. *International Journal of Control*, 74(9):889–904, 2001.
36. Q.G. Wang and Y. Zhang. Robust identification of continuous systems with dead-time from step response. *Automatica*, 37(3):377–390, 2001.
37. N. Wiener. *Modern Mathematics for the Engineers*, chapter 'The theory of prediction'. McGraw-Hill, Scarborough, CA, USA, 1956.
38. P.C. Young. In flight dynamic checkout-A discussion. *IEEE Transactions on Aerospace*, 2:1106–1111, 1964.
39. P.C. Young. An instrumental variable method for real time identification of noisy process. *Automatica*, 6:271–287, 1970.

40. P.C. Young. Optimal IV identification and estimation of continuous-time TF models. *15th IFAC World Congress*, Barcelona, Spain, pages 337–358, July 2002.

12

Iterative Methods for Identification of Multiple-input Continuous-time Systems with Unknown Time Delays

Zi-Jiang Yang

Kyushu University, Japan

12.1 Introduction

Many practical systems such as thermal processes, chemical processes and biological systems, *etc.*, have inherent time delay. If the time delay used in the system model for controller design does not coincide with the actual process time delay, a closed-loop system may be unstable or exhibit unacceptable transient response characteristics. Therefore, the problem of identifying such a system is of great importance for analysis, synthesis and prediction.

Numerous identification methods of time-delay systems based on discrete-time models have been proposed [3,32]. However, since physical systems considered in science and engineering are usually continuous in time, it is sometimes desirable to obtain a continuous-time model [12,24,29]. In [9], a general framework of the technique of step-invariant transform from the z- to s-domain was proposed, and it was pointed out that the technique may also be applicable to a system with fractional delay that is not an integral multiple of sampling period.

There have been some typical approaches to identification of continuous-time models with unknown delay. One approach is based on the approximation of the time delay in the frequency domain by a rational transfer function or the Padé approximation [1,4]. This approach requires estimation of more parameters because the order of the approximated system model is increased. And it is not easy to separate the parameters concerned with the time delay from those concerned with the transfer function of the system. Another problem in this method is that it may introduce an unacceptable approximation error when the system has a large time delay.

Another approach is based on the non-linear optimisation method like the non-linear least squares (LS) method that searches for the optimum by using a gradient-following technique. In [5,23], some variations of pure continuous-time on-line non-linear LS methods were studied.

The off-line separable non-linear least squares (SEPNLS) method was studied in [11], where the time delays were searched exhaustively with spline interpolation of the quadratic cost function, instead of the gradient-following-based iterative search algorithms.

A major drawback of the iterative non-linear search algorithms is that the solutions often converge to local optima. For the single-input, single-output, (SISO) continuous-time systems, since there exists only one non-linear parameter (the time delay) in the estimation problem, the problem of initial setting is considered to be relatively simple. Several trials are enough. For multiple-input, single-output (MISO) systems with multiple unknown time delays that may differ from each other, the problem is much more difficult. Although the genetic algorithm is considered to be a powerful approach to this problem [25,26], the algorithm is usually computationally demanding and requires a very long execution time.

It should be mentioned here that, on the other hand, both continuous-time and discrete-time transfer function models with time delay can also be identified from frequency-domain data. By using the maximum likelihood method, consistent estimates can be obtained in the presence of measurement noise [7,17,18]. In this chapter, however, we confine our focus to time-domain identification.

We consider the identification problem of MISO continuous-time systems with multiple unknown time delays from sampled input/output data. An approximated discrete-time estimation model is first derived, in which the system parameters remain in their original form and the time delays need not be an integral multiple of the sampling period. Then, an iterative SEPNLS method that estimates the time delays and transfer function parameters separably is first derived. Furthermore, we propose an iterative global SEPNLS (GSEPNLS) method to address the problem of convergence to a local minimum of the SEPNLS method, by using the stochastic global-optimisation techniques. In particular, we apply the stochastic approximation with convolution smoothing (SAS) technique [22] to the SEPNLS method. This results in the GSEPNLS method.

In high measurement noise situations, the LS method will yield biased estimates and hence some procedures for removing the bias errors will be required. The bootstrap instrumental variable (IV) method that employs iteratively the estimated noise-free output as IVs is one of the most useful techniques to yield consistent estimates in the presence of high measurement noise [8,21,28–31]. Although the IV technique is usually applied to the linear parameter estimation problems, we try to extend this technique to the non-linear LS problem. By using the bootstrap IV technique, the GSEPNLS method is further modified to a novel global separable non-linear instrumental variable (GSEPNIV) method to yield consistent estimates if the algorithm converges to the global minimum.

Finally, simulation results are included to show the performance of the proposed algorithms.

12.2 Statement of the Problem

Consider a strictly stable MISO continuous-time system with unknown time delays governed by the following differential equation

$$\sum_{i=0}^{n} a_i s^{n-i} x(t) = \sum_{j=1}^{r} \sum_{i=1}^{m_j} b_{ji} s^{m_j - i} u_j(t - \tau_j), \quad a_0 = 1, \ b_{j1} \neq 0 \quad (12.1)$$

where s is the differential operator, $u_j(t)$ is the jth input with time delay τ_j, $x(t)$ is the deterministic output. n and m_j are assumed to be known ($n \geq m_j$).

Consider the situation that the input signals are generated by digital computers. It is assumed that a zero-order hold is used such that

$$u_j(t) = \bar{u}_j(t_k) \quad (k-1)T_s \leq t < kT_s \quad (12.2)$$

where T_s is the sampling period.

Practically the discrete-time measurement of the output variable is corrupted by a stochastic measurement noise

$$y(t_k) = x(t_k) + \xi(t_k) \quad (12.3)$$

where $y(t_k), x(t_k), \xi(t_k)$ denote $y(kT_s), x(kT_s), \xi(kT_s)$, respectively.

Our goal is to identify the unknown time delays $\tau_j (j = 1, \cdots, r)$ and the transfer function parameters $a_i (i = 1, \cdots, n)$ and $b_{ji} (j = 1, \cdots, r, \ i = 1, \cdots, m_j)$ from sampled data of the inputs and the noisy output. To clarify the problem setting, some assumptions are made here.

Assumption 12.1 *The system under study is open loop strictly stable and strictly proper.*

Assumption 12.2 *The system order n is known, and the relative degree $n - m_j + 1$ ($n \geq m_j$) with respect to each input signal is known and well defined, that is, $b_{j1} \neq 0$ ($j = 1, \cdots, r$). Notice that this assumption is a necessary condition of identifiability. See Proposition 12.1 in Section 12.4.*

Assumption 12.3 *Each zero-order hold input $\bar{u}_j(t_k)$ is a quasi-stationary deterministic or random signal and the noise $\xi(t_k)$ is a quasi-stationary zero-mean random signal uncorrelated with each input such that*

$$\lim_{N \to \infty} \frac{1}{N} \sum_{k=1}^{N} \bar{u}_j(t_k) \xi(t_k) = 0$$

12.3 Approximate Discrete-time Model Estimation

The ordinary differential equation (12.1) in continuous time itself may not be suitable for identification when the discrete-time sampled input–output data are provided. Since differential operations may amplify the measurement noise as well as the round-off noise, it is inappropriate to identify the parameters using direct approximations of differentiations [12, 24, 29]. Our objective here is to introduce a digital low-pass prefilter that would reduce the noise effects sufficiently. Then, we can obtain an approximated discrete-time estimation model in which the transfer function parameters remain in their original form [25, 26]. A straightforward candidate of the low-pass prefilter $L(s)$ is

$$L(s) = \frac{1}{(\alpha s + 1)^n} \tag{12.4}$$

where α is the time constant that determines the passband of $L(s)$.

The prefilter was first suggested in [27]. Subsequently, the suitable choice of prefilters for discrete- and continuous-time models was discussed in [28–31] and it was suggested that if the prefilter is designed so that its passband matches that of the system closely the noise effects can be significantly reduced. Also, it was suggested that if the denominator of the prefilter is given by $A(s)$, which is the denominator of the system transfer function, the estimates can be made less vulnerable to higher measurement noise levels [28–31]. In this approach, since $A(s)$ is not known in advance, the prefilter has to be constructed iteratively by the estimate of $A(s)$, that is,

$$L(s) = \frac{1}{\hat{A}(s)}, \quad \hat{A}(s) = \sum_{i=0}^{n} \hat{a}_i s^{n-i} \tag{12.5}$$

Multiplying both sides of (12.1) by $L(s)$ and using the bilinear transformation based on the block-pulse functions [6], we obtain the following approximated discrete-time estimation model [15, 25, 26]

$$\bar{y}_{f_0}(t_k) + \sum_{i=1}^{n} a_i \bar{y}_{f_i}(t_k) = \sum_{j=1}^{r} \sum_{i=1}^{m_j} b_{ji} \bar{u}_{jf_{n-m_j+i}}(t_k - \tau_j) + r(t_k) \tag{12.6}$$

where

$$r(t_k) = \sum_{i=0}^{n} a_i \bar{\xi}_{f_i}(t_k)$$

$$\bar{u}_{jf_i}(t_k) = L_d(q^{-1}) \left(\frac{T_s}{2}\right)^i (1 + q^{-1})^i (1 - q^{-1})^{n-i} \bar{u}_j(t_k)$$

$$\bar{y}_{f_i}(t_k) = L_d(q^{-1}) \left(\frac{T_s}{2}\right)^i (1 + q^{-1})^i (1 - q^{-1})^{n-i} \bar{y}(t_k) \tag{12.7}$$

$$\bar{\xi}_{f_i}(t_k) = L_d(q^{-1}) \left(\frac{T_s}{2}\right)^i (1 + q^{-1})^i (1 - q^{-1})^{n-i} \bar{\xi}(t_k)$$

where q^{-1} is the backward shift operator, $\bar{\xi}(t_k) = (1+q^{-1})\xi(t_k)/2$, $\bar{y}(t_k) = (1+q^{-1})y(t_k)/2$, and

$$L_d(q^{-1}) = \begin{cases} \left[\alpha(1-q^{-1}) + \dfrac{T_s}{2}(1+q^{-1})\right]^{-n} & \text{if } L(s) = 1/(\alpha s+1)^n \\ \left[\displaystyle\sum_{i=0}^{n} \hat{a}_i \left(\dfrac{T_s}{2}\right)^i (1+q^{-1})^i (1-q^{-1})^{n-i}\right]^{-1} & \text{if } L(s) = 1/\hat{A}(s) \end{cases}$$

(12.8)

Notice that τ_j can be expressed as

$$\tau_j = l_j T_s + \Delta_j \qquad (12.9)$$

where $0 \leq \Delta_j < T_s$ and l_j is a non-negative integer.

Remark 12.1. In the approximated discrete-time estimation model, the time delays may be fractional, that is, $\Delta_j \neq 0$. In this case, we can get $\bar{u}_{jf_{n-m_j+i}}(t_k - \tau_j)$ by linear interpolation between $\bar{u}_{jf_{n-m_j+i}}(t_k - l_j T_s)$ and $\bar{u}_{jf_{n-m_j+i}}(t_k - l_j T_s - 1)$.

Remark 12.2. In the case where the system is excited by a band-limited continuous input signal $u_j(t)$ [19] instead of the piecewise-constant signal given in (12.2), we can define $\bar{u}_j(t_k) = (1+q^{-1})u_j(t_k)/2$. In this case, the bilinear transformation corresponds to the trapezoidal integration law [15].

Equation (12.6) can be written in vector form

$$\begin{aligned}
\bar{y}_{f_0}(t_k) &= \varphi^T(t_k, \tau)\theta + r(t_k) \\
\varphi^T(t_k, \tau) &= [-\varphi_{\bar{y}}^T(t_k), \varphi_{\bar{u}_1}^T(t_k - \tau_1), \cdots, \varphi_{\bar{u}_r}^T(t_k - \tau_r)] \\
\varphi_{\bar{y}}^T(t_k) &= [\bar{y}_{f_1}(t_k), \cdots, \bar{y}_{f_n}(t_k)] \\
\varphi_{\bar{u}_j}^T(t_k - \tau_j) &= [\bar{u}_{jf_{n-m_j+1}}(t_k - \tau_j), \cdots, \bar{u}_{jf_n}(t_k - \tau_j)] \\
\theta^T &= [\mathbf{a}^T, \mathbf{b}_1^T, \cdots, \mathbf{b}_r^T] \\
\tau^T &= [\tau_1, \cdots, \tau_r] \\
\mathbf{a}^T &= [a_1, \cdots, a_n] \\
\mathbf{b}_j^T &= [b_{j1}, \cdots, b_{jm_j}]
\end{aligned}$$

(12.10)

12.4 SEPNLS Method

Given a set of filtered data $\{\bar{y}_{f_0}(t_k), \varphi_{\bar{y}}^T(t_k), \varphi_{\bar{u}_1}^T(t_k), \cdots, \varphi_{\bar{u}_r}^T(t_k)\}_{k=1}^N$, the off-line parameter estimates are defined as the minimising arguments of the following quadratic cost function

$$V_N(\theta, \tau) = \frac{1}{N - k_s} \sum_{k=k_s+1}^{N} \frac{1}{2}\epsilon^2(t_k, \theta, \tau) \qquad (12.11)$$

$$\epsilon(t_k, \theta, \tau) = \bar{y}_{f_0}(t_k) - \varphi^T(t_k, \tau)\theta$$

such that

$$\left[\hat{\boldsymbol{\theta}}_N^T, \hat{\boldsymbol{\tau}}_N\right]^T = \arg\min_{\boldsymbol{\theta},\boldsymbol{\tau}} V_N(\boldsymbol{\theta}, \boldsymbol{\tau}) \qquad (12.12)$$

Notice that $(N - k_s)$ is the data length for identification, and k_s should be chosen such that $t_{k_s} \geq \max(\tau_1, \cdots, \tau_r)$, that is, the delayed signals should be casual.

For detailed descriptions of the iterative search algorithms applied to system identification, the readers are referred to [8, 10, 20].

The unseparable non-linear least squares (UNSEPNLS) method estimates $\boldsymbol{\theta}$ and $\boldsymbol{\tau}$ simultaneously by minimising the above quadratic cost function. Starting at a set of initial estimates $\hat{\boldsymbol{\theta}}_N^{(0)}$ and $\hat{\boldsymbol{\tau}}_N^{(0)}$, the minimising problem can be described by the following iteration (Gauss-Newton method).

$$\hat{\boldsymbol{\Theta}}_N^{(j+1)} = \hat{\boldsymbol{\Theta}}_N^{(j)} - \mu^{(j)} \left[\boldsymbol{R}_N\left(\hat{\boldsymbol{\Theta}}_N^{(j)}\right)\right]^{-1} \boldsymbol{V}_N'\left(\hat{\boldsymbol{\Theta}}_N^{(j)}\right) \qquad (12.13)$$

where $\boldsymbol{\Theta} = \left[\boldsymbol{\theta}^T, \boldsymbol{\tau}^T\right]^T$, $\mu^{(j)}$ is the step size that assures that $V_N(\boldsymbol{\Theta})$ decreases and that $\hat{\boldsymbol{\tau}}_N$ stays in a preassigned interval [8, 10, 20].

$\boldsymbol{V}_N'(\boldsymbol{\Theta})$ and $\boldsymbol{R}_N(\boldsymbol{\Theta})$ are, respectively, the gradient and the estimate of the Hessian of the quadratic cost function [8, 10, 20]

$$\begin{aligned}
\boldsymbol{V}_N'(\boldsymbol{\Theta}) &= -\frac{1}{N-k_s} \sum_{k=k_s+1}^{N} \boldsymbol{\gamma}(t_k, \boldsymbol{\Theta}) \epsilon(t_k, \boldsymbol{\Theta}) \\
\boldsymbol{R}_N(\boldsymbol{\Theta}) &= \frac{1}{N-k_s} \sum_{k=k_s+1}^{N} \boldsymbol{\gamma}(t_k, \boldsymbol{\Theta}) \boldsymbol{\gamma}^T(t_k, \boldsymbol{\Theta})
\end{aligned} \qquad (12.14)$$

where

$$\boldsymbol{\gamma}(t_k, \boldsymbol{\Theta}) = -\frac{\partial \epsilon(t_k, \boldsymbol{\Theta})}{\partial \boldsymbol{\Theta}}$$

$$= \begin{bmatrix} -\boldsymbol{\varphi}_{\bar{y}}^T(t_k) \\ \boldsymbol{\varphi}_{\bar{u}_1}^T(t_k - \tau_1) \\ \vdots \\ \boldsymbol{\varphi}_{\bar{u}_r}^T(t_k - \tau_r) \\ -\frac{\partial \epsilon(t_k, \boldsymbol{\Theta})}{\partial \tau_1} \\ \vdots \\ -\frac{\partial \epsilon(t_k, \boldsymbol{\Theta})}{\partial \tau_r} \end{bmatrix} = \begin{bmatrix} -\boldsymbol{\varphi}_{\bar{y}}^T(t_k) \\ \boldsymbol{\varphi}_{\bar{u}_1}^T(t_k - \tau_1) \\ \vdots \\ \boldsymbol{\varphi}_{\bar{u}_r}^T(t_k - \tau_r) \\ -\sum_{i=1}^{m_1} b_{1i} \bar{u}_{jf_{n-m_j+i-1}}(t_k - \tau_j) \\ \vdots \\ -\sum_{i=1}^{m_r} b_{ri} \bar{u}_{jf_{n-m_r+i-1}}(t_k - \tau_r) \end{bmatrix} \qquad (12.15)$$

Notice that $-\partial \epsilon(t_k)/\partial \tau_j (j = 1, \cdots, r)$ is derived as the following by replacing the differential operation with respect to time t by the bilinear transformation

$$
\begin{aligned}
-\frac{\partial \epsilon(t_k, \boldsymbol{\Theta})}{\partial \tau_j} &= \sum_{i=1}^{m_j} b_{ji} \frac{\partial}{\partial \tau_j} \bar{u}_{jf_{n-m_j+i}}(t-\tau_j) \bigg|_{t=kT_s} \\
&= -\sum_{i=1}^{m_j} b_{ji} \frac{\mathrm{d}}{\mathrm{d}t} \bar{u}_{jf_{n-m_j+i}}(t-\tau_j) \bigg|_{t=kT_s} \\
&= -\sum_{i=1}^{m_j} b_{ji} Q_d(q^{-1}) \left(\frac{T_s}{2}\right)^{n-m_j+i-1} \quad (12.16) \\
&\quad \times (1+q^{-1})^{n-m_j+i-1} (1-q^{-1})^{m_j-i+1} \bar{u}_j(t_k-\tau_j) \\
&= -\sum_{i=1}^{m_j} b_{ji} \bar{u}_{jf_{n-m_j+i-1}}(t_k-\tau_j)
\end{aligned}
$$

To avoid the ill-conditioned problem, $\boldsymbol{R}_N\left(\hat{\boldsymbol{\Theta}}_N^{(j)}\right)$ should not be singular or near singular. This requires that $b_{j1} \neq 0$ ($j = 1, \cdots, r$), since if $b_{j1} = 0$ for any j, $-\partial \epsilon(t_k, \boldsymbol{\Theta})/\partial \tau_j$ can be expressed by a linear combination of the elements of $\boldsymbol{\varphi}_{\bar{u}_j}^T(t_k - \tau_j)$ such that $\boldsymbol{R}_N\left(\hat{\boldsymbol{\Theta}}_N^{(j)}\right)$ is rank deficient. To summarise, we have the following proposition

Proposition 12.1. *If the relative degree of system (12.1) is not well defined, that is, $b_{j1} \approx 0$ for any $j = 1, \cdots, r$, then the LS minimising problem (12.12) with respect to unknown $\boldsymbol{\Theta} = \begin{bmatrix} \boldsymbol{\theta}^T, \boldsymbol{\tau}^T \end{bmatrix}^T$ becomes ill-conditioned.*

This means that Assumption 12.2 is necessary to obtain a unique solution of the LS minimising problem (12.12).

In contrast to the UNSEPNLS method, the SEPNLS method estimates the time delay vector $\boldsymbol{\tau}$ and the linear parameter vector $\boldsymbol{\theta}$ in a separable manner. When the time delays are known, the linear parameters can be estimated by the linear LS method as

$$
\begin{aligned}
\hat{\boldsymbol{\theta}}_N(\boldsymbol{\tau}) &= \boldsymbol{R}^{-1}(N, \boldsymbol{\tau}) \boldsymbol{f}(N, \boldsymbol{\tau}) \\
\boldsymbol{R}(N, \boldsymbol{\tau}) &= \frac{1}{N - k_s} \sum_{k=k_s+1}^{N} \boldsymbol{\varphi}(t_k, \boldsymbol{\tau}) \boldsymbol{\varphi}^T(t_k, \boldsymbol{\tau}) \quad (12.17) \\
\boldsymbol{f}(N, \boldsymbol{\tau}) &= \frac{1}{N - k_s} \sum_{k=k_s+1}^{N} \boldsymbol{\varphi}(t_k, \boldsymbol{\tau}) \bar{y}_{f_0}(t_k)
\end{aligned}
$$

Then the quadratic cost function $V_N(\boldsymbol{\theta}, \boldsymbol{\tau})$ becomes the following so that the time delays can be estimated separably

$$
\check{V}_N(\boldsymbol{\tau}) = \frac{1}{N - k_s} \sum_{k=k_s+1}^{N} \frac{1}{2} \check{\varepsilon}^2(t_k, \boldsymbol{\tau}) \quad (12.18)
$$

where

$$\check{\varepsilon}(t_k, \boldsymbol{\tau}) = \bar{y}_{f_0}(t_k) - \boldsymbol{\varphi}^T(t_k, \boldsymbol{\tau})\boldsymbol{R}^{-1}(N, \boldsymbol{\tau})\boldsymbol{f}(N, \boldsymbol{\tau}) \quad (12.19)$$

A non-linear LS problem is called separable if one set of parameters enter linearly and another set nonlinearly in the model for parameter estimation [10, 14]. The time delay vector $\boldsymbol{\tau}$ and the linear parameter vector $\boldsymbol{\theta}$ can be estimated separably according to the following theorem. See [10, 14] for proof and more detailed explanations.

Theorem 12.4. *Let $\boldsymbol{\theta} = \hat{\boldsymbol{\theta}}_N(\boldsymbol{\tau}) = \boldsymbol{R}^{-1}(N, \boldsymbol{\tau})\boldsymbol{f}(N, \boldsymbol{\tau})$ denote one solution of the quadratic cost function (12.11). Then*

$$\left[\hat{\boldsymbol{\theta}}_N^T, \hat{\boldsymbol{\tau}}_N^T\right]^T = \underset{\boldsymbol{\theta}, \boldsymbol{\tau}}{\mathrm{argmin}}\, V_N(\boldsymbol{\theta}, \boldsymbol{\tau}) = \underset{\boldsymbol{\tau}}{\mathrm{argmin}}\, \check{V}_N(\boldsymbol{\tau}) \quad (12.20)$$

Since the separable iterative research algorithm for the time delays is not found in the literature, the derivation of the algorithm will be described in detail.

The estimate of time delays can be obtained as

$$\hat{\boldsymbol{\tau}}_N = \underset{\boldsymbol{\tau}}{\mathrm{argmin}}\, \check{V}_N(\boldsymbol{\tau}) \quad (12.21)$$

through the following iterative search algorithm

$$\hat{\boldsymbol{\tau}}_N^{(j+1)} = \hat{\boldsymbol{\tau}}_N^{(j)} - \mu^{(j)} \left[\check{\boldsymbol{R}}_N\left(\hat{\boldsymbol{\tau}}_N^{(j)}\right)\right]^{-1} \check{\boldsymbol{V}}_N'\left(\hat{\boldsymbol{\tau}}_N^{(j)}\right) \quad (12.22)$$

where $\mu^{(j)}$ is the step size that assures that $\check{V}_N(\boldsymbol{\tau})$ decreases and that each element of $\hat{\boldsymbol{\tau}}_N$ stays in a preassigned interval, that is,

$$\hat{\tau}_{Ni}^{(j+1)} \in \Omega_{\tau_i} = \left\{\hat{\tau}_{Ni}^{(j+1)}\,\middle|\,0 \leq \hat{\tau}_{Ni}^{(j+1)} \leq \bar{\tau}_i\right\}, \quad i = 1, \cdots, r$$

$\check{\boldsymbol{V}}_N'(\boldsymbol{\tau})$ and $\check{\boldsymbol{R}}_N(\boldsymbol{\tau})$ are respectively the gradient and the estimate of the Hessian of the quadratic cost function (12.18)

$$\begin{aligned}\check{\boldsymbol{V}}_N'(\boldsymbol{\tau}) &= -\frac{1}{N-k_s}\sum_{k=k_s+1}^{N}\boldsymbol{\psi}(t_k, \boldsymbol{\tau})\check{\varepsilon}(t_k, \boldsymbol{\tau}) \\ \check{\boldsymbol{R}}_N(\boldsymbol{\tau}) &= \frac{1}{N-k_s}\sum_{k=k_s+1}^{N}\boldsymbol{\psi}(t_k, \boldsymbol{\tau})\boldsymbol{\psi}^T(t_k, \boldsymbol{\tau})\end{aligned} \quad (12.23)$$

$\boldsymbol{\psi}(t_k, \boldsymbol{\tau}) = [\psi_1(t_k, \boldsymbol{\tau}), \cdots, \psi_r(t_k, \boldsymbol{\tau})]^T$ can be obtained through tedious but straightforward calculations as follows, for $j = 1, \cdots, r$

$$\psi_j(t_k, \boldsymbol{\tau}) = -\frac{\partial \check{\varepsilon}(t_k, \boldsymbol{\tau})}{\partial \tau_j}$$
$$= \boldsymbol{\varphi}_{\tau_j}^T(t_k, \boldsymbol{\tau})\boldsymbol{R}^{-1}(N, \boldsymbol{\tau})\boldsymbol{f}(N, \boldsymbol{\tau}) + \boldsymbol{\varphi}^T(t_k, \boldsymbol{\tau})\boldsymbol{R}^{-1}(N, \boldsymbol{\tau})\boldsymbol{f}_{\tau_j}(N, \boldsymbol{\tau})$$
$$- \boldsymbol{\varphi}^T(t_k, \boldsymbol{\tau})\boldsymbol{R}^{-1}(N, \boldsymbol{\tau})\big[\boldsymbol{R}_{\tau_j}(N, \boldsymbol{\tau}) + \boldsymbol{R}_{\tau_j}^T(N, \boldsymbol{\tau})\big]\boldsymbol{R}^{-1}(N, \boldsymbol{\tau})\boldsymbol{f}(N, \boldsymbol{\tau})$$
$$\tag{12.24}$$

where

$$\boldsymbol{R}_{\tau_j}(N, \boldsymbol{\tau}) = \frac{1}{N - k_s} \sum_{k=k_s+1}^{N} \boldsymbol{\varphi}_{\tau_j}(t_k, \boldsymbol{\tau})\boldsymbol{\varphi}^T(t_k, \boldsymbol{\tau})$$

$$\boldsymbol{f}_{\tau_j}(N, \boldsymbol{\tau}) = \frac{1}{N - k_s} \sum_{k=k_s+1}^{N} \boldsymbol{\varphi}_{\tau_j}(t_k, \boldsymbol{\tau})\bar{y}_{f_0}(t_k)$$

$$\boldsymbol{\varphi}_{\tau_j}(t_k, \boldsymbol{\tau}) = \frac{\partial \boldsymbol{\varphi}(t_k, \boldsymbol{\tau})}{\partial \tau_j}$$
$$= \big[\boldsymbol{0}_{1 \times n}, \boldsymbol{0}_{1 \times m_1}, \cdots, \boldsymbol{0}_{1 \times m_{j-1}},$$
$$\boldsymbol{\varphi}_{\tau_j, u_j}^T(t_k - \tau_j), \boldsymbol{0}_{1 \times m_{j+1}}, \cdots, \boldsymbol{0}_{1 \times m_r}\big]^T$$
$$\boldsymbol{\varphi}_{\tau_j, u_j}^T(t_k - \tau_j) = [-\bar{u}_{jf_{n-m_j}}(t_k - \tau_j),$$
$$-\bar{u}_{jf_{n-m_j+1}}(t_k - \tau_j), \cdots, -\bar{u}_{jf_{n-1}}(t_k - \tau_j)]$$
$$\tag{12.25}$$

Remark 12.3. Comparative discussions on separable and unseparable algorithms can be found in the literature [10, 14, 20]. In [14], it was shown that the computational burden per iteration is of the same order for the separable and unseparable algorithms. In [10, 20], it was pointed out through theoretical analysis that $\check{\boldsymbol{R}}_N$ in the separable algorithm is better-conditioned than $\boldsymbol{R}_N$ in the unseparable algorithm. Therefore, the separable algorithm is likely to converge faster than the unseparable one, especially in the ill-conditioned cases, and numerical examples confirmed this [10, 20]. Therefore, we will confine our attention to the separable algorithms only.

12.5 GSEPNLS Method

A major drawback of the iterative non-linear search algorithms is that the solutions often converge to local optima. And hence the results may be sensitive to the initial estimates. For MISO systems with multiple unknown time delays that differ from each other, the problem of initial setting is non-trivial. Although the genetic algorithm is considered to be a powerful approach to achieve the global solution [25, 26], the algorithm is usually computationally demanding and requires a very long execution time. In this section, we propose an iterative GSEPNLS method to address the problem of convergence to a local minimum of the SEPNLS method by using the stochastic global-optimisation techniques. In particular, we apply the SAS technique [22] to the SEPNLS method. This results in the GSEPNLS method.

The SAS is a global-optimisation algorithm for minimising a non-convex function
$$\operatorname*{argmin}_{\boldsymbol{\tau}} \check{V}_N(\boldsymbol{\tau}) \tag{12.26}$$
by smoothing operations on it. The smoothing process represents the convolution of $\check{V}_N(\boldsymbol{\tau})$ with a smoothing function $\check{h}(\eta, \beta)$, where $\eta \in \mathbb{R}^r$ is a random vector used to perturb $\boldsymbol{\tau}$, and β controls the degree of smoothing. This smoothed functional, described in [13], is given by

$$\begin{aligned}\check{V}_N(\boldsymbol{\tau}, \beta) &= \int_{-\infty}^{\infty} \check{h}(\eta, \beta) \check{V}_N(\boldsymbol{\tau} - \eta) \mathrm{d}\eta \\ &= \int_{-\infty}^{\infty} \check{h}(\boldsymbol{\tau} - \eta, \beta) \check{V}_N(\eta) \mathrm{d}\eta \end{aligned} \tag{12.27}$$

which represents an averaged version of $\check{V}_N(\boldsymbol{\tau})$ weighted by $\check{h}(\cdot, \beta)$. The objective of convolution smoothing is to smooth the non-convex objective function by convolving it with a noise probability density function (PDF). To yield a properly smoothed functional $\check{V}_N(\boldsymbol{\tau}, \beta)$, the kernel functional $\check{h}(\eta, \beta)$ must have the following properties [13]

1. $\check{h}(\eta, \beta) = (1/\beta^r) h(\eta/\beta)$ is piecewise-differentiable with respect to β;
2. $\lim_{\beta \to 0} \check{h}(\eta, \beta) = \delta(\eta)$; ($\delta(\eta)$ is the Dirac delta function);
3. $\lim_{\beta \to 0} \check{V}_N(\boldsymbol{\tau}, \beta) = \check{V}_N(\boldsymbol{\tau})$;
4. $\check{h}(\eta, \beta)$ is a PDF.

One of the possible choices for $h(\eta)$ is a Gaussian PDF [13], which leads to the following expression for $\check{h}(\eta, \beta)$

$$\check{h}(\eta, \beta) = \frac{1}{(2\pi)^{(r/2)} \beta^r} \exp\left[-\frac{1}{2} \sum_{i=1}^{r} \left(\frac{\eta_i}{\beta}\right)^2\right] \tag{12.28}$$

Under these conditions, we can rewrite (12.27) as the expectation with respect to η

$$\check{V}_N(\boldsymbol{\tau}, \beta) = \mathsf{E}\{\check{V}_N(\boldsymbol{\tau} - \eta)\} \tag{12.29}$$

In our case, $\check{h}(\eta, \beta)$ will be the sampled values of its PDF, which is convolved with the original objective function for smoothing. Gaussian, uniform, and Cauchy distributions satisfy the above properties. In this chapter, we will use the Gaussian distribution.

The value of β plays a dominant role in the smoothing process by controlling the variance of $\check{h}(\eta, \beta)$; see properties 2 and 3. To significantly reduce the possibility of convergence to a local minimum, β has to be large at the start

12 Identification Methods for Multiple-input Systems with Time Delays 349

of the optimisation process and is then reduced to approximately zero as the quadratic cost function becomes sufficiently small.

Our objective now is to solve the following SAS optimisation problem: minimise the smoothed functional $\check{V}_N(\tau,\beta)$ with $\beta \to 0$ as $\tau \to \tau^*$, where τ^* is considered to be the global minimiser of the original function $\check{V}_N(\tau)$ [22].

The application of this technique to the SEPNLS method requires a gradient operation on the functional $\check{V}_N(\tau,\beta)$, that is, $\check{V}'_N(\tau,\beta)$. In the case where only the gradient of $\check{V}_N(\cdot)$ is known, the consistent gradient estimate of the smoothed functional can be expressed as [13, 22]

$$\check{V}'_N(\tau,\beta) = \frac{1}{M}\sum_{i=1}^{M}\check{V}'_N(\tau-\beta\eta_i) \tag{12.30}$$

In (12.30) M points η_i are sampled with the PDF $h(\eta)$. Substituting $M = 1$ in (12.30) one obtains the one-sample gradient estimator usually used in the stochastic approximation algorithms [22]

$$\check{V}'_N(\tau,\beta) = \check{V}'_N(\tau-\beta\eta) \tag{12.31}$$

In [22], using $\check{V}'_N(\tau-\beta\eta)$ in (12.31), SAS was applied to the normalised steepest-descent method.

A simplification that involves expressing the gradient $\check{V}'_N(\tau-\beta\eta)$ as a Taylor series around the operating point was proposed in [2]

$$\check{V}'_N(\tau-\beta\eta) = \check{V}'_N(\tau) - \beta\check{V}''_N(\tau)\eta + \cdots \tag{12.32}$$

Additionally, $\check{V}''_N(\tau)$ in the above equation is approximated as an identity matrix and only the first two terms of the Taylor series are kept such that

$$\check{V}'_N(\tau-\beta\eta) \approx \check{V}'_N(\tau) - \beta\eta \tag{12.33}$$

$\check{V}'_N(\tau-\beta\eta)$ was then used to modify the least mean-squares algorithm for the adaptive IIR filtering problem [2].

In this study, we extend the idea in [2] to our SEPNLS method. Replacing $\check{V}'_N(\tau)$ in (12.22) by $\check{V}'_N(\tau) - \beta\eta$, we obtain the following result.

$$\hat{\tau}_N^{(j+1)} = \hat{\tau}_N^{(j)} - \mu^{(j)}\left[\check{R}_N\left(\hat{\tau}_N^{(j)}\right)\right]^{-1}\left(\check{V}'_N\left(\hat{\tau}_N^{(j)}\right) - \beta^{(j)}\eta\right) \tag{12.34}$$

This is our GSEPNLS method that modifies the SEPNLS method with a random perturbation term.

Remark 12.4. As suggested in [22], β has to be chosen large at the start of the iterations and is then decreased to approximately zero as the cost function is sufficiently small. And in [2], the sequence of $\beta^{(j)}$ was chosen as a discrete exponentially decaying function of iteration number j. However, in both studies β were chosen by trial and error. And we have not found in the literature any reliable policy telling us how to determine reliable and efficient values of β. In this chapter, however, based on empirical studies, we recommend the following choice: $\beta^{(j)} = \beta_0 \check{V}_N\left(\hat{\boldsymbol{\tau}}_N^{(j)}\right)$, where β_0 is a sufficiently large positive constant. And β_0 is chosen as 10^5 in this study. It can be understood that if $\check{V}_N\left(\hat{\boldsymbol{\tau}}_N^{(j)}\right)$ is far from the global minimum, $\beta^{(j)}$ is large, and if it becomes near the global minimum, $\beta^{(j)}$ becomes small. Also, it should be mentioned here that the results are not sensitive to the constant β_0.

The overall algorithm of the GSEPNLS method can be summarised as follows

1. Let $j = 0$. Set β_0, the initial estimate $\hat{\boldsymbol{\tau}}_N^{(0)}$ and considerable upper bound of time delays $\bar{\boldsymbol{\tau}}$.
2. Compute prefiltered signals $\{\bar{y}_{f_0}(t_k), \boldsymbol{\varphi}_{\bar{y}}^T(t_k), \boldsymbol{\varphi}_{\bar{u}_1}^T(t_k), \cdots, \boldsymbol{\varphi}_{\bar{u}_r}^T(t_k)\}_{k=1}^N$.
3. Set $\beta^{(j)} = \beta_0 \check{V}_N\left(\hat{\boldsymbol{\tau}}_N^{(j)}\right)$.
4. Perform the following
   a) Compute
   $$\Delta\hat{\boldsymbol{\tau}}_N^{(j+1)} = -\left[\check{\boldsymbol{R}}_N\left(\hat{\boldsymbol{\tau}}_N^{(j)}\right)\right]^{-1}\left(\check{\boldsymbol{V}}_N'\left(\hat{\boldsymbol{\tau}}_N^{(j)}\right) - \beta^{(j)}\boldsymbol{\eta}\right)$$
   b) Compute
   $$\hat{\boldsymbol{\tau}}_N^{(j+1)} = \hat{\boldsymbol{\tau}}_N^{(j)} + \Delta\hat{\boldsymbol{\tau}}_N^{(j+1)}$$
   c) Check if $0 \leq \hat{\tau}_{Ni}^{(j+1)} \leq \bar{\tau}_i$. If not, let $\Delta\hat{\boldsymbol{\tau}}_N^{(j+1)} = 0.5\Delta\hat{\boldsymbol{\tau}}_N^{(j+1)}$ and go back to b).
   d) Check if $\check{V}_N\left(\hat{\boldsymbol{\tau}}_N^{(j+1)}\right) \leq \check{V}_N\left(\hat{\boldsymbol{\tau}}_N^{(j)}\right)$. If not, let $\Delta\hat{\boldsymbol{\tau}}_N^{(j+1)} = 0.5\Delta\hat{\boldsymbol{\tau}}_N^{(j+1)}$ and go back to b).
5. Terminate the algorithm if the stopping condition is satisfied. Otherwise, let $j = j + 1$ and go back to step 3.

Finally, by substituting $\hat{\boldsymbol{\tau}}_N$ into (12.17), the linear parameter vector $\boldsymbol{\theta}$ can be estimated by the linear LS method (12.17).

Remark 12.5. If β_0 is chosen to be zero such that $\beta^{(j)} = 0$ $(j = 0, 1, 2, \cdots)$, the GSEPNLS method is reduced to the local SPENLS method without the random perturbation term.

Remark 12.6. Notice that $\mu^{(j)}$ in (12.22) or (12.34) is chosen to guarantee $\hat{\tau}_{Ni}^{(j+1)} \in \Omega_{\tau_i} (i = 1, \cdots, r)$ and $\check{V}_N\left(\hat{\boldsymbol{\tau}}_N^{(j+1)}\right) \leq \check{V}_N\left(\hat{\boldsymbol{\tau}}_N^{(j)}\right)$. Typically, one starts with $\mu^{(j)} = 1$, and tests if these requirements are met. If not, let $\mu^{(j)} = 0.5\mu^{(j)}$, and re-calculate $\hat{\boldsymbol{\tau}}_N^{(j+1)}$. This process continues iteratively until the requirements are satisfied [8, 10].

Remark 12.7. Owing to d) in step 4 of the GSEPNLS method, $\check{V}_N\left(\hat{\boldsymbol{\tau}}_N^{(j)}\right) j = 0, 1, \cdots$ is a decreasing sequence. Therefore, even if the global minimum of $\check{V}_N\left(\hat{\boldsymbol{\tau}}_N^{(j)}\right)$ and hence $\beta^{(j)}$ are not exactly zero, $\check{V}_N\left(\hat{\boldsymbol{\tau}}_N^{(j)}\right)$ does converge.

12.6 GSEPNIV Method

Although the GSEPNLS method is able to converge to the global minimum, the estimates are acceptable only in the case of low measurement noise. In the case of high measurement noise, the estimates are usually biased.

In the problem of linear parameter estimation, the IV method is a well-known approach that eliminates the estimate bias due to measurement noise through correlation techniques [8, 21, 28–31]. In this section, the IV method is extended to the problem of non-linear parameter estimation. To achieve consistent estimates in the case of high measurement noise, we modify the GSEPNLS method to the GSEPNIV method.

We first introduce the following IV vector $\zeta(t_k, \tau)$ by using the input signals $\bar{u}_j(t_k)$ and sampled noise-free output $x(t_k)$

$$\zeta^T(t_k, \tau) = [-\varphi_{\bar{x}}^T(t_k), \varphi_{\bar{u}_1}^T(t_k - \tau_1), \cdots, \varphi_{\bar{u}_r}^T(t_k - \tau_r)]$$
$$\varphi_{\bar{x}}^T(t_k) = [\bar{x}_{f_1}(t_k), \cdots, \bar{x}_{f_n}(t_k)] \tag{12.35}$$

where, similar to those in (12.7), $\bar{x}_{f_i}(t_k)$ is given as

$$\bar{x}_{f_i}(t_k) = L_d(q^{-1})\left(\frac{T_s}{2}\right)^i (1+q^{-1})^i (1-q^{-1})^{n-i} \bar{x}(t_k) \tag{12.36}$$

and $\bar{x}(t_k) = (1+q^{-1})x(t_k)/2$.

Remark 12.8. In practice, however, the noise-free output is never known. Therefore, a bootstrap scheme is usually used to generate the instrumental variables [8, 21, 28–31]. The estimated noise-free output $\hat{\bar{x}}(t_k)$ is obtained by discretising the estimated system model by the bilinear transformation

$$\hat{\bar{z}}_j(t_k) = \frac{\sum_{i=1}^{m_j} \hat{b}_{ji}\left(\frac{T}{2}\right)^{n-m_j+i} (1+q^{-1})^{n-m_j+i}(1-q^{-1})^{m_j-i}}{\sum_{i=0}^{n} \hat{a}_i \left(\frac{T}{2}\right)^i (1+q^{-1})^i (1-q^{-1})^{n-i}} \bar{u}_j(t_k) \tag{12.37}$$

$$\hat{\bar{x}}(t_k) = \sum_{j=1}^{r} \hat{\bar{z}}(t_k - \hat{\tau}_j)$$

The bootstrap approach is rather *ad hoc*, and strict analysis of convergence is still not available in the literature. However, empirical studies indicate the bootstrap algorithms converge quite well [8, 21, 28–31].

By using the IV vector, we can estimate the linear transfer function parameters by the linear IV method as

$$\hat{\boldsymbol{\theta}}_{IVN}(\boldsymbol{\tau}) = \boldsymbol{R}_{IV}^{-1}(N,\boldsymbol{\tau})\boldsymbol{f}_{IV}(N,\boldsymbol{\tau})$$

$$\boldsymbol{R}_{IV}(N,\boldsymbol{\tau}) = \frac{1}{N-k_s}\sum_{k=k_s+1}^{N}\boldsymbol{\zeta}(t_k,\boldsymbol{\tau})\boldsymbol{\varphi}^T(t_k,\boldsymbol{\tau}) \qquad (12.38)$$

$$\boldsymbol{f}_{IV}(N,\boldsymbol{\tau}) = \frac{1}{N-k_s}\sum_{k=k_s+1}^{N}\boldsymbol{\zeta}(t_k,\boldsymbol{\tau})\bar{y}_{f_0}(t_k)$$

In this case, the residual is given as

$$\check{\varepsilon}_{IV}(t_k,\boldsymbol{\tau}) = \bar{y}_{f_0}(t_k) - \boldsymbol{\varphi}^T(t_k,\boldsymbol{\tau})\boldsymbol{R}_{IV}^{-1}(N,\boldsymbol{\tau})\boldsymbol{f}_{IV}(N,\boldsymbol{\tau}) \qquad (12.39)$$

Then the SEPNLS method (12.22) is modified to the following SEPNIV method

$$\hat{\boldsymbol{\tau}}_{IVN}^{(j+1)} = \hat{\boldsymbol{\tau}}_{IVN}^{(j)} - \mu^{(j)}\left[\check{\boldsymbol{R}}_{IVN}\left(\hat{\boldsymbol{\tau}}_{IVN}^{(j)}\right)\right]^{-1}\check{\boldsymbol{V}}_{IVN}'\left(\hat{\boldsymbol{\tau}}_{IVN}^{(j)}\right) \qquad (12.40)$$

where

$$\check{\boldsymbol{V}}_{IVN}'(\boldsymbol{\tau}) = -\frac{1}{N-k_s}\sum_{k=k_s+1}^{N}\boldsymbol{\psi}_{IV}(t_k,\boldsymbol{\tau})\check{\varepsilon}_{IV}(t_k,\boldsymbol{\tau})$$

$$\check{\boldsymbol{R}}_{IVN}(\boldsymbol{\tau}) = \frac{1}{N-k_s}\sum_{k=k_s+1}^{N}\boldsymbol{\psi}_{IV}(t_k,\boldsymbol{\tau})\boldsymbol{\psi}_m^T(t_k,\boldsymbol{\tau}) \qquad (12.41)$$

$\boldsymbol{\psi}_m(t_k,\boldsymbol{\tau}) = [\psi_{m1}(t_k,\boldsymbol{\tau}),\cdots,\psi_{mr}(t_k,\boldsymbol{\tau})]^T$ is a slight modification of $\boldsymbol{\psi}(t_k,\boldsymbol{\tau})$ given in (12.24)

$$\psi_{mj}(t_k,\boldsymbol{\tau}) = \boldsymbol{\varphi}_{\tau_j}^T(t_k,\boldsymbol{\tau})\boldsymbol{R}_{IV}^{-1}(N,\boldsymbol{\tau})\boldsymbol{f}_{IV}(N,\boldsymbol{\tau})$$
$$+\boldsymbol{\varphi}^T(t_k,\boldsymbol{\tau})\boldsymbol{R}_{IV}^{-1}(N,\boldsymbol{\tau})\boldsymbol{f}_{\tau_j}(N,\boldsymbol{\tau})$$
$$-\boldsymbol{\varphi}^T(t_k,\boldsymbol{\tau})\boldsymbol{R}_{IV}^{-1}(N,\boldsymbol{\tau})\left[\boldsymbol{R}_{\tau_j}(N,\boldsymbol{\tau})+\boldsymbol{R}_{\tau_j}^T(N,\boldsymbol{\tau})\right]\boldsymbol{R}_{IV}^{-1}(N,\boldsymbol{\tau})\boldsymbol{f}_{IV}(N,\boldsymbol{\tau})$$
$$(12.42)$$

and $\boldsymbol{\psi}_{IV}(t_k,\boldsymbol{\tau}) = [\psi_{IV1}(t_k,\boldsymbol{\tau}),\cdots,\psi_{IVr}(t_k,\boldsymbol{\tau})]^T$ is the IV vector to make $\check{\boldsymbol{V}}_{IVN}'(\boldsymbol{\tau})$ and $\check{\boldsymbol{R}}_{IVN}(\boldsymbol{\tau})$ consistent

$$\psi_{IVj}(t_k,\boldsymbol{\tau}) = \boldsymbol{\varphi}_{\tau_j}^T(t_k,\boldsymbol{\tau})\boldsymbol{R}_{IV}^{-1}(N,\boldsymbol{\tau})\boldsymbol{f}_{IV}(N,\boldsymbol{\tau})$$
$$+\boldsymbol{\zeta}^T(t_k,\boldsymbol{\tau})\boldsymbol{R}_{IV}^{-1}(N,\boldsymbol{\tau})\boldsymbol{f}_{\tau_j}(N,\boldsymbol{\tau})$$
$$-\boldsymbol{\zeta}^T(t_k,\boldsymbol{\tau})\boldsymbol{R}_{IV}^{-1}(N,\boldsymbol{\tau})\left[\boldsymbol{R}_{\tau_j}(N,\boldsymbol{\tau})+\boldsymbol{R}_{\tau_j}^T(N,\boldsymbol{\tau})\right]\boldsymbol{R}_{IV}^{-1}(N,\boldsymbol{\tau})\boldsymbol{f}_{IV}(N,\boldsymbol{\tau})$$
$$(12.43)$$

It can be shown through correlation analysis that the solution by the SEPNIV method is equivalent to the noise-free solution by the SEPNLS method, if the data length is sufficiently large.

We first show the following results according to Assumption 12.3

12 Identification Methods for Multiple-input Systems with Time Delays

$$\lim_{N\to\infty} \boldsymbol{R}_{IV}(N,\boldsymbol{\tau}) = \lim_{N\to\infty} \frac{1}{N-k_\mathrm{s}} \sum_{k=k_\mathrm{s}+1}^{N} \boldsymbol{\zeta}(t_k,\boldsymbol{\tau})\boldsymbol{\varphi}^T(t_k,\boldsymbol{\tau})$$
$$= \frac{1}{N-k_\mathrm{s}} \sum_{k=k_\mathrm{s}+1}^{N} \boldsymbol{\zeta}(t_k,\boldsymbol{\tau})\boldsymbol{\zeta}^T(t_k,\boldsymbol{\tau}) \quad (12.44)$$

$$\lim_{N\to\infty} \boldsymbol{f}_{IV}(N,\boldsymbol{\tau}) = \lim_{N\to\infty} \frac{1}{N-k_\mathrm{s}} \sum_{k=k_\mathrm{s}+1}^{N} \boldsymbol{\zeta}(t_k,\boldsymbol{\tau})\bar{y}_{f_0}(t_k)$$
$$= \frac{1}{N-k_\mathrm{s}} \sum_{k=k_\mathrm{s}+1}^{N} \boldsymbol{\zeta}(t_k,\boldsymbol{\tau})\bar{x}_{f_0}(t_k) \quad (12.45)$$

Therefore $\hat{\boldsymbol{\theta}}_{IVN}(\boldsymbol{\tau})$ is consistent if a consistent estimate of $\boldsymbol{\tau}$ is used in (12.38). Furthermore, we have

$$\lim_{N\to\infty} \boldsymbol{R}_{\tau_j}(N,\boldsymbol{\tau}) = \lim_{N\to\infty} \frac{1}{N-k_\mathrm{s}} \sum_{k=k_\mathrm{s}+1}^{N} \boldsymbol{\varphi}_{\tau_j}(t_k,\boldsymbol{\tau})\boldsymbol{\varphi}^T(t_k,\boldsymbol{\tau})$$
$$= \frac{1}{N-k_\mathrm{s}} \sum_{k=k_\mathrm{s}+1}^{N} \boldsymbol{\varphi}_{\tau_j}(t_k,\boldsymbol{\tau})\boldsymbol{\zeta}^T(t_k,\boldsymbol{\tau}) \quad (12.46)$$

$$\lim_{N\to\infty} \boldsymbol{f}_{\tau_j}(N,\boldsymbol{\tau}) = \lim_{N\to\infty} \frac{1}{N-k_\mathrm{s}} \sum_{k=k_\mathrm{s}+1}^{N} \boldsymbol{\varphi}_{\tau_j}(t_k,\boldsymbol{\tau})\bar{y}_{f_0}(t_k)$$
$$= \frac{1}{N-k_\mathrm{s}} \sum_{k=k_\mathrm{s}+1}^{N} \boldsymbol{\varphi}_{\tau_j}(t_k,\boldsymbol{\tau})\bar{x}_{f_0}(t_k) \quad (12.47)$$

From (12.44)–(12.47), we can conclude that each element of $\boldsymbol{\psi}_{IV}(t_k,\boldsymbol{\tau})$ obtained by (12.43) also converges to its noise-free counterpart when the data length is sufficiently large.

By using $\boldsymbol{\psi}_m(t_k,\boldsymbol{\tau})$ and $\boldsymbol{\psi}_{IV}(t_k,\boldsymbol{\tau})$ given in (12.42) and (12.43), and through some correlation operations, we have the following results

$$\lim_{N\to\infty} \check{\boldsymbol{R}}_{IVN}(\boldsymbol{\tau}) = \lim_{N\to\infty} \frac{1}{N-k_\mathrm{s}} \sum_{k=k_\mathrm{s}+1}^{N} \boldsymbol{\psi}_{IV}(t_k,\boldsymbol{\tau})\boldsymbol{\psi}_{IV}^T(t_k,\boldsymbol{\tau}) \quad (12.48)$$

$$\lim_{N\to\infty} \check{\boldsymbol{V}}'_{IVN}(\boldsymbol{\tau})$$
$$= -\lim_{N\to\infty} \frac{1}{N-k_\mathrm{s}} \sum_{k=k_\mathrm{s}+1}^{N} \boldsymbol{\psi}_{IV}(t_k,\boldsymbol{\tau})\big(\bar{y}_{f_0}(t_k) - \boldsymbol{\varphi}^T(t_k,\boldsymbol{\tau})\hat{\boldsymbol{\theta}}_{IVN}(\boldsymbol{\tau})\big)$$
$$= -\lim_{N\to\infty} \frac{1}{N-k_\mathrm{s}} \sum_{k=k_\mathrm{s}+1}^{N} \boldsymbol{\psi}_{IV}(t_k,\boldsymbol{\tau})\big(\bar{x}_{f_0}(t_k) - \boldsymbol{\zeta}^T(t_k,\boldsymbol{\tau})\hat{\boldsymbol{\theta}}_{IVN}(\boldsymbol{\tau})\big) \quad (12.49)$$

The results of (12.48) and (12.49) imply that $\check{\boldsymbol{R}}_{IVN}(\boldsymbol{\tau})$ and $\check{\boldsymbol{V}}'_{IVN}(\boldsymbol{\tau})$ do not include any bias term due to noise. Therefore, $\hat{\boldsymbol{\tau}}_{IVN}$ given in (12.40) is also expected to converge to the noise-free SEPNLS estimate, when the data length is sufficiently large.

The GSEPNLS method (12.34) is therefore modified to the following GSEP-NIV method

$$\hat{\boldsymbol{\tau}}_{IVN}^{(j+1)} = \hat{\boldsymbol{\tau}}_{IVN}^{(j)} - \mu^{(j)} \left[\check{\boldsymbol{R}}_{IVN}\left(\hat{\boldsymbol{\tau}}_{IVN}^{(j)}\right) \right]^{-1} \left(\check{\boldsymbol{V}}'_{IVN}\left(\hat{\boldsymbol{\tau}}_{IVN}^{(j)}\right) - \beta^{(j)}\eta \right) \tag{12.50}$$

Remark 12.9. We should notice that the SEPNIV estimate does not minimise the quadratic cost function. However, if the estimate is consistent, it should minimise the mean squares of the output error

$$V_{IV}(\boldsymbol{\theta}_{IV}, \boldsymbol{\tau}_{IV}) = \frac{1}{N-k_s} \sum_{k=k_s+1}^{N} \left(\bar{y}(t_k) - \hat{\bar{x}}(t_k) \right)^2 \tag{12.51}$$

Remark 12.10. As mentioned previously, in practice, however, the noise-free output is never known. Therefore, a bootstrap scheme is usually used to generate the instrumental variables. Also, in order to improve statistical efficiency of the estimates so that the estimates are less vulnerable to higher measurement noise levels, the prefilter can be iteratively updated by the estimated denominator of the transfer function as in (12.5) [27–31]. The price paid for this effort is that the prefiltered signals have to be computed iteratively so that the computational burden is increased.

The algorithm of the GSEPNIV method is summarised as follows

1. Let $j = 0$. Set β_0, the initial estimates $\hat{\boldsymbol{\theta}}_{IVN}^{(0)}$ and $\hat{\boldsymbol{\tau}}_{IVN}^{(0)}$, and the considerable upper bound of time delays $\bar{\boldsymbol{\tau}}$. Generate the estimated noise-free output $\{\hat{\bar{x}}(t_k)\}_{k=1}^{N}$ by using $\hat{\boldsymbol{\theta}}_{IVN}^{(0)}$ and $\hat{\boldsymbol{\tau}}_{IVN}^{(0)}$, and compute prefiltered signals $\{\boldsymbol{\varphi}_{\hat{\bar{x}}}^T(t_k)\}_{k=1}^{N}$.
2. Compute prefiltered signals $\{\bar{y}_{f_0}(t_k), \boldsymbol{\varphi}_{\bar{y}}^T(t_k), \boldsymbol{\varphi}_{\bar{u}_1}^T(t_k), \cdots, \boldsymbol{\varphi}_{\bar{u}_r}^T(t_k)\}_{k=1}^{N}$.
3. Set $\beta^{(j)} = \beta_0 V_{IV}\left(\hat{\boldsymbol{\theta}}_{IVN}^{(j)}, \hat{\boldsymbol{\tau}}_{IVN}^{(j)}\right)$
4. Perform the following.
    a) Compute
    $$\Delta\hat{\boldsymbol{\tau}}_{IVN}^{(j+1)} = -\check{\boldsymbol{R}}_{IVN}^{-1}\left(\hat{\boldsymbol{\tau}}_{IVN}^{(j)}\right)\left(\check{\boldsymbol{V}}'_{IVN}\left(\hat{\boldsymbol{\tau}}_{IVN}^{(j)}\right) - \beta^{(j)}\eta\right)$$
    b) Compute
    $$\hat{\boldsymbol{\tau}}_{IVN}^{(j+1)} = \hat{\boldsymbol{\tau}}_{IVN}^{(j)} + \Delta\hat{\boldsymbol{\tau}}_{IVN}^{(j+1)}$$
    c) Check if $0 \leq \hat{\tau}_{IVNi}^{(j+1)} \leq \bar{\tau}_i$. If not, let $\Delta\hat{\boldsymbol{\tau}}_{IVN}^{(j+1)} = 0.5\Delta\hat{\boldsymbol{\tau}}_{IVN}^{(j+1)}$ and go back to b).
    d) Compute
    $$\hat{\boldsymbol{\theta}}_{IVN}^{(j+1)} = \boldsymbol{R}_{IV}^{-1}\left(N, \hat{\boldsymbol{\tau}}_{IVN}^{(j+1)}\right)\boldsymbol{f}_{IV}\left(N, \hat{\boldsymbol{\tau}}_{IVN}^{(j+1)}\right).$$
    e) Check if the estimated system model that generates the estimated noise-free output is stable. If not, let $\hat{\boldsymbol{\theta}}_{IVN}^{(j+1)} = \hat{\boldsymbol{\theta}}_{IVN}^{(j)}$.

12 Identification Methods for Multiple-input Systems with Time Delays

Table 12.1. Estimates of the GSEPNLS method in the case of NSR= 5%

α	$\hat{a}_1(3.0)$	$\hat{a}_2(2.0)$	$\hat{b}_{11}(1.0)$	$\hat{b}_{12}(2.0)$	$\hat{b}_{21}(2.0)$	$\hat{b}_{22}(2.0)$	$\hat{\tau}_1(8.83)$	$\hat{\tau}_2(2.32)$
0.1	2.7012	1.6518	1.0016	1.6884	1.9687	1.5568	8.8289	2.3177
	±0.1148	±0.1513	±0.0178	±0.1547	±0.0229	±0.1621	±0.0056	±0.0027
0.4	2.9439	1.9388	1.0012	1.9432	1.9882	1.9325	8.8293	2.3181
	±0.0607	±0.0713	±0.0110	±0.0710	±0.0180	±0.0717	±0.0029	±0.0025
0.8	2.9673	1.9642	1.0019	1.9666	1.9908	1.9607	8.8292	2.3186
	±0.0420	±0.0509	±0.0110	±0.0510	±0.0155	±0.0528	±0.0023	±0.0014
1.2	2.9585	1.9554	1.0030	1.9576	1.9856	1.9515	8.8292	2.3183
	±0.0470	±0.0560	±0.0117	±0.0568	±0.0174	±0.0593	±0.0025	±0.0015
1.6	2.9405	1.9347	1.0059	1.9373	1.9781	1.9300	8.8295	2.3178
	±0.0527	±0.0631	±0.0135	±0.0648	±0.0197	±0.0675	±0.0031	±0.0016
2.0	2.9187	1.9085	1.0099	1.9115	1.9693	1.9033	8.8300	2.3171
	±0.0596	±0.0729	±0.0169	±0.0748	±0.0231	±0.0779	±0.0041	±0.0020
2.4	2.8954	1.8804	1.0142	1.8837	1.9595	1.8750	8.8306	2.3162
	±0.0684	±0.0850	±0.0213	±0.0865	±0.0275	±0.0902	±0.0054	±0.0024

  f) Generate the estimated noise-free output $\{\hat{\bar{x}}(t_k)\}_{k=1}^{N}$ by using $\hat{\boldsymbol{\theta}}_{IVN}^{(j+1)}$ and $\hat{\boldsymbol{\tau}}_{IVN}^{(j+1)}$.

  g) Check if
  $$V_{IV}\left(\hat{\boldsymbol{\theta}}_{IVN}^{(j+1)},\hat{\boldsymbol{\tau}}_{IVN}^{(j+1)}\right) \leq V_{IV}\left(\hat{\boldsymbol{\theta}}_{IVN}^{(j)},\hat{\boldsymbol{\tau}}_{IVN}^{(j)}\right).$$
  If not, let $\Delta\hat{\boldsymbol{\tau}}_{IVN}^{(j+1)} = 0.5\Delta\hat{\boldsymbol{\tau}}_{IVN}^{(j+1)}$ and go back to b).

  h) Compute prefiltered signals $\{\boldsymbol{\varphi}_{\hat{\bar{x}}}^T(t_k)\}_{k=1}^{N}$.

5. Terminate the algorithm if the stopping condition is satisfied. Otherwise, let $j = j + 1$. Go back to step 3 if the prefilter is fixed; go back to step 2 if the prefilter is iteratively updated.

Remark 12.11. In some cases, the GSEPNIV algorithm using the bootstrap technique may exhibit worse convergence behaviour than the GSEPNLS algorithm. Therefore, it is recommended to initialise the GSEPNIV algorithm by the GSEPNLS estimates.

Remark 12.12. There are two versions of the GSEPNIV algorithm described above. One is the GSEPNIV method with a fixed prefilter (GSEPNIV-F), and the other is the GSEPNIV method with an iteratively updated prefilter (GSEPNIV-IU). The former requires less computational burden whereas the latter yields better statistical performance of the estimates [27–31].

12.7 Numerical Results

Consider the following MISO continuous-time system

Table 12.2. Estimates of the GSEPNIV-F method in the case of NSR= 30%

α	$\hat{a}_1(3.0)$	$\hat{a}_2(2.0)$	$\hat{b}_{11}(1.0)$	$\hat{b}_{12}(2.0)$	$\hat{b}_{21}(2.0)$	$\hat{b}_{22}(2.0)$	$\hat{\tau}_1(8.83)$	$\hat{\tau}_2(2.32)$
0.4	2.9814	1.9863	0.9941	1.9887	1.9956	1.9809	8.8260	2.3175
	±0.0934	±0.1077	±0.0308	±0.1102	±0.0456	±0.1175	±0.0083	±0.0053
0.8	2.9825	1.9828	0.9965	1.9878	1.9992	1.9808	8.8262	2.3176
	±0.0931	±0.1168	±0.0310	±0.1180	±0.0414	±0.1258	±0.0081	±0.0052
1.2	2.9806	1.9824	0.9970	1.9881	1.9990	1.9811	8.8260	2.3175
	±0.1101	±0.1471	±0.0350	±0.1447	±0.0416	±0.1579	±0.0082	±0.0054
1.6	2.9749	1.9766	0.9984	1.9830	1.9976	1.9761	8.8257	2.3178
	±0.1298	±0.1803	±0.0414	±0.1739	±0.0446	±0.1931	±0.0086	±0.0055
2.0	2.9630	1.9620	1.0017	1.9692	1.9947	1.9626	8.8255	2.3168
	±0.1536	±0.2152	±0.0488	±0.2059	±0.0503	±0.2293	±0.0091	±0.0059
2.4	2.9435	1.9373	1.0072	1.9453	1.9896	1.9388	8.8255	2.3163
	±0.1826	±0.2519	±0.0559	±0.2410	±0.0589	±0.2671	±0.0097	±0.0064

Table 12.3. Estimates of the GSEPNIV-IU method in the case of NSR= 30%

$\hat{a}_1(3.0)$	$\hat{a}_2(2.0)$	$\hat{b}_{11}(1.0)$	$\hat{b}_{12}(2.0)$	$\hat{b}_{21}(2.0)$	$\hat{b}_{22}(2.0)$	$\hat{\tau}_1(8.83)$	$\hat{\tau}_2(2.32)$
2.9822	1.9822	0.9965	1.9872	1.9989	1.9795	8.8262	2.3177
±0.0923	±0.1138	±0.0305	±0.1153	±0.0422	±0.1226	±0.0081	±0.0053

$$\ddot{x}(t) + a_1\dot{x}(t) + a_2 x(t) = b_{11}\dot{u}_1(t-\tau_1) + b_{12}u_1(t-\tau_1) \\ + b_{21}\dot{u}_2(t-\tau_2) + b_{22}u_2(t-\tau_2) \quad (12.52)$$

where

$$a_1 = 3.0, \ a_2 = 2.0, \ b_{11} = 1.0, \ b_{12} = 2.0 \\ b_{21} = 2.0, \ b_{22} = 2.0, \ \tau_1 = 8.83, \ \tau_2 = 2.32 \quad (12.53)$$

The corresponding transfer function model is

$$X(s) = \frac{e^{-8.83s}}{s+1}U_1(s) + \frac{2e^{-2.32s}}{s+2}U_2(s) \\ = \frac{e^{-8.83s}(s+2)U_1(s) + e^{-2.32s}(2s+2)U_2(s)}{s^2+3s+2} \quad (12.54)$$

Each input signal is output of a zero-order hold driven by a white signal filtered by a Butterworth filter

$$\frac{1}{(s/\omega_c)^2 + \sqrt{2}(s/\omega_c) + 1} \quad (\omega_c = 4.0) \quad (12.55)$$

which is discretised by the bilinear transformation.
The input and output signals were sampled with the sampling period $T_s = 0.05$ s. Since the system passband was not known prior to identification, several different values of α in the fixed low-pass prefilter $L(s)$ were chosen. The effects of the values of α will be discussed later based on the simulation results.

Table 12.4. Hit ratio of the GSEPNLS method in the case of NSR= 5%

α	0.4	0.8	1.2	1.6	2.0	2.4
Hit ratio (%)	94	95	92	96	95	93

Table 12.5. Hit ratio of the GSEPNIV method in the case of NSR= 30%

α	0.4	0.8	1.2	1.6	2.0	2.4	IU
Hit ratio (%)	92	96	99	98	98	98	98

12.7.1 GSEPNLS Method in the Case of Low Measurement Noise

First, for one fixed realisation of the random perturbation vector η, and one fixed realisation of the input signals, the GSEPNLS methods were implemented for 20 realisations of the low measurement noise (NSR (noise-to-signal ratio) was set as 5%) with data length of 1000. NSR was defined as the ratio of σ_ξ/σ_x, where σ_ξ and σ_x are the standard deviations of the measurement noise and of the noise-free output, respectively. The algorithm was terminated after 150 iterations. The time delays were searched in the range of $\tau_1, \tau_2 \in [0, 10]$. The initial estimates were set at $\hat{\boldsymbol{\tau}}_N^{(0)} = [1, 9]^T$. The results are shown in Table 12.1. The table includes the means and standard deviations of the estimates. It can be seen that the GSEPNLS method gives satisfactory estimates when $\alpha = 0.4, 0.8, 1.2, 1.6$. Also, it is found that the results by $\alpha = 0.1$ are not acceptable. And when $\alpha = 2.0, 2.4$, the results become worse. Notice that the passband of the prefilter $L(s)$ by $\alpha = 0.4, 0.8, 1.2, 1.6$ is relatively close to that of the system under study. On the other hand, the passband of the filter $L(s)$ by $\alpha = 0.1$ is relatively too broad, and the passband by $\alpha = 2.0, 2.4$ is relatively too narrow, compared to that of the system. These results reflect the suggestions in [16, 28–31]. However, it can be found that the results are not sensitive to the value of α. After several trials of the identification algorithm, we can find the suitable range of α where the results are not sensitive and hence are similar to each other. If necessary, we can reset the prefilter parameter α based on the passband of the identified system model and then run the algorithm again, to improve the statistical efficiency of the estimates. Also, the prefilter can be iteratively updated by the estimated denominator of the transfer function as in (12.5) [27–31]. The price is that the prefiltered signals have to be computed iteratively so that the computational burden is increased.

It might be interesting to investigate the hit ratio of the algorithm, that is, the percentage of the case that the algorithm converges to the neighbourhood of the global minimum. For each value of $\alpha = 0.4, 0.8, 1.2, 1.6, 2.0, 2.4$, the GSEPNLS method was implemented, respectively, for 100 different realisations of the input signals, measurement noise, random perturbation vector η, initial values randomly chosen within the range of $[0, 10]$. The data length, the maximum number of iterations and the noise level were chosen as the same

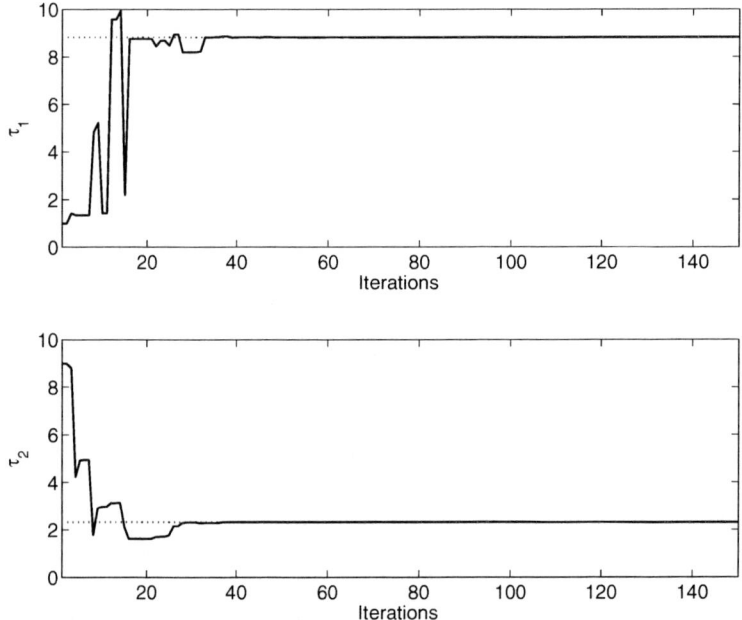

Fig. 12.1. Convergence behaviour of the time-delay estimates by the GSEPNLS method in the case of low measurement noise ($\alpha = 0.4$)

as the previous ones. The results are shown in Table 12.4. It can be seen that the GSEPNLS algorithm achieves a high hit ratio.

An example of the convergence behaviour of the estimates of the time delays is shown in Figure 12.1. The surface graph of the corresponding quadratic cost function (12.18) depending on τ_1 and τ_2 is shown in Figure 12.2, which implies that by the local SEPNLS method the estimate $\hat{\tau}$ can be easily stuck at a local minimum if the start point is not near the globally optimal solution. Figure 12.3 shows the locus of $\hat{\tau}$ on the contour of the quadratic cost function by the GSEPNLS method. It can be found that $\hat{\tau}$, which started at $\hat{\tau}_N^{(0)} = [1,9]^T$, reached the globally optimal solution very easily.

12.7.2 GSEPNIV Method

Through extensive simulation studies, we found that the GSEPNIV method converges to the global minimum in most cases of various combinations of the initial estimates, realisations of the inputs, random perturbation vector and measurement noise.

For one fixed realisation of η, and one fixed realisation of the input signals, the algorithms were implemented for 20 realisations of a high measurement noise

12 Identification Methods for Multiple-input Systems with Time Delays   359

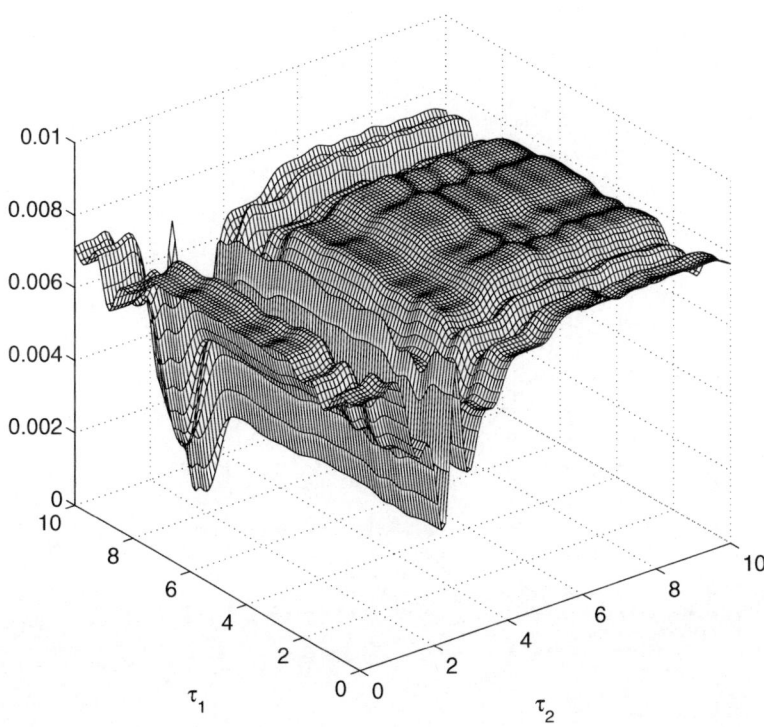

Fig. 12.2. Surface graph of the quadratic cost function in the case of low measurement noise ($\alpha = 0.4$)

of NSR=30%. The data length was chosen as 4000. The initial estimates were set at $\widehat{\boldsymbol{\tau}}_N^{(0)} = [1,9]^T$. The results by the GSEPNLS methods are not acceptable and hence are not shown here.

For the GSEPNIV method, the first 50 iterations were performed by the GSEPNLS method and the GSEPNLS estimates were provided as the initial values for the left 100 iterations of the GSEPNIV method.

The results by the GSEPNIV method with a fixed prefilter (GSEPNIV-F) are shown in Table 12.2, and the results by the GSEPNIV method with an iteratively updated prefilter (GSEPNIV-IU) are shown in Table 12.3. It can be seen that the GSEPNIV-F method yields satisfactory results for $\alpha = 0.4, 0.8, 1.2, 1.6$. When $\alpha = 2.0, 2.4$, however, the results become worse. The results by $\alpha = 0.1$ are not satisfactory and hence are not shown here. Also, it is found that the GSEPNIV-IU method yields satisfactory results that are similar to those by the best choice of α. The results reflect the claim

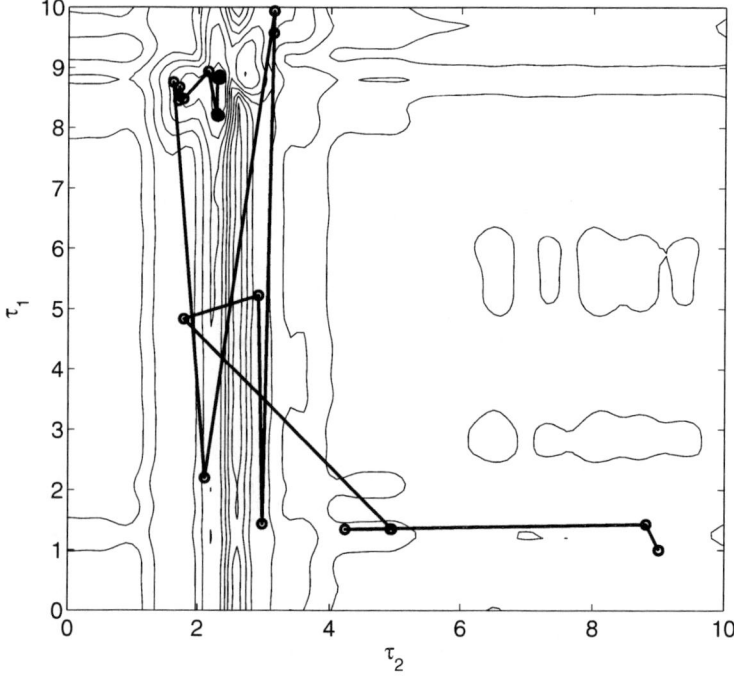

Fig. 12.3. Locus of the time-delay estimates on the contour of the quadratic cost function in the case of low measurement noise ($\alpha = 0.4$)

in Remark 12.10.

To investigate the hit ratio of the GSEPNIV algorithm, the GSEPNIV-F method for each value of $\alpha = 0.4, 0.8, 1.2, 1.6, 2.0, 2.4$ and the GSEPNIV-IU method were implemented respectively for 100 different realisations of the input signals, measurement noise, random perturbation vector η, initial values randomly chosen within the range of $[0, 10]$. The data length, the maximum number of iterations and the noise level were chosen as the same as the previous ones. The results are shown in Table 12.5, where IU means the case of iterative updated filter. It can be seen that the GSEPNIV algorithm achieves a high hit ratio in the case of high measurement noise.

12.8 Conclusions

In this chapter, we considered the identification problem of MISO continuous-time systems with multiple unknown time delays from sampled input–output data. The GSEPNLS method that estimates the time delays and transfer function parameters separably was proposed, by using the stochastic global-optimisation techniques to reduce the possibility of convergence to a local minimum. And then the GSEPNLS method was modified to a novel GSEP-NIV method to yield consistent estimates in the presence of high measurement noise if the algorithm converges to the global minimum. Through numerical studies, we found that the estimates by the proposed identification algorithms converge to the globally optimal solutions quite well. The GSEPNLS method yields acceptable estimates in the case of low measurement noise, and the GSEPNIV method yields consistent estimates in the presence of high measurement noise. Also, we found that the results are not sensitive to the initial estimates, although the estimation problem is strongly non-linear and multi-modal.

References

1. M. Agarwal and C. Canudas. On-line estimation of time delay and continuous-time process parameters. *International Journal of Control*, 46(1):295–311, 1987.
2. W. Edmonson, J. Principe, K. Srinivasan and C. Wang. A global least mean square algorithm for adaptive IIR filtering. *IEEE Transactions on Circuits and Systems II*, 45(3):379-384, 1998.
3. G. Ferreti, C. Maffezzoni and R. Scattolini. Recursive estimation of time delay in sampled systems. *Automatica*, 27(4): 653-661, 1991.
4. P.J. Gawthrop and M.T. Nihtilä. Identification of time-delays using a polynomial identification method. *System and Control Letters*, 5(4):267-271, 1985.
5. P.J. Gawthrop, M.T. Nihtilä and A.B. Rad. Recursive parameter estimation of continuous-time systems with unknown time delay. *Control Theory and Advanced Technology*, 5(3), 227-248, 1989.
6. Z.H. Jiang and W. Schaufelberger. *Block-Pulse Function and Their Applications in Control Systems*. Springer-Verlag, Berlin, Germany, 1992.
7. I. Kollar. *Frequency Domain System Identification Toolbox for MATLAB®*. Gamax Inc, 2002-2006.
8. L. Ljung. *System Identification. Theory for the User*. 2nd edition, Prentice Hall, Upper Saddle River, USA, 1999.
9. G. Nemeth and I. Kollar. Step-invariant transform from z- to s-domain–a general framework. *IEEE Instrumentation and Measurement Technology Conference*, pages 902–907, Baltimore, MD, 2000.
10. L.S.H. Ngia. Separable nonlinear least-squares methods for efficient off-line and on-line modeling of systems using Kautz and Laguerre filters. *IEEE Transactions on Circuits and Systems II*, 48(6):562-579, 2001.
11. A.E. Pearson and C.Y. Wuu. Decoupled delay estimation in the identification of differential delay systems. *Automatica*, 20(6):761-772, 1984.
12. G.P. Rao and H. Unbehauen. Identification of continuous-time systems. *IEE Proceedings, Control Theory and Applications*, 153(2):185-220, 2006.
13. R.Y. Rubinstein. *Simulation and the Monte Carlo Method*. John Wiley, New York, USA, 1981.
14. A. Ruhe and P.A. Wedin. Algorithms for separable nonlinear least squares problems. *SIAM Review*, 22(3):318-337, 1980.

15. S. Sagara, Z.J. Yang, K. Wada and T. Tsuji. Parameter identification and adaptive control of continuous systems with zero-order hold. *13th IFAC World Congress*, Sydney, Australia, 1993.
16. S. Sagara, Z.J. Yang and K. Wada. Identification of continuous systems using digital low-pass filters. *International Journal of Systems Science*, 22(7):1159-1176, 1991.
17. J. Schoukens and R. Pintelon. *Identification of Linear Systems: A Practical Guideline to Accurate Modeling*. Pergamon Press, Oxford, UK, 1991.
18. R. Pintelon and J. Schoukens. *System Identification: A Frequency Domain Approach*. IEEE Press, Piscataway, NJ, USA, 2001.
19. J. Schoukens, R. Pintelon and H. Van Hamme. Identification of linear dynamic systems using piecewise constant excitations: use, misuse and alternatives. *Automatica*, 30(7): 1152-1169, 1994.
20. J. Sjöberg and M. Viberg. Separable non-linear least-squares minimization–possible improvements for neural net fitting. *IEEE Workshop in Neural Networks for Signal Processing*, 1997.
21. T. Söderström and P. Stoica. *Instrumental Variable Methods for System Identification*. Springer Verlag, Berlin, Germany, 1983.
22. M.A. Styblinski and T.S. Tang. Experiments in nonconvex optimization: stochastic approximation with function smoothing and simulated annealing. *Neural Networks*, 3(5):467-483, 1990.
23. J. Tuch, A. Feuer and Z.J. Palmor. Time delay estimation in continuous linear time-invariant systems. *IEEE Transactions on Automatic Control*, 39(4):823-827, 1994.
24. H. Unbehauen and G.P. Rao. Continuous-time approaches to system identification–a survey. *Automatica*, 26(1):23-35, 1990.
25. Z.J. Yang, T. Hachino, T. Tsuji and S. Sagara. Identification of parameters and time delays of continuous systems using the genetic algorithm. *10th IFAC Symposium on System Identification*, Copenhagen, Denmark, July 1994.
26. Z.J. Yang, T. Hachino and T. Tsuji. On-line identification of continuous time-delay systems combining the least-squares techniques with the genetic algorithm. *International Journal of Control*, 66(1):23-42, 1997.
27. P.C. Young. In-flight dynamic checkout. *IEEE Transactions on Aerospace*, 2(3):1106-1111, 1964.
28. P.C. Young. Some observations on instrumental variable methods of time-series analysis. *International Journal of Control*, 23(5):593-612, 1976.
29. P.C. Young. Parameter estimation for continuous-time models–a survey. *Automatica*, 17(1):23-39, 1981.
30. P.C. Young. The instrumental variable method: A practical approach to identification and system parameter estimation. In H.A. Barker and P.C. Young (eds), *Identification and System Parameter Estimation*, pages 1–16, Pergamon Press, Oxford, UK, 1985.
31. P.C. Young and A.J. Jakeman. Refined instrumental variable methods of recursive time-series analysis. Part I, SISO systems. *International Journal of Control*, 29(1):1-30, 1979.
32. W.X. Zheng and C.B. Feng. Identification of stochastic time lag systems in the presence of colored noise. *Automatica*, 26:769-779, 1990.

13

Closed-loop Parametric Identification for Continuous-time Linear Systems via New Algebraic Techniques

Michel Fliess[1] and Hebertt Sira-Ramírez[2]

[1] Projet ALIEN, INRIA Futurs & Équipe MAX, LIX, France
[2] Cinvestav-IPN, México

13.1 Introduction

A few years ago the present authors launched a new approach to parametric identification of linear continuous-time systems [11]. Its main features may be summarised as follows:

- closed-loop identification is permitted thanks to the real-time identification scheme;
- the robustness with respect to noisy data is obtained without knowing the statistical properties of the corrupting noises.

This chapter is devoted to a new exposition of those methods and to their illustration via three examples and their computer simulations. Our mathematical techniques are quite different from those in the huge literature on this subject. We are mainly employing algebraic tools:

1. the module-theoretic approach to linear systems,
2. elementary non-commutative ring theory,
3. operational calculus.

The chapter is organised as follows. Section 13.2 gives a short summary of the module-theoretic setting for continuous-time linear systems[3]. Section 13.3 defines *linear identifiability*, which is sufficient for most practical cases, by introducing the notion of *algebraic derivatives* and the corresponding non-commutative ring theory. Section 13.4 discusses two kinds of perturbations:

1. The *structured perturbations*, which satisfy time-varying linear differential equations, are annihilated by suitable linear differential operators.

---

[3] This module-theoretic presentation of linear systems started in [4]. See [2] for an excellent introduction and related references. This standpoint provides a most useful way for synthesising model-based predictive control, which employs concepts stemming from *flatness*-based control [9, 28].

2. The *unstructured perturbations* are considered as highly fluctuating, or oscillating, phenomena. They are attenuated by suitable low-pass filters, like iterated time integrals.

We thus arrive at estimators that are robust with respect to a large class of noises. Our numerical simulations in Sections 13.5 and 13.6 deal, respectively, with the open-loop dragging of an unknown mass and the feedback control of a first-order system. They are compared to standard adaptive methods, which seem to be less efficient. We note also in Section 13.5 that the rather complex notion of persistently exciting signal becomes quite pointless in our setting. For the double-bridge buck converter of Section 13.7, which is more realistic (see [30]) than the previous examples, we are able to achieve a rather successful closed-loop parametric identification. A short conclusion relates our work to others.

Let us add that we tried to write the examples in such a way that they might be grasped without the necessity of reading the sections on the algebraic background. Our standpoint on parametric identification should therefore be accessible to most engineers.

13.2 A Module-theoretic Approach to Linear Systems: a Short Summary

13.2.1 Some Basic Facts about Modules over Principal Ideal Rings

Let k be a given field[4]. Write $k[s]$ the ring of polynomials $\sum_{\text{finite}} a_\nu s^\nu$, $a_\nu \in k$, in the indeterminate s. It is well known that $k[s]$ is a *principal ideal ring*, *i.e.*, any *ideal* of $k[s]$ may be generated by a single element. A $k[s]$-module M is said to be *finitely generated*, or of *finite type*, if, and only if, $M = \text{span}_{k[s]}(S)$, where S is a finite set. Module M is said to be *free*[5] if, and only if, there exists S whose elements are $k[s]$-linearly independent; S is then called a *basis*. The cardinalities, *i.e.*, the numbers of elements, of two bases are equal. Any submodule of a finitely generated (resp. free) $k[s]$-module is again finitely generated (resp. free). Any quotient module of a finitely generated $k[s]$-module is again finitely generated.

An element $x \in M$ is said to be *torsion* if, and only if, there exists $\varpi \in k[s]$, $\varpi \neq 0$, such that $\varpi x = 0$. The set of all torsion elements of M is a submodule M^{tor}, $\{0\} \subseteq M^{\text{tor}} \subseteq M$, which is called the *torsion submodule*. If $M^{\text{tor}} = M$, M is said to be *torsion*. If $M^{\text{tor}} = \{0\}$, M is said to be *torsion free*. Any free module is of course torsion free. As is well known, the converse holds true for finitely generated torsion-free $k[s]$-module M. The quotient module M/M^{tor} is free. The next property of a finitely generated $k[s]$-module M is crucial

---

[4] See, *e.g.*, [17] for a classic and well-written introduction to commutative algebra.
[5] By convention, the trivial module $\{0\}$, generated by the empty set $\emptyset$, is free.

$$M = M^{\text{tor}} \oplus F \tag{13.1}$$

where the free module $F = M/M^{\text{tor}}$ is defined up to isomorphism.

A module that is generated by a single element g is finite-dimensional, when viewed as a k-vector space, if, and only if, g is torsion. The extension to a finitely generated module M is immediate: M is torsion if, and only if, the dimension $\dim_k(M)$ of M, viewed as a k-vector space, is finite.

Example 13.1. Consider the set of $k[s]$-linear equations

$$\sum_{\kappa=1}^{\mu} a_{\iota\kappa}\xi_\kappa = 0, \quad a_{\iota\kappa} \in k[s], \ \iota = 1,\ldots,\nu \tag{13.2}$$

in the unknowns $\xi_1,\ldots,\xi_\mu$. Let F be the free $k[s]$-module with basis $(f_1,\ldots,f_\mu)$. Let E be the submodule generated by $e_\iota = \sum_{\kappa=1}^{\mu} a_{\iota\kappa}f_\kappa$, $\iota = 1,\ldots,\nu$. Then, the module corresponding to equations (13.2) is $M = F/E$. Equations (13.2) may be written in the following matrix form

$$P_M \begin{pmatrix} \xi_1 \\ \vdots \\ \xi_\mu \end{pmatrix} = 0 \tag{13.3}$$

$P_M \in k[s]^{\nu \times \mu}$ is a *presentation matrix* of Λ.

13.2.2 Formal Laplace Transform

Let $k(s)$ be the quotient field of $k[s]$, *i.e.*, the field of rational functions over k in the indeterminate s. Let M be a finitely generated $k[s]$-module. The elements of the tensor product $\hat{M} = k(s) \otimes_{k[s]} M$ are finite sums of products $q^{-1}x$, $x \in M$, $q \in k[s]$, $q \neq 0$. It is a $k(s)$-vector space, called the *transfer vector space* of M. The $k[s]$-linear mapping $M \to \hat{M}$, $m \mapsto \hat{m} = 1 \otimes m$, is the *formal Laplace transform*[6]. Its kernel is the torsion submodule M^{tor}. The formal Laplace transform is thus injective if, and only if, M is free. By definition, the *rank* of M, which is written $\text{rk}\,(M)$, is $\text{rk}\,(M) = \dim_{k(s)}(\hat{M})$. It is clear that M is torsion if, and only if, $\text{rk}\,(M) = 0$. Take two modules M_1, M_2, $M_1 \subseteq M_2$. Then, $\text{rk}\,(M_2/M_1) = \text{rk}\,(M_2) - \text{rk}\,(M_1)$. Thus, $\text{rk}\,(M_1) = \text{rk}\,(M_2)$ if, and only if, the quotient module M_2/M_1 is torsion. For any set $\boldsymbol{x} = (x_1,\ldots,x_\alpha) \subset M$, the following equality is obvious

$$\text{rk}\,(\text{span}_{k[s]}(\boldsymbol{x})) = \dim(\text{span}_{k(s)}(\hat{\boldsymbol{x}})) \tag{13.4}$$

The next property is stating a useful matrix characterisation of torsion modules

---

[6] See [5] for more details.

Proposition 13.1. *The module corresponding to (13.3) is torsion if, and only if, $rk(P_M) = \mu$. If (13.3) is square, i.e., $\mu = \nu$, this condition is equivalent to $\det(P_M) \neq 0$.*

Proof. The formal Laplace transform yields

$$P_M \begin{pmatrix} \hat{\xi}_1 \\ \vdots \\ \hat{\xi}_\mu \end{pmatrix} = 0$$

The module is torsion if, and only if, $\hat{\xi}_1 = \ldots \hat{\xi}_n = 0$. This latter condition is equivalent to $\text{rk}(P_M) = \mu$.

Example 13.2. Let T be a finitely generated torsion $k[s]$-module. Then, $\dim_k(T) = n < \infty$. Pick up a basis $\boldsymbol{b} = (b_1, \ldots, b_n)$ of T viewed as a k-vector space. To the k-linear mapping $s : T \to T$, $\tau \mapsto s\tau$, corresponds the matrix $A \in k^{n \times n}$ with respect to $\boldsymbol{b}$. This is equivalent to saying that T is defined by the following matrix equation

$$s \begin{pmatrix} b_1 \\ \vdots \\ b_n \end{pmatrix} = A \begin{pmatrix} b_1 \\ \vdots \\ b_n \end{pmatrix} \tag{13.5}$$

It is clear that $\det(s - A) \neq 0$.

13.2.3 Basic System-theoretic Definitions

A *k-linear system* is a finitely generated free $k[s]$-module Λ where we have distinguished a finite subset of *perturbation*, or *disturbance*, variables $\boldsymbol{\pi} = (\underline{\pi}_1, \ldots, \underline{\pi}_r)$.

Remark 13.1. Set $k = \mathbb{C}$. Consider the operational equation $a\chi = 0$, $a \in \mathbb{C}[s]$, $a \neq 0$, in the unknown χ. Its unique solution is $\chi = 0$. This means that any torsion element would be trivial. Note, moreover, that the linear differential equation $\dot{x} = 0$, which corresponds to a torsion $\mathbb{C}[\frac{d}{dt}]$-module, yields the operational equation $s\hat{x} - x(0) = 0$, which corresponds to a free $\mathbb{C}[s]$-module with basis $\{\hat{x}\}$. See [10] for a thorough discussion.

The *nominal*, or *unperturbed*, system Λ^{nom} is defined by the quotient module

$$\Lambda^{\text{nom}} = \Lambda/\text{span}_{k[s]}(\boldsymbol{\pi})$$

The canonical image of any $\lambda \in \Lambda$ is written $\lambda^{\text{nom}} \in \Lambda^{\text{nom}}$. We might sometimes call Λ a *perturbed* system. Note, moreover, that Λ^{nom} is not necessarily free.

Example 13.3. The module corresponding to $s\underline{x} = \underline{\pi}$, where $\underline{\pi} \in \mathrm{span}_{k[s]}(\underline{\pi})$, $\underline{\pi} \neq 0$, is free[7]. The module corresponding to the nominal system $s\underline{x}^{\mathrm{nom}} = 0$ is torsion.

A *k-linear dynamics* is a k-linear system Λ, which is equipped with a finite set $\boldsymbol{u} = (u_1, \ldots, u_m)$ of *control* variables, such that

- the control variables do not interact with the perturbation variables

$$\mathrm{span}_{k[s]}(\boldsymbol{u}) \cap \mathrm{span}_{k[s]}(\underline{\pi}) = \{0\} \qquad (13.6)$$

- the quotient module $\Lambda^{\mathrm{nom}}/\mathrm{span}_{k[s]}(\boldsymbol{u}^{\mathrm{nom}})$ is torsion.

If $\boldsymbol{u} = \emptyset$, this last condition implies that Λ^{nom} is torsion. The control variables are said to be *independent* if, and only if, $u_1, \ldots, u_m$ are $k[s]$-linearly independent.

The set of *output* variables is a finite subset $\boldsymbol{y} = (y_1, \ldots, y_p) \subset \Lambda$. A dynamics Λ with output variables is called a *k-linear input/output system*. A system is said to be *mono-variable* if, and only if, $m = p = 1$. If not, it is said to be *multi-variable*.

13.2.4 Transfer Matrices

Consider the nominal dynamics Λ^{nom} with control variables

$$\boldsymbol{u}^{\mathrm{nom}} = (u_1^{\mathrm{nom}}, \ldots, u_m^{\mathrm{nom}})$$

The transfer $k(s)$-vector space (see Section 13.2.2) $\hat{\Lambda}^{\mathrm{nom}}$ is spanned by $\hat{\boldsymbol{u}}^{\mathrm{nom}}$, which is a basis if the control variables are independent. It yields, with nominal output variables $\boldsymbol{y}^{\mathrm{nom}} = (y_1^{\mathrm{nom}}, \ldots, y_p^{\mathrm{nom}})$,

$$\begin{pmatrix} \hat{y}_1^{\mathrm{nom}} \\ \vdots \\ \hat{y}_p^{\mathrm{nom}} \end{pmatrix} = T \begin{pmatrix} \hat{u}_1^{\mathrm{nom}} \\ \vdots \\ \hat{u}_m^{\mathrm{nom}} \end{pmatrix} \qquad (13.7)$$

where $T \in k(s)^{p \times m}$ is the *(nominal) transfer matrix*, which is uniquely defined if, and only if, the control variables are independent. If $m = p = 1$, T is called a *(nominal) transfer function*. Matrix T is said to be *proper* (resp. *strictly proper*) if, and only if, its entries are proper (resp. strictly proper) rational functions.

---

[7] The initial condition $x(0)$ should be considered as a perturbation (see [10]).

13.3 Identifiability

13.3.1 Uncertain Parameters

Let the field k be a finite algebraic extension[8] of $k_0(\Theta)$, where

- k_0 is a given ground field;
- $\Theta = (\theta_1, \ldots, \theta_\tau)$ is a finite set of *uncertain*, or *unknown*, *parameters*.

13.3.2 The Algebraic Derivative and a New Module Structure

Call[9] with [21, 22, 32] the derivation $\frac{d}{ds}$ with respect to s the *algebraic derivative*[10]. Introduce a new commutative field K of *constants*, i.e., $\forall \xi \in K, \frac{d\xi}{ds} = 0$. The ring $K[s, \frac{d}{ds}]$ of linear differential operators $\sum_{\text{finite}} a_\nu \frac{d^\nu}{ds^\nu}$, $a_\nu \in K[s]$, with polynomial coefficients, is called the *Weyl algebra* (see, e.g., [20]). It is non-commutative, as shown by the commutator $[\frac{d}{ds}, s]$

$$[\frac{d}{ds}, s] = \frac{d}{ds}s - s\frac{d}{ds} = 1$$

Introduce the over-ring $K(s)[\frac{d}{ds}]$ of linear differential operators $\sum_{\text{finite}} b_\nu \frac{d^\nu}{ds^\nu}$, $b_\nu \in K(s)$, with rational coefficients. It is a non-commutative left and right principal ideal ring (see, e.g., [20]), i.e., any left (resp. right) ideal of $K(s)[\frac{d}{ds}]$ may be generated by a single element. Take again system Λ, i.e., a finitely generated free $k[s]$-module. Elements of the tensor product $\Lambda_{k(s)[\frac{d}{ds}]} = k_0(s)[\frac{d}{ds}] \otimes_{k[s]} \Lambda$ are (see, e.g., [20]) finite sums of products $r\lambda$, $r \in k_0(s)[\frac{d}{ds}]$, $\lambda \in \Lambda$. This means that $\Lambda_{k(s)[\frac{d}{ds}]}$ may be endowed with a structure of left $k(s)[\frac{d}{ds}]$-module.

13.3.3 Linear Identifiability

The uncertain parameters $\Theta = (\theta_1, \ldots, \theta_\tau)$ are said to be *linearly identifiable* if, and only if,

$$P \begin{pmatrix} \theta_1 \\ \vdots \\ \theta_\tau \end{pmatrix} = Q + R \tag{13.8}$$

where

---

[8] A *field extension* L/K is given by two fields K and L such that $K \subseteq L$ (see, e.g., [17]). It is a *finite algebraic* extension if, and only if, the dimension of L viewed as a vector space over K is finite. Then, any element of L is *algebraic* over K, i.e., satisfies a polynomial equation with coefficients in K.

[9] See [8, 10, 11] for more details.

[10] Remember (see, e.g., [3, 21, 22, 25, 32]) that $\frac{d}{ds}$ corresponds in the time domain to the multiplication by $-t$.

- the entries of the matrices P and Q, of respective sizes $\tau \times \tau$ and $\tau \times 1$, belong to $\mathrm{span}_{k_0(s)[\frac{d}{ds}]}(\boldsymbol{u}, \boldsymbol{y})$,
- $\det(P) \neq 0$;
- R is a $\tau \times 1$ matrix with entries in $\mathrm{span}_{k(s)[\frac{d}{ds}]}(\boldsymbol{\pi})$.

The uncertain parameters $\boldsymbol{\Theta}$ are said to be *projectively linearly identifiable* if, and only if,

- it is known that $\theta_\iota \neq 0$ for some ι, $1 \leq \iota \leq \tau$;
- the quantities $\{\frac{\theta_1}{\theta_\iota}, \ldots, \frac{\theta_{\iota-1}}{\theta_\iota}, \frac{\theta_{\iota+1}}{\theta_\iota}, \ldots, \frac{\theta_\tau}{\theta_\iota}\}$ are linearly identifiable[11].

The uncertain parameters $\boldsymbol{\Theta}$ are said to be *weakly linearly identifiable* if, and only if, there exists a set $\boldsymbol{\Theta}' = \{\theta'_1, \ldots, \theta'_\tau\}$ of linearly identifiable quantities such that the elements of Θ are algebraic over $k_0(\boldsymbol{\Theta}')$.

13.3.4 An Elementary Example

Set $k_0 = \mathbf{Q}$ and $\boldsymbol{\Theta} = \{a_1, \ldots, a_n, b_0, \ldots, b_m\}$. Consider the SISO system

$$\left(\frac{d^n}{dt^n} + a_1 \frac{d^{n-1}}{dt^{n-1}} + \cdots + a_n\right) y(t) = \left(b_0 \frac{d^m}{dt^m} + \cdots + b_m\right) u(t) \quad (13.9)$$

which reads with operational notations

$$\left(s^n + a_1 s^{n-1} + \cdots + a_n\right) Y = (b_0 s^m + \cdots + b_m) U + \mathcal{I}(s) \quad (13.10)$$

where $\mathcal{I}(s)$ is a polynomial over k in the indeterminate s, of degree $\max(m,n) - 1$, the coefficients of which depend on the initial conditions. By applying $\frac{d^{\max(m,n)}}{ds^{\max(m,n)}}$ to both sides of (13.10), we remove those conditions. The linear identifiability follows at once from the linear equations

$$\frac{d^\alpha}{ds^\alpha}\left(s^n + a_1 s^{n-1} + \cdots + a_n\right) Y = \frac{d^\alpha}{ds^\alpha}(b_0 s^m + \cdots + b_m) U$$

for $\max(m,n) \leq \alpha \leq \max(m,n) + m + n$.

Remark 13.2. See [31] for most interesting calculations via moments that bear some similarity with the ones above.

Remark 13.3. Replace (13.9) by

$$\left(a_0 \frac{d^n}{dt^n} + a_1 \frac{d^{n-1}}{dt^{n-1}} + \cdots + a_n\right) y(t) = \left(b_0 \frac{d^m}{dt^m} + \cdots + b_m\right) u(t)$$

where we have introduced the coefficient a_0. If we assume for instance that $a_0 \neq 0$, the set $\{a_0, a_1, \ldots, a_n, b_0, \ldots, b_m\}$ is obviously not linearly identifiable but projectively linearly identifiable.

---

[11] If $\iota = 1$ (resp. $\iota = \tau$), $\{\frac{\theta_2}{\theta_1}, \ldots, \frac{\theta_\tau}{\theta_1}\}$ (resp. $\{\frac{\theta_1}{\theta_\tau}, \ldots, \frac{\theta_{\tau-1}}{\theta_\tau}\}$) are linearly identifiable.

13.4 Perturbations

13.4.1 Structured Perturbations

For dealing with specific perturbations, introduce the left $k(s)[\frac{d}{ds}]$-module

$$\mathbf{L} = \Lambda_{k(s)[\frac{d}{ds}]}/M$$

where M is a submodule of $\text{span}_{k(s)[\frac{d}{ds}]}(\boldsymbol{\pi})$. Call again *perturbation*, or *disturbance*, variables the canonical image $\boldsymbol{\pi} = (\pi_1, \ldots, \pi_q) \subset \mathbf{L}$ of $\boldsymbol{\pi}$. A subset $S \subseteq \boldsymbol{\pi}$ is said to be *structured* if, and only if, the module $\text{span}_{k(s)[\frac{d}{ds}]}(S)$ is torsion. This means, in other words, that for any $\sigma \in \text{span}_{k(s)[\frac{d}{ds}]}(S)$, there exists $\varpi \in k(s)[\frac{d}{ds}]$, $\varpi \neq 0$, such that $\varpi \sigma = 0$. We say that the linear differential operator ϖ is *annihilating* the structured perturbation σ. The differential operator ϖ is also called an *annihilator* of σ.

Example 13.4. Set $k = \mathbb{C}$. The perturbation $\kappa \frac{e^{-Ls}}{s^n}$, $L \geq 0$, $\kappa \in \mathbb{C}$, $n \geq 0$, which is annihilated by $(\frac{d}{ds} + L)s^n = s^n(\frac{d}{ds} + L) + ns^{n-1}$, is structured. Note that the annihilating differential operator contains L, but not κ.

Example 13.5. The perturbation $\frac{a}{b}$, $a, b \in k[s]$, $b \neq 0$, which is annihilated by $\frac{d^\nu}{ds^\nu} b$, for ν large enough, is structured. In particular $\frac{\alpha s + \beta}{s^2 + \omega^2}$, $\alpha, \beta, \omega \in k$, is annihilated by $\frac{d^2}{ds^2}(s^2 + \omega^2) = 2 + 2s\frac{d}{ds} + (s^2 + \omega^2)\frac{d^2}{ds^2}$, which contains the 'frequency' ω, but not α and β.

The set of annihilators of any $\sigma \in \text{span}_{k(s)[\frac{d}{ds}]}(S)$ is a left ideal of $k(s)[\frac{d}{ds}]$. Any generator ϖ_0 of this principal ideal is said to be a *minimal* annihilator. Take two minimal annihilators ϖ_0 and ϖ_1. Then, $\varpi_1 = \varrho \varpi_0$, where $\varrho \in k(s)$, $\varrho \neq 0$.

13.4.2 Unstructured Perturbations

Perturbations that are not structured are said to be *unstructured*. Such noises are viewed as highly fluctuating, or oscillatory, signals, which may be attenuated by low-pass filters, like iterated time integrals.

Remark 13.4. See [6] for a precise mathematical foundation, which is based on *non-standard analysis*. A highly fluctuating function of zero mean is then defined by saying that its integral over a finite time interval is *infinitesimal*, i.e., 'very small'. Let us emphasise once more that this approach, which has been confirmed by numerous computer simulations and several laboratory experiments, is independent of any probabilistic setting. No knowledge of the statistical properties of the noises is required.

13.4.3 Linear Identifier

Equation (13.8) may be rewritten as

$$P \begin{pmatrix} \theta_1 \\ \vdots \\ \theta_\tau \end{pmatrix} = Q + R^{\text{struc}} + R^{\text{unstruc}}$$

where the components of the column matrix R^{struc} (resp. R^{unstruc}) are structured (resp. unstructured) perturbations. The set $\text{ann}(R^{\text{struc}})$ of differential polynomials $\omega \in k(s)[\frac{d}{ds}]$ annihilating R^{struc}, i.e., such that $\omega R^{\text{struc}} = 0$, is a left ideal. Two generators Δ_1, Δ_2 of this ideal are related by $\Delta_2 = \rho \Delta_1$, $\rho \in k(s)$, $\rho \neq 0$. Pick up a generator Δ

$$\Delta P \begin{pmatrix} \theta_1 \\ \vdots \\ \theta_\tau \end{pmatrix} = \Delta Q + \Delta R^{\text{unstruc}} \qquad (13.11)$$

If $\det(\Delta P) \neq 0$, then (13.11), where the structured perturbations have been eliminated, is called a *linear identifier* of the unknown parameters.

13.4.4 Robustness

By multiplying both sides of (13.11) by a suitable strictly proper transfer function in $k(s)$, we may ensure that any entry of the matrices is a k-linear combination of terms of the form $r \frac{d^\alpha}{ds^\alpha}(a)$, where

- $r \in k(s)$ is strictly proper;
- $\alpha = 0, 1, 2, \ldots$;
- a is either a control, an output, or an unstructured perturbation variable.

Denoising, *i.e.*, the attenuation of unstructured perturbations, is achieved by choosing appropriate low-pass filters, like iterated time integrals, which give rise to what we may call *invariant filtering*.

13.5 First Example: Dragging an Unknown Mass in Open Loop

13.5.1 Description and First Results

Consider the problem of dragging an unknown mass along a frictionless horizontal straight line. The model is given by

$$m\ddot{x}(t) = u(t)$$

where

- $x(t)$ is the mass displacement, perfectly measured from some reference point, or origin, labeled by 0;
- $u(t)$ is the applied force.

To make the problem simple, let us assume that $u(t)$ is a known, non-zero, open-loop control force at our disposal. The entire purpose of applying such a force to the mass is to gather some input/output information so that we can identify the unknown mass parameter m. The mass is initially, at time $t = 0$, placed at a distance x_0 of the origin and moves with unknown velocity $\dot{x}_0$. Operational calculus yields

$$m\left[s^2 X - s x_0 - \dot{x}_0\right] = U$$

The dependence of this expression upon the initial conditions is eliminated by differentiating both sides twice with respect to s

$$m\left[2X + 4s\frac{dX}{ds} + s^2\frac{d^2 X}{ds^2}\right] = \frac{d^2 U}{ds^2}$$

Time differentiation is avoided by multiplying both sides by s^{-2}

$$m\left[2s^{-2}X + 4s^{-1}\frac{dX}{ds} + \frac{d^2 X}{ds^2}\right] = s^{-2}\frac{d^2 U}{ds^2}$$

This reads in the time domain[12]

$$m\left[2\int_0^t \int_0^{\sigma_1} x(\sigma_2)d\sigma_2 d\sigma_1 - 4\int_0^t \sigma_1 x(\sigma_1)d\sigma_1 + t^2 x(t)\right]$$
$$= \int_0^t \int_0^{\sigma_1} u(\sigma_2)d\sigma_2 d\sigma_1$$

This expression has the advantage of being completely independent of the initial conditions and it only requires the measurement of the input force $u(t)$ and of the displacement output $x(t)$, in order to compute m. Set $1/m = \frac{n(t)}{d(t)}$, where

$$n(t) = t^2 x(t) - 4\int_0^t \sigma_1 x(\sigma_1) d\sigma_1 + 2\int_0^t \int_0^{\sigma_1} x(\sigma_2)d\sigma_2 d\sigma_1$$
$$d(t) = \int_0^t \int_0^{\sigma_1} \sigma_2^2 u(\sigma_2) d\sigma_2 d\sigma_1$$

At time $t = 0$, both the numerator and the denominator are 0: the quotient is undetermined. We must, therefore, begin to evaluate the formula not at time 0 but at a later time, say $\epsilon \gneq 0$, ϵ being small. Set for the estimate $1/m_e$ of $1/m$

---

[12] Remember (cf. Section 13.3.2) that the *algebraic derivative* $\frac{d}{ds}$ corresponds in the time domain to multiplication by $-t$.

13 Closed-loop CT Model Identification via New Algebraic Techniques 373

$$\frac{1}{m_e} = \begin{cases} \text{arbitrary for } t \in [0, \epsilon) \\ \dfrac{n(t)}{d(t)} \quad \text{for } t > \epsilon \end{cases} \qquad (13.12)$$

The evaluation of the quotient is, of course, valid as long as the denominator does not go through zero.

In order to easily implement the calculations on a digital computer, and given that time integrations are needed to synthesise the numerator and the denominator expressions, we would like to give to these two quantities the character of outputs of certain dynamic systems involving differential equations. We propose then the following linear time-varying 'filters'

$$\begin{cases} n(t) = t^2 x(t) + z_1 \\ \dot{z}_1 = -4tx(t) + z_2 \\ \dot{z}_2 = 2x(t) \end{cases} \qquad \begin{cases} d(t) = \eta_1 \\ \dot{\eta}_1 = \eta_2 \\ \dot{\eta}_2 = t^2 u(t) \end{cases} \qquad (13.13)$$

with $z_1(0) = z_2(0) = 0$ and $\eta_1(0) = \eta_2(0) = 0$.

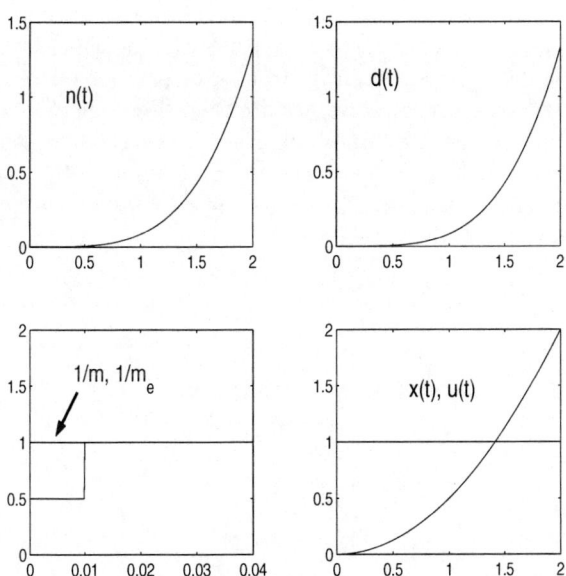

Fig. 13.1. Identification of the inverse-mass parameter

Figure 13.1 depicts the involved signals, *i.e.*, the numerator $n(t)$, the denominator $d(t)$, the input $u(t)$, which is here a constant force, the output $y(t) = x(t)$, and the estimate $1/m_e$ of $1/m$. The wrong, or arbitrary, guess for the parameter value, during the time interval $[0, \epsilon)$, was taken to be $1/m_e = 0.5$, as can be seen from the figure. We have set, in this case, $\epsilon = 0.01$

s, but still a smaller real value could have certainly been used. Also, we have let: $u(t) = 1$ for all t. For the simulations, the actual value of the mass was set to $m = 1$ kg.

Several distinctive features emerge from the simulations of this rather simple example:

1. The estimation of the mass parameter can be reliably achieved in quite a short amount of time that only depends on the arithmetic processor precision in being able to carry out the quotient of two very small quantities, the numerator and denominator signals.
2. The test input signal $u(t)$ being used does not necessarily exhibit the classical 'persistency of excitation' requirement.
3. The estimator of the inverse mass parameter is comprised of unstable signals in both the numerator and the denominator.

Regarding the first observation above, we should remark that the accurate precision with which we have obtained the mass parameter is not at all surprising, due to the fact that the used formula is as exact as the model and, very importantly, because we have not included any measurement noise in our simulations. This last feature may compromise not only the precision of the computation but, also, the fast character of the identification. The second feature of not needing a persistently exciting signal is certainly an unchallenged advantage. The last negative feature regarding our internally unstable scheme may be overcome in a simple mannor by prescribing the need to, at least temporarily, 'switch off' the estimator immediately after the precise parameter estimation is obtained. The noise-related aspects are quite essential. We propose below a possible approach.

13.5.2 Denoising

The expression

$$\frac{1}{m} = \frac{n(t)}{d(t)}$$

becomes

$$\frac{1}{m} = \frac{\mathcal{G} \star n(t)}{\mathcal{G} \star d(t)}$$

where

- $\mathcal{G}$ is a low-pass filter with rational transfer function $G(s)$;
- $\star$ denotes the convolution product.

According to Sections 13.4.2 and 13.4.4 such an invariant filtering permits to attenuation of zero-mean highly fluctuating noises, such as the plant noise $\zeta(t)$ and the measurement noise $\xi(t)$ occurring in

$$m\ddot{x}(t) = u(t) + \zeta(t), \quad y(t) = x(t) + \xi(t)$$

Corresponding to the inverse-mass parameter estimation we propose then the following time-varying filters, with second-order integration low-pass filtered outputs

$$\begin{cases} n(t) = z_1 \\ \dot{z}_1 = z_2 \\ \dot{z}_2 = (t)^2 y(t) + z_3 \\ \dot{z}_3 = -4ty(t) + z_4 \\ \dot{z}_4 = 2y(t) \end{cases} \qquad \begin{cases} d(t) = \eta_1 \\ \dot{\eta}_1 = \eta_2 \\ \dot{\eta}_2 = \eta_3 \\ \dot{\eta}_3 = \eta_4 \\ \dot{\eta}_4 = t^2 u(t) \end{cases} \qquad (13.14)$$

with $z_1(0) = z_2(0) = z_3(0) = z_4(0) = 0$ and $\eta_1(0) = \eta_2(0) = \cdots = \eta_4(0) = 0$. We set $\xi(t) = 0.02(\text{rect}(t) - 0.5)$ and $\zeta(t) = (\text{rect}(t) - 0.5)$, where $\text{rect}(t)$ is a computer-generated random process consisting of piecewise-constant random variables uniformly distributed in the interval $[0, 1]$ of the real line.

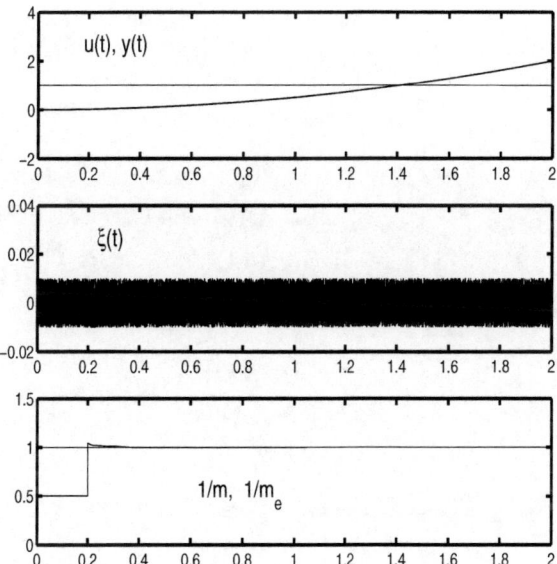

Fig. 13.2. Identification of the inverse mass parameter under noise measurements and using invariant filtering

Figure 13.2 depicts the outcome of the invariant filtering modification of our previously proposed parameter estimation scheme for the unknown dragged mass. We note that a larger ϵ parameter was used in this instance ($\epsilon = 0.2$) to allow for a reliable quotient yielding the inverse mass, once the signal-to-noise ratio becomes important in the numerator.

13.5.3 A Comparison with an Adaptive-observer Approach

For the purposes of comparison, and given that the parameter estimation problem has been cast, so far, into an open-loop problem, we choose now an observer approach for the estimation of the unknown parameter (see, e.g., [24]).
Consider the state-variable representation of the mass-dragging problem with a constant but unknown mass

$$\dot{x}_1 = x_2$$
$$\dot{x}_2 = x_3 u$$
$$\dot{x}_3 = 0$$
$$y = x_1 \tag{13.15}$$

where the state x_3 represents the inverse value of the unknown mass m. Here, u will be assumed to be, as before, a constant C. An adaptive observer is represented by the following certainty equivalence observer

$$\dot{x}_{1e} = x_{2e} + \lambda_3(y - x_{1e})$$
$$\dot{x}_{2e} = x_{3e}u + \lambda_2(y - x_{1e})$$
$$\dot{x}_{3e} = \lambda_1(y - x_{1e}) \tag{13.16}$$

The estimation error dynamics is given by

$$\dot{e}_1 = e_2 - \lambda_3 e_1$$
$$\dot{e}_2 = e_3 u - \lambda_2 e_1$$
$$\dot{e}_3 = -\lambda_1 e_1 \tag{13.17}$$

with $e_j = x_j - x_{je}$, $j = 1, 2, 3$. Thus,

$$\ddot{e}_1 + \lambda_3 \dot{e}_1 + \lambda_2 e_1 + \lambda_1 u \int_0^t e_1(\sigma) d\sigma = 0 \tag{13.18}$$

Clearly, for $u = C$, the characteristic polynomial of the estimation error dynamics is given by

$$p(s) = s^3 + \lambda_3 s^2 + \lambda_2 s + \lambda_1 C = 0 \tag{13.19}$$

Evidently the adaptive observer approach is limited, in this case, to those dragging maneuvers for which the constant value of C is strictly positive. Note that if C is to be negative, as in a 'pushing task', changing the sign of λ_1 to a negative value (so that the term $\lambda_1 C$ in the characteristic polynomial becomes strictly positive) simply destabilises e_1. This fact severely limits the applications in the context of trajectory-tracking problems, where u is not constant, and also in those situations in which the steady-state value of the control input is to become strictly negative, regardless of how small. Another limitation, as depicted in the simulation below, is the relatively slow convergence to the actual value of the inverse mass on the part of the estimate of x_3.

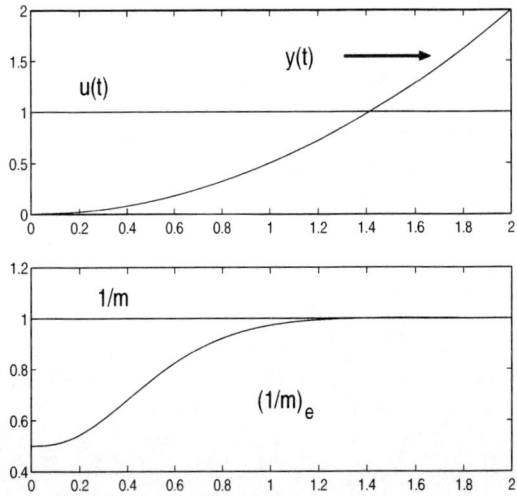

Fig. 13.3. Identification of the inverse-mass parameter using an adaptive observer

13.6 Second Example: A Perturbed First-order System

13.6.1 Presentation

We now turn our attention to a case where more than one unknown parameter is present in the system and, also, where the need arises for a closed-loop identification. As before, regarding the control part, we resort to the certainty equivalent control method. As for how to handle several parameters, we will have to generate as many algebraic equations as unknown parameters there may be.

Consider the linear parameter-uncertain, perturbed, first-order system

$$\dot{y}(t) = ay(t) + bu(t) + \kappa + \xi(t)$$

where

- a, b are uncertain parameters;
- κ is an unknown constant bias;
- $\xi(t)$ is a zero-mean highly fluctuating noise[13].

---

[13] One might replace κ and $\xi(t)$ by a highly fluctuating noise of constant but unknown mean (see [6]).

We would like to specify a feedback control law such that the following problem finds a solution

Devise a feedback control law that forces the output signal y to follow a given reference trajectory $y^*(t)$, in spite of the lack of knowledge about the plant parameters a, b, the uncertainty about the constant perturbation κ and the presence of the zero mean, rapidly varying, plant perturbation noise.

13.6.2 A Certainty Equivalence Controller

If the parameters a and b were perfectly known, and if there existed no plant perturbation noise, i.e., if $\xi(t) \equiv 0$, the following classical proportional integral controller may be used given its known robustness with respect to a constant, but unknown, perturbation

$$u = \frac{1}{b}\left[\dot{y}^*(t) - ay - c_1(y - y^*(t)) - c_0 \int_0^t (y(\sigma) - y^*(\sigma))\,d\sigma\right] \quad (13.20)$$

Set $e = y - y^*(t)$. The characteristic polynomial of the closed-loop tracking error dynamics is given by $p(s) = s^2 + c_1 s + c_0$. With a suitable choice of the design parameters c_0, c_1, the roots of $p(s)$ are all strictly located in the left portion of the complex plane. The tracking task is asymptotically accomplished.

We must therefore concentrate our efforts on obtaining the right values of a and b.

13.6.3 Parameter Identification

Assume again that $\xi(t) \equiv 0$. We try to generate a linear system of equations for the unknown parameters a and b. This system should be independent of the plant initial condition, and also, of course, independent of the constant perturbation, κ and, moreover, it should rely only on knowledge of the input u and the output y signals.

Operational calculus yields

$$sY - y_0 = aY + bU + \frac{\kappa}{s}$$

Multiply both sides by s

$$s^2 Y - s y_0 = asY + bsU + \kappa$$

Differentiating twice with respect to s removes the presence of the initial condition and, also, of the influence of the unknown parameter κ. We obtain, after some algebraic manipulations

$$a\left[2\frac{dY}{ds} + s\frac{d^2Y}{ds^2}\right] + b\left[2\frac{dU}{ds} + s\frac{d^2U}{ds^2}\right] = 2Y + 4s\frac{dY}{ds} + s^2\frac{d^2Y}{ds^2}$$

13 Closed-loop CT Model Identification via New Algebraic Techniques

Multiplying by s^{-2} to avoid time differentiations in the time domain, we get

$$a\left[2s^{-2}\frac{dY}{ds} + s^{-1}\frac{d^2Y}{ds^2}\right] + b\left[2s^{-2}\frac{dU}{ds} + s^{-1}\frac{d^2U}{ds^2}\right]$$
$$= 2s^{-2}Y + 4s^{-1}\frac{dY}{ds} + \frac{d^2Y}{ds^2}$$

This reads, in the time domain

$$\left[\int_0^t \sigma_1^2 y(\sigma_1)d\sigma_1 - 2\int_0^t \int_0^{\sigma_1} \sigma_2 y(\sigma_2)d\sigma_2 d\sigma_1\right]a$$
$$+ \left[\int_0^t \sigma_1^2 u(\sigma_1)d\sigma_1 - 2\int_0^t \int_0^{\sigma_1} \sigma_2 u(\sigma_2)d\sigma_2 d\sigma_1\right]b$$
$$= t^2 y(t) - 4\int_0^t \sigma_1 y(\sigma_1)d\sigma_1 + 2\int_0^t \int_0^{\sigma_1} y(\sigma_2)d\sigma_2 d\sigma_1$$

Integrating once more, we obtain a linear system for the constant parameters a and b. We arrive at a linear time-varying equation

$$P(t)\begin{bmatrix}a\\b\end{bmatrix} = q(t)$$

The 2×2 matrix $P(t)$ reads

$$P(t) = \begin{bmatrix}p_{11}(t) & p_{12}(t)\\p_{21}(t) & p_{22}(t)\end{bmatrix}$$

where

$$p_{11}(t) = \int_0^t \sigma^2 y d\sigma - 2\int_0^t \int_0^\sigma \lambda y d\lambda d\sigma$$
$$p_{12}(t) = \int_0^t \sigma^2 u d\sigma - 2\int_0^t \int_0^\sigma \lambda u d\lambda d\sigma$$
$$p_{21}(t) = \int_0^t \int_0^\sigma \lambda^2 y d\lambda d\sigma - 2\int_0^t \int_0^\sigma \int_0^\lambda \rho y d\rho d\lambda d\sigma$$
$$p_{22}(t) = \int_0^t \int_0^\sigma \lambda^2 u d\lambda d\sigma - 2\int_0^t \int_0^\sigma \int_0^\lambda \rho u d\rho d\lambda d\sigma$$

The column vector $q(t)$ is given by

$$q(t) = \begin{bmatrix}t^2 y - 4\int_0^t \sigma y d\sigma + 2\int_0^t \int_0^\sigma y d\lambda d\sigma \\ \int_0^t \sigma^2 y d\sigma - 4\int_0^t \int_0^\sigma \lambda y d\lambda d\sigma + 2\int_0^t \int_0^\sigma \int_0^\lambda y d\lambda d\sigma d\rho\end{bmatrix}$$

The matrix $P(t)$ and the vector $q(t)$ are equal to 0 at time $t = 0$. Nevertheless, it is easy to verify that the matrix $P(t)$ is, indeed, invertible at a small time, $t = \epsilon > 0$.

Under the noise-free circumstances, we may, then compute a and b *exactly*, at time $t = \epsilon > 0$, regardless of the constant perturbation input κ, and, moreover, for any initial condition on the plant output y.
as an $r \times r$ diagonal
A certainty equivalence controller, of the form (13.20), is proposed as follows

$$u = \frac{1}{b_e}\left[\ddot{y}^*(t) - a_e y - k_1(y - y^*(t)) - k_0 \int_0^t (y(\sigma) - y^*(\sigma))\,d\sigma\right]$$

with

$$\begin{bmatrix} a_e \\ b_e \end{bmatrix} = \begin{cases} \text{arbitrary, with } b_e \neq 0 & \text{for } t \in [0, \epsilon) \\ P^{-1}(t)q(t) & \text{for } t \in [\epsilon, +\infty) \end{cases}$$

13.6.4 Noise-free Simulation Results

Figure 13.4 depicts the fast adaptation system response in a rest-to-rest trajectory-tracking task. As can be seen, the determination of the system parameters happens quite fast, in approximately 4×10^{-3} s. The absence of measurement and plant noises certainly makes the algebraic estimation task quite precise and rather fast. The integral action on the proposed certainty equivalence controller annihilates the effects of the unknown constant perturbation input, while our estimation technique is shown to be totally independent of the constant perturbation input amplitude.

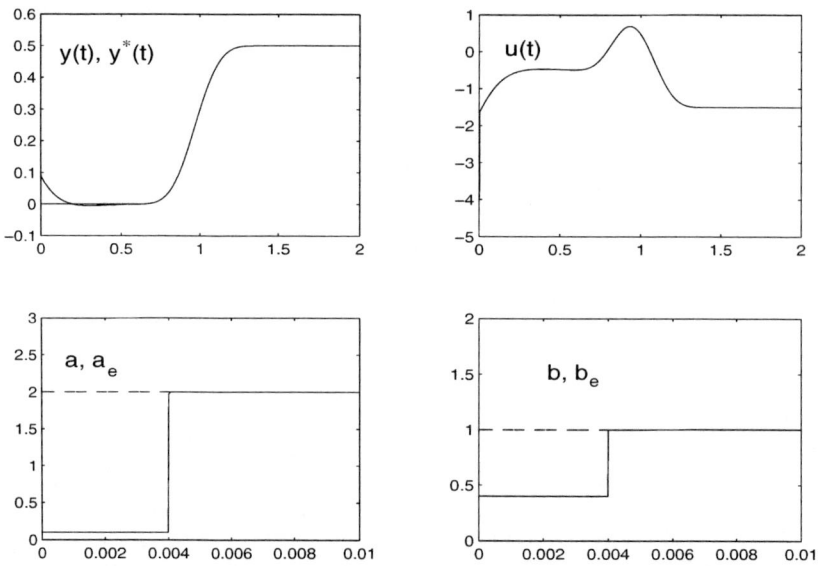

Fig. 13.4. System response, parameter determination, control input

13.6.5 Noisy Measurements and Plant Perturbations

To carry out our previously proposed algebraic parameter estimation approach to fast adaptive control, we considered the following intimately related perturbed system

$$\dot{x} = ax + bu + k + \eta(t), \quad y(t) = x(t) + \xi(t)$$

where $\eta(t)$ and $\xi(t)$ are zero-mean computer-generated noises consisting of a sequence of piecewise-constant random variables uniformly distributed in the interval $[-0.5R, 0.5R]$.

For the case of measurement noises, the same computational algorithm was used but now including an invariant filtering strategy. We low-pass filtered both members of each one of the algebraic equations derived before for the on-line calculation of the parameters. A second-order integration was used in each case. In the simulation shown in the Figure 13.5, we have assumed a zero-mean computer-generated measurement noise of significant amplitude. The output signal, and the control input signal do exhibit the influence of the measurement noise but the parameter estimates converge quite precisely and fast enough to the actual value of the parameters. The computation time is substantially increased in the noisy case. Nevertheless, the estimation of the unknown parameters is still quite accurate.

13.6.6 Simulation Results with Noises

Figure 13.5 depicts the performance of the algebraic parameter identifier including invariant filtering along with the systems response in a rest-to-rest trajectory-tracking task and the evolution of the applied feedback control input. For the measurement noise $\xi(t)$ we have chosen R to be 0.01 and for the plant system noise η, the corresponding R value was set to be 0.1.

13.6.7 Comparison with Adaptive Control

Adaptive control is usually approached from the viewpoint of Lyapunov stability theory via the synthesis of a suitable parameter update law derived on the basis of the behaviour around a closed-loop trajectory of the time derivative of a Lyapunov function that includes a quadratic parameter estimation error term. The feedback law is proposed as a *certainty equivalent controller*. The adaptation mechanism is derived by enforcing asymptotic stability of the closed-loop controlled system. The literature on the topic of adaptive control using Lyapunov arguments is certainly overwhelming. For further details, we refer the reader to popular references, like [1, 14, 23, 27].

We deal with the same system as in the previous section

$$\dot{y} = ay + bu + \kappa$$

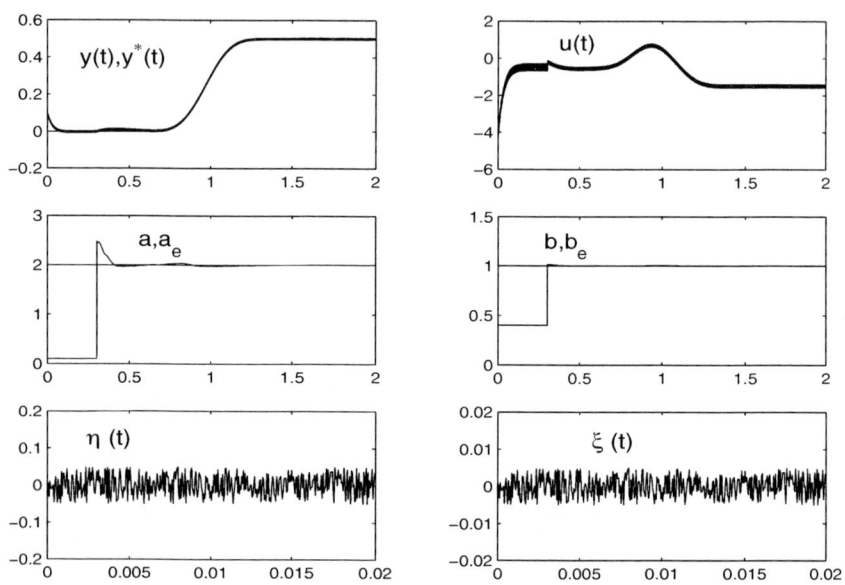

Fig. 13.5. System response, parameter determination and control input

A nominal desired trajectory for the system $y^*(t)$ demands the existence of a nominal control input $u^*(t)$ that satisfies the dynamics of the unperturbed system.

$$\dot{y}^*(t) = ay^*(t) + bu^*(t)$$

The tracking-error dynamics is then given by

$$\dot{e} = ae + be_u + \kappa$$

where $e = y - y^*(t)$ and $e_u = u^*(t)$. A certainty equivalence control, using estimated values of the unknown parameters, is given by

$$e_u = u - u^*(t) = \frac{1}{b_e}\left[-a_e e - k_1 e - k_0 \int_0^t e(\sigma)d\sigma\right]$$

where a_e and b_e are the estimated values of a and b.
The closed-loop system, after some algebraic manipulations, results in

$$\dot{e} + k_1 e + k_0 \rho = (a - a_e)e - \frac{1}{b_e}(b - b_e)\left[(k_1 - a_e)e + k_0 \int_0^t e(\sigma)d\sigma\right]$$

$$\dot{\rho} = e(\sigma), \quad \rho(0) = -\frac{\kappa}{k_0}$$

Taking as a Lyapunov function candidate the following positive-definite function

$$V(e,\rho,a-a_e,b-b_e) = \frac{1}{2}e^2 + \frac{k_0}{2}\rho^2 + \frac{1}{2\gamma}(a-a_e)^2 + \frac{1}{\beta}(b-b_e)^2$$

we find that the choice of the estimated values of a and b according to the following parameter update law

$$\dot{a}_e = \gamma e^2$$

$$\frac{\mathrm{d}}{\mathrm{d}t}[b_e]^2 = \beta \left[(k_1 - a_e)e^2 + k_0 e \int_0^t e(\sigma)\mathrm{d}\sigma \right]$$

leads to the following expression for the time derivative of $V(e)$ along the trajectories of the controlled system

$$\dot{V}(e) = -k_1 e^2 \leq 0$$

The non-positivity of $\dot{V}(e)$ implies that $V(e)$ is bounded. It is also clear that $\dot{V}(e)$ is absolutely continuous. It follows, according to Barabarat's lemma, that $\dot{V}(e)$ asymptotically converges to zero. Hence e tends to zero. It is also clear that the convergence of a_e and b_e to their actual values cannot be guaranteed. As a consequence of this, the value of k_1 must be chosen sufficiently large so that a_e does not cause an instability in the dynamics of b_e^2. But this in turn depends on the transient of the tracking error e. The approach may suffer severe limitations in trajectory-tracking tasks.

13.6.8 Simulations for the Adaptive Scheme

Figure 13.6 depicts the performance of the designed adaptive feedback control law in the same trajectory-tracking task of the previous algebraic approach example. Although the scheme manages to accomplish the trajectory-tracking task with rather low quality, the scheme fails to produce an accurate estimate of the unknown parameters. The values of the parameter update gains were chosen to be $\gamma = 100$ and $\beta = 1.25$. The values of a and b used in the simulations were the same as before $a = 2$ and $b = 1$. If the rest-to-rest maneuver entitles a higher final equilibrium value, say of 1, the scheme completely fails.

13.7 Third Example: A Double-bridge Buck Converter

The several electronic switches in Figure 13.7 take position values according to
$$\begin{cases} u = 1 , S_1 = ON, S_2 = ON, S_3 = OFF, S_4 = OFF \\ u = 0 , S_1 = OFF, S_2 = OFF, S_3 = ON, S_4 = ON \end{cases}$$
Consider the following (average) model of a double-bridge buck converter[14]

---

[14] See [30] for further details on this and other power converters. For a closely related on-line adaptive identification case on the same converter, see [29].

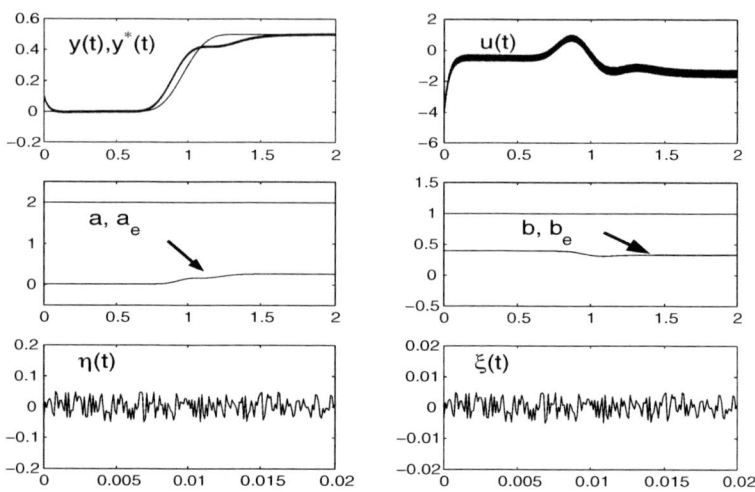

Fig. 13.6. Performance of adaptive-control approach in a trajectory-tracking task for the uncertain system

$$L\dot{x}_1 = -x_2 + \mu E$$
$$C\dot{x}_2 = x_1 - \frac{x_2}{R}$$
$$y = x_2$$

where

- x_1 is the inductor current;
- x_2 represents the capacitor voltage.

The average control input μ is assumed to take values in the closed interval $[-1, 1]$. This variable actually represents, in absolute value, the *duty ratio* of the switch positions. The parameters L, C, R and E are assumed to be unknown.

13.7.1 An Input–Output Model

Eliminating the state variable x_1 yields

$$\ddot{y} + \gamma_1 \dot{y} + \gamma_0 y = \gamma \mu$$

where the parameters $\gamma_1 = \frac{1}{RC}$, $\gamma_0 = \frac{1}{LC}$, $\gamma = \frac{E}{LC}$ are linearly identifiable according to Section 13.3.4. Estimating those parameters permits us to control the system without knowing the values of L, C, R and E.

Remark 13.5. It is straightforward to check that L, C, R, E are not simultaneously identifiable.

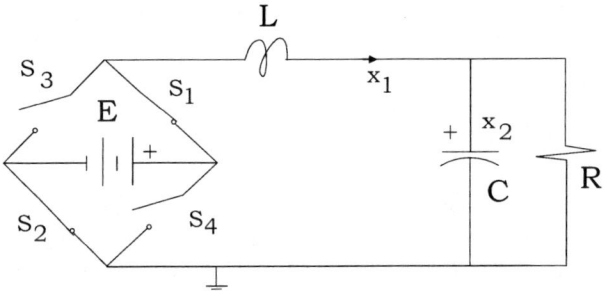

Fig. 13.7. The double-bridge buck converter

13.7.2 Problem Formulation

It is required to design an output feedback controller, possibly of dynamic nature, which induces in the uncertain system, representing the double-bridge buck converter average model, an exponentially asymptotic convergence of the output signal y towards the desired reference signal $y^*(t)$. In other words, we want

$$y \to y^*(t) \quad \text{exponentially}$$

13.7.3 A Certainty Equivalence Controller

We proceed to design the controller as if these parameters were all perfectly known. We propose the following certainty equivalence generalised GPI controller[15]

$$\mu = \mu^*(t) - \mathcal{G} \star (y - y^*(t)) \tag{13.21}$$

where

- $\star$ denotes the convolution product;
- the transfer function of $\mathcal{G}$ is

$$\frac{1}{\gamma} \left\{ \frac{[\gamma_1(\gamma_1 - c_1) + c_0 - \gamma_0]\, s + \gamma_0(\gamma_1 - c_1) + c_{-1}}{s + (c_1 - \gamma_1)} \right\}$$

13.7.4 Closed-loop Behaviour

The closed-loop behaviour of the tracking error, were the parameters perfectly known, is given by the following linear dynamics

---

[15] GPI controllers were introduced in [10] for linear systems in terms of *integral reconstructors* yielding states in terms of iterated integrals of inputs and outputs. It can be shown, with some work, that such controllers are also equivalent to classical compensation networks of which (13.21) is just an example.

$$\ddot{e}_y + c_1\dot{e}_y + c_0 e_y + c_{-1}\int_0^t e_y(\sigma)\mathrm{d}\sigma = 0$$

where $e_y(t) = y(t) - y^*(t)$ is the trajectory-tracking error.

The appropriate choice of the coefficients $\{c_1, c_0, c_{-1}\}$, in the characteristic polynomial of the tracking error dynamics, turns it into a Hurwitz polynomial with the associated asymptotically exponentially stable nature of the origin of coordinates of the natural tracking error state space $\{e_y = 0, \dot{e}_y = 0, \ddot{e}_y = 0\}$. Under the assumption of perfect parameter knowledge we obtain, modulo control input saturations,

$$e_y(t) \to 0 \quad \text{exponentially}$$

The problem now becomes one of accurate determination of the unknown parameters of the system *as required by the proposed GPI controller.*

13.7.5 Algebraic Determination of the Unknown Parameters

Consider the average input–output model of the converter system

$$\ddot{y} + \gamma_1 \dot{y} + \gamma_0 y = \gamma\mu$$

In the notation of operational calculus, we have

$$s^2 Y - sy_0 - \dot{y}_0 + \gamma_1(sY - y_0) + \gamma_0 Y = \gamma U$$

Taking derivatives with respect to s, twice, we obtain

$$(s^2 + \gamma_1 s + \gamma_0)\frac{\mathrm{d}^2 Y}{\mathrm{d}s^2} + (4s + 2\gamma_1)\frac{\mathrm{d}Y}{\mathrm{d}s} + 2Y = \gamma\frac{\mathrm{d}^2 U}{\mathrm{d}s^2}$$

This last expression may be rewritten as follows

$$\left[s\frac{\mathrm{d}^2 Y}{\mathrm{d}s^2} + 2\frac{\mathrm{d}Y}{\mathrm{d}s}\right]\gamma_1 + \left[\frac{\mathrm{d}^2 Y}{\mathrm{d}s^2}\right]\gamma_0 - \left[\frac{\mathrm{d}^2 U}{\mathrm{d}s^2}\right]\gamma =$$
$$-s^2\frac{\mathrm{d}^2 Y}{\mathrm{d}s^2} - 4s\frac{\mathrm{d}Y}{\mathrm{d}s} - 2Y$$

Multiplying out by a sufficient power of s^{-1}, say by s^{-4}, so that an invariant filtering effect is obtained, we also eliminate possible derivations in the time domain. We obtain

$$\left[s^{-3}\frac{\mathrm{d}^2 Y}{\mathrm{d}s^2} + 2s^{-4}\frac{\mathrm{d}Y}{\mathrm{d}s}\right]\gamma_1 + \left[s^{-4}\frac{\mathrm{d}^2 Y}{\mathrm{d}s^2}\right]\gamma_0 - \left[s^{-4}\frac{\mathrm{d}^2 U}{\mathrm{d}s^2}\right]\gamma =$$
$$-\left[s^{-2}\frac{\mathrm{d}^2 Y}{\mathrm{d}s^2} + 4s^{-3}\frac{\mathrm{d}Y}{\mathrm{d}s} + 2s^{-4}Y\right]$$

Reverting the previous expression to the time domain, we obtain a linear equation, with time-varying coefficients, in three unknowns $\{\gamma_1, \gamma_0, \gamma\}$. We write such an equation as

$$p_{11}(t)\gamma_1 + p_{12}(t)\gamma_0 + p_{13}(t)\gamma = q_1(t) \qquad (13.22)$$

We conform a system of three equations in three unknowns by simply adjoining to the previous equation its first integral and its iterated integral, *i.e.*,

$$p_{11}(t)\gamma_1 + p_{12}(t)\gamma_0 + p_{13}(t)\gamma = q_1(t)$$

$$\left(\int p_{11}(t)\right)\gamma_1 + \left(\int p_{12}(t)\right)\gamma_0 + \left(\int p_{13}(t)\right)\gamma = \left(\int q_1(t)\right)$$

$$\left(\int^{(2)} p_{11}(t)\right)\gamma_1 + \left(\int^{(2)} p_{12}(t)\right)\gamma_0 + \left(\int^{(2)} p_{13}(t)\right)\gamma = \left(\int^{(2)} q_1(t)\right)$$

This linear system of equation allows us to determine γ_1, γ_0 and γ for $t \geq \epsilon$ with ϵ being a very small positive real number.

13.7.6 Simulation Results

We considered the average model of a double-bridge buck converter with the following (unknown) parameters

$$R = 39.52 \; \Omega, \quad L = 1 \text{ mH}, \quad C = 1 \; \mu\text{F}, \quad E = 30 \text{ Volts}$$

It is desired that the average output voltage signal tracks a rest-to-rest trajectory starting at 21.0 V and landing at 9.0 V in approximately 0.474 ms. The tracking maneuver is to start at $t_{init} = 0.158$ ms and it ends at $t_f = 0.632$ ms. It was assumed that the output voltage could be measured through an additive noise process simulated with a computer generated sequence of random variables uniformly distributed in the interval $A[-0.5, 0.5]$ with the factor A taken to be $A = 0.3$.

Figure 13.8 depicts the simulated closed-loop performance of the GPI controlled double-bridge buck converter along with the performance of the proposed algebraic parameter estimator. The value of ϵ used to avoid the singularity of the formulae at time $t = 0$ was taken to be 22 μs. The three parameters are identified, rather accurately, in approximately 30 μs. Once identified, the value of the parameters is immediately substituted on the GPI feedback control law.

We tested our fast adaptive estimation algorithm now using a switched control signal for the control input whose average coincides with the previous control input. This is achieved using a double-sided sigma-delta modulator described by the following discontinuous dynamics

$$\dot{z} = \mu - u, \quad u = \frac{1}{2}\left[\text{sign}(\mu) + \text{sign}(z)\right]$$

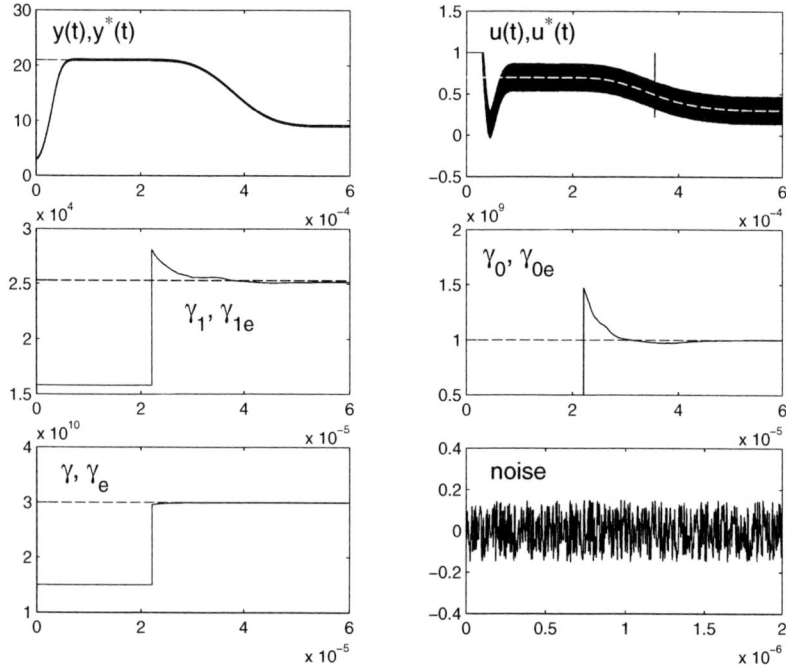

Fig. 13.8. Closed-loop average converter response with on-line identification of all linearly identifiable system parameters

The actual control input signal u being used in the identification algorithm is now a high-frequency signal actively switching and taking values in the discrete set $\{-1, 0, +1\}$. As can be inferred from Figure 13.9 the algebraic identifier works perfectly well with this input. In this instance, the maneuver entitled a trajectory-tracking task, with the same time-duration constraints as before, taking the output voltage from an initial equilibrium of 21 Volts towards a final equilibrium of –9 V.

Figure 13.9 shows the closed-loop response for the trajectory-tracking task of the switched input model as well as the precision and rapidity of the unknown parameter estimation process. The output voltage signal is also assumed to be measured through an additive noisy means. A sample of the noise process is also depicted in the figure. The actual bang-bang control input is shown along with the nominal value of the average control input.

13.8 Conclusion

It is a delicate matter to compare our theoretical techniques and results with today's parametric identification of linear continuous-time systems (see, e.g., [13, 15, 18, 19, 26, 33] and the references therein), which is perhaps less

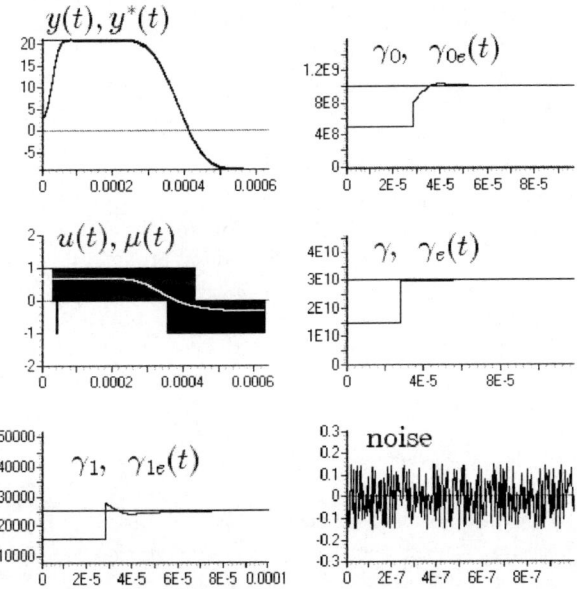

Fig. 13.9. Closed-loop switched converter response with on-line identification of all linearly identifiable system parameters

developed than its discrete-time counterpart, but nevertheless also makes generally a heavy utilisation of statistical methods[16]. Let us stress that all those approaches seem to rest on standpoints and therefore on mathematical tools that are rather far from ours. It is our belief that the only fair way for achieving such a comparison is provided by examples. We do hope that the readers will be convinced by the numerous case studies examined in this chapter and in [11].

The above techniques and results may be generalised to linear state reconstructors [12], to linear diagnosis [8], and to parametric identification of linear discrete-time systems [7]. See, *e.g.*, the references in [6] for their extensions to non-linear systems as well as to signal processing.

References

1. K.J. Åström and B. Wittenmark. *Adaptive Control.* Addison Wesley, New York, 2nd edition, 1995.
2. H. Bourlès. *Systèmes Linéaires: de la Modélisation à la Commande.* Hermès, Paris, 2006.
3. G. Doetsch. *Introduction to the Theory and Application of the Laplace Transform.* Translated from the German. Springer, Berlin, 1974.

---

[16] *Interval analysis* (see, *e.g.*, [16]) is a major exception to this trend.

4. M. Fliess. Some basic structural properties of generalized linear systems. *Systems Control Letters*, 15:391–396, 1990.
5. M. Fliess. Une interprétation algébrique de la transformation de Laplace et des matrices de transfert. *Linear Algebra and its Applications*, 203-204:429–442, 1994.
6. M. Fliess. Analyse non standard du bruit. *Comptes rendus de l'Académie des sciences, Paris Série I*, 342:797–802, 2006.
7. M. Fliess, S. Fuchshumer, K. Schlacher, and H. Sira-Ramírez. Discrete-time linear parametric identification: An algebraic approach. 2es *Journées Identification et Modélisation Expérimentale, JIME'2006*, Poitiers, 2006 (available at http://hal.inria.fr/inria-00105673).
8. M. Fliess, C. Join, and H. Sira-Ramírez. Robust residual generation for linear fault diagnosis: an algebraic setting with examples. *International Journal of Control*, 77:1223–1242, 2004.
9. M. Fliess and R. Marquez. Continuous-time linear predictive control and flatness: A module-theoretic setting with examples. *International Journal of Control*, 73:606–623, 2000.
10. M. Fliess, R. Marquez, E. Delaleau, and H. Sira-Ramírez. Correcteurs proportionnels-intégraux généralisés. *ESAIM: Control, Optimisation and Calculus of Variations*, 7:23–41, 2002.
11. M. Fliess and H. Sira-Ramírez. An algebraic framework for linear identification. *ESAIM: Control, Optimisation and Calculus of Variations*, 9:151–168, 2003.
12. M. Fliess and H. Sira-Ramírez. Reconstructeurs d'état. *Comptes rendus de l'Académie des sciences, Paris Série I*, 338:91–96, 2004.
13. H. Garnier, M. Mensler, and A. Richard. Continuous-time model identification from sampled data: implementation issues and performance evaluation. *International Journal of Control*, 76:1337–1357, 2003.
14. G. Goodwin and D. Mayne. A parameter estimation perspective to continuous time model reference adaptive control. *Automatica*, 23:57–70, 1987.
15. P. Van den Hof. Closed-loop issues in system identification. *Annual Reviews in Control*, 22:173–1186, 1998.
16. L. Jaulin, M. Kieffer, O. Didrit, and E. Walter. *Applied Interval: with Examples in Parameter and State Estimation, Robust Control and Robotics*. Springer, Berlin, 2001.
17. S. Lang. *Algebra*. Springer, Berlin, 3rd edition, 2002.
18. E.K. Larsson, M. Mossberg, and T. Soderstrom. An overview of important practical aspects of continuous-time ARMA system identification. *Circuits, Systems & Signal Processing*, 25:17–46, 2006.
19. K. Mahata and H. Garnier. Identification of continuous-time errors-in-variable models. *Automatica*, 42:1470–1490, 2006.
20. J. McConnel and J. Robson. *Noncommutative Noetherian Rings*. Amer. Math. Soc., Providence, RI, 2000.
21. J. Mikusinski. *Operational Calculus*. Vol. 1, PWN & Pergamon, Warsaw & Oxford, 2nd edition, 1983.
22. J. Mikusinski and T. Boehme. *Operational Calculus*. Vol. 2, PWN & Pergamon, Warsaw & Oxford, 2nd edition, 1987.
23. K. Narendra and A. Annaswamy. *Stable Adaptive Systems*. Prentice Hall, Englewood Cliffs, 1989.

24. M. Phan, L. Horta, and R. Longman. Linear system identification via an asymptotically stable observer. *Journal of Optimization Theory and Applications*, 79:59–86, 1993.
25. B. van der Pol and H. Bremmer. *Operational Calculus Based on the Two-Sided Laplace Integral.* Cambridge University Press, Cambridge, 2nd edition, 1955.
26. T.I. Salsbury. Continuous-time model identification for closed loop control performance assessment. *Control Engineering Practice*, 15:109–121, 2007.
27. S. Sastry and M. Bodson. *Adaptive Systems: Stability, Convergence and Robustness.* Prentice Hall, Englewood Cliffs, 1989.
28. H. Sira-Ramírez and S.K. Agrawal. *Differentially Flat Systems.* Marcel Dekker, New York, 2004.
29. H. Sira-Ramírez, E. Fossas, and M. Fliess. An algebraic, on-line, parameter identification approach to uncertain dc-to-ac power conversion. *41st IEEE Conference on Decision and Control*, Las Vegas, Nevada, USA, 2002.
30. H. Sira-Ramírez and R. Silva-Ortigoza. *Control Design Techniques in Power Electronics Devices.* Springer, London, 2006.
31. J.-C. Trigeassou. *Identification et commande des processus mono-entrée mono-sortie par la méthode des moments - Expérimentation sur calculatrice programmable.* Thèse 3e cycle, Université de Nantes, 1980.
32. K. Yosida. *Operational Calculus.* Springer, New York, 1984.
33. P.C. Young. Parameter estimation for continuous-time models - a survey. *Automatica*, 17:23–39, 1981.

14

Continuous-time Model Identification Using Spectrum Analysis with Passivity-preserving Model Reduction

Rolf Johansson

Lund University, Sweden

14.1 Introduction

This chapter deals with state-space model approximation of linear systems derived from linear regression and spectrum analysis – a problem that can be viewed as a problem of system identification. System identification deals with the problem of fitting mathematical models to time series of input–output data [16]. Important subproblems are the extraction both of a 'deterministic' subsystem—*i.e.*, computation of an input–output model—and a 'stochastic' subsystem that is usually modeled as a linear time-invariant system with white noise inputs and outputs that represent the misfit between model and data.
A pioneering effort in continuous-time model identification was Wiener's formulation of a Laguerre filter expansion [41]. As for the early literature on continuous-time model identification involving approaches with pseudo-linear regression, correlation and gradient search methods, there are algorithmic contributions [7, 15, 32, 41, 43, 44]; with surveys of Young [44], Unbehauen and Rao [13, 34, 35]; stochastic model estimation aspects [17]; software and algorithmic aspects [11, 40].
As is well known, linear regression methods are sensitive to coloured noise and need modification to provide unbiased parameter estimation. Such modifications may include restriction to finite impulse response (FIR) or moving average (MA) models, prewhitening filters, weighted least squares estimation, Markov estimates, approximate or iterative Markov estimates, or pseudo-linear regression.
Alternative methods are maximum likelihood (ML) methods (relying on numerical optimisation) and subspace-based methods (which may give poor results for low signal-to-noise ratios). The optimality of maximum likelihood estimates depends on the relevance and correctness of the assumptions on the underlying probability distribution functions—usually the normal distribution—and the numerical optimisation method that may perform poorly in cases of non-unique model parametrisation or when a unique ML

optimum fails to exist, *e.g.*, continuous-time and discrete-time multi-input, multi-output ARMAX models as well as overparameterised models.

State-space model identification has proved effective in determination of input–output relationships in the form of state-space models with early contributions reviewed in the comprehensive publications [29, 36–39]. As for subspace-based identification, continuous-time model identification methods were presented in [19]. Whereas subspace-based identification for high-to-moderate signal-to-noise ratios often provides good transfer function models and stochastic disturbance models, poor results may be obtained for high noise levels.

Evaluation of model misfit is often determined as an innovations sequence of a Kalman filter model that, in turn, also permits covariance matrix factorisation [37]. The related problem of stochastic realisation was approached by Ho and Kalman [14], Anderson and Moylan [4,5], Faurre [9,10], Akaike [2,3], Desai and Pal [8], Larimore [24,25], Lindquist and Picci [26,27], Juang and Pappa [21].

An important observation pointed out in [36] is that state-space model identification algorithms based on stochastic realisation algorithms may fail to provide a positive-definite solution of the Riccati equation that, in turn, brings attention to the problem of 'positive real sequences', their bias and variance [36, p. 85 ff.], one solution being provided when the innovations model fails to exist [20]. An issue in spectrum estimation as well as system identification is model-order determination and model-order reduction by means of balanced model reduction [1, 6, 12, 28, 30]. Attempts towards model approximation by means of balanced model-order reduction, however, may give rise to reduced-order state-space models that fail to satisfy the passivity condition. Such resulting state-space models will also fail in covariance analysis and spectral interpretation. In an important recent paper, Sorensen showed that passivity-preserving model approximation may be obtained as a by-product of a certain eigenvalue problem involving the state-space realisation matrices [33].

Here, we shall deal with model reduction applied to linear regression models in continuous-time model identification. After preliminaries on continuous-time model identification, positivity in spectrum-analysis model reduction, the main results will be presented.

14.2 Preliminaries

14.2.1 Continuous-time Model Identification

Consider a multi-input, multi-output continuous-time system with input $u \in \mathbb{R}^m$, output $y \in \mathbb{R}^p$ and disturbance $v \in \mathbb{R}^p$ related according to the linear signal model expressed as the Laplace transform relationship

$$Y(s) = G(s)U(s) + V(s) \tag{14.1}$$

which, in turn, is described by the left matrix fraction description

$$\mathcal{S}: \quad A_L(s)Y(s) = B_L(s)U(s) + C_L(s)W(s), \quad \det A_L(s) \neq 0 \quad (14.2)$$
$$A_L(s) \in \mathbb{R}^{p \times p}[s], \quad B_L(s) \in \mathbb{R}^{p \times m}[s], \quad C_L(s) \in \mathbb{R}^{p \times p}[s]$$
$$G(s) = A_L^{-1}(s)B_L(s), \quad G(s) \in \mathbb{R}^{p \times m}(s) \quad (14.3)$$
$$V(s) = H(s)W(s) \quad H(s) = A_L^{-1}(s)C_L(s), \quad H(s) \in \mathbb{R}^{p \times p}(s)$$

with p white noise inputs $W(s)$ ($w \in \mathbb{R}^p$), the transfer functions $G(s)$ and $H(s)$ describing the input–output transfer function and the coloured noise properties of $V(s)$, respectively. The polynomial matrices of the left matrix fraction description are

$$A_L(s) = s^n I_{p \times p} + A_1 s^{n-1} + \cdots + A_n, \quad A_1, \ldots, A_n \in \mathbb{R}^{p \times p} \quad (14.4)$$
$$B_L(s) = B_1 s^{n-1} + \cdots + B_{n-1} s + B_n, \quad B_1, \ldots, B_n \in \mathbb{R}^{p \times m} \quad (14.5)$$
$$C_L(s) = s^n I_{p \times p} + C_1 s^{n-1} + \cdots + C_n, \quad C_1, \ldots, C_n \in \mathbb{R}^{p \times p} \quad (14.6)$$

The continuous-time system identification problem involves estimation of $G(s)$, $H(s)$ (or the related generating polynomial matrices $A_L(s), B_L(s), C_L(s)$) from N samples of uniformly sampled input–output data $\{u_k\}_{k=1}^N, \{y_k\}_{k=1}^N$ with sample period T_s. Following [17], the following operator transformation is introduced

$$\lambda(s) = \frac{1}{1+s\tau}, \quad s = \frac{1-\lambda}{\tau\lambda}, \quad \tau > 0 \quad (14.7)$$

which permits the formulation of linear regression models in the spectral domain or in the time domain

$$A_\lambda(\lambda)Y(s) = B_\lambda(\lambda)U(s) + C_\lambda(\lambda)W(s) \quad (14.8)$$

with a one-to-one transformation from the polynomial matrices $\{A_L, B_L, C_L\}$ to $\{A_\lambda, B_\lambda, C_\lambda\}$ and with

$$A_\lambda(\lambda) = I_{p \times p} + A_\tau^{(1)} \lambda + \cdots + A_\tau^{(n)} \lambda^n, \quad A_\tau^{(1)}, \ldots, A_\tau^{(n)} \in \mathbb{R}^{p \times p}$$
$$B_\lambda(\lambda) = B_\tau^{(1)} \lambda + \cdots + B_\tau^{(n-1)} \lambda^{n-1} + B_\tau^{(n)} \lambda^n, \quad B_\tau^{(1)}, \ldots, B_\tau^{(n)} \in \mathbb{R}^{p \times m}$$
$$C_\lambda(\lambda) = I_{p \times p} + C_\tau^{(1)} \lambda + \cdots + C_\tau^{(n)} \lambda^n, \quad C_\tau^{(1)}, \ldots, C_\tau^{(n)} \in \mathbb{R}^{p \times p} \quad (14.9)$$

using the following regressors for $j = 1 \ldots n$

$$y^{(j)}(t) = [\lambda^j y](t), \quad u^{(j)}(t) = [\lambda^j u](t) \quad j = 1 \ldots n \quad (14.10)$$

For discrete-time input–output data, uniformly sampled with the sampling period T_s, linear regression applies to the discretised regressors

$$y_k^{(j)} = [[\lambda^j y](t)]_{t=kT_s}, \quad u_k^{(j)} = [[\lambda^j u](t)]_{t=kT_s} \quad (14.11)$$

Two problems arise for estimation of the stochastic disturbance properties. As the stochastic disturbance process ($\nu(\cdot)$ or $\{\nu_k\}$) is unknown, there is no immediate way to formulate regressors for estimation of the components of the polynomials $C_\lambda(\lambda)$.

For the purpose of least squares identification, then, it is suitable to organise model, data and notation according to linear regression

$$y_k = -A_\tau^{(1)} y_k^{(1)} - \cdots - A_\tau^{(n)} y_k^{(n)} + B_\tau^{(1)} u_k^{(1)} + \cdots + B_\tau^{(n)} u_k^{(n)} + \nu_k \quad (14.12)$$

$$y^T(t) = \varphi_\tau^T(t)\boldsymbol{\theta}_\tau + \nu(t) \tag{14.13}$$

$$\boldsymbol{\theta} = \begin{bmatrix} A_1 \ldots A_n \ B_1 \ldots B_n \end{bmatrix}^T, \tag{14.14}$$

$$y_k \in \mathbb{R}^p, \ \nu_k \in \mathbb{R}^p, \ \boldsymbol{\theta} \in \mathbb{R}^{n(m+p)\times p}$$

with a continuous-time noise process $\nu(t)$ with a discrete-time representation $\{\nu_k\}$ and regressor

$$\boldsymbol{\varphi}_k = \begin{bmatrix} -(y_k^{(1)})^T \cdots -(y_k^{(n)})^T \ (u_k^{(1)})^T \cdots (u_k^{(n)})^T \end{bmatrix}^T, \quad \boldsymbol{\varphi}_k \in \mathbb{R}^{n(m+p)}$$

which suggests the linear regression model

$$\mathcal{Y}_N = \boldsymbol{\Phi}_N \boldsymbol{\theta}, \quad \mathcal{Y}_N = \begin{bmatrix} y_1^T \\ y_2^T \\ \vdots \\ y_N^T \end{bmatrix}, \quad \boldsymbol{\Phi}_N = \begin{bmatrix} \varphi_1^T \\ \varphi_2^T \\ \vdots \\ \varphi_N^T \end{bmatrix} \tag{14.15}$$

Whereas least squares estimation based on noise-free data $\{u_k\}_{k=1}^N$, $\{y_k\}_{k=1}^N$ applied to the linear regression model (including some noise representation $\mathcal{V}_N$)

$$\mathcal{Y}_N = \boldsymbol{\Phi}_N \boldsymbol{\theta} + \mathcal{V}_N \tag{14.16}$$

effectively provides estimation of $G(s)$ (or $A_L(s), B_L(s)$), more elaborate algorithms are needed to estimate the noise spectrum $H(s)$ (or $A_L(s), C_L(s)$).

14.2.2 Spectrum Analysis and Positivity

Consider the signal models

$$Y(s) = G(s)U(s) + V(s) \tag{14.17}$$

with noise input V and measurable input U, output Y for the continuous-time and discrete-time models, respectively. Assume that the inputs U, V are uncorrelated. Then, the corresponding s-spectrum (continuous-time) are

$$S_{yy}(s) = G(s)S_{uu}(s)G^*(-s^*) + S_{vv}(s) \tag{14.18}$$

and the cross-spectrum fulfills [22]

$$S_{yu}(s) = G(s)S_{uu}(s), \quad S_{yu}(s) = S_{uy}^T(-s^*) \quad (14.19)$$

Based on standard spectrum analysis justified by (14.19), the transfer function and the noise spectrum may be estimated from the spectra as

$$\widehat{G}(s) = S_{yu}(s)S_{uu}^{-1}(s), \quad (14.20)$$

$$\widehat{S}_{vv}(s) = S_{yy}(s) - S_{yu}(s)S_{uu}^{-1}(s)S_{uy}^*(-s^*) \quad (14.21)$$

from which a noise state-space model may be determined by means of a stochastic realisation algorithm [8, 22].

A condition for interpretation of a function $S(s)$ as a spectral density is that $S(s)$ be positive real for $s = i\omega$ with the same conditions appearing in passivity analysis [18]. Spectrum analysis using state-space analysis is a time-honoured problem referred to as the partial realisation problem [26, 27] that starts with properties of the spectrum

$$S_{yy}(s) = S(s) = S_+(s) + S_+^*(-s) \quad (14.22)$$

$$S_+(s) = \frac{1}{2}S_0 + S_1 s + S_2 s^2 + \cdots \quad (14.23)$$

A condition for $S(s)$ to be spectral density is that $S_+(s)$ be positive real on the imaginary axis—i.e., for

$$S_+(i\omega) + S_+^T(-i\omega) \geq 0, \quad \forall \omega \in (-\infty, \infty) \quad (14.24)$$

If this passivity condition is not fulfilled, the stochastic realisation algorithm will fail [10, 27], [22, App. 8D], [20]. An example elaborating a system failing to exhibit positivity is given in [20, Ex. 1, p. 989].

Lemma (Kalman–Yakubovich–Popov [23, 31, 42], [22, p.307])

Consider the state-space system

$$\frac{dx}{dt} = Ax(t) + v_x(t), \quad x \in \mathbb{R}^n \quad (14.25)$$

$$y(t) = Cx(t) + v_y(t), \quad y \in \mathbb{R}^p \quad (14.26)$$

with stochastic inputs v_x, v_y with properties

$$\mathsf{E}\left(\begin{bmatrix} v_x(t) \\ v_y(t) \end{bmatrix}\right) = 0, \quad \mathsf{E}\left(\begin{bmatrix} v_x(t_1) \\ v_y(t_1) \end{bmatrix}\begin{bmatrix} v_x(t_2) \\ v_y(t_2) \end{bmatrix}\right) = Q\delta(t_1 - t_2) \quad (14.27)$$

where $Q = Q^T \geq 0$.

Let A be stable, (A, C) be an observable pair and assume that A does not have any eigenvalues on the imaginary axis with $A \in \mathbb{R}^{n \times n}$, $C \in \mathbb{R}^{p \times n}$.

$$S(s) = F(s)F^T(-s) = S_+(s) + S_+^T(-s), \quad F(s) = \left[C(sI-A)^{-1} \; I\right] Q^{1/2}$$

$$Q = Q^{1/2}(Q^{1/2})^T = \begin{bmatrix} Q_{11} & Q_{12} \\ Q_{12}^T & Q_{22} \end{bmatrix} \geq 0 \tag{14.28}$$

Then, the following statements are equivalent:

- $S_{ee}(i\omega) \geq 0$ for all $\omega \in (-\infty, \infty)$
- there exists a Hermitian matrix P such that

$$\begin{bmatrix} Q_{11} + AP + PA^T & Q_{12} + PC^T \\ Q_{12}^T + CP & Q_{22} \end{bmatrix} \geq 0 \tag{14.29}$$

- there exists a Hermitian matrix P such that

$$0 \leq \begin{bmatrix} Q_{11} + AP + PA^T & Q_{12} + PC^T \\ Q_{12}^T + CP & Q_{22} \end{bmatrix} = \begin{bmatrix} K \\ I \end{bmatrix} Q_{22} \begin{bmatrix} K \\ I \end{bmatrix}^T \tag{14.30}$$

The procedure to find a stable rational matrix function $F(s)$ of $S(s)$ is called spectral factorisation and the s_i such that $\det F(s_i) = 0$ are the spectral zeros of $S(s)$ [22].

A closely related result or corollary oriented towards rational function properties and known as the positive real lemma is the following

Positive Real Lemma
(Kalman–Yakubovich–Popov [23, 31, 42], [22, p. 308])

Let $G(s) = C(sI-A)^{-1}B + D$ be the transfer function of the state-space realisation

$$\frac{dx}{dt} = Ax + Bu, \quad x \in \mathbb{R}^n, \quad u \in \mathbb{R}^p \tag{14.31}$$

$$y = Cx + Du, \quad y \in \mathbb{R}^p \tag{14.32}$$

The following statements are equivalent:

1. $G(s)$ is positive real;
2. There exists an $n \times n$ matrix P, a $p \times p$ matrix R, and an $n \times p$ matrix K such that

$$0 \leq \begin{bmatrix} -AP - PA^T & B - PC^T \\ B^T - CP & D + D^T \end{bmatrix} = \begin{bmatrix} K \\ I_p \end{bmatrix} R \begin{bmatrix} K \\ I_p \end{bmatrix}^T \tag{14.33}$$

The following two statements are equivalent:

i. $G(s)$ is strictly positive real (SPR)—i.e., $G(i\omega) + G^T(-i\omega) > 0$;
ii. There exists a unique non-negative-definite solution P of the continuous-time Riccati equation

$$0 = AP + PA^T + KRK^T, \quad R = D + D^T,$$
$$K = (B - PC^T)R^{-1} \tag{14.34}$$

such that $A - KC$ is stable.

14.2.3 Spectral Factorisation and Positivity

Based on the Kalman–Yakubovich–Popov Lemma, there is one spectral factor of particular interest, namely the innovations model and its spectral factor $H(s)$

$$\begin{cases} \dfrac{\mathrm{d}x}{\mathrm{d}t} = Ax(t) + Kw(t) \\ y(t) = Cx(t) + w(t) \end{cases} \quad \begin{cases} H(s) = I + C(sI - A)^{-1}K \\ H^{-1}(s) = I - C(sI - A + KC)^{-1}K \end{cases} \quad (14.35)$$

so that

$$S(s) = H(s)QH^T(-s) \tag{14.36}$$

The stable spectral factor inverse $H^{-1}(s)$ provides the basis for the Kalman filter with the state-space realisation

$$\begin{aligned} \dfrac{\mathrm{d}\widehat{x}}{\mathrm{d}t} &= (A - KC)\widehat{x}(t) + Ky(t) \\ \widehat{y}(t) &= C\widehat{x}(t) \\ \widehat{w}(t) &= y(t) - \widehat{y}(t) = -C\widehat{x}(t) + y(t) \end{aligned} \tag{14.37}$$

14.2.4 Balanced Model Reduction

Optimal Hankel-norm model approximation has important applications for a wide variety of linear multi-variable systems [6, 12, 28, 30].

Consider application of model reduction to the state-space system

$$\dfrac{\mathrm{d}x}{\mathrm{d}t} = Ax(t) + Bu(t), \quad x \in \mathbb{R}^n, u \in \mathbb{R}^m \tag{14.38}$$

$$y(t) = Cx(t), \quad y \in \mathbb{R}^p \tag{14.39}$$

using Hankel singular values involves computation of the system Gramians

$$P = \int_0^\infty \mathrm{e}^{At}BB^T\mathrm{e}^{A^Tt}\mathrm{d}t, \quad Q = \int_0^\infty \mathrm{e}^{A^Tt}C^TC\mathrm{e}^{At}\mathrm{d}t \tag{14.40}$$

The Gramians may be computed by solving the Lyapunov equations

$$AP + PA^T = -BB^T, \quad P > 0 \tag{14.41}$$
$$A^TQ + QA = -C^TC, \quad Q > 0 \tag{14.42}$$

and the Hankel singular values are obtained from the eigenvalues $\{\sigma_k\}$

$$\sigma_k = \sqrt{\lambda_k(PQ)}, \quad k = 1, 2, \ldots, n \tag{14.43}$$

Under similarity transformation $z = Tx$ of a state-space system

$$P_z = TPT^T, \qquad Q_z = T^{-T}QT^{-1}, \qquad P_zQ_z = T(PQ)T^{-1} \qquad (14.44)$$

the balancing transformation T is found as the matrix diagonalising PQ to Σ^2. Let Q have the Cholesky factorisation

$$Q = Q_1^T Q_1, \quad Q_1 P Q_1^T = U\Sigma^2 U^T, \text{ with } U^T U = I, \quad \Sigma = \Sigma_1^T \Sigma_1 \qquad (14.45)$$
$$T = \Sigma_1^{-1} U^T Q_1 \qquad (14.46)$$

Model reduction of the balanced state-space system from model order n to model order r can be made as the balanced truncation

$$\mathcal{S} = \left[\begin{array}{c|c} TAT^{-1} & TB \\ \hline CT^{-1} & D \end{array}\right] = \left[\begin{array}{cc|c} A_{11} & A_{12} & B_1 \\ A_{21} & A_{22} & B_2 \\ \hline C_1 & C_2 & D \end{array}\right] \rightarrow \mathcal{S}_r = \left[\begin{array}{c|c} A_{11} & B_1 \\ \hline C_1 & D \end{array}\right] \qquad (14.47)$$

or

$$\mathcal{S}_r = \left[\begin{array}{c|c} E^T TAT^{-1}E & E^T TB \\ \hline CT^{-1}E & D \end{array}\right] = \left[\begin{array}{c|c} A_{11} & B_1 \\ \hline C_1 & D \end{array}\right], \quad E = \begin{bmatrix} I_r \\ 0 \end{bmatrix} \qquad (14.48)$$

Model reduction by means of balanced truncation preserves stability with the existence of a global error bound

$$\sigma_{r+1} = \|\Sigma - \Sigma_r\|_\infty \leq 2(\sigma_{r+1} + \cdots + \sigma_n) \qquad (14.49)$$

14.3 Problem Formulation

As the stochastic disturbance model included in the estimated model may be positive real but perhaps of high order, it is of interest to find a continuous-time model of reduced model order, while preserving stability and passivity properties for the continuous-time model.

A high-order model estimate with inputs u and w of the form of

$$\mathcal{S}_o = \left[\begin{array}{c|cc} A & B & K \\ \hline C & D & I \end{array}\right] \qquad (14.50)$$

may be balanced with respect to the input–output map $(u \rightarrow y)$ as

$$\mathcal{S} = \left[\begin{array}{c|cc} TAT^{-1} & TB & TK \\ \hline CT^{-1} & D & I \end{array}\right] = \left[\begin{array}{cc|cc} A_{11} & A_{12} & B_1 & K_1 \\ A_{21} & A_{22} & B_2 & K_2 \\ \hline C_1 & C_2 & D & I \end{array}\right] \qquad (14.51)$$

Balanced truncation suggests a reduced-order model of model order r on innovations form

$$\mathcal{S}_r = \left[\begin{array}{c|cc} A_{11} & B_1 & K_1 \\ \hline C_1 & D & I \end{array}\right] \qquad (14.52)$$

where the reduced-order spectral factor is $H(s) = C_1(sI_r - A_{11})^{-1}K_1 + I$.

In the following, we will consider the problem of such structure-preserving model reduction applied to least squares estimation in continuous-time model identification.

14.4 Main Results

The linear regression suggests a state-space model on innovations form

$$y(t) = -A_\tau^{(1)} y^{(1)} - \cdots - A_\tau^{(n)} y^{(n)} + B_\tau^{(1)} u^{(1)} + \cdots + B_\tau^{(n)} u^{(n)} + w \quad (14.53)$$

$$y_k = -A_\tau^{(1)} y_k^{(1)} - \cdots - A_\tau^{(n)} y_k^{(n)} + B_\tau^{(1)} u_k^{(1)} + \cdots + B_\tau^{(n)} u_k^{(n)} + w_k \quad (14.54)$$

which lends itself to least squares estimation using discrete-time input–output data [17]. Whereas an estimated regression model of (14.53) provides interpretations as an ARX or ARMAX model with immediate conversion to a transfer function model, we will now explore the benefit of an intermediate calculation of state-space models that lend themselves to model reduction. For this purpose, we introduce the non-minimal state-space description composed of the regressor components of (14.10).

$$x = \begin{bmatrix} y^{(1)} \\ \vdots \\ y^{(n)} \\ u^{(1)} \\ \vdots \\ u^{(n)} \end{bmatrix} \quad (14.55)$$

with dynamics according to the estimated state-space model

$$\frac{dx}{dt} = \frac{1}{\tau} \left(\begin{bmatrix} -I & 0 & 0 & 0 & 0 & 0 & \cdots & 0 \\ I & -I & \ddots & & \vdots & 0 & & \vdots \\ 0 & \ddots & \ddots & 0 & \vdots & & & 0 \\ 0 & & I & -I & 0 & \cdots & & \\ 0 & \cdots & & 0 & -I & 0 & 0 & 0 \\ 0 & & & 0 & I & -I & \ddots & \vdots \\ \vdots & & & & \vdots & 0 & \ddots & 0 \\ 0 & \cdots & & 0 & 0 & \cdots & I & -I \end{bmatrix} x + \begin{bmatrix} I \\ 0 \\ \vdots \\ 0 \\ 0 \\ 0 \\ \vdots \\ 0 \end{bmatrix} \hat{\boldsymbol{\theta}}_\tau^T x + \begin{bmatrix} 0 \\ 0 \\ \vdots \\ 0 \\ I \\ 0 \\ \vdots \\ 0 \end{bmatrix} u + \begin{bmatrix} I \\ 0 \\ \vdots \\ 0 \\ 0 \\ 0 \\ \vdots \\ 0 \end{bmatrix} w \right)$$

$$y = \hat{\boldsymbol{\theta}}_\tau^T x + w \quad (14.56)$$

Obviously, this non-minimal model is an innovations form and provides a spectral factor, and the non-minimality lends itself to model reduction. Thus, it is desirable to accomplish model reduction with preservation of the innovation form for the noise model. It follows that the model-reduction procedure of Section 14.3 has immediate application to model reduction of the system of (14.56).

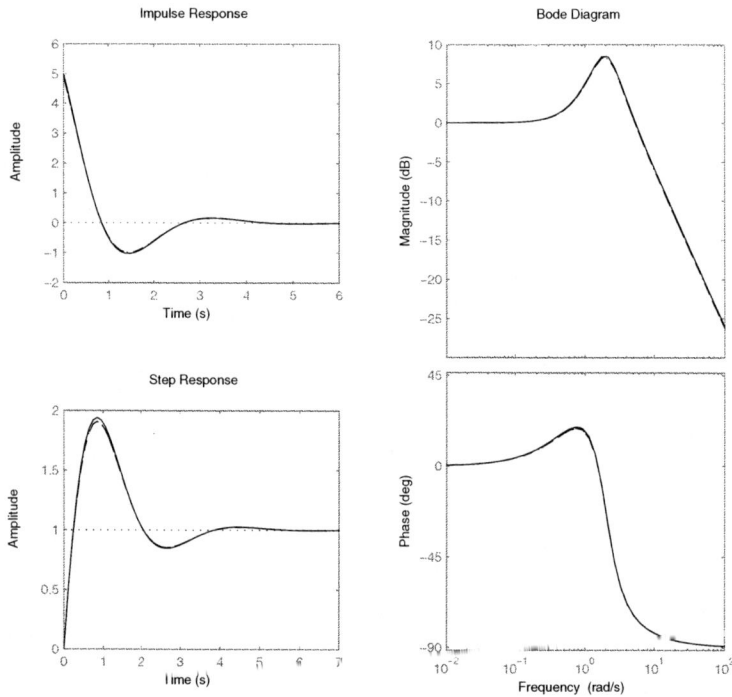

Fig. 14.1. True (*solid*) and estimated (*dashed*) impulse responses of $G(s)$ (*upper left*). True and estimated step responses of $G(s)$ (*lower left*). True and estimated Bode diagrams of $G(s)$ (*right*).

Example—Simulation Study

Consider the system

$$\mathcal{S}: \quad \frac{dx}{dt} = \begin{bmatrix} -2 & -4 \\ 1 & 0 \end{bmatrix} x(t) + \begin{bmatrix} 1 \\ 0 \end{bmatrix} u(t) + \begin{bmatrix} 1 \\ 0 \end{bmatrix} w(t) \tag{14.57}$$

$$y(t) = \begin{bmatrix} 5 & 4 \end{bmatrix} x(t) + w(t) \tag{14.58}$$

with the transfer function description

$$Y(s) = \frac{5s+4}{s^2+2s+4} U(s) + \frac{s^2+7s+8}{s^2+2s+4} W(s) \tag{14.59}$$

The input was generated from uniformly distributed pseudo-random numbers (see Figure 14.3). The input–output signals were sampled with a sampling period $T_s = 0.1$ s for measurement duration of $T = 100$ s for a signal-to-noise

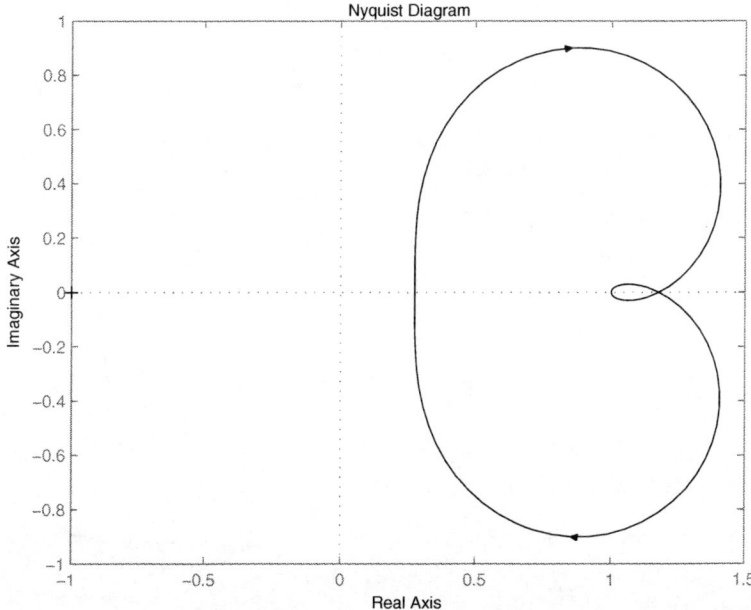

Fig. 14.2. Nyquist diagram of the spectral factors for the non-minimal estimated model (*solid*) and the reduced-order model (*dashed*)

ratio of $\|u\|/\|w\| = 10$. For $\tau = 1$, least squares identification was made for $n = 2$

$$\mathcal{M} : \frac{\mathrm{d}x}{\mathrm{d}t} = \begin{bmatrix} -0.6957 & -2.9238 & 4.2864 & -0.6657 \\ 1.00 & -1.00 & 0 & 0 \\ 0 & 0 & -1.00 & 0 \\ 0 & 0 & 1.00 & -1.00 \end{bmatrix} x + \begin{bmatrix} 0 \\ 0 \\ 1 \\ 0 \end{bmatrix} u + \begin{bmatrix} 1 \\ 0 \\ 0 \\ 0 \end{bmatrix} w$$

$$y = \begin{bmatrix} 0.3043 & -2.924 & 4.286 & -0.6657 \end{bmatrix} x + w \tag{14.60}$$

with good approximation properties as witnessed by Figure 14.1. A balancing transformation was made by means of balanced truncation with a similarity transformation matrix T and singular values Σ

$$T = \begin{bmatrix} 0.0227 & 1.4710 & -2.1163 & 0.3349 \\ -0.8027 & -0.4312 & 0.4388 & -0.0981 \\ -0.0095 & 0.0086 & 0.0118 & 0.0001 \\ 0.0001 & -0.0009 & -0.0004 & 0.0016 \end{bmatrix}, \quad \Sigma = \begin{bmatrix} 1.4088 \\ 0.9089 \\ 0.0000 \\ 0.0000 \end{bmatrix} \tag{14.61}$$

Based on the singular-value pattern, model reduction was made with reduction to a second-order model

$$\mathcal{M}_r : \frac{dx}{dt} = \begin{bmatrix} -1.59 & -1.858 \\ 1.858 & -0.1061 \end{bmatrix} x(t) + \begin{bmatrix} -2.116 \\ 0.439 \end{bmatrix} u(t) + \begin{bmatrix} 0.02275 \\ -0.8028 \end{bmatrix} w(t)$$
$$y(t) = \begin{bmatrix} -2.116 & -0.439 \end{bmatrix} x(t) + w(t) \quad (14.62)$$

The reduced-order model including a spectral factor (Figure 14.2) of the reduced-order continuous-time stochastic disturbance model provides a Kalman filter according to (14.37) as

$$\widehat{\mathcal{M}}_r : \frac{d\widehat{x}}{dt} = \begin{bmatrix} -1.5919 & -1.848 \\ 0.1593 & -0.4585 \end{bmatrix} \widehat{x}(t) + \begin{bmatrix} -2.116 \\ 0.439 \end{bmatrix} u(t) + \begin{bmatrix} 0.02275 \\ -0.8028 \end{bmatrix} y(t)$$
$$\widehat{y}(t) = \begin{bmatrix} -2.116 & -0.439 \end{bmatrix} \widehat{x}(t)$$
$$\widehat{w}(t) = y(t) - \widehat{y}(t) \quad (14.63)$$

As witnessed by Figure 14.3, the continuous-time Kalman filter provides a good predictor.

Spectral Factorisation and Positivity

In a compact format, the positive real lemma may be summarised in the Lyapunov equation

$$\mathcal{AP} + \mathcal{PA}^T = -\mathcal{Q}, \quad \mathcal{Q} = \begin{bmatrix} K \\ -I \end{bmatrix} R \begin{bmatrix} K \\ -I \end{bmatrix}^T,$$

$$\mathcal{P} = \begin{bmatrix} P & 0 \\ 0 & I \end{bmatrix}, \quad \mathcal{A} = \begin{bmatrix} A & B \\ -C & -D \end{bmatrix} \quad (14.64)$$

A necessary requirement for $G(p)$ to be positive real is that $\mathcal{A}$ be stable [18].

Finally, the simultaneous spectrum properties of stability and positivity are linked to the Lyapunov equation. Let

$$E = \begin{bmatrix} I_n & 0 \\ 0 & 0_p \end{bmatrix}, \quad G_{12}(s) = C^T + (-sI_n - A^T)P(sI_n - A)^{-1}B$$
$$L(s) = \begin{bmatrix} I_n & 0 \\ C(sI_n - A)^{-1} & I_p \end{bmatrix}, \quad R(s) = \begin{bmatrix} I_n & -(sI_n - A)^{-1}B \\ 0 & I_p \end{bmatrix} \quad (14.65)$$

Then

$$sE - \mathcal{A} = L(s) \begin{bmatrix} sI_n - A & 0 \\ 0 & G(s) \end{bmatrix} R(s) \quad (14.66)$$

$$0 \le \mathcal{Q} = -\mathcal{AP} - \mathcal{PA}^T = (sE - \mathcal{A})\mathcal{P} + \mathcal{P}(-sE - \mathcal{A}^T)$$
$$= R^T(s) \begin{bmatrix} -(AP + PA^T) & G_{12}(s) \\ G_{12}^T(s) & G(s) + G^T(-s) \end{bmatrix} R(s) \quad (14.67)$$

where the diagonal matrix blocks represents the stability and positivity, respectively, with non-negative semi-definite matrix properties. Rank deficit appears only at the transmission zeros—i.e., the spectral zeros.

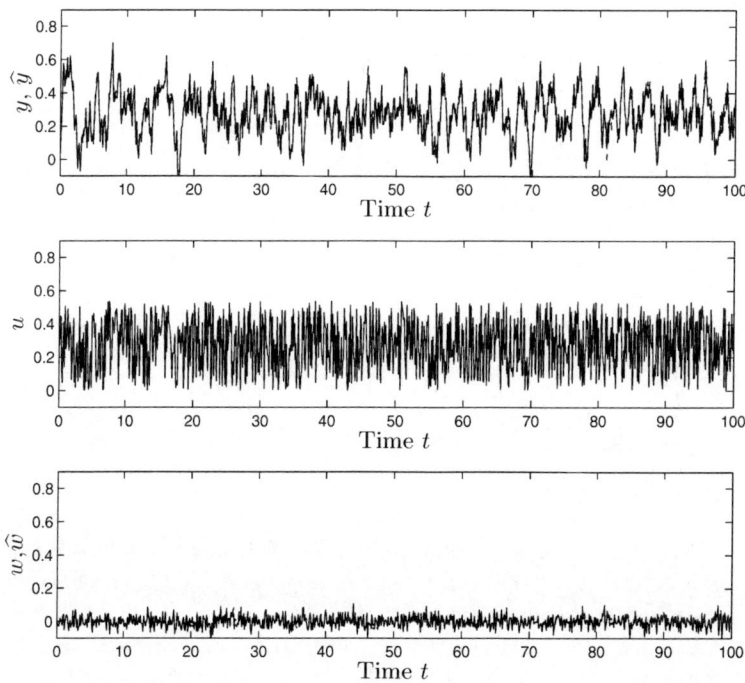

Fig. 14.3. Input-output data (*solid*) and predicted (*dashed*) by means of Kalman filter estimated from the spectral factor of the reduced-order model. Lower graph shows the sequence of true (*solid*) and estimated (*dashed*) innovations.

14.5 Discussion

This approach also lends itself to least squares identification without any autoregressive part—*i.e.*, the counterpart to moving average or finite impulse response models with particular interest to preclude bias resulting from correlation among stochastic disturbances and regressor components. As compared to numerical approaches to ML optimisation, the method presented has no specific problem with overparameterised models or multi-input, multi-output models.

Separate model reduction applied to the input–output model and the stochastic disturbance model will provide a reduced-order model of a form reminiscent of the Box–Jenkins models [16, p. 109].

The model identification may also be applied to linear regression models other than ARX-like or ARMAX-like structures of (14.12), one option being provided by data from spectrum analysis. For a noise process v uncorrelated with the input u, an unbiased estimate of $G(s)$ can be found as

$$\widehat{G}(s) = S_{yu}(s)S_{uu}^{-1}(s) = A_L^{-1}(s)B_L(s) \tag{14.68}$$
$$A_L(s)S_{yu}(s) = B_L(s)S_{uu}(s) \tag{14.69}$$

which suggests a linear regression model using spectral estimates $\widehat{S}_{yu}(s)$, $\widehat{S}_{uu}(s)$ as regressor components

$$A_L(s)\widehat{S}_{yu}(s) = B_L(s)\widehat{S}_{uu}(s) \tag{14.70}$$

A least squares spectrum-based estimate of $A_L(s)$, $B_L(s)$ permits use of the model-reduction algorithm to find the spectral factor of the stochastic disturbance model.

14.6 Conclusions

A two-stage continuous-time linear model identification problem is presented. The first stage provides discrete-time spectral estimation with an unbiased estimate of the input–output transfer function in the case of uncorrelated noise and control input. Note that the first stage provides an unbiased, overparameterised continuous-time linear model. Finally, the second stage of the algorithm provides passivity-preserving model reduction, resulting in a reduced-order continuous-time state-space model maintaining spectral properties and interpretations. The identification method combining linear regression and model reduction also provides an effective and interesting approach to tuning of continuous-time Kalman filters.

References

1. V.M. Adamjan, D.Z. Arov, and M.G. Krein. Analytic properties of Schmidt pairs for a Hankel operator and the generalized Schur-Takagi problem. *Math USSR Sbornik*, 15(1):31–73, 1971.
2. H. Akaike. Stochastic theory of minimal realization. *IEEE Transactions on Automatic Control*, 19:667–674, 1974.
3. H. Akaike. Markovian representation of stochastic processes by canonical variables. *SIAM Journal on Control and Optimization*, 13:162–173, 1975.
4. B.D.O. Anderson and J.B. Moore. *Optimal Filtering*. Prentice Hall, Englewood Cliffs, NJ, USA, 1979.
5. B.D.O. Anderson and P.J. Moylan. Spectral factorization of a finite-dimensional nonstationary matrix covariance. *IEEE Transactions on Automatic Control*, AC-19(6):680–692, 1974.
6. A.C. Antoulas. *Approximation of Large-scale Dynamical Systems*. SIAM, Philadelphia, 2005.

7. A.B. Clymer. Direct system synthesis by means of computers, Part I. *Transactions AIEE*, 77:798–806, 1959.
8. U.B. Desai and D. Pal. A realization approach to stochastic model reduction and balanced stochastic realizations. *IEEE Conference on Decision and Control*, pages 1105–1112, Orlando, Florida, USA, 1982.
9. P. Faurre. Stochastic realization algorithms. R.K. Mehra and D. Lainiotis (eds), *System Identification: Advances and Case Studies*. Academic Press, New York, USA,1976.
10. P. Faurre. *Opérateurs Rationnels Positifs*. Bordas, Paris, France, 1979.
11. H. Garnier, M. Mensler, and A. Richard. Continuous-time model identification from sampled data: Implementation issues and performance evaluation. *International Journal of Control*, 76(13):1337–1357, 2003.
12. K. Glover. All optimal Hankel-norm approximations of linear multivariable systems and their L^∞−error bounds. *International Journal of Control*, 39:1115–1193, 1984.
13. R. Haber and H. Unbehauen. Structure identification of nonlinear dynamic systems—A survey on input/output approaches. *Automatica*, 26:651–677, 1990.
14. B.L. Ho and R.E. Kalman. Effective construction of linear state-variable models from input/output functions. *Regelungstechn.*, 14:545–548, 1966.
15. R. Johansson. Identification of continuous-time dynamic systems. *25th IEEE Conference on Decision and Control*, pages 1653–1658, Athens, Greece, 1986.
16. R. Johansson. *System Modeling and Identification*. Prentice Hall, Englewood Cliffs, NJ, 1993.
17. R. Johansson. Identification of continuous-time models. *IEEE Transactions on Signal Processing*, 4:887–897, 1994.
18. R. Johansson and A. Robertsson. The Yakubovich-Kalman-Popov lemma and stability analysis of dynamic output feedback systems. *International Journal of Robust and Nonlinear Control*, 16:45–69, January 2006.
19. R. Johansson, M. Verhaegen, and C.T. Chou. Stochastic theory of continuous-time state-space identification. *IEEE Transactions on Signal Processing*, 47:41–51, January 1999.
20. R. Johansson, M. Verhaegen, C.T. Chou, and A. Robertsson. Residual models and stochastic realization in state-space system identification. *International Journal of Control*, 74:988–995, 2001.
21. J.N. Juang and R.S. Pappa. An eigensystem realization algorithm for modal parameter identification and model reduction. *Journal of Guidance, Control and Dynamics*, 8:620–627, 1985.
22. T. Kailath, A.H. Sayed, and B. Hassibi. *Linear Estimation*. Prentice Hall, Upper Saddle River, NJ, 2000.
23. R.E. Kalman. Lyapunov functions for the problem of Lur'e in automatic control. *National Academy of Sciences*, 49(2), 1963.
24. W. Larimore. Predictive inference, sufficiency, entropy and an asymptotic likelihood principle. *Biometrika*, 70:175–181, 1983.
25. W. Larimore. Canonical variate analysis in identification, filtering and adaptive control. *29th IEEE Conference on Decision and Control*, pages 596–604, Hawaii, USA, 1990.
26. A. Lindquist and G. Picci. Realization theory for multivariate stationary Gaussian processes. *SIAM Journal of Control and Optimization*, 23(6):809–857, 1985.

27. A. Lindquist and G. Picci. Canonical correlation analysis, approximate covariance extension and identification of stationary time series. *Automatica*, 32(5):709–733, 1996.
28. B.C. Moore. Principal component analysis in linear systems: controllability, observability, and model reduction. *IEEE Transactions on Automatic Control*, AC-26:17–32, 1981.
29. P. Van Overschee and B. De Moor. N4SID: subspace algorithm for the identification of combined deterministic-stochastic systems. *Automatica*, 30:75–93, 1994.
30. L. Pernebo and L.M. Silverman. Model reduction via balanced state-space realizations. *IEEE Transactions on Automatic Control*, pages 382–387, 1982.
31. V.M. Popov. Absolute stability of nonlinear systems of automatic control. *Avtomatika i Telemekhanika*, 22(8), 1961.
32. D.C. Saha and G.P. Rao. *Identification of Continuous Dynamical Systems—The Poisson Moment Functional (PMF) Approach*, volume 56 of *LNCIS*. Springer-Verlag, Berlin, 1983.
33. D. Sorensen. Passivity preserving model reduction via interpolation of spectral zeros. Technical Report TR02-15, Rice University, Dept. Comp. Appl. Math., 2002.
34. H. Unbehauen and G.P. Rao. *Identification of Continuous-time Systems*. North-Holland, Amsterdam, 1987.
35. H. Unbehauen and G.P. Rao. Continuous-time approaches to system identification—A survey. *Automatica*, 26:23–35, 1990.
36. P. Van Overschee and B. de Moor. *Subspace Identification for Linear Systems—Theory, Implementation, Applications*. Kluwer Academic Publishers, Boston-London-Dordrecht, 1996.
37. M. Verhaegen. Identification of the deterministic part of MIMO state space models given in innovation form from input-output data. *Special Issue on Statistical Signal Processing and Control of Automatica*, 30(1):61–74, 1994.
38. M. Verhaegen and P. Dewilde. Subspace model identification—Analysis of the elementary output-error state-space model identification algorithm. *International Journal of Control*, 56:1211–1241, 1992.
39. M. Verhaegen and P. Dewilde. Subspace model identification—The output-error state-space model identification class of algorithms. *International Journal of Control*, 56:1187–1210, 1992.
40. L. Wang. Continuous time model predictive control design using orthonormal functions. *International Journal of Control*, 74:1588–1600, 2001.
41. N. Wiener. *Nonlinear Problems in Random Theory*. The MIT Press, Cambridge, MA, 1958.
42. V.A. Yakubovich. Solution of certain matrix inequalities occuring in the theory of automatic controls. *Doklady Akademii Nauk SSSR*, 143(6), 1962.
43. P.C. Young. Applying parameter estimation to dynamic systems. *Control Engineering*, 16:Oct 119–125, Nov 118–124, 1969.
44. P.C. Young. Parameter estimation for continuous-time models. A survey. *Automatica*, 17:23–29, 1981.

Index

adaptive
   control, 381
   observer, 376
Akaike information criterion (AIC), 120, 147
aliasing effect, 69, 70
anti-alias filters, 219
arbitrary excitations, 228
ARMA model, 36, 95, 136, 255

bandwidth of validity, 84
bench-scale dryer, 331
BFGS method, 203, 209
bias-compensated least squares, 50, 133
block-pulse functions, 252
bond-graph approach, 201
Brownian motion, 220

cancer cells, 278
CAR, 34, 79, 82, 85
CARARX, 145
CARMA, 34, 94
CARMAX, 34, 241
CARX, 34, 144
central difference operator, 45
closed-loop identification, 133, 203, 227, 363, 377
COE, 101, 151, 254
coefficient of determination, 108, 179, 257
coloured noise, 94
computational
   load, 59, 340
   time, 59, 340

consistent estimates, 49, 102
continuous stirred tank, 332
continuous-time
   output error, 112, 254
   stochastic models, 38
   white noise, 34, 79, 94
covariance matrix, 54, 101, 104, 139, 140, 151
Cramér–Rao lower bound, 54, 104, 141, 231

data-based mechanistic model, 109
data-compression, 189
delta
   backward operator, 42
   forward operator, 42
   operator, 41, 68
deterministic models, 75
direct methods, 4, 40, 271
discrete Fourier transform, 73, 223
dynamics of human standing, 204

empirical transfer function estimate (ETFE), 196, 277
equality constraints, 198
errors-in-variables, 219, 236
expectation maximisation, 314

filter
   invariant, 371, 374
   Kalman, 399
   time-varying, 375
finite impulse response (FIR), 190, 324
Fisher information matrix, 54

flight flutter, 240
Fourier functions, 252
Fourier modulating functions (FMF), 252
fractional delay, 339
fractional derivatives, 227
frequency domain maximum likelihood, 72, 75, 229
frequency-sampling filter (FSF) model, 189, 191

generalised least squares, 195
genetic algorithm, 347
global circulation model, 120
global warming, 119
global-optimisation algorithm, 348
Gramians, 399

Hartley modulating functions, 252
heteroscedasticity, 127
hot air-flow device, 331
hybrid Box–Jenkins model, 95, 149, 227, 255
hybrid modelling, 135, 226, 242

impulse invariant transformation, 219
indirect methods, 3, 40, 271
inequality constraints, 198
initial conditions, 320
input inter-sample behaviour, 146, 216, 266, 324
instrumental product matrix, 108, 161, 166
instrumental variable
  auxiliary model, 97, 165, 262, 322, 340, 351
  basic, 138
  delayed, 84
  estimate, 48, 99
  extended, 104, 139
  for closed-loop systems, 138
  multi-step, 144
  optimal, 103, 104, 141
  refined, 105, 147, 255
  simplified refined, 101, 151, 254
  tailor-made, 133
inverted pendulum, 203
iterative procedure, 103, 149, 322, 346
  initiation, 324

IVSVF method, 262

leakage error, 224
least absolute-value algorithms, 229
least mean-squares, 349
least squares estimate, 48
left matrix fraction description, 395
lifting technique, 314
linear filter method, 315
linear identifiability, 368
linear integral filter, 252
local minimum, 347
LSSVF method, 111, 262
Lyapunov equation, 56, 399, 404

measurement setup
  band-limited, 217
  zero-order-hold, 217
mixing process, 332
model order identification, 108, 161, 257, 274
model order reduction, 394, 399
module-theoretic approach, 363
Monte Carlo simulation, 59, 80, 109, 154, 184, 199, 329
multi-rate data, 314
multi-sine, 277, 329
multiple-input systems, 103, 340
multiple-model structure, 166
multivariable, 283

noise modelling, 220
non-commutative ring theory, 363
non-linear instrumental variable
  global separable (GSEPNIV), 340, 354, 358
  global separable algorithm, 354
  separable (SEPNIV), 352
non-linear least squares, 202, 339
  global separable (GSEPNLS), 340, 347, 357
  global separable algorithm, 350
  problem, 346
  separable (SEPNLS), 339
  unseparable (UNSEPNLS), 344
non-minimum phase system, 323
numerical conditioning, 230

Octave, 203

operational amplifier, 237
operational calculus, 363
optimisation methods, 203
orthogonal basis-function model, 325
  Kautz model, 325
  Laguerre model, 325
outliers, 229, 314

partially known physical model, 211
periodic excitation, 241
persistently exciting signal, 374
photosensitising drug, 278
physically plausible model, 201
Poisson moment functionals, 252
polynomials
  Chebychev, 252
  Hermite, 252
  Laguerre, 252
  Legendre, 252
  orthogonal, 252
power converter, 383
PRBS, 113, 154, 179, 258, 261
PRESS statistic, 175, 178
principal ideal ring, 364

quadratic programming problem, 199

re-initialised partial moments, 252
recursive estimation, 259
relative degree, 341
residual alias error, 224
Riccati equation, 398
RIVC algorithm, 100
robot arm, 275
robustness issues, 67

sample maximum likelihood estimator, 234
sampled-data model, 68
sampling, 34
  equidistant, 36, 314
  frequency, 112, 274
  instantaneous, 36
  integrated, 52
  non-uniformly, 52, 103, 260, 266, 279, 314, 324
  of CARMA systems, 35

very rapid, 272
zeros, 37, 71, 80
shifted least squares method, 50
signal-to-noise ratio, 43, 113, 154, 262
Slepian–Bang formula, 55
smoothing function, 348
spectrum analysis, 393
SRIVC algorithm, 102
state-space model
  approximation, 393
  canonical form, 256
  fully parameterised form, 256, 292
  non-minimal description, 401
state-variable filter, 154, 162, 252, 342
Steiglitz and McBride algorithm, 103
step-invariant transformation, 221, 339
stochastic realisation algorithm, 397
strictly positive real (SPR), 398
subspace methods, 256, 291

time delay, 93, 251, 313, 339
toolbox
  ARMASA, 242
  CAPTAIN, 129
  CONTSID, 128, 242, 249
  FDIDENT, 242
  FREQID, 242
  SID, 242
  UNIT, 205
torsion module, 365
transient term, 224
trapezoidal-pulse functions, 252

UDV factorisation, 173
undermodelling errors, 75
unmodelled dynamics, 82
unstable plants, 230

Walsh functions, 252
Wiener process, 220
winding process, 122, 283

Young information criterion (YIC), 108, 257

zero-order hold, 35, 216, 341

Other titles published in this series (continued):

Soft Sensors for Monitoring and Control of Industrial Processes
Luigi Fortuna, Salvatore Graziani, Alessandro Rizzo and Maria G. Xibilia

Adaptive Voltage Control in Power Systems
Giuseppe Fusco and Mario Russo

Advanced Control of Industrial Processes
Piotr Tatjewski

Process Control Performance Assessment
Andrzej W. Ordys, Damien Uduehi and Michael A. Johnson (Eds.)

Modelling and Analysis of Hybrid Supervisory Systems
Emilia Villani, Paulo E. Miyagi and Robert Valette

Process Control
Jie Bao and Peter L. Lee

Distributed Embedded Control Systems
Matjaž Colnarič, Domen Verber and Wolfgang A. Halang

Precision Motion Control (2nd Ed.)
Tan Kok Kiong, Lee Tong Heng and Huang Sunan

Optimal Control of Wind Energy Systems
Iulian Munteanu, Antoneta Iuliana Bratcu, Nicolaos-Antonio Cutululis and Emil Ceangă

Model-based Process Supervision
Arun K. Samantaray and Belkacem Ould Bouamama

Dry Clutch Control for Automated Manual Transmission Vehicles
Pietro J. Dolcini, Carlos Canudas-de-Wit and Hubert Béchart
Publication due May 2008

Real-time Iterative Learning Control
Xu Jian-Xin, Sanjib K. Panda and Lee Tong Heng
Publication due July 2008

Model Predictive Control Design and Implementation Using MATLAB®
Liuping Wang
Publication due July 2008

Magnetic Control of Tokamak Plasmas
Marco Ariola and Alfredo Pironti
Publication due July 2008

Design of Fault-tolerant Control Systems
Hassan Noura, Didier Theilliol, Jean-Christophe Ponsart and Abbas Chamseddine
Publication due October 2008